Communications and Control Engineering

More information about this series at http://www.springer.com/series/61

Lars Grüne · Jürgen Pannek

Nonlinear Model Predictive Control

Theory and Algorithms

Second Edition

Springer

Lars Grüne
Mathematisches Institut
Universität Bayreuth
Bayreuth
Germany

Jürgen Pannek
Bremer Institut für Produktion und Logistik
(BIBA)
Universität Bremen
Bremen
Germany

Additional material to this book can be downloaded from http://extras.springer.com.

MATLAB® and Simulink® are registered trademarks of The MathWorks, Inc., 3 Apple Hill Drive, Natick, MA 01760-2098, USA, http://www.mathworks.com.
MAPLE™ is a trademark of Waterloo Maple Inc., 615 Kumpf Drive, Waterloo, ON Canada, N2V 1K8, http://www.maplesoft.com.

ISSN 0178-5354 ISSN 2197-7119 (electronic)
Communications and Control Engineering
ISBN 978-3-319-46023-9 ISBN 978-3-319-46024-6 (eBook)
DOI 10.1007/978-3-319-46024-6

Library of Congress Control Number: 2016950897

Mathematics Subject Classification (2010): 93-02, 92C10, 93D15, 49M37

Printed on acid-free paper

This Springer imprint is published by Springer Nature
The registered company is Springer International Publishing AG
The registered company address is: Gewerbestrasse 11, 6330 Cham, Switzerland

For Brigitte, Florian and Carla
LG

For Sabina, Alina, Laura and Timo
JP

Preface to the Second Edition

The continuing research activities in the field of Nonlinear Model Predictive Control have resulted in various new developments which were not represented in the first edition of this book. Five years after its publication we therefore proposed to write a second, significantly updated edition and we were glad that the Springer-Verlag immediately approved this proposal.

In order to maintain the character of the book as a mixture between a research monograph and an advanced textbook, we decided to limit the new material. We added two topics which have reached a certain level of maturity and to which we ourselves have made contributions—and which received widespread interest in recent years. This is on the one hand the new paradigm of Economic Nonlinear Model Predictive Control, in which more general optimal control problems than those penalizing the distance to a desired reference solution are considered. In Chap. 8 we provide a detailed presentation of the most important properties of the resulting closed-loop solution: averaged optimality, asymptotic stability of an optimal equilibrium, and finite time or transient optimality, both for formulations with and without stabilizing terminal conditions. On the other hand, Chap. 9 gives an introduction into Distributed Nonlinear Model Predictive Control. Here, we first develop an abstract framework for distributed schemes and their feasibility and stability, and then show how a number of schemes proposed in the literature fit into this framework. Beyond these two new chapters, the material in the remaining chapters has been updated and all typos and inaccuracies we were aware of have been corrected. Most importantly, with Proposition 6.19 we have provided an alternative simplified version of Proposition 6.18 as the main building block for the analysis of schemes without terminal constraints. While being more conservative, Proposition 6.19 allows for a considerably simpler proof than Proposition 6.18, which is thus also suitable for presentation in lectures.

Like the first edition, the present second edition also benefitted from the help of many people. We would like to thank Frank Allgöwer, David Angeli, Bob Bitmead, Andrea Boccia, Philipp Braun, Moritz Diehl, Timm Faulwasser, Rolf Findeisen, Arthur Fleig, Matthias Gerdts, Achim Ilchmann, Chris Kellett, David Mayne, Matthias A. Müller, Vryan Gil Palma, Simon Pirkelmann, Jim Rawlings, Marcus

Reble, Willi Semmler, Tobias Sprodowski, Marleen Stieler, Andy Teel, Steve Weller, Karl Worthmann, and Mario Zanon for valuable discussions and critical comments on the material presented in this book, for joint research leading to new results presented or at least mentioned in this book, for pointing out inaccuracies and suggesting improvements and for proofreading the new chapters. LG particularly thanks Chris Kellett and Steve Weller for their fantastic hospitality during his sabbatical at the University of Newcastle, Australia. Finally, we thank Oliver Jackson from Springer for his continuing excellent support.

Bayreuth, Germany — Lars Grüne
Bremen, Germany — Jürgen Pannek
June 2016

Preface to the First Edition

The idea for this book grew out of a course given at a winter school of the International Doctoral Program "Identification, Optimization and Control with Applications in Modern Technologies" in Schloss Thurnau in March 2009. Initially, the main purpose of this course was to present results on stability and performance analysis of nonlinear model predictive control algorithms, which had at that time recently been obtained by ourselves and coauthors. However, we soon realized that both the course and even more the book would be inevitably incomplete without a comprehensive coverage of classical results in the area of nonlinear model predictive control and without the discussion of important topics beyond stability and performance, like feasibility, robustness, and numerical methods.

As a result, this book has become a mixture between a research monograph and an advanced textbook. On the one hand, the book presents original research results obtained by ourselves and coauthors during the last 5 years in a comprehensive and self-contained way. On the other hand, the book also presents a number of results—both classical and more recent—of other authors. Furthermore, we have included a lot of background information from mathematical systems theory, optimal control, numerical analysis, and optimization to make the book accessible to graduate students—on PhD and Master level—from applied mathematics and control engineering alike. Finally, via our web page www.nmpc-book.com we provide MATLAB and C++ software for all examples in this book which enables the reader to perform his or her own numerical experiments. For reading this book, we assume a basic familiarity with control systems, their state space representation as well as with concepts like feedback and stability as provided, e.g., in undergraduate courses on control engineering or in courses on mathematical systems and control theory in an applied mathematics curriculum. However, no particular knowledge of nonlinear systems theory is assumed. Substantial parts of the systems theoretic chapters of the book have been used by us for a lecture on nonlinear model predictive control for master students in applied mathematics and we believe that the book is well suited for this purpose. More advanced concepts like time varying formulations or peculiarities of sampled data systems can be easily skipped if only time invariant problems or discrete time systems shall be treated.

The book centers around two main topics: systems theoretic properties of nonlinear model predictive control schemes on the one hand and numerical algorithms on the other hand; for a comprehensive description of the contents we refer to Sect. 1.3. As such, the book is somewhat more theoretical than engineering or application-oriented monographs on nonlinear model predictive control, which are furthermore often focused on linear methods.

Within the nonlinear model predictive control literature, distinctive features of this book are the comprehensive treatment of schemes without stabilizing terminal constraints and the in-depth discussion of performance issues via infinite horizon suboptimality estimates, both with and without stabilizing terminal constraints. The key for the analysis in the systems theoretic part of this book is a uniform way of interpreting both classes of schemes as relaxed versions of infinite horizon optimal control problems. The *relaxed dynamic programming* framework developed in Chap. 4 is thus a cornerstone of this book, even though we do not use dynamic programming for actually solving nonlinear model predictive control problems; for this task we prefer direct optimization methods as described in the last chapter of this book since they also allow for the numerical treatment of high-dimensional systems.

There are many people whom we have to thank for their help in one or the other way. For pleasant and fruitful collaboration within joint research projects and on joint papers—of which many have been used as the basis for this book—we are grateful to Frank Allgöwer, Nils Altmüller, Rolf Findeisen, Marcus von Lossow, Dragan Nešić, Anders Rantzer, Martin Seehafer, Paolo Varutti, and Karl Worthmann. For enlightening talks, inspiring discussions, for organizing workshops and minisymposia (and inviting us) and, last but not least, for pointing out valuable references to the literature we would like to thank David Angeli, Moritz Diehl, Knut Graichen, Peter Hokayem, Achim Ilchmann, Andreas Kugi, Daniel Limón, Jan Lunze, Lalo Magni, Manfred Morari, Davide Raimondo, Saša Raković, Jörg Rambau, Jim Rawlings, Markus Reble, Oana Serea and Andy Teel, and we apologize to everyone who is missing in this list although he or she should have been mentioned. Without the proofreading of Nils Altmüller, Robert Baier, Thomas Jahn, Marcus von Lossow, Florian Müller, and Karl Worthmann the book would contain even more typos and inaccuracies than it probably does—of course, the responsibility for all remaining errors lies entirely with us and we appreciate all comments on errors, typos, missing references, and the like. Beyond proofreading, we are grateful to Thomas Jahn for his help with writing the software supporting this book and to Karl Worthmann for his contributions to many results in Chaps. 6 and 10, most importantly the proof of Proposition 6.18. Finally, we would like to thank Oliver Jackson and Charlotte Cross from Springer-Verlag for their excellent support.

Bayreuth, Perth
April 2011

Lars Grüne
Jürgen Pannek

Contents

Chapter 1
Introduction

1.1 What Is Nonlinear Model Predictive Control?

Nonlinear model predictive control (henceforth abbreviated as NMPC) is an optimization-based method for the feedback control of nonlinear systems. Its primary applications are *stabilization* and *tracking* problems, which we briefly introduce in order to describe the basic idea of model predictive control.

Suppose we are given a controlled process whose state $x(n)$ is measured at discrete time instants t_n, $n = 0, 1, 2, \ldots$. "Controlled" means that at each time instant we can select a control input $u(n)$ which influences the future behavior of the state of the system. In tracking control, the task is to determine the control inputs $u(n)$ such that $x(n)$ follows a given *reference* $x^{\mathrm{ref}}(n)$ as good as possible. This means that if the current state is far away from the reference then we want to control the system towards the reference and if the current state is already close to the reference then we want to keep it there. In order to keep this introduction technically simple, we consider $x(n) \in X = \mathbb{R}^d$ and $u(n) \in U = \mathbb{R}^m$, furthermore we consider a reference which is constant and equal to $x_* = 0$, i.e., $x^{\mathrm{ref}}(n) = x_* = 0$ for all $n \geq 0$. With such a constant reference, the tracking problem reduces to a stabilization problem; in its full generality the tracking problem will be considered in Sect. 3.3.

Since we want to be able to react to the current deviation of $x(n)$ from the reference value $x_* = 0$, we would like to have $u(n)$ in *feedback form*, i.e., in the form $u(n) = \mu(x(n))$ for some map μ mapping the state $x \in X$ into the set U of control values.

The idea of model predictive control—linear or nonlinear—is now to utilize a model of the process in order to predict and optimize the future system behavior. In this book, we will use models of the form

$$x^+ = f(x, u) \tag{1.1}$$

where $f : X \times U \to X$ is a known and in general nonlinear map which assigns to a state x and a control value u the successor state x^+ at the next time instant. Starting

L. Grüne and J. Pannek, *Nonlinear Model Predictive Control*,
Communications and Control Engineering, DOI 10.1007/978-3-319-46024-6_1

from the current state $x(n)$, for any given control sequence $u(0), \ldots, u(N-1)$ with *horizon length* $N \geq 2$, we can now iterate (1.1) in order to construct a prediction trajectory x_u defined by

$$x_u(0) = x(n), \quad x_u(k+1) = f(x_u(k), u(k)), \quad k = 0, \ldots, N-1. \tag{1.2}$$

Proceeding this way, we obtain predictions $x_u(k)$ for the state of the system $x(n+k)$ at time t_{n+k} in the future. Hence, we obtain a prediction of the behavior of the system on the discrete interval $t_n, \ldots, t_{n+N}$ depending on the chosen control sequence $u(0), \ldots, u(N-1)$.

Now we use optimal control in order to determine $u(0), \ldots, u(N-1)$ such that x_u is as close as possible to $x_* = 0$. To this end, we measure the distance between $x_u(k)$ and $x_* = 0$ for $k = 0, \ldots, N-1$ by a function $\ell(x_u(k), u(k))$. Here, we not only allow for penalizing the deviation of the state from the reference but also—if desired—the distance of the control values $u(k)$ to a reference control u_*, which here we also choose as $u_* = 0$. A common and popular choice for this purpose is the quadratic function

$$\ell(x_u(k), u(k)) = \|x_u(k)\|^2 + \lambda \|u(k)\|^2,$$

where $\|\cdot\|$ denotes the usual Euclidean norm and $\lambda \geq 0$ is a weighting parameter for the control, which could also be chosen as 0 if no control penalization is desired. The optimal control problem now reads

$$\text{minimize} \quad J(x(n), u(\cdot)) := \sum_{k=0}^{N-1} \ell(x_u(k), u(k))$$

with respect to all admissible[1] control sequences $u(0), \ldots, u(N-1)$ with x_u generated by (1.2).

Let us assume that this optimal control problem has a solution which is given by the minimizing control sequence $u^\star(0), \ldots, u^\star(N-1)$, i.e.,

$$\min_{u(0),\ldots,u(N-1)} J(x(n), u(\cdot)) = \sum_{k=0}^{N-1} \ell(x_{u^\star}(k), u^\star(k)).$$

In order to get the desired feedback value $\mu(x(n))$, we now set $\mu(x(n)) := u^\star(0)$, i.e., we apply the first element of the optimal control sequence. This procedure is sketched in Fig. 1.1.

At the following time instants $t_{n+1}, t_{n+2}, \ldots$ we repeat the procedure with the new measurements $x(n+1), x(n+2), \ldots$ in order to derive the feedback values $\mu(x(n+1)), \mu(x(n+2)), \ldots$. In other words, we obtain the feedback law μ by

[1] The meaning of "admissible" will be defined in Sect. 3.2.

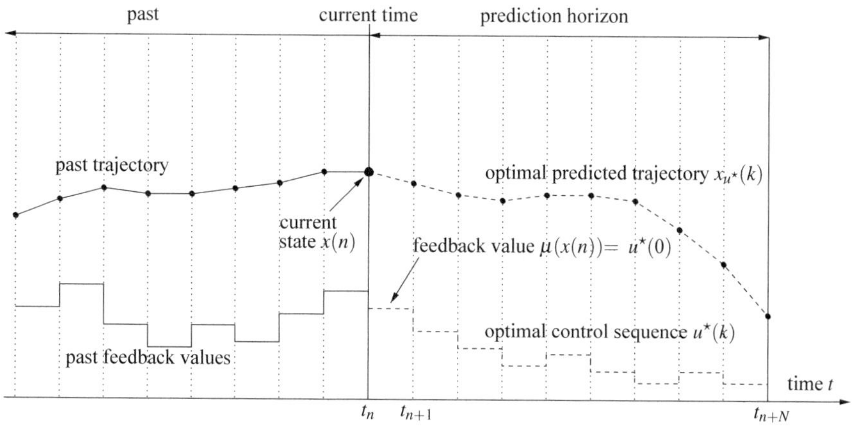

Fig. 1.1 Illustration of the NMPC step at time t_n

an *iterative online optimization* over the predictions generated by our model (1.1).[2] This is the first key feature of model predictive control.

From the prediction horizon point of view, proceeding this iterative way the trajectories $x_u(k), k = 0, \ldots, N$ provide a prediction on the discrete interval $t_n, \ldots, t_{n+N}$ at time t_n, on the interval $t_{n+1}, \ldots, t_{n+N+1}$ at time t_{n+1}, on the interval $t_{n+2}, \ldots, t_{n+N+2}$ at time t_{n+2}, and so on. Hence, the prediction horizon is moving and this *moving horizon* is the second key feature of model predictive control.

Regarding terminology, another term which is often used alternatively to *model predictive control* is *receding horizon control*. While the former expression stresses the use of model-based predictions, the latter emphasizes the moving horizon idea. Despite these slightly different literal meanings, we prefer and follow the common practice to use these names synonymously. The additional term *nonlinear* indicates that our model (1.1) need not be a linear map.

1.2 Where Did NMPC Come From?

Due to the vast amount of literature, the brief history of NMPC we provide in this section is inevitably incomplete and focused on those references in the literature from which we ourselves learned about the various NMPC techniques. Furthermore, we focus on the systems theoretic aspects of NMPC and on the academic development; some remarks on numerical methods specifically designed for NMPC can be found

[2] Attentive readers may already have noticed that this description is mathematically idealized since we neglected the computation time needed to solve the optimization problem. In practice, when the measurement $x(n)$ is provided to the optimizer the feedback value $\mu(x(n))$ will only be available after some delay. For simplicity of exposition, throughout our theoretical investigations we will assume that this delay is negligible. We will come back to this problem in Sect. 10.6.

in Sect. 12.7. Information about the use of linear and nonlinear MPC in practical applications can be found in many articles, books and proceedings volumes, e.g., in [9, 16, 23].

Nonlinear model predictive control grew out of the theory of optimal control which had been developed in the middle of the twentieth century with seminal contributions like the maximum principle of Pontryagin, Boltyanskii, Gamkrelidze and Mishchenko [21] and the dynamic programming method developed by Bellman [2]. The first paper we are aware of in which the central idea of model predictive control—for discrete time linear systems—is formulated was published by Propoĭ [22] in the early 1960s. Interestingly enough, in this paper neither Pontryagin's maximum principle nor dynamic programming is used in order to solve the optimal control problem. Rather, the paper already proposed the method which is predominant nowadays in NMPC, in which the optimal control problem is transformed into a static optimization problem, in this case a linear one. For nonlinear systems, the idea of model predictive control can be found in the book by Lee and Markus [15] from 1967 on p. 423:

> One technique for obtaining a feedback controller synthesis from knowledge of open-loop controllers is to measure the current control process state and then compute very rapidly for the open-loop control function. The first portion of this function is then used during a short time interval, after which a new measurement of the process state is made and a new open-loop control function is computed for this new measurement. The procedure is then repeated.

Due to the fact that neither computer hardware nor software for the necessary "very rapid" computation were available at that time, for a while this observation had little practical impact.

In the late 1970s, due to the progress in algorithms for solving constrained linear and quadratic optimization problems, MPC for linear systems became popular in control engineering. Richalet, Rault, Testud, and Papon [25] and Cutler and Ramaker [6] were among the first to propose this method in the area of process control, in which the processes to be controlled are often slow enough in order to allow for an online optimization, even with the computer technology available at that time. It is interesting to note that in [25] the method was described as a "new method of digital process control" and earlier references were not mentioned; it appears that the basic MPC principle was reinvented several times. Systematic stability investigations appeared a little bit later; an account of early results in that direction for linear MPC can, e.g., be found in the survey paper of García, Prett, and Morari [11] or in the monograph by Bitmead, Gevers and Wertz [3]. Many of the techniques which later turned out to be useful for NMPC, like Lyapunov function-based stability proofs or stabilizing terminal conditions were in fact first developed for linear MPC and later carried over to the nonlinear setting.

The earliest paper we were able to find which analyzes an NMPC algorithm similar to the ones used today is an article by Chen and Shaw [4] from 1982. In this paper, stability of an NMPC scheme with equilibrium terminal constraint in continuous time is proved using Lyapunov function techniques; however, the whole optimal control function on the optimization horizon is applied to the plant, as opposed to only the first part as in our NMPC paradigm. For NMPC algorithms meeting this paradigm, first comprehensive stability studies for schemes with equilibrium terminal constraint were given in 1988 by Keerthi and Gilbert [14] in discrete time and in 1990 by Mayne and Michalska [18] in continuous time. The fact that for nonlinear systems equilibrium terminal constraints may cause severe numerical difficulties subsequently motivated the investigation of alternative techniques. Terminal conditions, which consist of regional terminal constraints in combination with appropriate terminal costs, turned out to be a suitable tool for this purpose and in the second half of the 1990s there was a rapid development of such techniques with contributions by De Nicolao, Magni, and Scattolini [7, 8], Magni and Sepulchre [17] or Chen and Allgöwer [5], both in discrete and continuous time. This development eventually led to the formulation of a widely accepted "axiomatic" stability framework for NMPC schemes with stabilizing terminal conditions as formulated in discrete time in the survey article by Mayne, Rawlings, Rao, and Scokaert [19] in 2000, which is also an excellent source for more detailed information on the history of various NMPC variants not mentioned here. This framework also forms the core of our stability analysis of such schemes in Chap. 5 of this book. A continuous time version of such a framework was given by Fontes [10] in 2001.

All stability results discussed so far add terminal conditions including terminal costs and additional state constraints to the finite horizon optimization in order to ensure stability. Among the first who provided a rigorous stability result of an NMPC scheme without such constraints were Parisini and Zoppoli [20] and Alamir and Bornard [1], both in 1995 and for discrete time systems. Parisini and Zoppoli [20], however, still needed a terminal cost with specific properties similar to the one used in [5]. Alamir and Bonnard [1] were able to prove stability without such terminal cost by imposing a rank condition on the linearization on the system. Under less restrictive conditions, stability results were provided in 2005 by Grimm, Messina, Tuna, and Teel [12] for discrete time systems and by Jadbabaie and Hauser [13] for continuous time systems. The results presented in Chap. 6 of this book are qualitatively similar to these references but use slightly different assumptions and a different proof technique which allows for quantitatively tighter results; for more details we refer to the discussions in Sects. 6.1 and 6.9.

After the basic systems theoretic principles of NMPC had been clarified, more advanced topics such as robustness of stability and feasibility under perturbations, performance estimates and efficiency of numerical algorithms, and extensions to optimal control problems not a priori related to stabilization or tracking were addressed. For a discussion of these more recent issues including a number of references, we refer to the final sections of the respective chapters of this book.

1.3 How Is This Book Organized?

The book consists of two main parts, which cover systems theoretic aspects of NMPC in Chaps. 2–10 on the one hand and numerical and algorithmic aspects in Chaps. 11–12 on the other hand. These parts are, however, not strictly separated; in particular, many of the theoretical and structural properties of NMPC developed in the first part are used when looking at the performance of numerical algorithms.

The basic theme of the first part of the book is the systems theoretic analysis of stability, performance, feasibility, and robustness of NMPC schemes. This part starts with the introduction of the *class of systems* and the presentation of *background material* from Lyapunov stability theory in Chap. 2 and proceeds with a *detailed description of different NMPC algorithms* as well as related background information on dynamic programming in Chap. 3.

A distinctive feature of this book is that both schemes with stabilizing terminal conditions as well as schemes without such conditions are considered and treated in a uniform way. This "uniform way" consists of interpreting both classes of schemes as relaxed versions of *infinite horizon optimal control*. To this end, Chap. 4 first develops the theory of infinite horizon optimal control and shows by means of dynamic programming and Lyapunov function arguments that infinite horizon optimal feedback laws are actually asymptotically stabilizing feedback laws. The main building block of our subsequent analysis is the development of a *relaxed dynamic programming* framework in Sect. 4.3. Roughly speaking, Theorems 4.11 and 4.14 in this section extract the main structural properties of the infinite horizon optimal control problem, which ensure

- asymptotic or practical asymptotic stability of the closed loop,
- admissibility, i.e., maintaining the imposed state constraints,
- a guaranteed bound on the infinite horizon performance of the closed loop,
- applicability to NMPC schemes with and without stabilizing terminal conditions.

The application of these theorems does not necessarily require that the feedback law to be analyzed is close to an infinite horizon optimal feedback law in some quantitative sense. Rather, it requires that the two feedback laws share certain properties which are sufficient in order to conclude asymptotic or practical asymptotic stability and admissibility for the closed loop. While our approach allows for investigating the infinite horizon performance of the closed loop for most schemes under consideration—which we regard as an important feature of the approach in this book—we would like to emphasize that near optimal infinite horizon performance is not needed for ensuring stability and admissibility.

The results from Sect. 4.3 are then used in the subsequent Chaps. 5 and 6 in order to analyze stability, admissibility and infinite horizon performance properties for stabilizing NMPC schemes with and without stabilizing terminal conditions, respectively. Here, the results for *NMPC schemes with stabilizing terminal conditions* in Chap. 5 can by now be considered as classical and thus mainly summarize what can be found in the literature, although some results—e.g., Theorems 5.21 and 5.22—generalize known results. In contrast to this, the results for *NMPC schemes without stabilizing terminal conditions* in Chap. 6 were mainly developed by ourselves and coauthors and have not been presented before in this way.

In Chap. 7 we discuss *feasibility and robustness* issues. In particular, in Sects. 7.1–7.3 we present feasibility results for NMPC schemes without terminal constraints and without imposing viability assumptions on the state constraints. The results are reformulations of results from the literature, however, some of them so far were only established for linear MPC. These results finish our study of the properties of stabilizing NMPC schemes in the nominal case, which is why it is followed by a comparative discussion of the advantages and disadvantages of the various stabilizing NMPC schemes presented in this book in Sect. 7.4. The remaining sections in Chap. 7 address the robustness of the stability of the NMPC closed loop with respect to additive perturbations and measurement errors. Here we decided to present a selection of results we consider representative, partially from the literature and partially based on our own research.

Chapter 8 investigates a variant of NMPC known as *Economic NMPC*. Despite the name, these schemes are not necessarily connected to problems from economy. The name rather refers to a class of NMPC schemes in which the stage cost does not penalize the distance to a desired reference solution, but models different optimization objectives like minimization of energy consumption or maximization of output. As these objectives are often motivated by economic considerations, the name Economic NMPC emerged. In Chap. 8, we investigate some of the important properties of the resulting closed-loop solutions, such as averaged optimality, asymptotic stability of an optimal equilibrium, and finite time or transient optimality, both for formulations with and without stabilizing terminal conditions. In both cases, the notion of strict dissipativity plays an important role. The results in this chapter are partially developed by us and partially by other researchers. The unified presentation of schemes with and without terminal conditions is again a particular feature of this book.

Chapter 9 gives an introduction into *Distributed NMPC*. In this variant of NMPC, the overall system is divided into subsystems and the optimization problem to be solved in each step of the NMPC scheme is solved individually in each subsystem. The approach is motivated by practical applications in which a centralized NMPC controller cannot be realized for various reasons, either because of the inherent distributed nature of the overall system as, e.g., in traffic control, or because of the sheer size of the system as, e.g., in large production plants or logistic networks. Of course, there exist many different ways to divide a system into subsystems and to solve optimization problems on the level of the individual subsystems. During the optimization, the subsystems may communicate with each other or not, and they may try to minimize a common optimization objective together or focus on their individual objectives. For this reason, the chapter starts with an abstract framework, in which the problem formulation and basic goals like feasibility and stability are formulated independent of the concrete communication structures and couplings. Popular algorithms from the literature like the scheme by Richards and How and the ADMM algorithm are then discussed as particular examples in this framework.

While most of the results in this book are formulated and proved in a mathematically rigorous way, Chap. 10 deviates from this practice and presents a couple of *variants and extensions* of the basic NMPC schemes considered before in a more survey like manner. Here, proofs are occasionally only sketched with appropriate references to the literature. These extensions finish the systems theoretic part of the book.

The numerical part of the book covers two central questions in NMPC: how can we numerically compute the predicted trajectories needed in NMPC for finite-dimensional sampled data systems and how is the optimization in each NMPC step performed numerically? The first issue is treated in Chap. 11, in which we start by giving an overview on *numerical one step methods*, a classical numerical technique for solving ordinary differential equations. After having looked at the convergence analysis and adaptive step size control techniques, we discuss some implementational issues for the use of this methods within NMPC schemes. Finally, we investigate how the numerical approximation errors affect the closed-loop behavior, using the robustness results from Chap. 7.

The last Chap. 12 is devoted to numerical algorithms for solving nonlinear finite horizon optimal control problems. We concentrate on so-called *direct methods* which form the currently by far preferred class of algorithms in NMPC applications. In these methods, the optimal control problem is transformed into a static optimization problem which can then be solved by nonlinear programming algorithms. We describe different ways of how to do this transformation and then give a detailed introduction into some popular nonlinear programming algorithms for constrained optimization. The focus of this introduction is on explaining how these algorithms work rather than on a rigorous convergence theory and its purpose is twofold: on the one hand, even though we do not expect our readers to implement such algorithms, we still think that some background knowledge is helpful in order to understand the opportunities and limitations of these numerical methods. On the other hand, we want to highlight the key features of these algorithms in order to be able to explain how they can be efficiently used within an NMPC scheme. This is the topic of the final Sects. 12.4–12.6, in which several issues regarding efficient implementation, warm start and feasibility are investigated. Like Chap. 10 and in contrast to the other chapters in the book, Chap. 12 has in large parts a more survey like character, since a comprehensive and rigorous treatment of these topics would easily fill an entire book. Still, we hope that this chapter contains valuable information for those readers who are interested not only in systems theoretic foundations but also in the practical numerical implementation of NMPC schemes.

Last but not least, for all examples presented in this book we offer either MATLAB® or C++ code in order to reproduce our numerical results. This code is available from the web page

www.nmpc-book.com

Both our MATLAB NMPC routine—which is suitable for smaller problems—and our C++ NMPC package—which can also handle larger problems with reasonable computing time—can also be modified in order to perform simulations for problems not treated in this book. In order to facilitate both the usage and the modification, the Appendix A contains brief descriptions of our routines.

Beyond numerical experiments, almost every chapter contains a small selection of problems related to the more theoretical results. Solutions for these problems are available from the authors upon request by email. Attentive readers will note that several of these problems—as well as some of our examples—are actually linear problems. Even though all theoretical and numerical results apply to general nonlinear systems, we have decided to include such problems and examples, because nonlinear problems hardly ever admit analytical solutions, which are needed in order to solve problems or to work out examples without the help of numerical algorithms.

Let us finally say a few words on the class of systems and NMPC problems considered in this book. Most results are formulated for discrete time systems on arbitrary metric spaces, which in particular cover finite- and infinite-dimensional sampled data systems. The discrete time setting has been chosen because of its notational and conceptual simplicity compared to a continuous time formulation. Still, since sampled data continuous time systems form a particularly important class of systems, we have made considerable effort in order to highlight the peculiarities of this system class whenever appropriate. This concerns, among other topics, the relation between sampled data systems and discrete time systems in Sect. 2.2, the derivation of continuous time stability properties from their discrete time counterparts in Sect. 2.4 and Remark 4.13, the transformation of continuous time NMPC schemes into the discrete time formulation in Sect. 3.5 and the numerical solution of ordinary differential equations in Chap. 11. Readers or lecturers who are interested in NMPC in a pure discrete time framework may well skip these parts of the book.

The settings discussed in this book include the asymptotic tracking problem in which the goal is to asymptotically stabilize a time-varying reference $x^{\mathrm{ref}}(n)$. This leads to a time-varying NMPC formulation; in particular, the optimal control problem to be solved in each step of the NMPC algorithm explicitly depends on the current time. All of the fundamental results in Chaps. 2–4 explicitly take this time dependence into account. However, in order to be able to concentrate on concepts rather than on technical details, in the subsequent chapters we often decided to simplify the setting. To this end, many results in Chaps. 5–7 are first formulated for time-invariant problems $x^{\mathrm{ref}} \equiv x_*$—i.e., for stabilizing an x_*—and the necessary modifications for the time-varying case are discussed afterwards.

1.4 What Is Not Covered in This Book?

The area of NMPC has grown so rapidly over the last two decades that it is virtually impossible to cover all developments in detail. In order not to overload this book, we have decided to omit several topics, despite the fact that they are certainly important and useful in a variety of applications. We end this introduction by giving a brief overview over some of these topics.

For this book, we decided to concentrate on NMPC schemes with online optimization only, thus leaving out all approaches in which part of the optimization is carried out offline. Some of these methods, which can be based on both infinite horizon and finite horizon optimal control and are often termed *explicit MPC*, are briefly discussed in Sects. 3.5 and 4.4. Furthermore, we will not discuss special classes of nonlinear systems, e.g., piecewise linear systems often considered in the explicit MPC literature.

Regarding robustness of NMPC controllers under perturbations, we have restricted our attention to schemes in which the optimization is carried out for a nominal model, i.e., in which the perturbation is not explicitly taken into account in the optimization objective, cf. Sects. 7.5–7.9. Some variants of model predictive control in which the perturbation is explicitly taken into account, like min–max MPC schemes building on game theoretic ideas or tube-based MPC schemes relying on set oriented methods are briefly discussed in Sect. 7.10.

At the very heart of each NMPC algorithm is a mathematical model of the systems dynamics, which leads to the discrete time dynamics f in (1.1). While we will explain in detail in Sect. 2.2 and Chap. 11 how to obtain such a discrete time model from a differential equation, we will not address the question of how to obtain a suitable differential equation or how to identify the parameters in this model. Both modeling and parameter identification are serious problems in their own right which cannot be covered in this book. It should, however, be noted that optimization methods similar to those used in NMPC can also be used for parameter identification; see, e.g., Schittkowski [26].

A somewhat related problem stems from the fact that NMPC inevitably leads to a feedback law in which the full state $x(n)$ needs to be measured in order to evaluate the feedback law, i.e., a state feedback law. In most applications, this information is not available; instead, only output information $y(n) = h(x(n))$ for some output map h is at hand. This implies that the state $x(n)$ must be reconstructed from the output $y(n)$ by means of a suitable observer. While there is a variety of different techniques for this purpose, it is interesting to note that an idea which is very similar to NMPC can be used for this purpose: in the so-called *moving horizon state estimation* approach the state is estimated by iteratively solving optimization problems over a moving time horizon, analogous to the repeated minimization of $J(x(n), u(\cdot))$ described above. However, instead of minimizing the future deviations of the predictions from the reference value, here the past deviations of the trajectory from the measured output values are minimized. More information on this topic can be found, e.g., in Rawlings and Mayne [24, Chap. 4] and the references therein.

References

1. Alamir, M., Bornard, G.: Stability of a truncated infinite constrained receding horizon scheme: the general discrete nonlinear case. Automatica **31**(9), 1353–1356 (1995)
2. Bellman, R.: Dynamic Programming. Princeton University Press, Princeton (1957). Reprinted in 2010
3. Bitmead, R.R., Gevers, M., Wertz, V.: Adaptive Optimal Control: The Thinking Man's GPC. International Series in Systems and Control Engineering. Prentice-Hall, New York (1990)
4. Chen, C.C., Shaw, L.: On receding horizon feedback control. Automatica **18**(3), 349–352 (1982)
5. Chen, H., Allgöwer, F.: Nonlinear model predictive control schemes with guaranteed stability. In: Berber, R., Kravaris, C. (eds.) Nonlinear Model Based Process Control, pp. 465–494. Kluwer Academic, Dodrecht (1999)

6. Cutler, C.R., Ramaker, B.L.: Dynamic matrix control – a computer control algorithm. In: Proceedings of the Joint Automatic Control Conference, pp. 13–15 (1980)
7. De Nicolao, G., Magni, L., Scattolini, R.: Stabilizing nonlinear receding horizon control via a nonquadratic terminal state penalty. In: CESA'96 IMACS Multiconference: Computational Engineering in Systems Applications, Lille, France, pp. 185–187 (1996)
8. De Nicolao, G., Magni, L., Scattolini, R.: Stabilizing receding-horizon control of nonlinear time-varying systems. IEEE Trans. Automat. Control **43**(7), 1030–1036 (1998)
9. Del Re, L., Allgöwer, F., Glielmo, L., Guardiola, C., Kolmanovsky, I. (eds.): Automotive Model Predictive Control – Models, Methods and Applications. Lecture Notes in Control and Information Sciences. Springer, Berlin (2010)
10. Fontes, F.A.C.C.: A general framework to design stabilizing nonlinear model predictive controllers. Syst. Control Lett. **42**(2), 127–143 (2001)
11. García, C.E., Prett, D.M., Morari, M.: Model predictive control: Theory and practice - a survey. Automatica **25**(3), 335–348 (1989)
12. Grimm, G., Messina, M.J., Tuna, S.E., Teel, A.R.: Model predictive control: for want of a local control Lyapunov function, all is not lost. IEEE Trans. Automat. Control **50**(5), 546–558 (2005)
13. Jadbabaie, A., Hauser, J.: On the stability of receding horizon control with a general terminal cost. IEEE Trans. Automat. Control **50**(5), 674–678 (2005)
14. Keerthi, S.S., Gilbert, E.G.: Optimal infinite-horizon feedback laws for a general class of constrained discrete-time systems: stability and moving-horizon approximations. J. Optim. Theory Appl. **57**(2), 265–293 (1988)
15. Lee, E.B., Markus, L.: Foundations of Optimal Control Theory. Wiley, New York (1967)
16. Maciejowski, J.M.: Predictive Control with Constraints. Prentice Hall, New York (2002)
17. Magni, L., Sepulchre, R.: Stability margins of nonlinear receding-horizon control via inverse optimality. Syst. Control Lett. **32**(4), 241–245 (1997)
18. Mayne, D.Q., Michalska, H.: Receding horizon control of nonlinear systems. IEEE Trans. Automat. Control **35**(7), 814–824 (1990)
19. Mayne, D.Q., Rawlings, J.B., Rao, C.V., Scokaert, P.O.M.: Constrained model predictive control: stability and optimality. Automatica **36**(6), 789–814 (2000)
20. Parisini, T., Zoppoli, R.: A receding-horizon regulator for nonlinear systems and a neural approximation. Automatica **31**(10), 1443–1451 (1995)
21. Pontryagin, L.S., Boltyanskii, V.G., Gamkrelidze, R.V., Mishchenko, E.F.: The Mathematical Theory of Optimal Processes. Translated by D. E. Brown. Pergamon/Macmillan, New York (1964)
22. Propoĭ, A.I.: Application of linear programming methods for the synthesis of automatic sampled-data systems. Avtom. Telemeh. **24**, 912–920 (1963)
23. Qin, S.J., Badgwell, T.A.: A survey of industrial model predictive control technology. Control Eng. Pract. **11**, 733–764 (2003)
24. Rawlings, J.B., Mayne, D.Q.: Model Predictive Control: Theory and Design. Nob Hill Publishing, Madison (2009)
25. Richalet, J., Rault, A., Testud, J.L., Papon, J.: Model predictive heuristic control: applications to industrial processes. Automatica **14**, 413–428 (1978)
26. Schittkowski, K.: Numerical Data Fitting in Dynamical Systems. Applied Optimization, vol. 77. Kluwer Academic, Dordrecht (2002)

Chapter 2
Discrete Time and Sampled Data Systems

2.1 Discrete Time Systems

In this book, we investigate model predictive control for discrete time nonlinear control systems of the form

$$x^+ = f(x, u). \tag{2.1}$$

Here, the *transition map* $f : X \times U \to X$ assigns the state $x^+ \in X$ at the next time instant to each pair of state $x \in X$ and control value $u \in U$. The *state space* X and the *control value space* U are arbitrary metric spaces, i.e., sets in which we can measure distances between two elements $x, y \in X$ or $u, v \in U$ by metrics $d_X(x, y)$ or $d_U(u, v)$, respectively. Readers who are less familiar with metric spaces may think of $X = \mathbb{R}^d$ and $U = \mathbb{R}^m$ for $d, m \in \mathbb{N}$ with the Euclidean metrics $d_X(x, y) = \|x - y\|$ and $d_U(u, v) = \|u - v\|$ induced by the usual Euclidean norm $\|\cdot\|$, although some of our examples use different spaces. While most of the systems we consider possess continuous transition maps f, we do not require continuity in general.

The *set of finite control sequences* $u(0), \dots, u(N-1)$ for $N \in \mathbb{N}$ will be denoted by U^N and the *set of infinite control sequences* $u(0), u(1), u(2), \dots$ by U^∞. Note that we may interpret the control sequences as functions $u : \{0, \dots, N-1\} \to U$ or $u : \mathbb{N}_0 \to U$, respectively. For either type of control sequences we will briefly write $u(\cdot)$ or simply u if there is no ambiguity. With $\mathbb{N}_\infty$ we denote the natural numbers including ∞ and with $\mathbb{N}_0$ the natural numbers including 0.

A *trajectory* of (2.1) is obtained as follows: given an initial value $x_0 \in X$ and a control sequence $u(\cdot) \in U^K$ for $K \in \mathbb{N}_\infty$, we define the trajectory $x_u(k)$ iteratively via

$$x_u(0) = x_0, \qquad x_u(k+1) = f\big(x_u(k), u(k)\big), \tag{2.2}$$

for all $k \in \mathbb{N}_0$ if $K = \infty$ and for $k = 0, 1, \dots, K-1$ otherwise. Whenever we want to emphasize the dependence on the initial value we write $x_u(k, x_0)$.

An important basic property of the trajectories is the *cocycle property*: given an initial value $x_0 \in X$, a control $u \in U^N$, and time instants $k_1, k_2 \in \{0, \dots, N-1\}$

L. Grüne and J. Pannek, *Nonlinear Model Predictive Control*,
Communications and Control Engineering, DOI 10.1007/978-3-319-46024-6_2

with $k_1 \le k_2$ the solution trajectory satisfies

$$x_u(k_2, x_0) = x_{u(\cdot+k_1)}(k_2 - k_1, x_u(k_1, x_0)). \tag{2.3}$$

Here, the *shifted* control sequence $u(\cdot + k_1) \in U^{N-k_1}$ is given by

$$u(\cdot + k_1)(k) := u(k + k_1), \quad k \in \{0, \dots, N - k_1 - 1\}, \tag{2.4}$$

i.e., if the sequence u consists of the N elements $u(0), u(1), \dots, u(N-1)$, then the sequence $\tilde{u} = u(\cdot + k_1)$ consists of the $N - k_1$ elements $\tilde{u}(0) = u(k_1), \tilde{u}(1) = u(k_1 + 1), \dots, \tilde{u}(N - k_1 - 1) = u(N-1)$. With this definition, the identity (2.3) is easily proved by induction using (2.2).

We illustrate our class of models by three simple examples—the first two being in fact linear.

Example 2.1 One of the simplest examples of a control system of type (2.1) is given by $X = U = \mathbb{R}$ and

$$x^+ = x + u =: f(x, u).$$

This system can be interpreted as a very simple model of a vehicle on an infinite straight road in which $u \in \mathbb{R}$ is the traveled distance in the period until the next time instant. For $u > 0$ the vehicle moves right and for $u < 0$ it moves left.

Example 2.2 A slightly more involved version of Example 2.1 is obtained if we consider the state $x = (x_1, x_2)^\top \in X = \mathbb{R}^2$, where x_1 represents the position and x_2 the velocity of the vehicle. With the dynamics

$$\begin{pmatrix} x_1^+ \\ x_2^+ \end{pmatrix} = \begin{pmatrix} x_1 + x_2 + u/2 \\ x_2 + u \end{pmatrix} =: f(x, u)$$

on an appropriate time scale the control $u \in U = \mathbb{R}$ can be interpreted as the (constant) acceleration in the period until the next time instant. For a formal derivation of this model from a continuous time system see Example 2.6, below.

Example 2.3 Another variant of Example 2.1 is obtained if we consider the vehicle on a road which forms an ellipse, cf. Fig. 2.1, in which half of the ellipse is shown.

Here, the set of possible states is given by

$$X = \left\{ x \in \mathbb{R}^2 \,\middle|\, \left\| \begin{pmatrix} x_1 \\ 2x_2 \end{pmatrix} \right\| = 1 \right\}.$$

Since X is a compact subset of $\mathbb{R}^2$ (more precisely a submanifold, but we will not need this particular geometric structure) we can use the metric induced by the Euclidean norm on $\mathbb{R}^2$, i.e., $d_X(x, y) = \|x - y\|$. Defining the dynamics

$$\begin{pmatrix} x_1^+ \\ x_2^+ \end{pmatrix} = \begin{pmatrix} \sin(\vartheta(x) + u) \\ \cos(\vartheta(x) + u)/2 \end{pmatrix} =: f(x, u)$$

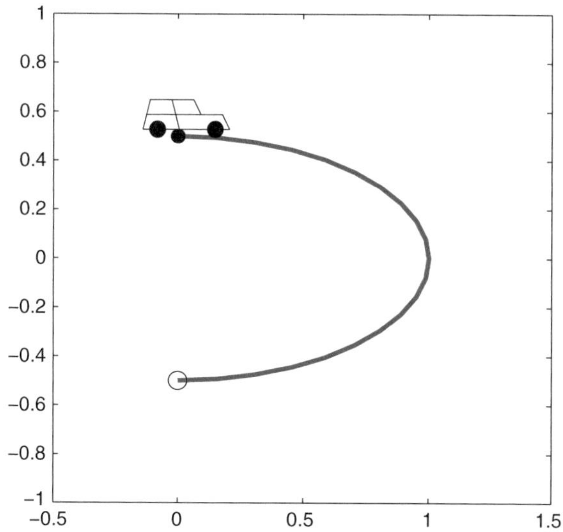

Fig. 2.1 Illustration of Example 2.3

with $u \in U = \mathbb{R}$ and

$$\vartheta(x) = \begin{cases} \arccos 2x_2, & x_1 \geq 0 \\ 2\pi - \arccos 2x_2, & x_1 < 0 \end{cases}$$

the vehicle moves on the ellipse with traveled distance $u \in U = \mathbb{R}$ in the next time step, where the traveled distance is now expressed in terms of the angle ϑ. For $u > 0$ the vehicle moves clockwise and for $u < 0$ it moves counterclockwise.

The main purpose of these very simple examples is to provide test cases which we will use in order to illustrate various effects in model predictive control. Due to their simplicity, we can intuitively guess what a reasonable controller should do and often even analytically compute different optimal controllers. This enables us to compare the behavior of the NMPC controller with our intuition and other controllers. More sophisticated models will be introduced in the next section.

As outlined in the introduction, the model (2.1) will serve for generating the predictions $x_u(k, x(n))$ which we need in the optimization algorithm of our NMPC scheme, i.e., (2.1) will play the role of the model (1.1) used in the introduction. Clearly, in general we cannot expect that this mathematical model produces exact predictions for the trajectories of the real process to be controlled. Nevertheless, during the Chaps. 3–6 and 8–10 and in Sects. 7.1–7.4 of this book we will suppose this idealized assumption. In other words, given the NMPC feedback law $\mu : X \to U$, we assume that the resulting closed-loop system satisfies

$$x^+ = f(x, \mu(x)) \tag{2.5}$$

with f from (2.1). We will refer to (2.5) as the *nominal closed-loop system*.

There are several good reasons for using this idealized assumption: First, satisfactory behavior of the nominal NMPC closed loop is a natural necessary condition for the correctness of our controller—if we cannot ensure proper functioning in the absence of modeling errors we can hardly expect the method to work under real life conditions. Second, the assumption that the prediction is based on an exact model of the process considerably simplifies the analysis and thus allows us to derive sufficient conditions under which NMPC works in a simplified setting. Last, based on these conditions for the nominal model (2.5), we can investigate additional robustness conditions which ensure satisfactory performance also for the realistic case in which (2.5) is only an approximate model for the real closed-loop behavior. This issue will be treated in Sects. 7.5–7.9.

2.2 Sampled Data Systems

Most models of real-life processes in technical and other applications are given as continuous time models, usually in form of differential equations. In order to convert these models into the discrete time form (2.1) we introduce the concept of sampling.

Let us assume that the control system under consideration is given by a finite-dimensional ordinary differential equation

$$\dot{x}(t) = f_c\big(x(t), v(t)\big) \tag{2.6}$$

with *vector field* $f_c : \mathbb{R}^d \times \mathbb{R}^m \to \mathbb{R}^d$, *control function* $v : \mathbb{R} \to \mathbb{R}^m$, and unknown function $x : \mathbb{R} \to \mathbb{R}^d$, where $\dot{x}$ is the usual short notation for the derivative dx/dt and $d, m \in \mathbb{N}$ are the dimensions of the state and the control vector. Here, we use the slightly unusual symbol v for the control function in order to emphasize the difference between the continuous time control function $v(\cdot)$ in (2.6) and the discrete time control sequence $u(\cdot)$ in (2.1).

Caratheodory's Theorem (see, e.g., [15, Theorem 54]) states conditions on f_c and v under which (2.6) has a unique solution. For its application we need the following assumption.

Assumption 2.4 The vector field $f_c : \mathbb{R}^d \times \mathbb{R}^m \to \mathbb{R}^d$ is continuous and Lipschitz in its first argument in the following sense: for each $r > 0$ there exists a constant $L > 0$ such that the inequality

$$\big\| f_c(x, v) - f_c(y, v)\big\| \le L\|x - y\|$$

holds for all $x, y \in \mathbb{R}^d$ and all $v \in \mathbb{R}^m$ with $\|x\| \le r$, $\|y\| \le r$ and $\|v\| \le r$.

Under Assumption 2.4, Caratheodory's Theorem yields that for each initial value $x_0 \in \mathbb{R}^d$, each initial time $t_0 \in \mathbb{R}$ and each locally Lebesgue integrable control function $v : \mathbb{R} \to \mathbb{R}^m$ Eq. (2.6) has a unique solution $x(t)$ with $x(t_0) = x_0$ defined for all times t contained in some open interval $I \subseteq \mathbb{R}$ with $t_0 \in I$. We denote this solution by $\varphi(t, t_0, x_0, v)$.

We further denote the space of locally Lebesgue integrable control functions mapping $\mathbb{R}$ into $\mathbb{R}^m$ by $L^\infty(\mathbb{R}, \mathbb{R}^m)$. For a precise definition of this space see, e.g., [15, Sect. C.1]. Readers who are not familiar with Lebesgue measure theory may always think of v being piecewise continuous, which is the approach taken in [7, Chap. 3]. Since the space of piecewise continuous functions is a subset of $L^\infty(\mathbb{R}, \mathbb{R}^m)$, existence and uniqueness holds for these control functions as well. Note that if we consider (2.6) only for times t from an interval $[t_0, t_1]$ then it is sufficient to specify the control function v for these times $t \in [t_0, t_1]$, i.e., it is sufficient to consider $v \in L^\infty([t_0, t_1], \mathbb{R}^m)$. Furthermore, note that two Caratheodory solutions $\varphi(t, t_0, x_0, v_1)$ and $\varphi(t, t_0, x_0, v_2)$ for $v_1, v_2 \in L^\infty(\mathbb{R}, \mathbb{R}^m)$ coincide if v_1 and v_2 coincide for almost all $\tau \in [t_0, t]$, where *almost all* means that $v_1(\tau) \neq v_2(\tau)$ may hold for $\tau \in \mathscr{T} \subset [t_0, t]$ where $\mathscr{T}$ is a set with zero Lebesgue measure. Since, in particular, sets $\mathscr{T}$ with only finitely many values have zero Lebesgue measure, this implies that for any $v \in L^\infty(\mathbb{R}, \mathbb{R}^m)$ the solution $\varphi(t, t_0, x_0, v)$ does not change if we change the value of $v(\tau)$ for finitely many times $\tau \in [t_0, t]$.[1]

The idea of sampling consists of defining a discrete time system (2.1) such that the trajectories of this discrete time system and the continuous time system coincide at the sampling times $t_0 < t_1 < t_2 < \cdots < t_N$, i.e.,

$$\varphi(t_n, t_0, x_0, v) = x_u(n, x_0), \quad n = 0, 1, 2, \ldots, N, \tag{2.7}$$

provided the continuous time control function $v : \mathbb{R} \to \mathbb{R}^m$ and the discrete time control sequence $u(\cdot) \in U^N$ are chosen appropriately. Before we investigate how this appropriate choice can be done, cf. Theorem 2.7, below, we need to specify the discrete time system (2.1) which allows for such a choice.

Throughout this book we use equidistant sampling times $t_n = nT$, $n \in \mathbb{N}_0$, with *sampling period* $T > 0$. For this choice, we claim that

$$x^+ = f(x, u) := \varphi(T, 0, x, u) \tag{2.8}$$

for $x \in \mathbb{R}^d$ and $u \in L^\infty([0, T], \mathbb{R}^m)$ is the desired discrete time system (2.1) for which (2.7) can be satisfied. Clearly, $f(x, u)$ is only well defined if the solution $\varphi(t, 0, x, u)$ exists for the time $t = T$. Unless explicitly stated otherwise, we will tacitly assume that this is the case whenever using $f(x, u)$ from (2.8).

Before we explain the precise relation between u in (2.8) and $u(\cdot)$ and $v(\cdot)$ in (2.7), cf. Theorem 2.7, below, we first look at possible choices of u in (2.8). In general, u in (2.8) may be any function in $L^\infty([0, T], \mathbb{R}^m)$, i.e., any measurable continuous time control function defined on one sampling interval. This suggests that we should use

---

[1] Strictly speaking, L^∞ functions are not even defined pointwise but rather via equivalence classes which identify all functions $v \in L^\infty(\mathbb{R}, \mathbb{R}^m)$ which coincide for almost all $t \in \mathbb{R}$. However, in order not to overload the presentation with technicalities we prefer the slightly heuristic explanation given here.

$U = L^\infty([0,T],\mathbb{R}^m)$ in (2.1) when f is defined by (2.8). However, other—much simpler—choices of U as appropriate subsets of $L^\infty([0,T],\mathbb{R}^m)$ are often possible and reasonable. This is illustrated by the following examples and discussed after Theorem 2.7 in more detail.

Example 2.5 Consider the continuous time control system

$$\dot{x}(t) = v(t)$$

with $n = m = 1$. It is easily verified that the solutions of this system are given by

$$\varphi(t,0,x_0,v) = x_0 + \int_0^t v(\tau)\,d\tau.$$

Hence, for $U = L^\infty([0,T],\mathbb{R})$ we obtain (2.8) as

$$x^+ = f(x,u) = x + \int_0^T u(\tau)\,d\tau.$$

If we restrict ourselves to constant control functions $u(t) \equiv u \in \mathbb{R}$ (for ease of notation we use the same symbol u for the function and for its constant value), which corresponds to choosing $U = \mathbb{R}$, then f simplifies to

$$f(x,u) = x + Tu.$$

If we further specify $T = 1$, then this is exactly Example 2.1.

Example 2.6 Consider the continuous time control system

$$\begin{pmatrix} \dot{x}_1(t) \\ \dot{x}_2(t) \end{pmatrix} = \begin{pmatrix} x_2(t) \\ v(t) \end{pmatrix}$$

with $n = 2$ and $m = 1$. In this model, if we interpret $x_1(t)$ as the position of a vehicle at time t, then $x_2(t) = \dot{x}_1(t)$ is its velocity and $v(t) = \dot{x}_2(t)$ its acceleration.

Again, one easily computes the solutions of this system with initial value $x_0 = (x_{01}, x_{02})^\top$ as

$$\varphi(t,0,x_0,v) = \begin{pmatrix} x_{01} + \int_0^t x_2(\tau)\,d\tau \\ x_{02} + \int_0^t v(\tau)\,d\tau \end{pmatrix} = \begin{pmatrix} x_{01} + \int_0^t \left(x_{02} + \int_0^\tau v(s)\,ds\right)\,d\tau \\ x_{02} + \int_0^t v(\tau)\,d\tau \end{pmatrix}.$$

Hence, for $U = L^\infty([0,T],\mathbb{R})$ and $x = (x_1, x_2)^\top$ we obtain (2.8) as

$$x^+ = f(x,u) = \begin{pmatrix} x_1 + Tx_2 + \int_0^T \int_0^t u(s)\,ds\,dt \\ x_2 + \int_0^T u(t)\,dt. \end{pmatrix}.$$

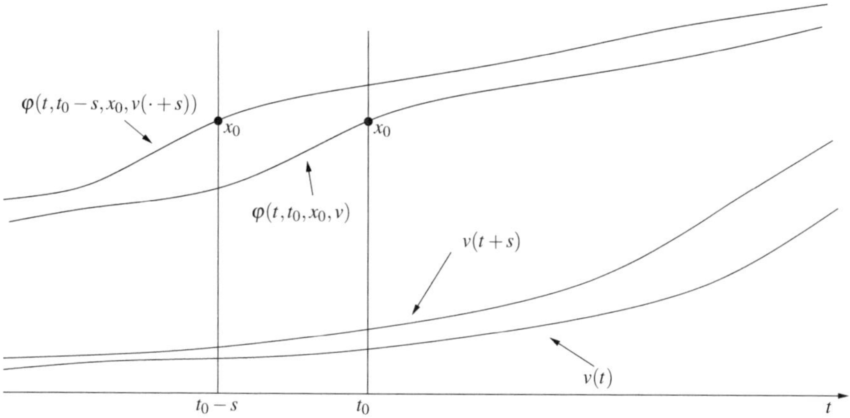

Fig. 2.2 Illustration of equality (2.9)

If we restrict ourselves to constant control functions $u(t) \equiv u \in \mathbb{R}$ (again using the same symbol u for the function and for its constant value), i.e., $U = \mathbb{R}$, then f simplifies to

$$f(x,u) = \begin{pmatrix} x_1 + Tx_2 + T^2u/2 \\ x_2 + Tu \end{pmatrix}$$

If we further specify $T = 1$, then this is exactly Example 2.2.

In order to see how the control inputs $v(\cdot)$ in (2.6) and $u(\cdot)$ in (2.8) need to be related such that (2.8) ensures (2.7), we use that the continuous time trajectories satisfy the identity

$$\varphi(t, t_0, x_0, v) = \varphi\big(t - s, t_0 - s, x_0, v(\cdot + s)\big) \tag{2.9}$$

for all $t, s \in \mathbb{R}$, provided, of course, the solutions exist for the respective times. Here $v(\cdot + s) : \mathbb{R} \to \mathbb{R}^m$ denotes the shifted control function, i.e., $v(\cdot + s)(t) = v(t + s)$, see also (2.4). This identity is illustrated in Fig. 2.2: changing $\varphi(t, t_0 - s, x_0, v(\cdot + s))$ to $\varphi(t - s, t_0 - s, x_0, v(\cdot + s))$ implies a shift of the upper graph by s to the right after which the two graphs coincide.

Identity (2.9) follows from the fact that $x(t) = \varphi(t - s, t_0 - s, x_0, v(\cdot + s))$ satisfies

$$\begin{aligned} \dot{x}(t) &= \frac{d}{dt}\varphi\big(t - s, t_0 - s, x_0, v(\cdot + s)\big) \\ &= f\big(\varphi\big(t - s, t_0 - s, x_0, v(\cdot + s)\big), v(\cdot + s)(t - s)\big) = f(x(t), v(t)) \end{aligned}$$

and

$$x(t_0) = \varphi\big(t_0 - s, t_0 - s, x_0, v(\cdot + s)\big) = x_0.$$

Hence, both functions in (2.9) satisfy (2.6) with the same control function and fulfill the same initial condition. Consequently, they coincide by uniqueness of the solution.

Using a similar uniqueness argument one sees that the solutions φ satisfy the *cocycle property*

$$\varphi(t, t_0, x_0, v) = \varphi\big(t, s, \varphi(s, t_0, x_0, v), v\big) \tag{2.10}$$

for all $t, s \in \mathbb{R}$, again provided all solutions in this equation exist for the respective times. This is the continuous time version of the discrete time cocycle property (2.3). Note that in (2.3) we have combined the discrete time counterparts of (2.9) and (2.10) into one equation since by (2.2) the discrete time trajectories always start at time 0.

With the help of (2.9) and (2.10) we can now prove the following theorem.

Theorem 2.7 *Assume that* (2.6) *satisfies Assumption 2.4 and let $x_0 \in \mathbb{R}^d$ and $v \in L^\infty([t_0, t_N], \mathbb{R}^m)$ be given such that $\varphi(t_n, t_0, x_0, v)$ exists for all sampling times $t_n = nT$, $n = 0, \ldots, N$ with $T > 0$. Define the control sequence $u(\cdot) \in U^N$ with $U = L^\infty([0, T], \mathbb{R}^m)$ by*

$$u(n) = v|_{[t_n, t_{n+1}]}(\cdot + t_n), \quad n = 0, \ldots, N-1, \tag{2.11}$$

where $v|_{[t_n, t_{n+1}]}$ denotes the restriction of v onto the interval $[t_n, t_{n+1}]$. Then

$$\varphi(t_n, t_0, x_0, v) = x_u(n, x_0) \tag{2.12}$$

holds for $n = 0, \ldots, N$ and the trajectory of the discrete time system (2.1) *defined by* (2.8).

Conversely, given $u(\cdot) \in U^N$ with $U = L^\infty([0, T], \mathbb{R}^m)$, then Eq. (2.12) *holds for $n = 0, \ldots, N$ for any $v \in L^\infty([t_0, t_N], \mathbb{R}^m)$ satisfying*

$$v(t) = u(n)(t - t_n) \quad \textit{for almost all } t \in [t_n, t_{n+1}] \textit{ and all } n = 0, \ldots, N-1, \tag{2.13}$$

provided $\varphi(t_n, t_0, x_0, v)$ exists for all sampling times $t_n = nT$, $n = 0, \ldots, N$.

Proof We prove the assertion by induction over n. For $n = 0$ we can use the initial conditions to get

$$x_u(t_0, u) = x_0 = \varphi(t_0, t_0, x_0, v).$$

For the induction step $n \to n+1$ assume (2.12) for t_n as induction assumption. Then by definition of x_u we get

$$\begin{aligned} x_u(n+1, x_0) &= f\big(x_u(n, x_0), u(n)\big) \;=\; \varphi\big(T, 0, x_u(n, x_0), u(n)\big) \\ &= \varphi\big(T, 0, \varphi(t_n, t_0, x_0, v), v(\cdot + t_n)\big) \\ &= \varphi\big(t_{n+1}, t_n, \varphi(t_n, t_0, x_0, v), v\big) \\ &= \varphi\big(t_{n+1}, t_0, x_0, v\big), \end{aligned}$$

where we used the induction assumption in the third equality, (2.9) in the fourth equality and (2.10) in the last equality.

The converse statement follows by observing that applying (2.11) for any v satisfying (2.13) yields a sequence of control functions $u(0), \dots, u(N-1)$ whose elements coincide with the original ones for almost all $t \in [0, T]$. □

Remark 2.8 At a first glance it may seem that the condition on v in (2.13) is not well defined at the sampling times t_n: from (2.13) for $n-1$ and $t = t_n$ we obtain $v(t_n) = u(n-1)(t_n - t_{n-1})$ while (2.13) for n and $t = t_n$ yields $v(t_n) = u(n)(0)$ and, of course, the values $u(n-1)(t_n - t_{n-1})$ and $u(n)(0)$ need not coincide. However, this does not pose a problem because the set of sampling times t_n in (2.13) is finite and thus the solutions $\varphi(t, t_0, x_0, v)$ do not depend on the values $v(t_n), n = 0, \dots, N-1$, cf. the discussion after Assumption 2.4. Formally, this is reflected in the words *almost all* in (2.13) which in particular imply that (2.13) is satisfied regardless of how $v(t_n)$, $n = 0, \dots, N-1$ is chosen.

Theorem 2.7 shows that we can reproduce every continuous time solution at the sampling times if we choose $U = L^\infty([0, T], \mathbb{R}^m)$. Although this is a nice property for our subsequent theoretical investigations, usually this is not a good choice for practical purposes in an NMPC context: recall from the introduction that in NMPC we want to optimize over the sequence $u(0), \dots, u(N-1) \in U^N$ in order to determine the feedback value $\mu(x(n)) = u(0) \in U$. Using $U = L^\infty([0, T], \mathbb{R}^m)$, each element of this sequence and hence also $\mu(x(n))$ is an element from a very large infinite-dimensional function space. In practice, such a general feedback concept is impossible to implement. Furthermore, although theoretically it is well possible to optimize over sequences from this space, for practical algorithms we will have to restrict ourselves to finite-dimensional sets, i.e., to subsets $U \subset L^\infty([0, T], \mathbb{R}^m)$ whose elements can be represented by finitely many parameters.

A popular way to achieve this—which is also straightforward to implement in technical applications—is via *zero-order hold*, where we choose U to be the space of constant functions which we can identify with $\mathbb{R}^m$, cf. also the Examples 2.5 and 2.6. For $u(n) \in U$, the continuous time control functions v generated by (2.13) are then piecewise constant on the sampling intervals, i.e., $v(t) = u(n)$ for almost all $t \in [t_n, t_{n+1}]$, as illustrated in Fig. 2.3. Recall from Remark 2.8 that the fact that the sampling intervals overlap at the sampling instants t_n does not pose a problem.

Consequently, the feedback $\mu(x(n))$ is a single control value from $\mathbb{R}^m$ to be used as a constant control signal on the sampling interval $[t_n, t_{n+1}]$. This is also the choice we will use in Chap. 11 on numerical methods for solving (2.6) and which is implemented in our NMPC software, cf. Appendix A. In our theoretical investigations, we will nevertheless allow for arbitrary $U \subseteq L^\infty([0, T], \mathbb{R}^m)$.

Other possible choices of U can be obtained, e.g., by polynomials $u : [0, T] \to \mathbb{R}^m$ resulting in piecewise polynomial control functions v. Yet another choice can be obtained by multirate sampling, in which we introduce a smaller sampling period $\tau = T/K$ for some $K \in \mathbb{N}$, $K \geq 2$ and choose U to be the space of functions which are constant on the intervals $[j\tau, (j+1)\tau)$, $j = 0, \dots, K-1$. In all cases the time n

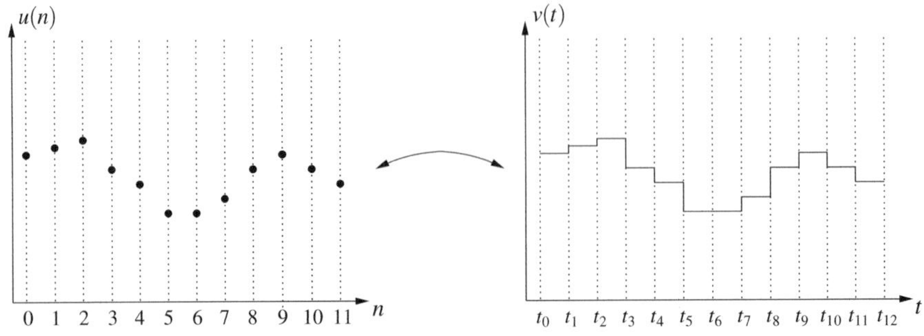

Fig. 2.3 Illustration of zero-order hold: the sequence $u(n) \in \mathbb{R}^m$ on the left corresponds to the piecewise constant control functions with $v(t) = u(n)$ for almost all $t \in [t_n, t_{n+1}]$ on the *right*

in the discrete time system (2.1) corresponds to the time $t_n = nT$ in the continuous time system.

Remark 2.9 The particular choice of U affects various properties of the resulting discrete time system. For instance, in Chap. 5 we will need the sets $\mathbb{X}_N$ which contain all initial values x_0 for which we can find a control sequence $u(\cdot)$ with $x_u(N, x_0) \in \mathbb{X}_0$ for some given set $\mathbb{X}_0$. Obviously, for sampling with zero-order hold, i.e., for $U = \mathbb{R}^m$, this set $\mathbb{X}_N$ will be smaller than for multirate sampling or for sampling with $U = L^\infty([0, T], \mathbb{R}^m)$. For this reason, we will formulate all assumptions needed in the subsequent chapters directly in terms of the discrete time system (2.1) rather than for the continuous time system (2.6), cf. also Remark 6.9.

When using sampled data models, the map f from (2.8) is usually not available in exact analytical form but only as a numerical approximation. We will discuss this issue in detail in Chap. 11.

We end this section by three further examples we will use for illustration purposes later in this book.

Example 2.10 A standard example in control theory is the *inverted pendulum on a cart* problem shown in Fig. 2.4.

This problem has two types of equilibria, the stable downright position and the unstable upright position. A typical task is to stabilize one of the unstable upright equilibria. Normalizing the mass of the pendulum to 1, the dynamics of this system can be expressed via the system of ordinary differential equations

$$\begin{aligned}
\dot{x}_1(t) &= x_2(t)\\
\dot{x}_2(t) &= -\frac{g}{l}\sin(x_1(t)) - u(t)\cos(x_1(t)) - \frac{k_L}{l}x_2(t)|x_2(t)| - k_R\operatorname{sgn}(x_2(t))\\
\dot{x}_3(t) &= x_4(t)\\
\dot{x}_4(t) &= u(t)
\end{aligned}$$

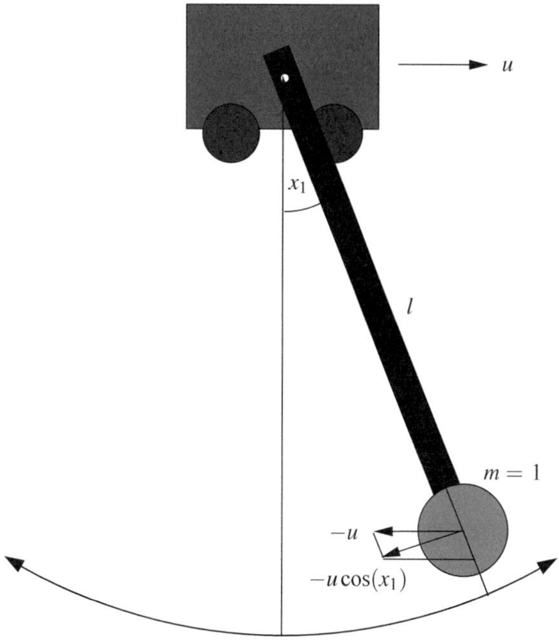

Fig. 2.4 Schematical sketch of the inverted pendulum on a cart problem: The pendulum (with unit mass $m = 1$) is attached to a cart which can be controlled using the acceleration force u. Through the joint, this force will have an effect on the dynamics of the pendulum

with gravitational force g, length of the pendulum l, air friction constant k_L, and rotational friction constant k_R. Here, x_1 denotes the angle of the pendulum, x_2 the angular velocity of the pendulum, x_3 the position, and x_4 the velocity of the cart. For this system the upright unstable equilibria are of the form $((2k+1)\pi, 0, 0, 0)^\top$ for $k \in \mathbb{Z}$.

Our model thus presented deviates from other variants often found in the literature, see, e.g., [2, 9], in terms of the types of friction we included. Instead of the linear friction model often considered, here we use a nonlinear air friction term $\frac{k_L}{l} x_2(t)|x_2(t)|$ and a rotational discontinuous Coulomb friction term $k_R \text{sgn}(x_2(t))$. The air friction term captures the fact that the force induced by the air friction grows quadratically with the speed of the pendulum mass. The Coulomb friction term is derived from first principles using Coulomb's law, see, e.g., [17] for an introduction and a description of the mathematical and numerical difficulties related to discontinuous friction terms. We consider this type of modeling as more appropriate in an NMPC context, since it describes the evolution of the dynamics more accurately, especially around the upright equilibria which we want to stabilize. For short time intervals, these nonlinear effect may be neglected, but within the NMPC design we have to predict the future development of the system for rather long periods, which may render the linear friction model inappropriate.

Unfortunately, these friction terms pose problems both theoretically and numerically.

$$\dot{x}_2(t) = -\frac{g}{l}\sin(x_1(t)) - u(t)\cos(x_1(t)) - \underbrace{\frac{k_L}{l}x_2(t)|x_2(t)|}_{\text{not } C^2} - \underbrace{k_R \operatorname{sgn}(x_2(t))}_{\text{discontinuous}}$$

The rotational Coulomb friction term is discontinuous in $x_2(t)$, hence Assumption 2.4, which is needed for Caratheodory's existence and uniqueness theorem, is not satisfied. In addition, the air friction term is only once continuously differentiable in $x_2(t)$, which poses problems when using higher order numerical methods for solving the ODE for computing the NMPC predictions, cf. the discussion before Theorem 11.5 in Chap. 11.

Hence, for the friction terms we use smooth approximations which allow us to approximate the behavior of the original equation.

$$\dot{x}_1(t) = x_2(t) \tag{2.14}$$

$$\begin{aligned}\dot{x}_2(t) = &-\frac{g}{l}\sin(x_1(t)) - \frac{k_L}{l}\arctan(1000x_2(t))x_2^2(t) - u(t)\cos(x_1(t)) \\ &- k_R\left(\frac{4ax_2(t)}{1+4(ax_2(t))^2} + \frac{2\arctan(bx_2(t))}{\pi}\right)\end{aligned} \tag{2.15}$$

$$\dot{x}_3(t) = x_4(t) \tag{2.16}$$

$$\dot{x}_4(t) = u(t). \tag{2.17}$$

In some examples in this book we will also use the linear variant of this system. To obtain it, a transformation of coordinates is applied which shifts one unstable equilibrium to the origin and then the system is linearized. Using a simplified set of parameters including only the gravitational constant g and a linear friction constant k, this leads to the linear control system

$$\dot{x}(t) = \begin{pmatrix} 0 & 1 & 0 & 0 \\ g & -k & 0 & 0 \\ 0 & 0 & 0 & 1 \\ 0 & 0 & 0 & 0 \end{pmatrix} x(t) + \begin{pmatrix} 0 \\ 1 \\ 0 \\ 1 \end{pmatrix} u(t). \tag{2.18}$$

Example 2.11 In contrast to the inverted pendulum example where our task was to stabilize one of the upright equilibria, the control task for the Arm/Rotor/Platform (ARP) model illustrated in Fig. 2.5 the different elements A, R, P, and B in the model is indicated in the description of this figure) is a digital redesign problem, see [4, 12].

Such problems consist of two separate steps: First, a continuous time control signal $v(t)$ derived from a continuous time feedback law is designed which—in the case considered here—solves a tracking problem. Since continuous time control laws may perform poorly under sampling, in a second step, the trajectory corresponding to $v(t)$ is used as a reference function to compute a digital control using NMPC such that the resulting sampled data closed loop mimics the behavior of the continuous time reference trajectory. Compared to a direct formulation of a tracking problem,

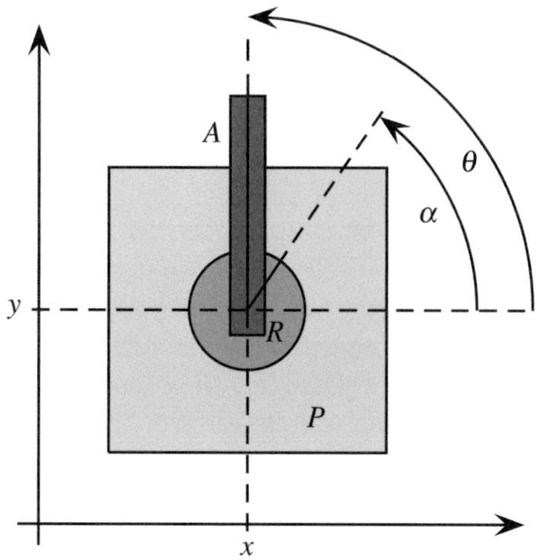

Fig. 2.5 Graphical illustration of the arm/rotor/platform (ARP) problem, see also [1, Sect. 7.3]: The arm (A) is driven by a motor (R) via a flexible joint. This motor is mounted on a platform (P) which is again flexibly connected to a fixed base (B). Moreover, we assume that there is no vertical force and that the rotational motion of the platform is not present

this approach is advantageous since the resulting NMPC problem is easier to solve. Here, we describe the model and explain the derivation of the continuous time control function $v(t)$. Numerical results for the corresponding NMPC controller are given in Example 10.21 in Chap. 10.

Using the Lagrange formalism and a change of coordinates detailed in [1, Sect. 7.3], the ARP model can be described by the differential equation system

$$\dot{x}_1(t) = x_2(t) + x_6(t)x_3(t) \tag{2.19}$$

$$\dot{x}_2(t) = -\frac{k_1}{M}x_1(t) - \frac{b_1}{M}x_2(t) + x_6(t)x_4(t) - \frac{mr}{M^2}b_1x_6(t) \tag{2.20}$$

$$\dot{x}_3(t) = -x_6(t)x_1(t) + x_4(t) \tag{2.21}$$

$$\dot{x}_4(t) = -x_6(t)x_2(t) - \frac{k_1}{M}x_3(t) - \frac{b_1}{M}x_4(t) + \frac{mr}{M^2}k_1 \tag{2.22}$$

$$\dot{x}_5(t) = x_6(t) \tag{2.23}$$

$$\dot{x}_6(t) = -a_1x_5(t) - a_2x_6(t) + a_1x_7(t) + a_3x_8(t) - p_1x_1(t) - p_2x_2(t) \tag{2.24}$$

$$\dot{x}_7(t) = x_8(t) \tag{2.25}$$

$$\dot{x}_8(t) = a_4x_5(t) + a_5x_6(t) - a_4x_7(t) - (a_5 + a_6)x_8(t) + \frac{1}{J}v(t) \tag{2.26}$$

where

$$a_1 = \frac{k_3M}{MI - (mr)^2} \qquad a_4 = \frac{k_3}{J} \qquad p_1 = \frac{mr}{MI - (mr)^2}k_1$$

$$a_2 = \frac{b_3M^2 - b_1(mr)^2}{M[MI - (mr)^2]} \qquad a_5 = \frac{b_3}{J} \qquad p_2 = \frac{mr}{MI - (mr)^2}b_1$$

$$a_3 = \frac{b_3 M}{MI - (mr)^2} \qquad\qquad a_6 = \frac{b_4}{J}$$

Here, M represents the total mass of arm, rotor, and platform and m is the mass of arm, r denotes the distance from the A/R joint to arm center of mass and I, J, and D are the moment of inertia of arm about A/R joint, of the rotor, and of the platform, respectively. Moreover, k_1, k_2, and k_3 denote the translational spring constant of P/B connection as well as the rotational spring constants of the P/B connection and the A/R joint. Last, b_1, b_2, b_3, and b_4 describe the translational friction coefficient of P/B connection as well as the rotational friction coefficients of P/B, A/R, and R/P connection, respectively. The coordinates x_1 and x_2 correspond to the (transformed) x position of P and its velocity of the platform in direction x, whereas x_3 and x_4 represent the (transformed) y position of P and the respective velocity. The remaining coordinates x_5 and x_7 denote the angles θ and α and the coordinates x_6 and x_8 the corresponding angular velocities.

Our design goal is to regulate the system such that the position of the arm relative to the platform, i.e., the angle x_5, tracks a given reference signal. Note that this task is not simple since both connections of the rotor are flexible. Here, we assume that the reference signal and its derivatives are known and available to the controller. Moreover, we assume that the relative positions and velocities x_5, x_6, x_7, and x_8 are supplied to the controller.

In order to derive the continuous time feedback, we follow the backstepping approach from [1] using the output

$$\zeta(t) = x_5(t) - \frac{a_3}{a_1 - a_2 a_3}\left[x_6(t) - a_3 x_7(t)\right]. \tag{2.27}$$

The output has relative degree 4, that is the control $v(t)$ appears explicitly within the fourth derivative of $\zeta(t)$. Expressing $\zeta^{(4)}(t)$ by the known data, we obtain the continuous time input signal[2]

$$\begin{aligned}
v(t) = & \frac{J}{a_1^2 + a_3[p] \cdot \left[\left[\frac{\partial F(x_6(t))}{\partial x_6(t)}\right] \cdot [\eta(t)] + \left[\frac{\partial G(x_6(t))}{\partial x_6(t)}\right]\right]} \\
& \Big(- \Big(- a_1 x_5(t) - a_2 x_6(t) + a_1 x_7(t) + a_3 x_8(t) - [p] \cdot [\eta(t)]\Big) \\
& \Big(- a_1^2 + a_1 a_2 (a_2 - a_3) + \Big(a_3 [p] \cdot [F(x_6(t)) - (a_1 + a_2 a_3)[p]]\Big) \\
& \left[\left[\frac{\partial F(x_6(t))}{\partial x_6(t)}\right] \cdot [\eta(t)] + \left[\frac{\partial G(x_6(t))}{\partial x_6(t)}\right]\right] \\
& + 2a_3[p]\left[\frac{\partial F(x_6(t))}{\partial x_6(t)}\right] \cdot [[F(x_6(t))] \cdot [\eta(t)] + [G(x_6(t))]] \Big)
\end{aligned}$$

[2]For details of the derivation see [13, Sect. 7.3].

$$
\begin{aligned}
&- \Big(a_4 x_5(t) + a_5 x_6(t) - a_4 x_7(t) - (a_5 + a_6) x_8(t)\Big) \\
&\Bigg(a_1^2 + a_3\Bigg(a_3[p]\cdot\Bigg[\bigg[\frac{\partial F(x_6(t))}{\partial x_6(t)}\bigg]\cdot[\eta(t)] + \bigg[\frac{\partial G(x_6(t))}{\partial x_6(t)}\bigg]\Bigg] - a_1(a_2 - a_3)\Bigg)\Bigg) \\
&- \Big(a_3[p]\cdot[F(x_6(t))] - a_1[p]\Big)\cdot[F(x_6(t))]\cdot[[F(x_6(t))]\cdot[\eta(t)] + [G(x_6(t))]] \\
&- \Big(-a_1(x_6(t) - x_8(t)) - [p]\cdot[[F(x_6(t))]\cdot[\eta(t)] + [G(x_6(t))]]\Big) \\
&\Bigg(-a_1(a_2 - a_3) + a_3[p]\cdot\Bigg[\bigg[\frac{\partial F(x_6(t))}{\partial x_6(t)}\bigg]\cdot[\eta(t)] + \bigg[\frac{\partial G(x_6(t))}{\partial x_6(t)}\bigg]\Bigg]\Bigg) \\
&+ (a_1 - a_2 a_3)\hat{v}(t)\Bigg)
\end{aligned}
\tag{2.28}
$$

where we used the abbreviations

$$
[\eta(t)] := \big(x_1(t)\ x_2(t)\ x_3(t)\ x_4(t)\big)^T, \quad [\chi(t)] := \big(x_5(t)\ x_6(t)\ x_7(t)\ x_8(t)\big)^T,
$$

$$
[F(x_6(t))] := \begin{pmatrix} 0 & 1 & x_6(t) & 0 \\ -\frac{k_1}{M} & -\frac{b_1}{M} & 0 & x_6(t) \\ -x_6(t) & 0 & 0 & 1 \\ 0 & -x_6(t) & -\frac{k_1}{M} & -\frac{b_1}{M} \end{pmatrix}, \quad [G(x_6(t))] := \begin{pmatrix} 0 \\ -\frac{mrb_1}{M^2} x_6(t) \\ 0 \\ \frac{mrk_1}{M^2} \end{pmatrix},
$$

$$
[A] := \begin{pmatrix} 0 & 1 & 0 & 0 \\ -a_1 & -a_2 & a_1 & a_3 \\ 0 & 0 & 0 & 1 \\ a_4 & a_5 & -a_4 & -(a_5 + a_6) \end{pmatrix}, \quad [E] := \begin{pmatrix} 0 & 0 \\ -p_1 & -p_2 \\ 0 & 0 \\ 0 & 0 \end{pmatrix}, \quad [B] := \begin{pmatrix} 0 \\ 0 \\ 0 \\ \frac{1}{J} \end{pmatrix}
$$

as well as the row vector $[p] := \big(p_1\ p_2\ 0\ 0\big)$. In (2.28), we added the function $\hat{v}(t)$ which we will now use as the new input. Given a desired reference $\zeta_{\text{ref}}(\cdot)$ for the output (2.27), we can track this reference by setting $\hat{v}$ in (2.28) as

$$
\begin{aligned}
\hat{v}(t) := \zeta_{\text{ref}}^{(4)}(t) - c_3(\zeta^{(3)}(t) - \zeta_{\text{ref}}^{(3)}(t)) - c_2(\ddot{\zeta}(t) - \ddot{\zeta}_{\text{ref}}(t)) \\
- c_1(\dot{\zeta}(t) - \dot{\zeta}_{\text{ref}}(t)) - c_0(\zeta(t) - \zeta_{\text{ref}}(t))
\end{aligned}
$$

with design parameters $c_i \in \mathbb{R}$, $c_i \geq 0$. These parameters are degrees of freedom within the design of the continuous time feedback which can be used as tuning parameters, e.g., to reduce the transient time or the overshoot.

Example 2.12 Another class of systems fitting our framework, which actually goes beyond the setting we used for introducing sampled data systems, are infinite-dimensional systems induced by partial differential equations (PDEs). In this example, we slightly change our notation in order to be consistent with the usual PDE notation.

In the following controlled parabolic PDE (2.29) the solution $y(t, x)$ with $y : \mathbb{R} \times \overline{\Omega} \to \mathbb{R}$ depends on time t as well as on a one-dimensional state variable $x \in \Omega = (0, L)$ for a parameter $L > 0$. Thus, the state of the system at each time t is now a continuous function $y(t, \cdot) : \overline{\Omega} \to \mathbb{R}$ and x becomes an independent variable. The control v in this example is a so-called distributed control, i.e., a measurable function $v : \mathbb{R} \times \Omega \to \mathbb{R}$. The evolution of the state is defined by the equation

$$y_t(t, x) = \theta y_{xx}(t, x) - y_x(t, x) + \rho \left(y(t, x) - y(t, x)^3\right) + v(t, x) \qquad (2.29)$$

for $x \in \Omega$ and $t \geq 0$ together with the initial condition $y(0, x) = y_0(x)$ and the boundary conditions $y(t, 0) = y(0, L) = 0$.

Here y_t and y_x denote the partial derivatives with respect to t and x, respectively and y_{xx} denotes the second partial derivative with respect to x. The parameters θ and ρ are positive constants. Of course, in order to ensure that (2.29) is well defined, we need to interpret this equation in an appropriate weak sense and make sure that for the chosen class of control functions a solution to (2.29) exists in appropriate function spaces. For details on these issues we refer to, e.g., [10] or [18]. As we will see later in Example 6.32, for suitable values of the parameters θ and ρ the uncontrolled equation, i.e., (2.29) with $v \equiv 0$, has an unstable equilibrium $y_* \equiv 0$ which can be stabilized by NMPC.

Using the letter z for the state of the discrete time system associated to the sampled-data solution of (2.29), we can abstractly write this system as

$$z^+ = f(z, u)$$

with z and z^+ being continuous functions from $\overline{\Omega}$ to $\mathbb{R}$. The function f maps $y_0 = z$ to the solution $y(T, x)$ of (2.29) at the sampling time T using the measurable control function $u = v : [0, T] \times \Omega \to \mathbb{R}$. Thus, it maps continuous functions to continuous functions; again we omit the exact details of the respective functions spaces.

As in the ordinary differential equation case, we can restrict ourselves to the zero-order hold situation, i.e., to control functions $u(t, x)$ which are constant in $t \in [0, T]$. The corresponding control functions v generated via (2.11) are again constant in t on each sampling interval $[t_n, t_{n+1})$. Note, however, that in our distributed control context both u and v are still arbitrary measurable—i.e., in particular nonconstant—functions in x.

For sampled data systems, the nominal closed-loop system (2.5) corresponds to the closed-loop sampled data system

$$\dot{x}(t) = f_c(x(t), \mu(x(t_n))(t - t_n)), \quad t \in [t_n, t_{n+1}), \; n = 0, 1, 2, \ldots \qquad (2.30)$$

whose solution with initial value $x_0 \in X$ we denote by $\varphi(t, t_0, x_0, \mu)$. Note that the argument "$(t - t_n)$" of $\mu(x(t_n))$ can be dropped in case of sampling with zero order hold when—as usual—we interpret the control value $\mu(x(t_n)) \in U = \mathbb{R}^m$ as a constant control function.

2.3 Stability of Discrete Time Systems

In the introduction, we already specified the main goal of model predictive control, namely to control the state $x(n)$ of the system toward a reference trajectory $x^{\text{ref}}(n)$ and then keep it close to this reference. In this section we formalize what we mean by "toward" and "close to" using concepts from stability theory of nonlinear systems.

We first consider the case where x^{ref} is constant, i.e., where $x^{\text{ref}} \equiv x_*$ holds for some $x_* \in X$. We assume that the states $x(n)$ are generated by a difference equation of the form

$$x^+ = g(x) \tag{2.31}$$

for a not necessarily continuous map $g : X \to X$ via the usual iteration $x(n + 1) = g(x(n))$. As before, we write $x(n, x_0)$ for the trajectory satisfying the initial condition $x(0, x_0) = x_0 \in X$. Allowing g to be discontinuous is important for our NMPC application, because g will later represent the nominal closed-loop system (2.5) controlled by the NMPC-feedback law μ, i.e., $g(x) = f(x, \mu(x))$. Since μ is obtained as an outcome of an optimization algorithm, in general we cannot expect μ to be continuous and thus g will in general be discontinuous, too.

Nonlinear stability properties can be expressed conveniently via so-called comparison functions which were first introduced by Hahn in 1967 [5] and popularized in nonlinear control theory during the 1990s by Sontag, particularly in the context of input-to-state stability [14]. Although we mainly deal with discrete time systems, we stick to the usual continuous time definition of these functions using the notation $\mathbb{R}_0^+ = [0, \infty)$.

Definition 2.13 (*Comparison functions*) We define the following classes of comparison functions.

$$\begin{aligned}
\mathcal{K} &:= \{\alpha : \mathbb{R}_0^+ \to \mathbb{R}_0^+ \mid \alpha \text{ is continuous \& strictly increasing with } \alpha(0) = 0\} \\
\mathcal{K}_\infty &:= \{\alpha : \mathbb{R}_0^+ \to \mathbb{R}_0^+ \mid \alpha \in \mathcal{K},\ \alpha \text{ is unbounded}\} \\
\mathcal{L} &:= \{\delta : \mathbb{R}_0^+ \to \mathbb{R}_0^+ \mid \delta \text{ is continuous \& strictly decreasing with } \lim_{t\to\infty} \delta(t) = 0\} \\
\mathcal{KL} &:= \{\beta : \mathbb{R}_0^+ \times \mathbb{R}_0^+ \to \mathbb{R}_0^+ \mid \beta \text{ is continuous, } \beta(\cdot, t) \in \mathcal{K},\ \beta(r, \cdot) \in \mathcal{L}\}.
\end{aligned}$$

The graph of a typical function $\beta \in \mathcal{KL}$ is shown in Fig. 2.6.

Using this function, we can now introduce the concept of asymptotic stability. Here, for arbitrary $x_1, x_2 \in X$ we denote the distance from x_1 to x_2 by

$$|x_1|_{x_2} := d_X(x_1, x_2).$$

Furthermore, we use the ball

$$\mathcal{B}_\eta(x_*) := \{x \in X \mid |x|_{x_*} < \eta\}$$

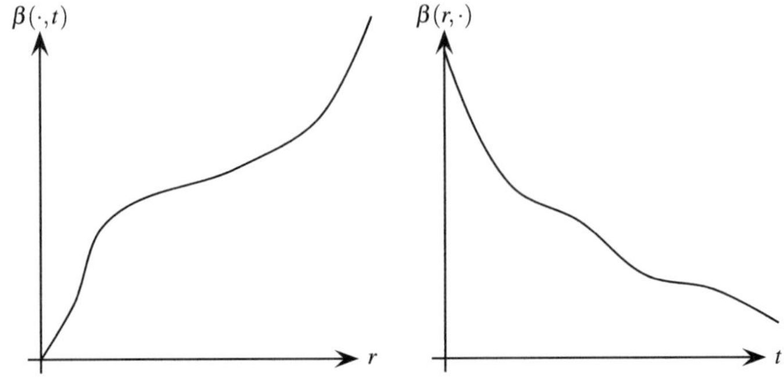

Fig. 2.6 Illustration of a typical class $\mathcal{KL}$ function

and we say that a set $Y \subseteq X$ is *forward invariant* for (2.31) if $g(x) \in Y$ holds for all $x \in Y$.

Definition 2.14 (*Asymptotic stability*) Let $x_* \in X$ be an equilibrium for (2.31), i.e., $g(x_*) = x_*$. Then we say that x_* is *locally asymptotically stable* if there exist $\eta > 0$ and a function $\beta \in \mathcal{KL}$ such that the inequality

$$|x(n, x_0)|_{x_*} \leq \beta(|x_0|_{x_*}, n) \tag{2.32}$$

holds for all $x_0 \in \mathcal{B}_\eta(x_*)$ and all $n \in \mathbb{N}_0$.

We say that x_* is *asymptotically stable on a forward invariant set* Y with $x_* \in Y$ if there exists $\beta \in \mathcal{KL}$ such that (2.32) holds for all $x_0 \in Y$ and all $n \in \mathbb{N}_0$ and we say that x_* is *globally asymptotically stable* if x_* is asymptotically stable on $Y = X$.

If one of these properties holds then β is called *attraction rate*.

Note that asymptotic stability on a forward invariant set Y implies local asymptotic stability if Y contains a ball $\mathcal{B}_\eta(x_*)$. However, we do not necessarily require this property.

Asymptotic stability thus defined consists of two main ingredients:

(i) The smaller the initial distance from x_0 to x_* is, the smaller the distance from $x(n)$ to x_* becomes for all future n, or formally: for each $\varepsilon > 0$ there exists $\delta > 0$ such that $|x(n, x_0)|_{x_*} \leq \varepsilon$ holds for all $n \in \mathbb{N}_0$ and all $x_0 \in Y$ (or $x_0 \in \mathcal{B}_\eta(x_*)$) with $|x_0|_{x_*} \leq \delta$.
This fact is easily seen by choosing δ so small that $\beta(\delta, 0) \leq \varepsilon$ holds, which is possible since $\beta(\cdot, 0) \in \mathcal{K}$. Since β is decreasing in its second argument, for $|x_0|_{x_*} \leq \delta$ from (2.32) we obtain

$$|x(n, x_0)|_{x_*} \leq \beta(|x_0|_{x_*}, n) \leq \beta(|x_0|_{x_*}, 0) \leq \beta(\delta, 0) \leq \varepsilon.$$

(ii) As the system evolves, the distance from $x(n, x_0)$ to x_* becomes arbitrarily small, or formally: for each $\varepsilon > 0$ and each $R > 0$ there exists $N > 0$ such that $|x(n, x_0)|_{x_*} \leq \varepsilon$ holds for all $n \geq N$ and all $x_0 \in Y$ (or $x_0 \in \mathcal{B}_\eta(x_*)$) with

$|x_0|_{x_*} \le R$. This property easily follows from (2.32) by choosing $N > 0$ with $\beta(R, N) \le \varepsilon$ and exploiting the monotonicity properties of β.

These two properties are known as (i) stability (in the sense of Lyapunov) and (ii) attraction. In the literature, asymptotic stability is often defined via these two properties. In fact, for continuous time (and continuous) systems (i) and (ii) are known to be equivalent to the continuous time counterpart of Definition 2.14, cf. [8, Sect. 3]. We conjecture that the arguments in this reference can be modified in order to prove that equivalence also holds for our discontinuous discrete time setting.

Asymptotic stability includes the desired properties of the NMPC closed loop described earlier: whenever we are already close to the reference equilibrium we want to stay close; otherwise we want to move toward the equilibrium.

Asymptotic stability also includes that eventually the distance of the closed-loop solution to the equilibrium x_* becomes arbitrarily small. Occasionally, this may be too demanding. In the following chapters, this is for instance the case if the system is subject to perturbations or modelling errors, cf. Sects. 7.5–7.9 or if in NMPC without stabilizing terminal constraints the system cannot be controlled to x_* sufficiently fast, cf. Sect. 6.7. In this case, one can relax the asymptotic stability definition to practical asymptotic stability as follows. Here we only consider the case of asymptotic stability on a forward invariant set Y.

Definition 2.15 (*P-practically asymptotic stability*) Let Y be a forward invariant set and let $P \subset Y$ be a subset of Y. Then we say that a point $x_* \in Y$ is *P-practically asymptotically stable on Y* if there exists $\beta \in \mathcal{KL}$ such that (2.32) holds for all $x_0 \in Y$ and all $n \in \mathbb{N}_0$ with $x(n, x_0) \notin P$.

Figure 2.7 illustrates practical asymptotic stability (on the right) as opposed to "usual" asymptotic stability (on the left).

This definition is typically used with P contained in a small ball around the equilibrium, i.e., $P \subseteq \mathcal{B}_\delta(x_*)$ for some small $\delta > 0$. In this case one obtains the estimate

$$|x(n, x_0)|_{x_*} \le \max\{\beta(|x_0|_{x_*}, n), \delta\} \tag{2.33}$$

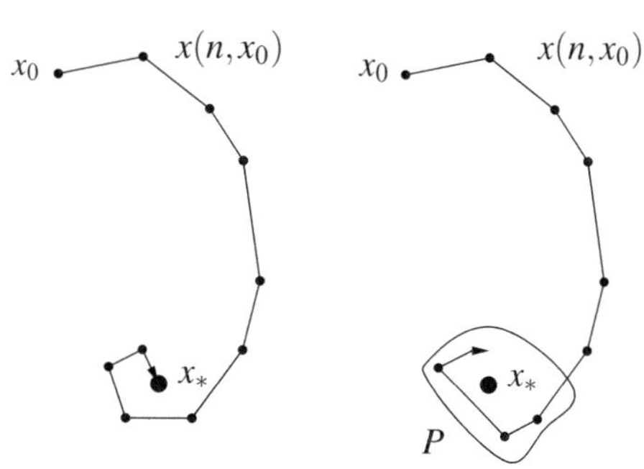

Fig. 2.7 Sketch of asymptotic stability (*left*) as opposed to practical asymptotic stability (*right*)

for all $x_0 \in Y$ and all $n \in \mathbb{N}_0$, i.e., the system behaves like an asymptotically stable system until it reaches the ball $\mathscr{B}_\delta(x_*)$. Note that x_* does not need to be an equilibrium in Definition 2.15.

For general nonconstant reference functions $x^{\text{ref}} : \mathbb{N}_0 \to X$ we can easily extend Definition 2.14 if we take into account that the objects under consideration become time varying in two ways: (i) the distance under consideration varies with n and (ii) the system (2.31) under consideration varies with n. While (i) is immediate, (ii) follows from the fact that with time varying reference also the feedback law μ is time varying, i.e., we obtain a feedback law of the type $\mu(n, x(n))$. Consequently, we now need to consider systems

$$x^+ = g(n, x) \tag{2.34}$$

with g of the form $g(n, x) = f(x, \mu(n, x))$. Furthermore, we now have to take the initial time n_0 into account: while the solutions of (2.31) look the same for all initial times n_0 (which is why we only considered $n_0 = 0$) now we need to keep track of this value. To this end, by $x(n, n_0, x_0)$ we denote the solution of (2.34) with initial condition $x(n_0, n_0, x_0) = x_0$ at time n_0. The appropriate modification of Definition 2.14 then looks as follows. Here we say that a time dependent family of sets $Y(n) \subseteq X$, $n \in \mathbb{N}_0$ is *forward invariant* if $g(n, x) \in Y(n + 1)$ holds for all $n \in \mathbb{N}_0$ and all $x \in Y(n)$.

Definition 2.16 (*Uniform asymptotic stability*) Let $x^{\text{ref}} : \mathbb{N}_0 \to X$ be a trajectory for (2.31), i.e., $x^{\text{ref}}(n+1) = g(x^{\text{ref}}(n))$ for all $n \in \mathbb{N}_0$. Then we say that x^{ref} is *locally uniformly asymptotically stable* if there exists $\eta > 0$ and a function $\beta \in \mathscr{KL}$ such that the inequality

$$|x(n, n_0, x_0)|_{x^{\text{ref}}(n)} \le \beta(|x_0|_{x^{\text{ref}}(n_0)}, n - n_0) \tag{2.35}$$

holds for all $x_0 \in \mathscr{B}_\eta(x^{\text{ref}}(n_0))$ and all $n_0, n \in \mathbb{N}_0$ with $n \ge n_0$.

We say that x_* is *uniformly asymptotically stable on a forward invariant family of sets* $Y(n)$ with $x^{\text{ref}}(n) \in Y(n)$ if there exists $\beta \in \mathscr{KL}$ such that (2.35) holds for all $n_0, n \in \mathbb{N}_0$ with $n \ge n_0$ and all $x_0 \in Y(n_0)$ and we say that x_* is *globally uniformly asymptotically stable* if x_* is asymptotically stable on $Y(n) = X$ for all $n_0 \in \mathbb{N}_0$.

If one of these properties hold then β is called *(uniform) attraction rate*.

The term "uniform" describes the fact that the bound $\beta(|x_0|_{x^{\text{ref}}(n_0)}, n - n_0)$ only depends on the elapsed time $n - n_0$ but not on the initial time n_0. If this were the case, i.e., if we needed different β for different initial times n_0, then we would call the asymptotic stability "nonuniform". For a comprehensive discussion of nonuniform stability notions and their representation via time-dependent $\mathscr{KL}$ functions we refer to [3].

As in the time-invariant case, asymptotic stability on a forward invariant family of sets $Y(n)$ implies local asymptotic stability if each $Y(n)$ contains a ball $\mathscr{B}_\eta(x^{\text{ref}}(n))$. Again, we do not necessarily require this property.

The time-varying counterpart of P-practical asymptotic stability is defined as follows.

Definition 2.17 (*Uniform P-practical asymptotic stability*) Let $Y(n)$ be a forward invariant family of sets and let $P(n) \subset Y(n)$ be subsets of $Y(n)$. Then we say that a reference trajectory x^{ref} with $x^{\text{ref}}(n) \in Y(n)$ is P-practically uniformly asymptotically stable on $Y(n)$ if there exists $\beta \in \mathcal{KL}$ such that (2.35) holds for all $x_0 \in Y(n_0)$ and all $n_0, n \in \mathbb{N}_0$ with $n \geq n_0$ and $x(n, n_0, x_0) \notin P(n)$.

Analogous to the time-invariant case, this definition is typically used with $P(n) \subseteq \mathcal{B}_\delta(x^{\text{ref}}(n))$ for some small value $\delta > 0$ which then yields

$$|x(n, n_0, x_0)|_{x^{\text{ref}}(n)} \leq \max\{\beta(|x_0|_{x^{\text{ref}}(n_0)}, n - n_0), \ \delta\}. \tag{2.36}$$

In order to verify that our NMPC controller achieves asymptotic stability, we will utilize the concept of Lyapunov functions. For constant reference $x^{\text{ref}} \equiv x_* \in X$ these functions are defined as follows.

Definition 2.18 (*Lyapunov function*) Consider a system (2.31), a point $x_* \in X$ and let $S \subseteq X$ be a subset of the state space. A function $V : S \to \mathbb{R}_0^+$ is called a *Lyapunov function* on S if the following conditions are satisfied:

(i) There exist functions $\alpha_1, \alpha_2 \in \mathcal{K}_\infty$ such that

$$\alpha_1(|x|_{x_*}) \leq V(x) \leq \alpha_2(|x|_{x_*}) \tag{2.37}$$

holds for all $x \in S$.

(ii) There exists a function $\alpha_V \in \mathcal{K}$ such that

$$V(g(x)) \leq V(x) - \alpha_V(|x|_{x_*}) \tag{2.38}$$

holds for all $x \in S$ with $g(x) \in S$.

The following theorem shows that the existence of a Lyapunov function ensures asymptotic stability.

Theorem 2.19 (Asymptotic stability using Lyapunov functions) *Let x_* be an equilibrium of* (2.31) *and assume there exists a Lyapunov function V on S. If S contains a ball $\mathcal{B}_\nu(x_*)$ with $g(x) \in S$ for all $x \in \mathcal{B}_\nu(x_*)$ then x_* is locally asymptotically stable with $\eta = \alpha_2^{-1} \circ \alpha_1(\nu)$. If $S = Y$ holds for some forward invariant set $Y \subseteq X$ containing x_* then x_* is asymptotically stable on Y. If $S = X$ holds then x_* is globally asymptotically stable.*

Proof The idea of the proof lies in showing that by (2.38) the function $V(x(n, x_0))$ is strictly decreasing in n and converges to 0. Then by (2.37) we can conclude that $x(n, x_0)$ converges to x_*. The function β from Definition 2.14 will be constructed from α_1, α_2 and α_V. In order to simplify the notation, throughout the proof we write $|x|$ instead of $|x|_{x_*}$.

First, if S is not forward invariant, define the value $\gamma := \alpha_1(\nu)$ and the set $\widetilde{S} := \{x \in S \mid V(x) < \gamma\}$. Then from (2.37) we get

$$x \in \widetilde{S} \Rightarrow \alpha_1(|x|) \leq V(x) < \gamma \Rightarrow |x| < \alpha_1^{-1}(\gamma) = \nu \Rightarrow x \in \mathscr{B}_\nu(x_*),$$

observing that each $\alpha \in \mathscr{K}_\infty$ is invertible with $\alpha^{-1} \in \mathscr{K}_\infty$.

Hence, for each $x \in \widetilde{S}$ inequality (2.38) applies and consequently $V(g(x)) \leq V(x) < \gamma$ implying $g(x) \in \widetilde{S}$. If $S = Y$ for some forward invariant set $Y \subseteq X$ we define $\widetilde{S} := S$. With these definitions, in both cases the set $\widetilde{S}$ becomes forward invariant.

Now we define $\alpha_V' := \alpha_V \circ \alpha_2^{-1}$. Note that concatenations of $\mathscr{K}$-functions are again in $\mathscr{K}$, hence $\alpha_V' \in \mathscr{K}$. Since $|x| \geq \alpha_2^{-1}(V(x))$, using monotonicity of α_V this definition implies

$$\alpha_V(|x|) \geq \alpha_V \circ \alpha_2^{-1}(V(x)) = \alpha_V'(V(x)).$$

Hence, along a trajectory $x(n, x_0)$ with $x_0 \in \widetilde{S}$, from (2.38) we get the inequality

$$V(x(n+1, x_0)) \leq V(x(n, x_0)) - \alpha_V(|x(n, x_0)|) \leq V(x(n, x_0)) - \alpha_V'(V(x(n, x_0))). \tag{2.39}$$

For the construction of β we need the last expression in (2.39) to be strictly increasing in $V(x(n, x_0))$. To this end we define

$$\tilde{\alpha}_V(r) := \min_{s \in [0,r]} \{\alpha_V'(s) + (r - s)/2\}.$$

Straightforward computations show that this function satisfies $r_2 - \tilde{\alpha}_V(r_2) > r_1 - \tilde{\alpha}_V(r_1) \geq 0$ for all $r_2 > r_1 \geq 0$ and $\min\{\alpha_V'(r/2), r/4\} \leq \tilde{\alpha}_V(r) \leq \alpha_V'(r)$ for all $r \geq 0$. In particular, (2.39) remains valid and we get the desired monotonicity when α_V' is replaced by $\tilde{\alpha}_V$.

We inductively define a function $\beta_1 : \mathbb{R}_0^+ \times \mathbb{N}_0 \to \mathbb{R}_0^+$ via

$$\beta_1(r, 0) := r, \quad \beta_1(r, n+1) = \beta_1(r, n) - \tilde{\alpha}_V(\beta_1(r, n)). \tag{2.40}$$

By induction over n using the properties of $\tilde{\alpha}_V(r)$ and Inequality (2.39) one easily verifies the following inequalities:

$$\beta_1(r_2, n) > \beta_1(r_1, n) \geq 0 \text{ for all } r_2 > r_1 \geq 0 \text{ and all } n \in \mathbb{N}_0 \tag{2.41}$$

$$\beta_1(r, n_1) > \beta_1(r, n_2) > 0 \text{ for all } n_2 > n_1 \geq 0 \text{ and all } r > 0 \tag{2.42}$$

$$V(x(n, x_0)) \leq \beta_1(V(x_0), n) \text{ for all } n \in \mathbb{N}_0 \text{ and all } x_0 \in \widetilde{S} \tag{2.43}$$

From (2.42) it follows that $\beta_1(r, n)$ is monotone decreasing in n and by (2.41) it is bounded from below by 0. Hence, for each $r \geq 0$ the limit $\beta_1^\infty(r) = \lim_{n\to\infty} \beta_1(r, n)$

exists. We claim that $\beta_1^\infty(r) = 0$ holds for all r. Indeed, convergence implies $\beta_1(r,n) - \beta_1(r,n+1) \to 0$ as $n \to \infty$ which together with (2.40) yields $\tilde{\alpha}_V(\beta_1(r,n)) \to 0$. On the other hand, since $\tilde{\alpha}_V$ is continuous, we get $\tilde{\alpha}_V(\beta_1(r,n)) \to \tilde{\alpha}_V(\beta_1^\infty(r))$. This implies

$$\tilde{\alpha}_V(\beta_1^\infty(r)) = 0$$

which because of $\tilde{\alpha}_V(r) \geq \min\{\alpha_V(r/2), r/4\}$ and $\alpha_V \in \mathcal{K}$ is only possible if $\beta_1^\infty(r) = 0$.

Consequently, $\beta_1(r,n)$ has all properties of a $\mathcal{KL}$ function except that it is only defined for $n \in \mathbb{N}_0$. Defining the linear interpolation

$$\beta_2(r,t) := (n+1-t)\beta_1(r,n) + (t-n)\beta_1(r,n+1)$$

for $t \in [n, n+1)$ and $n \in \mathbb{N}_0$, we obtain a function $\beta_2 \in \mathcal{KL}$ which coincides with β_1 for $t = n \in \mathbb{N}_0$. Finally, setting

$$\beta(r,t) := \alpha_1^{-1} \circ \beta_2(\alpha_2(r), t)$$

we can use (2.43) in order to obtain

$$\begin{aligned} |x(n,x_0)| &\leq \alpha_1^{-1}(V(x(n,x_0)) \;\leq\; \alpha_1^{-1} \circ \beta_1(V(x_0), n) \\ &= \alpha_1^{-1} \circ \beta_2(V(x_0), n) \;\leq\; \alpha_1^{-1} \circ \beta_2(\alpha_2(|x_0|, n) \;=\; \beta(|x_0|, n), \end{aligned}$$

for all $x_0 \in \widetilde{S}$ and all $n \in \mathbb{N}_0$. This is the desired inequality (2.32). If $\widetilde{S} = S = Y$ this shows the claimed asymptotic stability on Y and global asymptotic stability if $Y = X$. If $\widetilde{S} \neq S$, then in order to satisfy the local version of Definition 2.14 it remains to show that $x \in \mathcal{B}_\eta(x_*)$ implies $x \in \widetilde{S}$. Since by definition of η and γ we have $\eta = \alpha_2^{-1}(\gamma)$, we get

$$x \in \mathcal{B}_\eta(x_*) \Rightarrow |x| < \eta = \alpha_2^{-1}(\gamma) \Rightarrow V(x) \leq \alpha_2(|x|) < \gamma \Rightarrow x \in \widetilde{S}.$$

This finishes the proof. □

Likewise, P-practical asymptotic stability can be ensured by a suitable Lyapunov function condition provided the set P is forward invariant.

Theorem 2.20 (P-practical asymptotic stability) *Consider forward invariant sets Y and $P \subset Y$ and a point $x_* \in P$. If there exists a Lyapunov function V on $S = Y \setminus P$ then x_* is P-practically asymptotically stable on Y.*

Proof The same construction of β as in the proof of Theorem 2.19 yields

$$|x(n,x_0)|_{x_*} \leq \beta(|x|_{x_*}, n) \tag{2.32}$$

for all $n = 0, \ldots, n^* - 1$, where $n^* \in \mathbb{N}_0$ is minimal with $x(n^*, x_0) \in P$. This follows with the same arguments as in the proof of Theorem 2.19 by restricting the times considered in (2.39) and (2.43) to $n = 0, \ldots, n^* - 2$ and $n = 0, \ldots, n^* - 1$, respectively.

Since forward invariance of P ensures $x(n, x_0) \in P$ for all $n \geq n^*$, the times n for which $x(n, x_0) \notin P$ holds are exactly $n = 0, \ldots, n^* - 1$. Since these are exactly the times at which (2.32) is required, this yields the desired P-practical asymptotic stability. □

In case of a time-varying reference x^{ref} we need to use the time varying asymptotic stability from Definition 2.16. The corresponding Lyapunov function concept is as follows.

Definition 2.21 (*Uniform time-varying Lyapunov function*) Consider a system (2.34), reference points $x^{\text{ref}}(n)$, subsets of the state space $S(n) \subseteq X$ and define $\mathcal{S} := \{(n, x) \mid n \in \mathbb{N}_0, x \in S(n)\}$. A function $V : \mathcal{S} \to \mathbb{R}_0^+$ is called a *uniform time-varying Lyapunov function* on $S(n)$ if the following conditions are satisfied:

(i) There exist functions $\alpha_1, \alpha_2 \in \mathcal{K}_\infty$ such that

$$\alpha_1(|x|_{x^{\text{ref}}(n)}) \leq V(n, x) \leq \alpha_2(|x|_{x^{\text{ref}}(n)}) \tag{2.44}$$

holds for all $n \in \mathbb{N}_0$ and all $x \in S(n)$.

(ii) There exists a function $\alpha_V \in \mathcal{K}$ such that

$$V(n + 1, g(n, x)) \leq V(n, x) - \alpha_V(|x|_{x^{\text{ref}}(n)}) \tag{2.45}$$

holds for all $n \in \mathbb{N}_0$ and all $x \in S(n)$ with $g(n, x) \in S(n + 1)$.

Theorem 2.22 (Asymptotic stability) *Let x^{ref} be a trajectory of* (2.34) *and assume there exists a uniform time-varying Lyapunov function V on $S(n)$. If each $S(n)$ contains a ball $\mathcal{B}_\nu(x^{ref}(n))$ with $g(n, x) \in S(n + 1)$ for all $x \in \mathcal{B}_\nu(x^{ref}(n))$ then x^{ref} is locally asymptotically stable with $\eta = \alpha_2^{-1} \circ \alpha_1(\nu)$. If the family of sets $S(n)$ is forward invariant in the sense stated before Definition 2.16, then x^{ref} is asymptotically stable on $S(n)$. If $S(n) = X$ holds for all $n \in \mathbb{N}_0$ then x^{ref} is globally asymptotically stable.*

Proof The proof is analogous to the proof of Theorem 2.19 with the obvious modifications to take $n \in \mathbb{N}_0$ into account. □

Indeed, the necessary modification in the proof are straightforward because the time-varying Lyapunov function is uniform, i.e., α_1, α_2 and α_V do not depend on n. For the more involved nonuniform case we again refer to [3].

The P-practical version of this statement is provided by the following theorem in which we assume forward invariance of the sets $P(n)$. Observe that here x^{ref} does not need to be a trajectory of the system (2.34).

Theorem 2.23 (P-practical asymptotic stability) *Consider forward invariant families of sets $Y(n)$ and $P(n) \subset Y(n)$, $n \in \mathbb{N}_0$, and reference points $x^{ref}(n) \in P(n)$.*

If there exists a uniform time-varying Lyapunov function V on $S(n) = Y(n) \setminus P(n)$ then x^{ref} is P-practically asymptotically stable on $Y(n)$.

Proof The proof is analogous to the proof of Theorem 2.20 with the obvious modifications. □

2.4 Stability of Sampled Data Systems

We now investigate the special case in which (2.31) represents the nominal closed-loop system (2.5) with f obtained from a sampled data system via (2.8). In this case, the solutions $x(n, x_0)$ of (2.31) and the solutions $\varphi(t_n, t_0, x_0, \mu)$ of the sampled data closed-loop system (2.30) satisfy the identity

$$x(n, x_0) = \varphi(t_n, t_0, x_0, \mu) \tag{2.46}$$

for all $n \in \mathbb{N}_0$. This implies that the stability criterion from Definition 2.14 (and analogous for the other stability definitions) only yields inequalities for the continuous state of the system at the sampling times t_n, i.e.,

$$|\varphi(t_n, t_0, x_0, \mu)|_{x_*} \leq \beta(|x_0|_{x_*}, n) \quad \text{for all } n = 0, 1, 2, \ldots \tag{2.47}$$

for a suitable $\beta \in \mathscr{KL}$. However, for a continuous time system it is in general desirable to ensure the existence of $\overline{\beta} \in \mathscr{KL}$ such that the continuous time asymptotic stability property

$$|\varphi(t, t_0, x_0, \mu)|_{x_*} \leq \overline{\beta}(|x_0|_{x_*}, t) \quad \text{for all } t \geq 0 \tag{2.48}$$

holds.

In the remainder of this chapter we will show that under a reasonable additional assumption (2.47) implies the existence of $\overline{\beta} \in \mathscr{KL}$ such that (2.48) holds. For simplicity, we restrict ourselves to local asymptotic stability and to the case of time-invariant reference $x^{\mathrm{ref}} \equiv x_*$. The arguments can be modified to cover the other cases, as well.

The necessary additional condition is the following boundedness assumption on the solutions in between two sampling instants.

Definition 2.24 Consider a sampled data closed-loop system (2.30) with sampling period $T > 0$. If there exists a function $\gamma \in \mathscr{K}$ and a constant $\eta > 0$ such that for all $x \in X$ with $|x|_{x_*} \leq \eta$, the solutions of (2.30) exist on $[0, T]$ and the solutions satisfy

$$|\varphi(t; 0, x, \mu)|_{x_*} \leq \gamma(|x|_{x_*})$$

for all $t \in [0, T]$ then the solutions of (2.30) are called *uniformly bounded over T*.

Effectively, this condition demands that in between two sampling times t_n and t_{n+1} the continuous time solution does not deviate too much from the solution at the sampling time t_n. Sufficient conditions for this property formulated directly in terms of the vector field f_c in (2.30) can be found in [11, Lemma 3]. A sufficient condition in our NMPC setting is discussed in Remark 4.13.

For the subsequent analysis we introduce the following class of $\mathcal{KL}$ functions which will allow us to deal with the inter sampling behavior of the continuous time solution.

Definition 2.25 A function $\beta \in \mathcal{KL}$ is called *uniformly incrementally bounded* if there exists $P > 0$ such that $\beta(r,k) \leq P\beta(r,k+1)$ holds for all $r \geq 0$ and all $k \in \mathbb{N}$.

Uniformly incrementally bounded $\mathcal{KL}$ functions exhibit a nice bounding property compared to standard $\mathcal{KL}$ functions which we will use the proof of Theorem 2.27. Before, we show that any $\mathcal{KL}$ function β—like the one in (2.47)—can be bounded from above by a uniformly incrementally bounded $\mathcal{KL}$ function.

Lemma 2.26 *For any $\beta \in \mathcal{KL}$ the function*

$$\tilde{\beta}(r,t) := \max_{\tau\in[0,t]} 2^{-\tau}\beta(r,t-\tau).$$

is a uniformly incrementally bounded $\mathcal{KL}$ function with $\beta(r,t) \leq \tilde{\beta}(r,t)$ for all $r \geq 0$ and all $t \geq 0$ and $P = 2$.

Proof The inequality $\beta \leq \tilde{\beta}$ follows immediately from the definition. Uniform incremental boundedness with $P = 2$ follows from the inequality

$$\begin{aligned}\tilde{\beta}(r,t) &= \max_{\tau\in[0,t]} 2^{-\tau}\beta(r,t-\tau) = \max_{\tau\in[1,t+1]} 2^{1-\tau}\beta(r,t-\tau+1)\\ &= 2\max_{\tau\in[1,t+1]} 2^{-\tau}\beta(r,t-\tau+1) \leq 2\max_{\tau\in[0,t+1]} 2^{-\tau}\beta(r,t-\tau+1) = 2\tilde{\beta}(r,t+1).\end{aligned}$$

It remains to show that $\tilde{\beta} \in \mathcal{KL}$.

Since $\beta \in \mathcal{KL}$ it follows that $\tilde{\beta}$ is continuous and $\tilde{\beta}(0,t) = 0$ for any $t \geq 0$. For any $r_2 > r_1 \geq 0$, $\beta \in \mathcal{KL}$ implies $2^{-\tau}\beta(r_2,t-\tau) > 2^{-\tau}\beta(r_1,t-\tau)$. This shows that $\tilde{\beta}(r_2,t) > \tilde{\beta}(r_1,t)$ and hence $\tilde{\beta}(\cdot,t) \in \mathcal{K}$.

Next we show that for any fixed $r > 0$ the function $t \mapsto \tilde{\beta}(r,t)$ is strictly decreasing to 0. To this end, in the following we use that for all $t \geq s \geq q \geq 0$ and all $r \geq 0$ the inequality

$$\max_{\tau\in[q,s]} 2^{-\tau}\beta(r,t-\tau) \leq 2^{-q}\beta(r,t-s)$$

holds. In order to show the strict decrease property for $r > 0$, let $t_2 > t_1 \geq 0$. Defining $d := t_2 - t_1$ we obtain

$$\begin{aligned}
\tilde{\beta}(r,t_2) &= \max_{\tau\in[0,t_2]} 2^{-\tau}\beta(r,t_2-\tau)\\
&= \max\left\{\max_{\tau\in[0,d/2]} 2^{-\tau}\beta(r,t_2-\tau),\ \max_{\tau\in[d/2,d]} 2^{-\tau}\beta(r,t_2-\tau),\right.\\
&\qquad\left.\max_{\tau\in[d,t_2]} 2^{-\tau}\beta(r,t_2-\tau)\right\}\\
&\le \max\left\{\beta(r,t_2-d/2),\ 2^{-d/2}\beta(r,t_2-d),\ \max_{\tau\in[0,t_1]} 2^{-\tau-d}\beta(r,t_1-\tau)\right\}\\
&= \max\left\{\beta(r,t_1+d/2),\ 2^{-d/2}\beta(r,t_1),\ 2^{-d}\tilde{\beta}(r,t_1)\right\}.
\end{aligned}$$

Now the strict monotonicity $\tilde{\beta}(r,t_2) < \tilde{\beta}(r,t_1)$ follows since $\beta(r,t_1+d/2) < \beta(r,t_1) \le \tilde{\beta}(r,t_1)$, $2^{-d/2}\beta(r,t_1) < \beta(r,t_1) \le \tilde{\beta}(r,t_1)$ and $2^{-d}\tilde{\beta}(r,t_1) < \tilde{\beta}(r,t_1)$.

Finally, we prove $\lim_{t\to\infty}\tilde{\beta}(r,t) = 0$ for any $r > 0$. Since

$$\begin{aligned}
\tilde{\beta}(r,t) &\le \max\left\{\max_{\tau\in[0,t/2]} 2^{-\tau}\beta(r,t-\tau),\ \max_{\tau\in[t/2,t]} 2^{-\tau}\beta(r,t-\tau)\right\}\\
&\le \max\left\{\beta(r,t/2), 2^{-t/2}\beta(r,0)\right\} \to 0 \quad \text{as } t\to\infty
\end{aligned}$$

the assertion follows. □

Now, we are ready to prove the final stability result.

Theorem 2.27 (Asymptotic stability and uniform boundedness over T) *Consider the sampled data closed-loop system* (2.30) *with sampling period* $T > 0$ *and the corresponding discrete time closed-loop system* (2.5) *with* f *from* (2.8). *Then* (2.30) *is locally asymptotically stable, i.e., there exists* $\eta > 0$ *and* $\overline{\beta} \in \mathcal{KL}$ *such that* (2.48) *holds for all* $x \in \mathcal{B}_\eta(x_*)$, *if and only if* (2.5) *is locally asymptotically stable and the solutions of* (2.30) *are uniformly bounded over* T.

Proof If (2.30) is locally asymptotically stable with some $\overline{\beta} \in \mathcal{KL}$, then by (2.46) it immediately follows that the discrete time system (2.5) is asymptotically stable with $\beta(r,k) = \overline{\beta}(r,kT)$ and that the solutions of (2.30) are uniformly bounded with $\gamma(r) = \overline{\beta}(r,0)$.

Conversely, assume that (2.5) is locally asymptotically stable and that the solutions of (2.30) are uniformly bounded over T. Denote the values $\eta > 0$ from Definitions 2.14 and 2.24 by η^s and η^b, respectively. These two properties imply that there exist $\beta \in \mathcal{KL}$ and $\gamma \in \mathcal{K}_\infty$ such that

$$|x|_{x_*} \le \eta^s \implies |\varphi(kT;0,x,\mu)|_{x_*} \le \beta(|x|_{x_*},k) \quad \text{for all } k \ge 0 \tag{2.49}$$

$$|x|_{x_*} \le \eta^b \implies |\varphi(t;0,x,\mu)|_{x_*} \le \gamma(|x|_{x_*}) \quad \text{for all } t \in [0,T]. \tag{2.50}$$

In order to show the assertion we have to construct $\eta > 0$ and $\overline{\beta} \in \mathcal{KL}$ with

$$|x|_{x_*} \le \eta \implies |\varphi(t;0,x,\mu)|_{x_*} \le \overline{\beta}(|x|_{x_*},t) \quad \text{for all } t \ge 0. \tag{2.51}$$

Define $\gamma_0(r) := \beta(r, 0)$ and let $\eta = \min\{\eta^s, \gamma_0^{-1}(\eta^b)\}$. This definition implies $\beta(\eta, 0) \le \eta^b$ and $\eta \le \eta^s$. In what follows we consider arbitrary $x \in X$ with $|x|_{x_*} \le \eta$. For these x, (2.49) and $\eta \le \eta^s$ yield

$$|\varphi(kT; 0, x, \mu)|_{x_*} \le \beta(\|x\|_{x_*}, k) \le \beta(\eta, 0) \le \eta^b \qquad \text{for all } k \ge 0. \tag{2.52}$$

For any $k \ge 0$ and $t \in [kT, (k+1)T]$ the definition of (2.30) implies

$$\varphi(t, 0, x, \mu) = \varphi(t - kT, 0, \varphi(kT, 0, x, \mu), \mu).$$

Since (2.52) implies $|\varphi(kT, 0, x, \mu)|_{x_*} \le \eta^b$ for all $k \ge 0$, (2.50) holds for $x = \varphi(kT, 0, x, \mu)$ and from (2.50) and (2.52) we obtain

$$|\varphi(t, 0, x, \mu)|_{x_*} \le \gamma(\|\varphi(kT, 0, x, \mu)\|) \le \gamma(\beta(\|x\|_{x_*}, k)) \tag{2.53}$$

for all $t \in [kT, (k+1)T]$ and all $k \ge 0$.

Now we define $\hat{\beta}(r, t) := \gamma(\beta(r, t))$. Clearly, $\hat{\beta} \in \mathcal{KL}$ and by Lemma 2.26 we can assume without loss of generality that $\hat{\beta}$ is uniformly incrementally bounded; otherwise we replace it by $\tilde{\beta}$ from this lemma.

Hence, for $k \in \mathbb{N}_0$ and $s \in [0, 1]$ we obtain

$$\hat{\beta}(r, k) \le P\hat{\beta}(r, k+1) \le P\hat{\beta}(r, k+s). \tag{2.54}$$

Now pick an arbitrary $t \ge 0$ and let $k \in \mathbb{N}_0$ be maximal with $k \le t/T$. Then (2.53) and (2.54) with $s = t/T - k \in [0, 1]$ imply

$$|\varphi(t, 0, x, \mu)|_{x_*} \le \hat{\beta}(\|x\|_{x_*}, k)) \le P\hat{\beta}(|x|_{x_*}, k + (t/T - k)) = P\hat{\beta}(|x|_{x_*}, t/T).$$

This shows the assertion with $\overline{\beta}(r, t) = P\hat{\beta}(r, t/T)$. □

Concluding, if we can compute an asymptotically stabilizing feedback law for the discrete time system induced by the sampled data system, then the resulting continuous time sampled data closed loop is also asymptotically stable provided its solutions are uniformly bounded over T.

2.5 Notes and Extensions

The general setting presented in Sect. 2.1 is more or less standard in discrete time control theory, except maybe for the rather general choice of the state space X and the control value space U which allows us to cover infinite-dimensional systems as illustrated in Example 2.12 and sampled data systems without the zero order hold assumption as discussed after Theorem 2.7.

This definition of sampled data systems is not so frequently found in the literature, where often only the special case of zero-order hold is discussed. While zero-order hold is usually the method of choice in practical applications and is also used in the numerical examples later in this book, for theoretical investigations the more general approach given in Sect. 2.2 is appealing, too.

The discrete time stability theory presented in Sect. 2.3 has a continuous time counterpart which is actually more frequently found in the literature. Introductory textbooks on this subject in a control theoretic setting are, e.g., the books by Khalil [7] and Sontag [15]. The proofs in this section are not directly taken from the literature, but they are based on standard arguments, which appear in many books and papers on the subject. Formulating asymptotic stability via $\mathcal{KL}$-function goes back to Hahn [5] and became popular in nonlinear control theory during the 1990s via the input-to-state stability (ISS) property introduced by Sontag in [14]. A good survey on this theory can be found in Sontag [16].

While here we only stated direct Lyapunov function theorems which state that the existence of a Lyapunov function ensures asymptotic stability, there is a rather complete converse theory, which shows that asymptotic stability implies the existence of Lyapunov functions. A collection of such results—again in a control theoretic setting—can be found in the PhD thesis of Kellett [6].

The final Sect. 2.4 on asymptotic stability of sampled data systems is based on the Paper [11] by Nešić, Teel and Sontag, in which this topic is treated in a more general setting. In particular, this paper also covers ISS results for perturbed systems.

Problems

1. Show that there exists no differential equation $\dot{x}(t) = f_c(x(t))$ (i.e., without control input) such that the difference equation $x^+ = f(x)$ with

$$f(x) = \begin{cases} \frac{x}{2}, & x \geq 0 \\ -x, & x < 0 \end{cases}$$

is the corresponding sampled data system.

2. (a) Show that $x^{\text{ref}}(n) = \sum_{k=0}^{n} \frac{1}{2^{n-k}} \sin(k)$ is a solution of the difference equation

$$x(n+1) = \frac{1}{2}x(n) + \sin(n).$$

(b) Prove that x^{ref} from (a) is uniformly asymptotically stable and derive a comparison function $\beta \in \mathcal{KL}$ such that (2.35) holds. Here it is sufficient to derive a formula for $\beta(r, n)$ for $n \in \mathbb{N}_0$.

(c) Show that $x^{\mathrm{ref}}(n) = \sum_{k=0}^{n} \frac{k+1}{n+1} \sin(k)$ is a solution of the difference equation

$$x(n+1) = \frac{n+1}{n+2} x(n) + \sin(n).$$

(d) Can you also prove uniform asymptotic stability for x^{ref} from (c)?

Hint for (b) and (d): One way to proceed is to derive a difference equation for $z(n) = x(n, n_0, x_0) - x^{\mathrm{ref}}(n)$ and look at the equilibrium $x_* = 0$ for this new equation.

3. Consider the two-dimensional difference equation

$$x^+ = (1 - ||x||) \begin{pmatrix} 0 & 1 \\ -1 & 0 \end{pmatrix} x$$

with $x = (x_1, x_2)^\top \in \mathbb{R}^2$.

(a) Prove that $V(x) = x_1^2 + x_2^2$ is a Lyapunov function for the equilibrium $x_* = 0$ on $S = \{x \in \mathbb{R}^2 \mid \|x\| \leq 1\}$.

(b) Is V also a Lyapunov function on $S = \mathbb{R}^2$?

(c) Solve (a) and (b) for the difference equation

$$x^+ = \frac{1}{1 + ||x||} \begin{pmatrix} 0 & 1 \\ -1 & 0 \end{pmatrix} x.$$

4. Consider a globally asymptotically stable difference equation (2.31) with equilibrium $x_* \in X$ and a Lyapunov function V on $S = X$ with $\alpha_1(r) = 2r^2, \alpha_2(r) = 3r^2$ and $\alpha_V(r) = r^2$.

Compute the rate of attraction $\beta \in \mathcal{KL}$ such that (2.32) holds. Here it is sufficient to derive a formula for $\beta(r, n)$ for $n \in \mathbb{N}_0$.

Hint: Follow the construction of β from the proof of Theorem 2.19. Why can you use $\tilde{\alpha}_V = \alpha_V'$ for this problem?

5. Consider a difference equation (2.31) with equilibrium $x_* \in X$ and a function $V : X \to \mathbb{R}_0^+$ which satisfies (2.37) but only

$$V(g(x)) \leq V(x)$$

instead of (2.38).

(a) Prove that there exists $\alpha_L \in \mathcal{K}_\infty$ such that the solutions of (2.1) satisfy the inequality

$$|x(n, x_0)|_{x_*} \leq \alpha_L(|x_0|).$$

(b) Conclude from (a) that the system is stable in the sense of Lyapunov, cf. the discussion after Definition 2.14.

References

1. Freeman, R.A., Kokotovic, P.V.: Robust Nonlinear Control Design: Systems and Control: Foundations and Applications. Birkhäuser, Boston (1996)
2. Giselsson, P., Åkesson, J., Robertsson, A.: Optimization of a pendulum system using optimica and modelica. In: 7th International Modelica Conference 2009. Modelica Association (2009)
3. Grüne, L., Kloeden, P.E., Siegmund, S., Wirth, F.R.: Lyapunov's second method for nonautonomous differential equations. Discret. Contin. Dyn. Syst. **18**(2–3), 375–403 (2007)
4. Grüne, L., Nešić, D., Pannek, J., Worthmann, K.: Redesign techniques for nonlinear sampled-data systems. at - Automatisierungstechnik **56**(1), 38–47 (2008)
5. Hahn, W.: Stability of Motion. Springer, Heidelberg (1967)
6. Kellett, C.M.: Advances in converse and control Lyapunov functions. PhD thesis University of California, Santa Barbara (2000)
7. Khalil, H.K.: Nonlinear Systems, 3rd edn. Prentice Hall, Upper Saddle River (2002)
8. Lin, Y., Sontag, E.D., Wang, Y.: A smooth converse Lyapunov theorem for robust stability. SIAM J. Control Optim. **34**(1), 124–160 (1996)
9. Magni, L., Scattolini, R., Aström, K.J.: Global stabilization of the inverted pendulum using model predictive control. In: Proceedings of the 15th IFAC World Congress (2002)
10. Neittaanmäki, P., Tiba, D.: Optimal control of nonlinear parabolic systems. In: Theory, Algorithms, and Applications, Monographs and Textbooks in Pure and Applied Mathematics, vol. 179. Marcel Dekker Inc., New York (1994)
11. Nešić, D., Teel, A.R., Sontag, E.D.: Formulas relating $\mathcal{KL}$ stability estimates of discrete-time and sampled-data nonlinear systems. Syst. Control Lett. **38**(1), 49–60 (1999)
12. Nešić, D., Grüne, L.: A receding horizon control approach to sampled-data implementation of continuous-time controllers. Syst. Control Lett. **55**, 660–672 (2006)
13. Pannek, J.: Receding Horizon Control: A Suboptimality-based Approach. PhD thesis, University of Bayreuth, Germany (2009)
14. Sontag, E.D.: Smooth stabilization implies coprime factorization. IEEE Trans. Automat. Control **34**(4), 435–443 (1989)
15. Sontag, E.D.: Mathematical Control Theory. Texts in Applied Mathematics, vol. 6, 2nd edn. Springer, New York (1998)
16. Sontag, E.D.: Input to state stability: basic concepts and results. In: Nonlinear and Optimal Control Theory, Lecture Notes in Mathematics, vol. 1932, pp. 163–220. Springer, Berlin (2008)
17. Stewart, D.E.: Rigid-body dynamics with friction and impact. SIAM Rev. **42**(1), 3–39 (2000)
18. Tröltzsch, F.: Optimal control of partial differential equations, Graduate Studies in Mathematics, vol. 112. American Mathematical Society, Providence, RI. Theory, methods and applications, Translated from the 2005 German original by Jürgen Sprekels (2010)

Chapter 3
Nonlinear Model Predictive Control

In this chapter, we introduce the nonlinear model predictive control algorithm in a rigorous way. We start by defining a basic NMPC algorithm for constant reference and continue by formalizing state and control constraints. Viability (or weak forward invariance) of the set of state constraints is introduced and the consequences for the admissibility of the NMPC-feedback law are discussed. After having introduced NMPC in a special setting, we describe various extensions of the basic algorithm, considering time varying reference solutions, terminal constraints, and costs and additional weights. Finally, we investigate the optimal control problem corresponding to this generalized setting and prove several properties, most notably the dynamic programming principle.

3.1 The Basic NMPC Algorithm

As already outlined in the introductory Chap. 1, the idea of the NMPC scheme is as follows: at each sampling instant n we optimize the predicted future behavior of the system over a finite time horizon $k = 0, \dots, N-1$ of length $N \geq 2$ and use the first element of the resulting optimal control sequence as a feedback control value for the next sampling interval. In this section we give a detailed mathematical description of this basic idea for a constant reference $x^{\mathrm{ref}} \equiv x_* \in X$. The time varying case as well as several other variants will then be presented in Sect. 3.3.

A prerequisite for being able to find a feedback law which stabilizes the system at x_* is that x_* is an equilibrium of the nominal closed-loop system (2.5), i.e., $x_* = f(x_*, \mu(x_*))$—this follows immediately from Definition 2.14 with $g(x) = f(x, \mu(x))$. A necessary condition for this is that there exists a control value $u_* \in U$ with

$$x_* = f(x_*, u_*), \tag{3.1}$$

which we will assume in the sequel. The cost function to be used in our optimization should penalize the distance of an arbitrary state $x \in X$ to x_*. In addition, it is often

L. Grüne and J. Pannek, *Nonlinear Model Predictive Control*,
Communications and Control Engineering, DOI 10.1007/978-3-319-46024-6_3

desired to penalize the control $u \in U$. This can be useful for computational reasons, because optimal control problems may be easier to solve if the control variable is penalized. On the other hand, penalizing u may also be desired for modeling purposes, e.g., because we want to avoid the use of control values $u \in U$ corresponding to expensive high energy. For these reasons, we choose our cost function to be of the form $\ell : X \times U \to \mathbb{R}_0^+$.

In any case, we require that if we are in the equilibrium x_* and use the control value u_* in order to stay in the equilibrium, then the cost should be 0. Outside the equilibrium, however, the cost should be positive, i.e.,

$$\ell(x_*, u_*) = 0 \quad \text{and} \quad \ell(x, u) > 0 \ \text{ for all } x \in X,\ u \in U \text{ with } x \neq x_*. \tag{3.2}$$

If our system is defined on Euclidean space, i.e., $X = \mathbb{R}^d$ and $U = \mathbb{R}^m$, then we may always assume $x_* = 0$ and $u_* = 0$ without loss of generality: if this is not the case we can replace $f(x, u)$ by $f(x + x_*, u + u_*) - x_*$ which corresponds to a simple linear coordinate transformation on X and U. Indeed, this transformation is always possible if X and U are vector spaces, even if they are not Euclidean spaces. In this case, a popular choice for ℓ meeting Condition (3.2) is the quadratic function

$$\ell(x, u) = \|x\|^2 + \lambda\|u\|^2,$$

with the usual Euclidean norms and a parameter $\lambda \geq 0$. In our general setting on metric spaces with metrics d_X and d_U on X and U, the analogous choice of ℓ is

$$\ell(x, u) = d_X(x, x_*)^2 + \lambda d_U(u, u_*)^2. \tag{3.3}$$

Note, however, that in both settings many other choices are possible and often reasonable, as we will see in the subsequent chapters. Moreover, we will introduce additional conditions on ℓ later, which we require for a rigorous stability proof of the NMPC closed loop.

In the case of sampled data systems we can take the continuous time nature of the underlying model into account by defining the stage cost ℓ as an integral over a continuous time running cost function $L : X \times U \to \mathbb{R}_0^+$ on a sampling interval. Using the continuous time solution φ from (2.8), we can define

$$\ell(x, u) := \int_0^T L(\varphi(t, 0, x, u), u(t))dt. \tag{3.4}$$

Defining ℓ this way, we can incorporate the intersampling behavior of the sampled data system explicitly into our optimal control problem. As we will see later in Remark 4.13, this enables us to derive rigorous stability properties not only for the sampled data closed-loop system (2.30). The numerical computation of the integral in (3.4) can be efficiently integrated into the numerical solution of the ordinary differential equation (2.6), see Sect. 11.4 for details.

Given such a cost function ℓ and a prediction horizon length $N \geq 2$, we can now formulate the basic NMPC scheme as an algorithm. In the optimal control problem ($\mathrm{OCP_N}$) within this algorithm we introduce a set of control sequences $\mathbb{U}^N(x_0) \subseteq U^N$ over which we optimize. This set may include constraints depending on the initial value x_0. Details about how this set should be chosen will be discussed in Sect. 3.2. For the moment we simply set $\mathbb{U}^N(x_0) := U^N$ for all $x_0 \in X$.

Algorithm 3.1 (*Basic NMPC algorithm for constant reference* $x^{\mathrm{ref}} \equiv x_*$) At each sampling time t_n, $n = 0, 1, 2 \ldots$

(1) Measure the state $x(n) \in X$ of the system
(2) Set $x_0 := x(n)$, solve the optimal control problem

$$\begin{array}{l} \text{minimize} \quad J_N(x_0, u(\cdot)) := \sum_{k=0}^{N-1} \ell(x_u(k, x_0), u(k)) \\ \text{with respect to } \ u(\cdot) \in \mathbb{U}^N(x_0), \quad \text{subject to} \\ x_u(0, x_0) = x_0, \quad x_u(k+1, x_0) = f(x_u(k, x_0), u(k)) \end{array} \tag{$\mathrm{OCP_N}$}$$

and denote the obtained optimal control sequence by $u^\star(\cdot) \in \mathbb{U}^N(x_0)$.
(3) Define the NMPC-feedback value $\mu_N(x(n)) := u^\star(0) \in U$ and use this control value in the next sampling period.

Observe that in this algorithm we have assumed that an optimal control sequence $u^\star(\cdot)$ exists. Sufficient conditions for this existence are briefly discussed after Definition 3.14, below.

The nominal closed-loop system resulting from Algorithm 3.1 is given by (2.5) with state feedback law $\mu = \mu_N$, i.e.,

$$x^+ = f(x, \mu_N(x)). \tag{3.5}$$

The trajectories of this system will be denoted by $x_{\mu_N}(n)$ or, if we want to emphasize the initial value $x_0 = x_{\mu_N}(0)$, by $x_{\mu_N}(n, x_0)$.

During our theoretical investigations we will neglect the fact that computing the solution of ($\mathrm{OCP_N}$) in Step (2) of the algorithm usually needs some computation time τ_c which—in the case when τ_c is relatively large compared to the sampling period T—may not be negligible in a real-time implementation. We will sketch a solution to this problem in Sect. 10.6.

In our abstract formulations of the NMPC Algorithm 3.1 only the first element $u^\star(0)$ of the respective minimizing control sequence is used in each step, the remaining entries $u^\star(1), \ldots, u^\star(N-1)$ are discarded. In the practical implementation, however, these entries play an important role because numerical optimization algorithms for solving ($\mathrm{OCP_N}$) (or its variants) usually work iteratively: starting from an initial guess $u^0(\cdot)$ an optimization algorithm computes iterates $u^i(\cdot)$, $i = 1, 2, \ldots$ converging to the minimizer $u^\star(\cdot)$ and a good choice of $u^0(\cdot)$ is crucial in order to obtain fast convergence of this iteration, or even to ensure convergence, at all. Here, the minimizing sequence from the previous time step can be efficiently used in order to construct such a good initial guess. Several different ways to implement this idea are discussed in Sect. 12.4.

3.2 Constraints

One of the main reasons for the success of NMPC (and MPC in general) is its ability to explicitly take constraints into account. Here, we consider constraints both on the control as well as on the state. To this end, we introduce a nonempty *state constraint set* $\mathbb{X} \subseteq X$ and for each $x \in \mathbb{X}$ we introduce a nonempty *control constraint set* $\mathbb{U}(x) \subseteq U$. Of course, $\mathbb{U}$ may also be chosen independent of x. The idea behind introducing these sets is that we want the trajectories to lie in $\mathbb{X}$ and the corresponding control values to lie in $\mathbb{U}(x)$. This is made precise in the following definition:

Definition 3.2 (*Admissibility*) Consider a control system (2.1) and the state and control constraint sets $\mathbb{X} \subseteq X$ and $\mathbb{U}(x) \subseteq U$.

(i) The states $x \in \mathbb{X}$ are called *admissible states* and the control values $u \in \mathbb{U}(x)$ are called *admissible control values for* x. The elements of the set $\mathbb{Y} := \{(x, u) \in X \times U \mid x \in \mathbb{X}, u \in \mathbb{U}(x)\}$ are called *admissible pairs*.

(ii) For $N \in \mathbb{N}$ and an initial value $x_0 \in \mathbb{X}$ we call a control sequence $u \in U^N$ and the corresponding trajectory $x_u(k, x_0)$ *admissible for* x_0 *up to time* N, if

$$(x_u(k, x_0), u(k)) \in \mathbb{Y} \quad \text{for all} \quad k = 0, \dots, N-1 \quad \text{and} \quad x_u(N, x_0) \in \mathbb{X}$$

holds. We denote the set of admissible control sequences for x_0 up to time N by $\mathbb{U}^N(x_0)$.

(iii) A control sequence $u \in U^\infty$ and the corresponding trajectory $x_u(k, x_0)$ are called *admissible for* x_0 if they are admissible for x_0 up to every time $N \in \mathbb{N}$. We denote the set of admissible control sequences for x_0 by $\mathbb{U}^\infty(x_0)$.

(iv) A (possibly time varying) feedback law $\mu : \mathbb{N}_0 \times X \to U$ is called *admissible* if $\mu(n, x) \in \mathbb{U}^1(x)$ holds for all $x \in \mathbb{X}$ and all $n \in \mathbb{N}_0$.

Whenever the reference to x or x_0 is clear from the context we will omit the additional "for x" or "for x_0".

Since we can (and will) identify control sequences with only one element with the respective control value, we can consider $\mathbb{U}^1(x_0)$ as a subset of U, which we already implicitly did in the definition of admissibility for the feedback law μ, above. However, in general $\mathbb{U}^1(x_0)$ does not coincide with $\mathbb{U}(x_0) \subseteq U$ because using $x_u(1, x) = f(x, u)$ and the definition of $\mathbb{U}^N(x_0)$ we get $\mathbb{U}^1(x) := \{u \in \mathbb{U}(x) \mid f(x, u) \in \mathbb{X}\}$. With this subtle difference in mind, one sees that our admissibility condition (iv) on μ ensures both $\mu(n, x) \in \mathbb{U}(x)$ and $f(x, \mu(n, x)) \in \mathbb{X}$ whenever $x \in \mathbb{X}$.

Furthermore, our definition of $\mathbb{U}^N(x)$ implies that even if $\mathbb{U}(x) = \mathbb{U}$ is independent of x the set $\mathbb{U}^N(x)$ may depend on x for some or all $N \in \mathbb{N}_\infty$.

Often, in order to be suitable for optimization purposes these sets are assumed to be compact and convex. For our theoretical investigations, however, we do not need any regularity requirements of this type except that these sets are nonempty. We will, however, frequently use the following assumption.

Assumption 3.3 (*Viability*) For each $x \in \mathbb{X}$ there exists $u \in \mathbb{U}(x)$ such that $f(x,u) \in \mathbb{X}$ holds.

The property defined in this assumption is called *viability* or *weak (or controlled) forward invariance* of $\mathbb{X}$. It excludes the situation that there are states $x \in \mathbb{X}$ from which the trajectory leaves the set $\mathbb{X}$ for all admissible control values. Hence, it ensures $\mathbb{U}^N(x_0) \neq \emptyset$ for all $x_0 \in \mathbb{X}$ and all $N \in \mathbb{N}_\infty$. This property is important to ensure the feasibility of (OCP$_\mathrm{N}$): the optimal control problem (OCP$_\mathrm{N}$) is called *feasible* for an initial value x_0 if the set $\mathbb{U}^N(x_0)$ over which we optimize is non empty. Viability of $\mathbb{X}$ thus implies that (OCP$_\mathrm{N}$) is feasible for each $x_0 \in \mathbb{X}$ and hence ensures that $\mu_N(x)$ is well defined for each $x \in \mathbb{X}$. Furthermore, a straightforward induction shows that under Assumption 3.3 any finite admissible control sequence $u(\cdot) \in \mathbb{U}^N(x_0)$ can be extended to an infinite admissible control sequence $\tilde{u}(\cdot) \in \mathbb{U}^\infty(x_0)$ with $u(k) = \tilde{u}(k)$ for all $k = 0, \ldots, N-1$.

In order to see that the construction of a constraint set $\mathbb{X}$ meeting Assumption 3.3 is usually a nontrivial task, we reconsider Example 2.2.

Example 3.4 Consider Example 2.2, i.e.,

$$x^+ = f(x,u) = \begin{pmatrix} x_1 + x_2 + u/2 \\ x_2 + u \end{pmatrix}.$$

Assume we want to constrain all variables, i.e., the position x_1, the velocity x_2 and the acceleration u to the interval $[-1,1]$. For this purpose one could define $\mathbb{X} = [-1,1]^2$ and $\mathbb{U}(x) = \mathbb{U} = [-1,1]$. Then, however, for $x = (1,1)^\top$, one immediately obtains

$$x_1^+ = x_1 + x_2 + u/2 = 2 + u/2 \geq 3/2$$

for all u, hence $x^+ \notin \mathbb{X}$ for all $u \in \mathbb{U}$. Thus, in order to find a viable set $\mathbb{X}$ we need to either tighten or relax some of the constraints. For instance, relaxing the constraint on u to $\mathbb{U} = [-2,2]$ the viability of $\mathbb{X} = [-1,1]^2$ is guaranteed, because then by elementary computations one sees that for each $x \in \mathbb{X}$ the control value

$$u = \begin{cases} 0, & x_1 + x_2 \in [-1,1] \\ 2 - 2x_1 - 2x_2, & x_1 + x_2 > 1 \\ -2 - 2x_1 - 2x_2, & x_1 + x_2 < -1 \end{cases}$$

is in $\mathbb{U}$ and satisfies $f(x,u) \in \mathbb{X}$. A way to achieve viability without changing $\mathbb{U}$ is by tightening the constraint on x_2 by defining

$$\mathbb{X} = \{(x_1,x_2)^T \in \mathbb{R}^2 \mid x_1 \in [-1,1], x_2 \in [-1,1] \cap [-3/2 - x_1, 3/2 - x_1]\}, \quad (3.6)$$

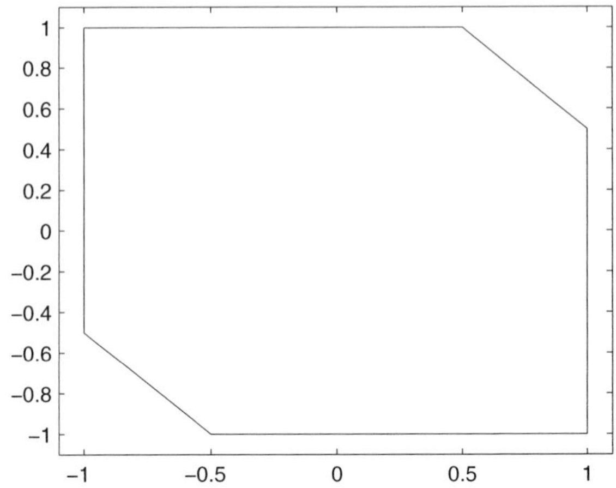

Fig. 3.1 Illustration of the set $\mathbb{X}$ from (3.6)

see Fig. 3.1. Again, elementary computations show that for each $x \in \mathbb{X}$ and

$$u = \begin{cases} 1, & x_2 < -1/2 \\ -2x_2, & x_2 \in [-1/2, 1/2] \\ -1, & x_2 > 1/2 \end{cases}$$

the desired properties $u \in \mathbb{U}$ and $f(x, u) \in \mathbb{X}$ hold.

This example shows that finding viable constraint sets $\mathbb{X}$ (and the corresponding $\mathbb{U}$ or $\mathbb{U}(x)$) is a tricky task already for very simple systems. Still, Assumption 3.3 significantly simplifies the subsequent analysis, cf. Theorem 3.5, below. For this reason we will impose this condition in our theoretical investigations for schemes without stabilizing terminal conditions in Chap. 6. Ways to relax this condition will be discussed in Sects. 7.1–7.3.

For schemes with stabilizing terminal constraints as featured in Chap. 5, we will not need this assumption, since for these schemes the region on which the NMPC controller is defined is by construction confined to feasible subsets $\mathbb{X}_N$ of $\mathbb{X}$, see Definition 3.9, below. Even if $\mathbb{X}$ is not viable, these feasible sets $\mathbb{X}_N$ turn out to be viable provided the terminal constraint set is viable, cf. Lemmas 5.2 and 5.10. For a more detailed discussion of these issues see also Part (iv) of the discussion in Sect. 7.4.

NMPC is well suited to handle constraints because these can directly be inserted into Algorithm 3.1. In fact, since we already formulated the corresponding optimization problem (OCP$_\mathrm{N}$) with state dependent control value sets, the constraints are readily included if we use $\mathbb{U}^N(x_0)$ from Definition 3.2(ii) in (OCP$_\mathrm{N}$). The following theorem shows that the viability assumption ensures that the NMPC closed-loop system obtained this way indeed satisfies the desired constraints.

Theorem 3.5 (Admissibility) *Consider Algorithm 3.1 using $\mathbb{U}^N(x_0)$ from Definition 3.2(ii) in the optimal control problem* (OCP$_\text{N}$) *for constraint sets $\mathbb{X} \subset X$, $\mathbb{U}(x) \subset U$, $x \in \mathbb{X}$, satisfying Assumption 3.3. Consider the nominal closed-loop system* (3.5) *and suppose that $x_{\mu_N}(0) \in \mathbb{X}$. Then the constraints are satisfied along the solution of* (3.5), *i.e.,*

$$(x_{\mu_N}(n), \mu_N(x_{\mu_N}(n))) \in \mathbb{Y} \tag{3.7}$$

for all $n \in \mathbb{N}$. Thus, the NMPC-feedback μ_N is admissible in the sense of Definition 3.2(iv).

Proof First, recall from the discussion after Assumption 3.3 that under this assumption the optimal control problem (OCP$_\text{N}$) is feasible for each $x \in \mathbb{X}$, hence $\mu_N(x)$ is well defined for each $x \in \mathbb{X}$.

We now show that $x_{\mu_N}(n) \in \mathbb{X}$ implies $\mu_N(x_{\mu_N}(n)) \in \mathbb{U}(x_{\mu_N}(n))$ and $x_{\mu_N}(n+1) \in \mathbb{X}$. Then the assertion follows by induction from $x_{\mu_n}(0) \in \mathbb{X}$.

The viability of $\mathbb{X}$ from Assumption 3.3 ensures that whenever $x_{\mu_N}(n) \in \mathbb{X}$ holds in Algorithm 3.1 then $x_0 \in \mathbb{X}$ holds for the respective optimal control problem (OCP$_\text{N}$). Since the optimization is performed with respect to admissible control sequences only, also the optimal control sequence $u^\star(\cdot)$ is admissible for $x_0 = x_{\mu_N}(n)$. This implies $\mu_N(x_{\mu_N}(n)) = u^\star(0) \in \mathbb{U}^1(x_{\mu_N}(n)) \subseteq \mathbb{U}(x_{\mu_N}(n))$ and thus also

$$x_{\mu_N}(n+1) = f(x_{\mu_N}(n), \mu_N(x_{\mu_N}(n))) = f(x(n), u^\star(0)) \in \mathbb{X},$$

i.e., $x_{\mu_N}(n+1) \in \mathbb{X}$. □

Theorem 3.5 in particular implies that if a state x is feasible for (OCP$_\text{N}$), which under Assumption 3.3 is equivalent to $x \in \mathbb{X}$ (cf. the discussion after Assumption 3.3), then its closed-loop successor state $f(x, \mu_N(x))$ is again feasible. This property is called *recursive feasibility* of $\mathbb{X}$.

In the case of sampled data systems, the constraints are only defined for the sampling times t_n but not for the intersampling times $t \neq t_n$. That is, for the sampled data closed-loop system (2.30) we can only guarantee

$$\varphi(t_n, t_0, x_0, \mu) \in \mathbb{X} \quad \text{for} \quad n = 0, 1, 2, \ldots$$

but in general not

$$\varphi(t, t_0, x_0, \mu) \in \mathbb{X} \quad \text{for} \quad t \neq t_n, \; n = 0, 1, 2, \ldots$$

Since we prefer to work within the discrete time framework, directly checking $\varphi(t, t_0, x_0, u) \in \mathbb{X}$ for all t does not fit our setting. If desired, however, one could implicitly include this condition in the definition of $\mathbb{U}(x)$, e.g., by defining new control constraint sets via

$$\widetilde{\mathbb{U}}(x) := \{u \in \mathbb{U}(x) \mid \varphi(t, 0, x, u) \in \mathbb{X} \text{ for all } t \in [0, T]\}.$$

In practice, however, this is often not necessary because continuity of φ in t ensures that the constraints are usually only "mildly" violated for $t \neq t_n$, i.e., $\varphi(t, t_0, x_0, \mu)$ will still be close to $\mathbb{X}$ at intersampling times. Still, one should keep this fact in mind when designing the constraint set $\mathbb{X}$.

In the underlying optimization algorithms for solving (OCP$_\mathrm{N}$), usually the constraints cannot be specified via sets $\mathbb{X}$ and $\mathbb{U}(x)$. Rather, one uses so-called *equality* and *inequality constraints* in order to specify $\mathbb{X}$ and $\mathbb{U}(x)$ according to the following definition.

Definition 3.6 Given functions $G_i^S : X \times U \to \mathbb{R}$, $i \in \mathcal{E}^S = \{1, \dots, p_g\}$ and $H_i^S : X \times U \to \mathbb{R}$, $i \in \mathcal{I}^S = \{p_g + 1, \dots, p_g + p_h\}$ with $p_g, p_h \in \mathbb{N}_0$, we define the constraint sets $\mathbb{X}$ and $\mathbb{U}(x)$ via

$$\mathbb{X} := \left\{ x \in X \,\middle|\, \begin{array}{l} \text{there exists } u \in U \text{ with } G_i^S(x, u) = 0 \text{ for all } i \in \mathcal{E}^S \\ \text{and } H_i^S(x, u) \geq 0 \text{ for all } i \in \mathcal{I}^S \end{array} \right\}$$

and for $x \in \mathbb{X}$

$$\mathbb{U}(x) := \left\{ u \in U \,\middle|\, \begin{array}{l} G_i^S(x, u) = 0 \text{ for all } i \in \mathcal{E}^S \text{ and} \\ H_i^S(x, u) \geq 0 \text{ for all } i \in \mathcal{I}^S \end{array} \right\}$$

Here, the functions G_i^S and H_i^S do not need to depend on both arguments. The functions G_i^S, H_i^S not depending on u are called *pure state constraints*, the functions G_i^S, H_i^S not depending on x are called *pure control constraints* and the functions G_i^S, H_i^S depending on both x and u are called *mixed constraints*.

Observe that if we do not have mixed constraints then $\mathbb{U}(x)$ is independent of x.

The reason for defining $\mathbb{X}$ and $\mathbb{U}(x)$ via these (in)equality constraints is purely algorithmic: the plain information "$x_u(k, x_0) \notin \mathbb{X}$" does not yield any information for the optimization algorithm in order to figure out how to find an admissible $u(\cdot)$, i.e., a $u(\cdot)$ for which "$x_u(k, x_0) \in \mathbb{X}$" holds. In contrast to that, an information of the form "$H_i^S(x_u(k, x_0), u(k)) < 0$" together with additional knowledge about H_i^S (provided, e.g., by the derivative of H_i^S) enables the algorithm to compute a "direction" in which $u(\cdot)$ needs to be modified in order to reach an admissible $u(\cdot)$. For more details on this we refer to Chap. 12.

In our theoretical investigations we will use the notationally more convenient set characterization of the constraints via $\mathbb{X}$ and $\mathbb{U}(x)$ or $\mathbb{U}^N(x)$. In the practical implementation of our NMPC method, however, we will use their characterization via the inequality constraints from Definition 3.6.

3.3 Variants of the Basic NMPC Algorithms

In this section we discuss some important variants and extensions of the basic NMPC Algorithm 3.1; several further variants will be briefly discussed in Sect. 3.5. We start by incorporating nonconstant references $x^{\mathrm{ref}}(n)$ and afterwards turn to including terminal constraints, terminal costs and weights.

If the reference x^{ref} is time varying, we need to take this fact into account in the formulation of the NMPC algorithm. Similar to the constant case where we assumed that x_* is an equilibrium of (2.1) for control value u_*, we now assume that x^{ref} is a trajectory of the system, i.e.,

$$x^{\mathrm{ref}}(n) = x_{u^{\mathrm{ref}}}(n, x_0)$$

for $x_0 = x^{\mathrm{ref}}(0)$ and some suitable admissible reference control sequence $u^{\mathrm{ref}}(\cdot) \in \mathbb{U}^\infty(x_0)$. In contrast to the constant reference case of Sect. 3.1, even for $X = \mathbb{R}^d$ and $U = \mathbb{R}^m$ we do not assume that these references are constantly equal to 0, because this would lead to time varying coordinate transformations in X and U. For this reason, we always need to take $x^{\mathrm{ref}}(\cdot)$ and $u^{\mathrm{ref}}(\cdot)$ into account when defining ℓ. As a consequence, ℓ becomes time varying, too, i.e., we use a function $\ell : \mathbb{N}_0 \times X \times U \to \mathbb{R}_0^+$. Furthermore, we need to keep track of the current sampling instant n in the optimal control problem.

Again, we require that the cost function ℓ vanishes if and only if we are exactly on the reference. In the time varying case (3.2) becomes

$$\begin{aligned} &\ell(n, x^{\mathrm{ref}}(n), u^{\mathrm{ref}}(n)) = 0 \ \text{ for all } n \in \mathbb{N}_0 \qquad \text{and} \\ &\ell(n, x, u) > 0 \ \text{ for all } n \in \mathbb{N}_0,\ x \in X,\ u \in U \text{ with } x \neq x^{\mathrm{ref}}(n). \end{aligned} \tag{3.8}$$

For $X = \mathbb{R}^d$, $U = \mathbb{R}^m$ with Euclidean norms, a quadratic distance function is now of the form

$$\ell(n, x, u) = \|x - x^{\mathrm{ref}}(n)\|^2 + \lambda \|u - u^{\mathrm{ref}}(n)\|^2$$

with $\lambda \geq 0$ and in the general case

$$\ell(n, x, u) = d_X(x, x^{\mathrm{ref}}(n))^2 + \lambda d_U(u, u^{\mathrm{ref}}(n))^2$$

is an example for ℓ meeting (3.8).

For sampled data systems, we can again define ℓ via an integral over a continuous time running cost function L analogous to (3.4). Note, however, that for defining L we will then need a continuous time reference.

For each $k = 0, \ldots, N-1$, the prediction $x_u(k, x_0)$ with $x_0 = x(n)$ used in the NMPC algorithm now becomes a prediction for the closed-loop state $x(n+k)$ which we would like to have close to $x^{\mathrm{ref}}(n+k)$. Consequently, in the optimal control problem at time n we need to penalize the distance of $x_u(k, x_0)$ to $x^{\mathrm{ref}}(n+k)$, i.e., we need to use the cost $\ell(n+k, x_u(k, x_0), u(k))$. This leads to the following algorithm where we minimize over the set of control sequences $\mathbb{U}^N(x_0)$ defined in Sect. 3.2.

Algorithm 3.7 (*Basic NMPC algorithm for time varying reference* x^{ref}) At each sampling time t_n, $n = 0, 1, 2\ldots$

(1) Measure the state $x(n) \in X$ of the system.
(2) Set $x_0 = x(n)$, solve the optimal control problem

$$\boxed{\begin{aligned} &\text{minimize} \quad J_N(n, x_0, u(\cdot)) := \sum_{k=0}^{N-1} \ell(n+k, x_u(k, x_0), u(k)) \\ &\text{with respect to } u(\cdot) \in \mathbb{U}^N(x_0), \quad \text{subject to} \\ &x_u(0, x_0) = x_0, \quad x_u(k+1, x_0) = f(x_u(k, x_0), u(k)) \end{aligned}} \qquad (\mathrm{OCP}_\mathrm{N}^\mathrm{n})$$

and denote the obtained optimal control sequence by $u^\star(\cdot) \in \mathbb{U}^N(x_0)$.
(3) Define the NMPC-feedback value $\mu_N(n, x(n)) := u^\star(0) \in U$ and use this control value in the next sampling period.

Note that Algorithm 3.7 and ($\mathrm{OCP}_\mathrm{N}^\mathrm{n}$) reduce to Algorithm 3.1 and (OCP_N), respectively, if ℓ does not depend on n.

The resulting nominal closed-loop system is now given by (2.5) with $\mu(x) = \mu_N(n, x)$, i.e.,

$$x^+ = f(x, \mu_N(n, x)). \tag{3.9}$$

As before, the trajectories of this system will be denoted by $x_{\mu_N}(n)$. Since the right hand side is now time varying, whenever necessary we include both the initial time and the initial value in the notation, i.e., for a given $n_0 \in \mathbb{N}_0$ we write $x_{\mu_N}(n, n_0, x_0)$ for the closed-loop solution satisfying $x_{\mu_N}(n_0, n_0, x_0) = x_0$. It is straightforward to check that Theorem 3.5 remains valid for Algorithm 3.7 when (3.7) is replaced by

$$(x_{\mu_N}(n), \mu_N(n, x_{\mu_N}(n))) \in \mathbb{Y}. \tag{3.10}$$

Remark 3.8 Observe that Algorithm 3.7 can be straightforwardly extended to the case when f and $\mathbb{X}$ depend on n, too. However, in order to keep the presentation simple, we do not explicitly reflect this possibility in our notation.

More often than not one can find variations of the basic NMPC Algorithms 3.1 and 3.7 in the literature in which the optimal control problem (OCP_N) or ($\mathrm{OCP}_\mathrm{N}^\mathrm{n}$) is changed in one way or another in order to improve the closed-loop performance. These techniques will be discussed in detail in Chap. 5 and in Sects. 10.1 and 10.2. We now introduce generalizations ($\mathrm{OCP}_\mathrm{N,e}$) and ($\mathrm{OCP}_\mathrm{N,e}^\mathrm{n}$) of ($\mathrm{OCP}_\mathrm{N}$) and ($\mathrm{OCP}_\mathrm{N}^\mathrm{n}$), respectively, that contain all the variants which we will investigate in these chapters and sections.

A typical choice for such a variant is an additional terminal constraint of the form

$$x_u(N, x(n)) \in \mathbb{X}_0 \text{ for a } \textit{terminal constraint set } \mathbb{X}_0 \subseteq \mathbb{X} \tag{3.11}$$

for the time-invariant case of (OCP_N) and

$$x_u(N, x(n)) \in \mathbb{X}_0(n+N) \text{ for } \textit{terminal constraint sets } \mathbb{X}_0(n) \subseteq \mathbb{X},\ n \in \mathbb{N}_0 \tag{3.12}$$

for the time varying problem ($\mathrm{OCP}_\mathrm{N}^\mathrm{n}$). Of course, in the practical implementation, the constraint sets $\mathbb{X}_0$ or $\mathbb{X}_0(n)$ are again expressed via (in)equalities of the form given in Definition 3.6.

When using terminal constraints, the NMPC-feedback law is only defined for those states x_0 for which the optimization problem within the NMPC algorithm is feasible also for these additional constraints, i.e., for which there exists an admissible control sequence with corresponding trajectory starting in x_0 and ending in the terminal constraint set. Such initial values are again called *feasible* and the set of all feasible initial values form the feasible set. This set along with the corresponding admissible control sequences is formally defined as follows.

Definition 3.9 (*Feasible set and admissible control sequences*)
(i) For $\mathbb{X}_0$ from (3.11) we define the *feasible set* for horizon $N \in \mathbb{N}$ by

$$\mathbb{X}_N := \{x_0 \in \mathbb{X} \mid \text{there exists } u(\cdot) \in \mathbb{U}^N(x_0) \text{ with } x_u(N, x_0) \in \mathbb{X}_0\}$$

and for each $x_0 \in \mathbb{X}_N$ we define the set of *admissible control sequences* by

$$\mathbb{U}^N_{\mathbb{X}_0}(x_0) := \{u(\cdot) \in \mathbb{U}^N(x_0) \mid x_u(N, x_0) \in \mathbb{X}_0\}.$$

(ii) For $\mathbb{X}_0(n)$ from (3.12) we define the *feasible set* for horizon $N \in \mathbb{N}$ at time $n \in \mathbb{N}_0$ by

$$\mathbb{X}_N(n) := \{x_0 \in \mathbb{X} \mid \text{there exists } u(\cdot) \in \mathbb{U}^N(x_0) \text{ with } x_u(N, x_0) \in \mathbb{X}_0(n+N)\}$$

and for each $x_0 \in \mathbb{X}_N(n)$ we define the set of *admissible control sequences* by

$$\mathbb{U}^N_{\mathbb{X}_0}(n, x_0) := \{u(\cdot) \in \mathbb{U}^N(x_0) \mid x_u(N, x_0) \in \mathbb{X}_0(n+N)\}.$$

Note that in (i) $\mathbb{X}_N = \mathbb{X}$ and $\mathbb{U}^N_{\mathbb{X}_0}(x) = \mathbb{U}^N(x)$ holds if $\mathbb{X}_0 = \mathbb{X}$, i.e., if no additional terminal constraints are imposed. Similarly, in case (ii) $\mathbb{X}_N(n) = \mathbb{X}$ and $\mathbb{U}^N_{\mathbb{X}_0}(n, x) = \mathbb{U}^N(x)$ holds if $\mathbb{X}_0(n) = \mathbb{X}$.

Another modification of the optimal control problems (OCP_N) and ($\mathrm{OCP}_\mathrm{N}^\mathrm{n}$), often used in conjunction with this terminal constraint is an additional terminal cost of the form $F(x_u(N, x(n)))$ with $F : X \to \mathbb{R}_0^+$ in the optimization objective. This function may also be time depending, i.e., it may be of the form $F(n+N, x_u(N, x(n)))$ with $F : \mathbb{N}_0 \times X \to \mathbb{R}_0^+$. As terminal constraints $\mathbb{X}_0$ and terminal costs F often come together, we will refer to their combination as *terminal conditions*.

An alternative to using terminal costs is to put weights on some summands of the objective, i.e., replacing $\ell(x_u(k, x_0), u(k))$ by $\omega_{N-k}\ell(x_u(k, x_0), u(k))$ for weights $\omega_1, \dots, \omega_N \geq 0$. Although for NMPC schemes we will only investigate the effect of the weight ω_1 in detail, cf. Sect. 10.2, here we introduce weights for all summands since this offers more flexibility and does not further complicate the subsequent analysis in this chapter. The need for the "backward" numbering of the ω_{N-k} will become clear in the proof of Theorem 3.15, below.

In the sequel, we will analyze schemes with terminal cost F and schemes with weights ω_{N-k} separately, cf. Sects. 5.3, 10.1, and 10.2. However, in order to reduce the number of variants of NMPC algorithms in this book we include both features in the optimization problems ($\mathrm{OCP}_{\mathrm{N,e}}$) and ($\mathrm{OCP}_{\mathrm{N,e}}^{\mathrm{n}}$) in the following NMPC algorithms extending the basic Algorithms 3.1 and 3.7, respectively. Note that compared to these basic algorithms only the optimal control problems are different, i.e., the part in the boxes in Step (2). We start by extending the time-invariant Algorithm 3.1.

Algorithm 3.10 (*Extended NMPC algorithm for constant reference $x^{\mathrm{ref}} \equiv x_*$*) At each sampling time t_n, $n = 0, 1, 2 \ldots$:

(1) Measure the state $x(n) \in X$ of the system.
(2) Set $x_0 := x(n)$, solve the optimal control problem

$$\begin{aligned}
&\text{minimize} \quad J_N(x_0, u(\cdot)) := \sum_{k=0}^{N-1} \omega_{N-k}\ell(x_u(k, x_0), u(k)) \\
&\qquad\qquad\qquad\qquad\qquad + F(x_u(N, x_0)) \\
&\text{with respect to } u(\cdot) \in \mathbb{U}^N_{\mathbb{X}_0}(x_0), \quad \text{subject to} \\
&x_u(0, x_0) = x_0, \quad x_u(k+1, x_0) = f(x_u(k, x_0), u(k))
\end{aligned} \qquad (\mathrm{OCP}_{\mathrm{N,e}})$$

and denote the obtained optimal control sequence by $u^\star(\cdot) \in \mathbb{U}^N_{\mathbb{X}_0}(x_0)$.
(3) Define the NMPC-feedback value $\mu_N(x(n)) := u^\star(0) \in U$ and use this control value in the next sampling period.

Similarly, we can extend the time-variant Algorithm 3.7.

Algorithm 3.11 (*Extended NMPC algorithm for time varying reference x^{ref}*) At each sampling time t_n, $n = 0, 1, 2 \ldots$:

(1) Measure the state $x(n) \in X$ of the system.

(2) Set $x_0 = x(n)$, solve the optimal control problem

$$\begin{aligned}
&\text{minimize} \quad J_N(n, x_0, u(\cdot)) := \sum_{k=0}^{N-1} \omega_{N-k}\ell(n+k, x_u(k, x_0), u(k)) \\
&\qquad\qquad\qquad\qquad\qquad + F(n+N, x_u(N, x_0)) \\
&\text{with respect to } u(\cdot) \in \mathbb{U}^N_{\mathbb{X}_0}(n, x_0), \quad \text{subject to} \\
&x_u(0, x_0) = x_0, \quad x_u(k+1, x_0) = f(x_u(k, x_0), u(k))
\end{aligned} \qquad (\mathrm{OCP}_{\mathrm{N,e}}^{\mathrm{n}})$$

and denote the obtained optimal control sequence by $u^\star(\cdot) \in \mathbb{U}^N_{\mathbb{X}_0}(n, x_0)$.

(3) Define the NMPC-feedback value $\mu_N(n, x(n)) := u^\star(0) \in U$ and use this control value in the next sampling period.

Observe that the terminal constraints (3.11) and (3.12) are included via the restrictions $u(\cdot) \in \mathbb{U}^N_{\mathbb{X}_0}(x_0)$ and $u(\cdot) \in \mathbb{U}^N_{\mathbb{X}_0}(n, x_0)$, respectively.

Algorithm 3.10 is a special case of Algorithm 3.11 if ℓ, F and $\mathbb{X}_0$ do not depend on n. Furthermore, Algorithm 3.1 is obtained from Algorithm 3.10 for $F \equiv 0$, $\omega_{N-k} = 1$, $k = 0, \ldots, N-1$ and $\mathbb{X}_0 = \mathbb{X}$. Likewise, we can derive Algorithm 3.7 from Algorithm 3.11 by setting $F \equiv 0$, $\omega_{N_k} = 1$, $k = 0, \ldots, N-1$ and $\mathbb{X}_0(n) = \mathbb{X}$, $n \in \mathbb{N}_0$. Consequently, all NMPC Algorithms in this book are special cases of Algorithm 3.11 and all optimal control problems included in these algorithms are special cases of $(\mathrm{OCP}^{\mathrm{n}}_{\mathrm{N,e}})$.

We end this section with two useful results on the sets of admissible control sequences from Definition 3.9 which we formulate for the general setting of Algorithm 3.11, i.e., for time varying terminal constraint set $\mathbb{X}_0(n)$.

Lemma 3.12 *Let $x_0 \in \mathbb{X}_N(n)$, $N \in \mathbb{N}$ and $K \in \{0, \ldots, N\}$ be given.*

(i) *For each $u(\cdot) \in \mathbb{U}^N_{\mathbb{X}_0}(n, x_0)$ we have $x_u(K, x_0) \in \mathbb{X}_{N-K}(n+K)$.*

(ii) *For each $u(\cdot) \in \mathbb{U}^N_{\mathbb{X}_0}(n, x_0)$ the control sequences $u_1 \in U^K$ and $u_2 \in U^{N-K}$ uniquely defined by the relation*

$$u(k) = \begin{cases} u_1(k), & k = 0, \ldots, K-1 \\ u_2(k-K), & k = K, \ldots, N-1 \end{cases} \tag{3.13}$$

satisfy $u_1 \in \mathbb{U}^K_{\mathbb{X}_{N-K}}(n, x_0)$ and $u_2 \in \mathbb{U}^{N-K}_{\mathbb{X}_0}(n+K, x_{u_1}(K, x_0))$.

(iii) *For each $u_1(\cdot) \in \mathbb{U}^K_{\mathbb{X}_{N-K}}(n, x_0)$ there exists $u_2(\cdot) \in \mathbb{U}^{N-K}_{\mathbb{X}_0}(n+K, x_{u_1}(K, x_0))$ such that $u(\cdot)$ from (3.13) satisfies $u \in \mathbb{U}^N_{\mathbb{X}_0}(n, x_0)$.*

Proof (i) Using (2.3) we obtain the identity

$$x_{u(K+\cdot)}(N-K, x_u(K, x_0))) = x_u(N, x_0) \in \mathbb{X}_0(n+N),$$

which together with the definition of $\mathbb{X}_{N-K}$ implies the assertion.

(ii) The relation (3.13) together with (2.3) implies

$$x_u(k, x_0) = \begin{cases} x_{u_1}(k, x_0), & k = 0, \ldots, K \\ x_{u_2}(k-K, x_{u_1}(K, x_0)), & k = K, \ldots, N \end{cases} \tag{3.14}$$

For $k = 0, \ldots, K-1$ this identity and (3.13) yield

$$u_1(k) = u(k) \in \mathbb{U}(x_u(k, x_0)) = \mathbb{U}(x_{u_1}(k, x_0))$$

and for $k = 0, \ldots, N-K-1$ we obtain

$$u_2(k) = u(k+K) \in \mathbb{U}(x_u(k+K, x_0)) = \mathbb{U}(x_{u_2}(k, x_{u_1}(K, x_0))),$$

implying $u_1 \in \mathbb{U}^K(x_0)$ and $u_2 \in \mathbb{U}^{N-K}(x_{u_1}(K, x_0))$. Furthermore, (3.14) implies the equation $x_{u_2}(N-K, x_{u_1}(K, x_0)) = x_u(N, x_0) \in \mathbb{X}_0(n+N)$

which proves $u_2 \in \mathbb{U}^{N-K}_{\mathbb{X}_0}(n+K, x_{u_1}(K, x_0))$. This, in turn, implies that $\mathbb{U}^{N-K}_{\mathbb{X}_0}(n+K, x_{u_1}(K, x_0))$ is nonempty, hence $x_{u_1}(K, x_0) \in \mathbb{X}_{N-K}(n+K)$ and consequently $u_1 \in \mathbb{U}^K_{\mathbb{X}_{N-K}}(n, x_0)$ follows.

(iii) By definition, for each $x \in \mathbb{X}_{N-K}(n+K)$ there exists $u_2 \in \mathbb{U}^{N-K}_{\mathbb{X}_0}(n+K, x)$. Choosing such a u_2 for $x = x_{u_1}(K, x_0) \in \mathbb{X}_{N-K}(n+K)$ and defining u via (3.13), similar arguments as in Part (ii), above, show the claim $u \in \mathbb{U}^N_{\mathbb{X}_0}(n, x_0)$. □

A straightforward corollary of this lemma is the following.

Corollary 3.13
(i) For each $x \in \mathbb{X}_N$ the NMPC-feedback law μ_N obtained from Algorithm 3.10 satisfies

$$f(x, \mu_N(x)) \in \mathbb{X}_{N-1}.$$

(ii) For each $n \in \mathbb{N}$ and each $x \in \mathbb{X}_N(n)$ the NMPC-feedback law μ_N obtained from Algorithm 3.11 satisfies

$$f(x, \mu_N(n, x)) \in \mathbb{X}_{N-1}(n+1).$$

Proof We show (ii) which contains (i) as a special case. Since $\mu_N(n, x)$ is the first element $u^\star(0)$ of the optimal control sequence $u^\star \in \mathbb{U}^N_{\mathbb{X}_0}(n, x)$ we get $f(x, \mu_N(n, x)) = x_{u^\star}(1, x)$. Now Lemma 3.12(i) yields the assertion. □

3.4 The Dynamic Programming Principle

In this section we provide one of the classical tools in optimal control, the *dynamic programming principle*. We will formulate and prove the results in this section for $(\mathrm{OCP}^{\mathrm{n}}_{\mathrm{N,e}})$, since all other optimal control problems which have been introduced above can be obtained as special cases of this problem. We will first formulate the principle for the open-loop control sequences in $(\mathrm{OCP}^{\mathrm{n}}_{\mathrm{N,e}})$ and then derive consequences for the NMPC-feedback law μ_N. The dynamic programming principle is often used as a basis for numerical algorithms, cf. Sect. 3.5. In contrast to this, in this book we will exclusively use the principle for analyzing the behavior of NMPC closed-loop systems, while for the actual numerical solution of $(\mathrm{OCP}^{\mathrm{n}}_{\mathrm{N,e}})$ we use different algorithms as described in Chap. 12. The reason for this is that the numerical effort of solving $(\mathrm{OCP}^{\mathrm{n}}_{\mathrm{N,e}})$ via dynamic programming usually grows exponentially with the dimension of the state of the system, see the discussion in Sect. 3.5. In contrast to this, the computational effort of the methods described in Chap. 12 scales much more moderately with the space dimension.

We start by defining some objects we need in the sequel.

Definition 3.14 Consider the optimal control problem ($OCP^{n}_{N,e}$) with initial value $x_0 \in \mathbb{X}$, time instant $n \in \mathbb{N}_0$ and optimization horizon $N \in \mathbb{N}_0$.

(i) The function

$$V_N(n, x_0) := \inf_{u(\cdot) \in \mathbb{U}^N_{\mathbb{X}_0}(x_0)} J_N(n, x_0, u(\cdot))$$

is called *optimal value function*.

(ii) A control sequence $u^\star(\cdot) \in \mathbb{U}^N_{\mathbb{X}_0}(x_0)$ is called *optimal control sequence* for x_0, if

$$V_N(n, x_0) = J_N(n, x_0, u^\star(\cdot))$$

holds. The corresponding trajectory $x_{u^\star}(\cdot, x_0)$ is called *optimal trajectory*.

In our NMPC Algorithm 3.11 and its variants we have assumed that an optimal control sequence $u^\star(\cdot)$ exists, cf. the comment after Algorithm 3.1. In general, this is not necessarily the case but under reasonable continuity and compactness conditions the existence of $u^\star(\cdot)$ can be rigorously shown. Examples of such theorems for a general infinite-dimensional state space can be found in Keerthi and Gilbert [10] or Doležal [7]. While for formulating and proving the dynamic programming principle we will not need the existence of $u^\star(\cdot)$, for all subsequent results we will assume that $u^\star(\cdot)$ exists, in particular when we derive properties of the NMPC-feedback law μ_N. While we conjecture that most of the results in this book can be generalized to the case when μ_N is defined via an approximately minimizing control sequence, we decided to use the existence assumption because it considerably simplifies the presentation of the results in this book.

The following theorem introduces the *dynamic programming principle*. It gives an equation which relates the optimal value functions for different optimization horizons N and for different points in space.

Theorem 3.15 (Dynamic programming principle) *Consider the optimal control problem* ($OCP^{n}_{N,e}$) *with* $x_0 \in \mathbb{X}_N(n)$ *and* $n, N \in \mathbb{N}_0$. *Then for all* $N \in \mathbb{N}$ *and all* $K = 1, \ldots, N$ *the equation*

$$V_N(n, x_0) = \inf_{u(\cdot) \in \mathbb{U}^K_{\mathbb{X}_{N-K}}(n, x_0)} \left\{ \sum_{k=0}^{K-1} \omega_{N-k} \ell(n + k, x_u(k, x_0), u(k)) + V_{N-K}(n + K, x_u(K, x_0)) \right\} \tag{3.15}$$

holds. If, in addition, an optimal control sequence $u^\star(\cdot) \in \mathbb{U}^N_{\mathbb{X}_0}(n, x_0)$ *exists for* x_0, *then we get the equation*

$$V_N(n, x_0) = \sum_{k=0}^{K-1} \omega_{N-k} \ell(n+k, x_{u^\star}(k, x_0), u^\star(k)) + V_{N-K}(n+K, x_{u^\star}(K, x_0)). \tag{3.16}$$

In particular, in this case the "inf" *in* (3.15) *is a* "min"*.*

Proof First observe that from the definition of J_N for $u(\cdot) \in \mathbb{U}^N_{\mathbb{X}_0}(n, x_0)$ we immediately obtain

$$J_N(n, x_0, u(\cdot)) = \sum_{k=0}^{K-1} \omega_{N-k} \ell(n+k, x_u(k, x_0), u(k)) \\ + J_{N-K}(n+K, x_u(K, x_0), u(\cdot + K)). \tag{3.17}$$

Since $u(\cdot + K)$ equals $u_2(\cdot)$ from Lemma 3.12(ii) we obtain $u(\cdot + K) \in \mathbb{U}^{N-K}_{\mathbb{X}_0}(n + K, x_u(K, x_0))$. Note that for (3.17) to hold we need the backward numbering of ω_{N-k}.

We now prove (3.15) by proving "$\geq$" and "$\leq$" separately. From (3.17) we obtain

$$\begin{aligned} J_N(n, x_0, u(\cdot)) &= \sum_{k=0}^{K-1} \omega_{N-k} \ell(n+k, x_u(k, x_0), u(k)) \\ &\quad + J_{N-K}(n+K, x_u(K, x_0), u(\cdot + K)) \\ &\geq \sum_{k=0}^{K-1} \omega_{N-k} \ell(n+k, x_u(k, x_0), u(k)) + V_{N-K}(n+K, x_u(K, x_0)). \end{aligned}$$

Since this inequality holds for all $u(\cdot) \in \mathbb{U}^N_{\mathbb{X}_0}(n, x_0)$, it also holds when taking the infimum on both sides. Hence we get

$$\begin{aligned} V_N(n, x_0) &= \inf_{u(\cdot) \in \mathbb{U}^N_{\mathbb{X}_0}(n, x_0)} J_N(n, x_0, u(\cdot)) \\ &\geq \inf_{u(\cdot) \in \mathbb{U}^N_{\mathbb{X}_0}(n, x_0)} \left\{ \sum_{k=0}^{K-1} \omega_{N-k} \ell(n+k, x_u(k, x_0), u(k)) \right. \\ &\qquad \left. + \; V_{N-K}(n+K, x_u(K, x_0)) \right\} \\ &= \inf_{u_1(\cdot) \in \mathbb{U}^K_{\mathbb{X}_{N-K}}(n, x_0)} \left\{ \sum_{k=0}^{K-1} \omega_{N-k} \ell(n+k, x_{u_1}(k, x_0), u(k)) \right. \\ &\qquad \left. + \; V_{N-K}(n+K, x_{u_1}(K, x_0)) \right\}, \end{aligned}$$

i.e., (3.15) with "$\geq$". Here in the last step we used the fact that by Lemma 3.12(ii) the control sequence u_1 consisting of the first K elements of $u(\cdot) \in \mathbb{U}^N_{\mathbb{X}_0}(n, x_0)$ lies in $\mathbb{U}^K_{\mathbb{X}_{N-K}}(n, x_0)$ and, conversely, by Lemma 3.12(iii) each control sequence in $u_1(\cdot) \in \mathbb{U}^K_{\mathbb{X}_{N-K}}(n, x_0)$ can be extended to a sequence in $u(\cdot) \in \mathbb{U}^N_{\mathbb{X}_0}(n, x_0)$. Thus, since the expression in braces does not depend on $u(K), \ldots, u(N-1)$, the infima coincide.

In order to prove "$\le$", fix $\varepsilon > 0$ and let $u^\varepsilon(\cdot)$ be an approximately optimal control sequence for the right hand side of (3.17), i.e.,

$$\sum_{k=0}^{K-1} \omega_{N-k}\ell(n+k, x_{u^\varepsilon}(k,x_0), u^\varepsilon(k)) + J_{N-K}(n+K, x_{u^\varepsilon}(K,x_0), u^\varepsilon(\cdot+K))$$

$$\le \inf_{u(\cdot)\in\mathbb{U}^N_{\mathbb{X}_0}(n,x_0)} \left\{ \sum_{k=0}^{K-1} \omega_{N-k}\ell(n+k, x_u(k,x_0), u(k)) \right.$$

$$\left. + \; J_{N-K}(n+K, x_u(K,x_0), u(\cdot+K)) \right\} + \varepsilon.$$

Now we use the decomposition (3.13) of $u(\cdot)$ into $u_1 \in \mathbb{U}^K_{\mathbb{X}_{N-K}}(n,x_0)$ and $u_2 \in \mathbb{U}^{N-K}_{\mathbb{X}_0}(n+K, x_{u_1}(K,x_0))$ from Lemma 3.12(ii). This way we can proceed

$$\inf_{u(\cdot)\in\mathbb{U}^N_{\mathbb{X}_0}(n,x_0)} \left\{ \sum_{k=0}^{K-1} \omega_{N-k}\ell(n+k, x_u(k,x_0), u(k)) \right.$$

$$\left. + \; J_{N-K}(n+K, x_u(K,x_0), u(\cdot+K)) \right\}$$

$$= \inf_{\substack{u_1(\cdot)\in\mathbb{U}^K_{\mathbb{X}_{N-K}}(n,x_0) \\ u_2(\cdot)\in\mathbb{U}^{N-K}_{\mathbb{X}_0}(n+K,x_{u_1}(K,x_0))}} \left\{ \sum_{k=0}^{K-1} \omega_{N-k}\ell(n+k, x_{u_1}(k,x_0), u_1(k)) \right.$$

$$\left. + \; J_{N-K}(n+K, x_{u_1}(K,x_0), u_2(\cdot)) \right\}$$

$$= \inf_{u_1(\cdot)\in\mathbb{U}^K_{\mathbb{X}_{N-K}}(n,x_0)} \left\{ \sum_{k=0}^{K-1} \omega_{N-k}\ell(n+k, x_{u_1}(k,x_0), u_1(k)) \right.$$

$$\left. + \; V_{N-K}(n+K, x_{u_1}(K,x_0)) \right\}$$

Now (3.17) yields

$$V_N(n,x_0) \le J(n,x_0,u^\varepsilon(\cdot))$$

$$= \sum_{k=0}^{K-1} \omega_{N-k}\ell(n+k, x_{u^\varepsilon}(k,x_0), u^\varepsilon(k)) + J_{N-K}(n+K, x_{u^\varepsilon}(K,x_0), u^\varepsilon(\cdot+K))$$

$$\le \inf_{u(\cdot)\in\mathbb{U}^K_{\mathbb{X}_{N-K}}(n,x_0)} \left\{ \sum_{k=0}^{K-1} \omega_{N-k}\ell(n+k, x_u(k,x_0), u(k)) \right.$$

$$\left. + \; V_{N-K}(n+K, x_u(K,x_0)) \right\} + \varepsilon.$$

Since the first and the last term in this inequality chain are independent of ε and since $\varepsilon > 0$ was arbitrary, this shows (3.15) with "$\le$" and thus (3.15).

In order to prove (3.16) we use (3.17) with $u(\cdot) = u^\star(\cdot)$. This yields

$$\begin{aligned}
V_N(n,x_0) &= J(n,x_0,u^\star(\cdot)) \\
&= \sum_{k=0}^{K-1} \omega_{N-k}\ell(n+k, x_{u^\star}(k,x_0), u^\star(k)) + J_{N-K}(n+K, x_{u^\star}(K,x_0), u^\star(\cdot + K)) \\
&\ge \sum_{k=0}^{K-1} \omega_{N-k}\ell(n+k, x_{u^\star}(k,x_0), u^\star(k)) + V_{N-K}(n+K, x_{u^\star}(K,x_0)) \\
&\ge \inf_{u(\cdot)\in\mathbb{U}^K_{\mathbb{X}_{N-K}}(n,x_0)} \left\{ \sum_{k=0}^{K-1} \omega_{N-k}\ell(n+k, x_u(k,x_0), u(k)) + V_{N-K}(n+K, x_u(K,x_0)) \right\} \\
&= V_N(n,x_0),
\end{aligned}$$

where we used the (already proven) equality (3.15) in the last step. Hence, the two "$\ge$" in this chain are actually "$=$" which implies (3.16). □

The following corollary states an immediate consequence of the dynamic programming principle. It shows that tails of optimal control sequences are again optimal control sequences for suitably adjusted optimization horizon, time instant, and initial value.

Corollary 3.16 *If $u^\star(\cdot)$ is an optimal control sequence for initial value $x_0 \in \mathbb{X}_N(n)$, time instant n and optimization horizon $N \ge 2$, then for each $K = 1, \ldots, N-1$ the sequence $u^\star_K(\cdot) = u^\star(\cdot + K)$, i.e.,*

$$u^\star_K(k) = u^\star(K+k), \quad k = 0, \ldots, N-K-1$$

is an optimal control sequence for initial value $x_{u^\star}(K,x_0)$, time instant $n+K$ and optimization horizon $N-K$.

Proof Inserting $V_N(n,x_0) = J_N(n,x_0,u^\star(\cdot))$ and the definition of $u^\star_k(\cdot)$ into (3.17) we obtain

$$V_N(n,x_0) = \sum_{k=0}^{K-1} \omega_{N-k}\ell(n+k, x_{u^\star}(k,x_0), u^\star(k)) + J_{N-K}(n+K, x_{u^\star}(K,x_0), u^\star_K(\cdot))$$

Subtracting (3.16) from this equation yields

$$0 = J_{N-K}(n+K, x_{u^\star}(K,x_0), u^\star_K(\cdot)) - V_{N-K}(n+K, x_{u^\star}(K,x_0))$$

which shows the assertion. □

The next theorem relates the NMPC-feedback law μ_N defined in the NMPC Algorithm 3.11 and its variants to the dynamic programming principle. Here we use the argmin operator in the following sense: for a map $a : U \to \mathbb{R}$, a nonempty subset $\widetilde{U} \subseteq U$ and a value $u^\star \in \widetilde{U}$ we write

$$u^\star = \operatorname*{argmin}_{u \in \widetilde{U}} a(u) \tag{3.18}$$

if and only if $a(u^\star) = \inf_{u \in \widetilde{U}} a(u)$ holds. Whenever (3.18) holds the existence of the minimum $\min_{u \in \widetilde{U}} a(u)$ follows. However, we do not require uniqueness of the minimizer $u^\star$. In case of uniqueness equation (3.18) can be understood as an assignment, otherwise it is just a convenient way of writing "$u^\star$ minimizes $a(u)$".

Theorem 3.17 (Dynamic programming and NMPC) *Consider the optimal control problem* ($\mathrm{OCP}^{\mathrm{n}}_{\mathrm{N,e}}$) *with* $x_0 \in \mathbb{X}_N(n)$ *and* $n, N \in \mathbb{N}_0$ *and assume that an optimal control sequence* $u^\star$ *exists. Then the NMPC-feedback law* $\mu_N(n, x_0) = u^*(0)$ *satisfies*

$$\mu_N(n, x_0) = \operatorname*{argmin}_{u \in \mathbb{U}^1_{\mathbb{X}_{N-1}}(n, x_0)} \{\omega_N \ell(n, x_0, u) + V_{N-1}(n+1, f(x_0, u))\} \tag{3.19}$$

and

$$V_N(n, x_0) = \omega_N \ell(n, x_0, \mu_N(n, x_0)) + V_{N-1}(n+1, f(x_0, \mu_N(n, x_0))) \tag{3.20}$$

where in (3.19) *we interpret* $\mathbb{U}^1_{\mathbb{X}_{N-1}}(n, x_0)$ *as a subset of* U, *i.e., we identify the one element sequence* $u = u(\cdot)$ *with its only element* $u = u(0)$.

Proof Equation (3.20) follows by inserting $u^\star(0) = \mu_N(n, x_0)$ and $x_{u^\star}(1, x_0) = f(x_0, \mu_N(n, x_0))$ into (3.16) for $K = 1$.

Inserting $x_u(1, x_0) = f(x_0, u)$ into the dynamic programming principle (3.15) for $K = 1$ we further obtain

$$V_N(n, x_0) = \inf_{u \in \mathbb{U}^1_{\mathbb{X}_{N-1}}(n, x_0)} \{\omega_N \ell(n, x_0, u) + V_{N-1}(n+1, f(x_0, u))\}\,. \tag{3.21}$$

This implies that the right-hand sides of (3.20) and (3.21) coincide. Thus, the definition of argmin in (3.18) with $a(u) = \omega_N \ell(n, x_0, u) + V_{N-1}(n+1, f(x_0, u))$ and $\widetilde{U} = \mathbb{U}^1_{\mathbb{X}_{N-1}}(n, x_0)$ yields (3.19). $\square$

Our final corollary in this section shows that we can reconstruct the whole optimal control sequence $u^\star(\cdot)$ using the feedback from (3.19).

Corollary 3.18 *Consider the optimal control problem* ($\mathrm{OCP}^{\mathrm{n}}_{\mathrm{N,e}}$) *with* $x_0 \in \mathbb{X}$ *and* $n, N \in \mathbb{N}_0$ *and consider admissible feedback laws* $\mu_{N-k} : \mathbb{N}_0 \times \mathbb{X} \to U$, $k = 0, \ldots, N-1$, *in the sense of Definition 3.2(iv). Denote the solution of the closed-loop system*

$$x(0) = x_0, \quad x(k+1) = f(x(k), \mu_{N-k}(n+k, x(k))), \; k = 0, \ldots, N-1 \qquad (3.22)$$

by $x_\mu(\cdot)$ *and assume that the* μ_{N-k} *satisfy* (3.19) *with horizon* $N-k$ *instead of* N, *time index* $n+k$ *instead of* n *and initial value* $x_0 = x_\mu(k)$ *for* $k = 0, \ldots, N-1$. *Then*

$$u^\star(k) = \mu_{N-k}(n+k, x_\mu(k)), \quad k = 0, \ldots, N-1 \qquad (3.23)$$

is an optimal control sequence for initial time n *and initial value* x_0 *and the solution of the closed-loop system* (3.22) *is a corresponding optimal trajectory.*

Proof Applying the control (3.23) to the dynamics (3.22) we immediately obtain

$$x_{u^\star}(n, x_0) = x_\mu(n), \quad n = 0, \ldots, N-1.$$

Hence, we need to show that

$$V_N(n, x_0) = J_N(n, x_0, u^\star) = \sum_{k=0}^{N-1} \omega_{N-k} \ell(n+k, x(k), u^\star(k)) + F(n+N, x(N)).$$

Using (3.23) and (3.20) for $N-k$ instead of N we get

$$V_{N-k}(n+k, x(k)) = \omega_{N-k} \ell(n+k, x(k), u^\star(k)) + V_{N-k-1}(n+k+1, x(k+1))$$

for $k = 0, \ldots, N-1$. Summing these equalities for $k = 0, \ldots, N-1$ and eliminating the identical terms $V_{N-k}(n+k, x_0)$, $k = 1, \ldots, N-1$ on both sides we obtain

$$V_N(n, x_0) = \sum_{k=0}^{N-1} \omega_{N-k} \ell(n+k, x(k), u^\star(k)) + V_0(n+N, x(N))$$

Since by definition of J_0 we have $V_0(n+N, x) = F(x)$, this shows the assertion. □

3.5 Notes and Extensions

The discrete time nonlinear model predictive control framework introduced in Sects. 3.1–3.3 covers most of the settings commonly found in the literature. For continuous time systems, one often also finds nonlinear model predictive control frameworks in explicit continuous time form. In these frameworks, the optimization in ($\mathrm{OCP}^{\mathrm{n}}_{\mathrm{N,e}}$) and its variants is carried out at times $t_0, t_1, t_2, \ldots$ minimizing an integral criterion along the continuous time solution of the form

$$J_{T_{\mathrm{opt}}}(x_0, v) = \int_0^{T_{\mathrm{opt}}} L(\varphi(t, x_0, v), v(t))dt + F(\varphi(T_{\mathrm{opt}}, N, x_0, v)).$$

The feedback law $\mu_{T_{\text{opt}}}$ computed at time t_n is then obtained by applying the first portion $v^\star|_{[0,t_{n+1}-t_n]}$ of the optimal control function $v^\star$ to the system, see, e.g., Alamir [1] or Findeisen [9]. Provided that $t_{n+1} - t_n = T$ holds for all n, this problem is equivalent to our setting if the sampled data system (2.8) and the integral criterion (3.4) with running cost L is used.

Regarding notation, in NMPC it is important to distinguish between the open-loop predictions and the NMPC closed loop. Here we have decided to denote the open-loop predictions by $x_u(k)$ or $x_u(k, x_0)$ and the NMPC closed-loop trajectories by either $x(n)$ or—more often—by $x_{\mu_N}(n)$ or $x_{\mu_N}(n, x_0)$. There are, however, various other notations commonly found in the literature. For instance, the prediction at time instant n is occasionally denoted as $x(k|n)$ in order to emphasize the dependence on the time instant n. In our notation, the dependence on n is implicitly expressed via the initial condition $x_0 = x(n)$ and the index n in ($\text{OCP}_\text{N}^\text{n}$) or ($\text{OCP}_\text{N,e}^\text{n}$). Whenever necessary, the value of n under consideration will be specified in the context. On the other hand, we decided to always explicitly indicate the dependence of open-loop solutions on the control sequence u. This notation enables us to easily distinguish between open-loop and closed-loop solutions and also for simultaneously considering open-loop solutions for different control sequences.

In linear discrete time MPC, the optimization at each sampling instant is occasionally performed over control sequences with predefined values $u(K), \ldots, u(N-1)$ for some $K \in \{1, \ldots, N-1\}$, i.e., only $u(0), \ldots, u(K-1)$ are used as optimization variables in ($\text{OCP}_\text{N,e}$) and its variants. For instance, if $x_* = 0$ and $u_* = 0$, cf. Sect. 3.1, then $u(K), \ldots, u(N-1) = 0$ is a typical choice. In this setting, K is referred to as *optimization horizon* (or *control horizon*) and N is referred to as *prediction horizon*. Since this variant is less common in nonlinear MPC, we do not consider it in this book; in particular, we use the terms optimization horizon and prediction horizon synonymously, while the term control horizon will receive a different meaning in Sect. 10.4. Still, most of the subsequent analysis could be extended to the case in which the optimization horizon and the prediction horizon do not coincide.

Regarding the cost function ℓ, the setting described in Sects. 3.1 and 3.3 is easily extended to the case in which a set instead of a single equilibrium or a time-variant family of sets instead of a single reference shall be stabilized. Indeed, if we are given a family of sets $X^{\text{ref}}(n) \subset X$ such that for each $x \in X^{\text{ref}}(n)$ there is a control u_x with $f(x, u_x) \in X^{\text{ref}}(n+1)$, then we can modify (3.8) to

$$\begin{aligned} &\ell(n, x, u_x) = 0 \text{ for all } x \in X^{\text{ref}}(n) \quad \text{and} \\ &\ell(n, x, u) > 0 \;\text{ for all } x \in X \setminus X^{\text{ref}}(n),\; u \in U. \end{aligned} \tag{3.24}$$

Similarly, we can modify (3.2) in the time-invariant case.

Another modification of ℓ, again often found in the linear MPC literature, are stage cost functions which include two consecutive control values, i.e., $\ell(x_u(k), u(k), u(k-1))$. Typically, this is used in order to penalize large changes in the control input by adding a term $\sigma\|u(k) - u(k-1)\|$ (assuming U to be a vector space with norm $\|\cdot\|$, for simplicity). Using the augmented state $\tilde{x}_u(k) = (x_u(k), u(k-1))$ this can be transformed into a cost function meeting our setting by defining $\widetilde{\ell}(\tilde{x}_u(k), u(k)) = \ell(x_u(k), u(k), u(k-1))$.

Yet another commonly used variant are stage costs in which only an output $y = h(x)$ instead of the whole state is taken into account. In this case, ℓ will usually no longer satisfy (3.2) or (3.8), i.e., ℓ will not be positive definite, anymore. We will discuss this case in Sect. 10.3. In this context it should be noted that even if the stage cost ℓ depends only on an output, the NMPC-feedback μ_N will nevertheless be a state feedback law. Hence, if only output data is available, suitable observers need to be used in order to reconstruct the state of the system. An even more general choice of ℓ is allowed in *economic NMPC*, which we will discuss in Chap. 8.

The term dynamic programming was introduced by Bellman [2] and due to his seminal contributions to this area the dynamic programming principle is often also called Bellman's principle of optimality. The principle is widely used in many application areas and a quite comprehensive account of its use in various different settings is given in the monographs by Bertsekas [4, 5]. For $K = 1$, the dynamic programming principle (3.15) simplifies to

$$V_N(n, x) = \inf_{u \in \mathbb{U}^1_{\mathbb{X}_{N-1}}(n,x)} \{\omega_N \ell(n+k, x, u) + V_{N-1}(n+1, f(x, u))\} \tag{3.25}$$

and in this form it can be used for recursively computing $V_1, V_2, \ldots V_N$ starting from $V_0(x) = F(x)$. Once V_N and V_{N-1} are known, the feedback law μ_N can be obtained from (3.19).

Whenever V_N can be expressed using simple functions this approach of computing V_N can be efficiently used. For instance, when the dynamics are linear and finite dimensional, the stage cost is quadratic and there are no constraints, then V_N can be expressed as $V_N(x) = x^\top P_N x$ for a matrix $P_N \in \mathbb{R}^{d \times d}$ and (3.25) reduces to the Riccati difference equation, see, e.g., Dorato and Levis [8].

For nonlinear systems with low-dimensional state space it is also possible to approximate V_N numerically using the backward recursion induced by (3.25) with approximations $\widetilde{V}_1 \approx V_1, \ldots, \widetilde{V}_N \approx V_N$. These approximations can then, in turn, be used in order to compute a numerical approximation of the NMPC-feedback law μ_N. This is, roughly speaking, the idea behind the so-called *explicit MPC* methods, see, e.g., Borrelli, Baotic, Bemporad and Morari [6], Bemporad and Filippi [3], Tøndel, Johansen and Bemporad [11], to mention but a few papers from this area in which often special problem structures like piecewise linear dynamics instead of general nonlinear models are considered. The main advantage of this approach is that $\widetilde{V}_N$ and the approximation of μ_N can be computed offline and thus the online computational effort of evaluating μ_N is very low. Hence, in contrast to conventional NMPC in which ($\mathrm{OCP}^{\mathrm{n}}_{\mathrm{N,e}}$) is entirely solved online, this method is also applicable to very fast systems which require fast sampling.

Unfortunately, for high-dimensional systems, the numerical effort of this approach becomes prohibitive since the computational complexity of computing $\widetilde{V}_N$ grows exponentially in the state dimension, unless one can exploit very specific problem structure. This fact—the so-called curse of dimensionality—arises because the approximation of V_N requires a global solution to ($\mathrm{OCP}^{\mathrm{n}}_{\mathrm{N,e}}$) or its variants for all initial values $x_0 \in \mathbb{X}$ or at least in the set of interest, which is typically a set of full dimension in state space. Consequently, dynamic programming method cannot be applied to high-dimensional systems. In contrast to this, the methods we will discuss in Chap. 12 solve ($\mathrm{OCP}^{\mathrm{n}}_{\mathrm{N,e}}$) for a single initial value x_0 only at each sampling instant, i.e., locally in space. Since this needs to be done online, these methods are in principle slower, but since the numerical effort scales much more moderate with the state dimension they are nevertheless applicable to systems with much higher state dimension.

Problems

1. Consider the control system

$$x^+ = f(x, u) = ax + bu$$

with $x \in X = \mathbb{R}$, $u \in U = \mathbb{R}$, constraints $\mathbb{X} = [-1, 1]$ and $\mathbb{U} = [-100, 100]$ and real parameters $a, b \in \mathbb{R}$.

   (a) For which parameters $a, b \in \mathbb{R}$ is the state constraint set $\mathbb{X}$ viable?
   (b) For those parameters for which $\mathbb{X}$ is not viable, determine a viable state constraint set contained in $\mathbb{X}$.

2. Compute an optimal trajectory for the optimal control problem ($\mathrm{OCP}_{\mathrm{N,e}}$)

$$\begin{aligned} \text{Minimize} \quad & \sum_{k=0}^{N-1} u(k)^2 \\ \text{subject to} \quad & x_1(k+1) = x_1(k) + 2x_2(k) \\ & x_2(k+1) = 2u(k) - x_2(k) \\ & x_1(0) = 0, \quad x_2(0) = 0 \\ & x_1(N) = 4, \quad x_2(N) = 0 \end{aligned}$$

with $N = 4$ via dynamic programming.

3. Consider the NMPC problem defined by the dynamics

$$x^+ = f(x, u) = x + u$$

with $x \in X = \mathbb{R}$, $u \in U = \mathbb{R}$ and stage costs

$$\ell(x, u) = x^2 + u^2.$$

(a) Compute the optimal value function V_2 and the NMPC-feedback law μ_2 by dynamic programming.
(b) Show that V_2 is a Lyapunov function for the closed loop and compute the functions α_1, α_2 and α_V in (2.37) and (2.38).
(c) Show that the NMPC closed loop is globally asymptotically stable without using the Lyapunov function V_2.

4. Consider an optimal trajectory $x_{u^\star}(\cdot, x_0)$ for the optimal control problem (OCP$_\mathrm{N}$) with initial value x_0 and optimization horizon $N \geq 2$. Prove that for any $K \in \{1, \dots, N-1\}$ the tail

$$x_{u^\star}(K, x_0), \dots, x_{u^\star}(N-1, x_0)$$

of the optimal trajectory along with the tail

$$u^\star(K), \dots, u^\star(N-1)$$

of the optimal control sequence are optimal for (OCP$_\mathrm{N}$) with new initial value $x_{u^\star}(K, x_0)$ and optimization horizon $N-K$, i.e., that

$$\sum_{k=K}^{N-1} \ell(x_{u^\star}(k, x_0), u^\star(k)) = V_{N-K}(x_{u^\star}(K, x_0))$$

holds.

5. After a lecture in which you presented the basic NMPC Algorithm 3.1, a student asks the following question:
"If I ride my bicycle and want to make a turn to the left, I first steer a little bit to the right to make my bicycle tilt to the left. Let us assume that this way of making a turn is optimal for a suitable problem of type (OCP$_\mathrm{N}$). This would mean that the optimal control sequence will initially steer to the right and later steer to the left. If we use this optimal control sequence in an NMPC algorithm, only the first control action will be implemented. As a consequence, we will always steer to the right, and we will make a turn to the right instead of a turn to the left. Does this mean that NMPC does not work for controlling my bicycle?"
What do you respond?

References

1. Alamir, M.: Stabilization of nonlinear systems using receding-horizon control schemes. Lecture Notes in Control and Information Sciences, vol. 339. Springer, London (2006)
2. Bellman, R.: Dynamic programming. Princeton University Press. Reprinted in 2010 (1957)
3. Bemporad, A., Filippi, C.: Suboptimal explicit MPC via approximate multiparametric quadratic programming. In: Proceedings of the 40th IEEE Conference on Decision and Control - CDC 2001, pp. 4851–4856. Orlando, Florida, USA (2001)

4. Bertsekas, D.P.: Dynamic Programming and Optimal Control, vol. I, 3rd edn. Athena Scientific, Belmont (2005)
5. Bertsekas, D.P.: Dynamic Programming and Optimal Control, vol. II, 2nd edn. Athena Scientific, Belmont (2001)
6. Borrelli, L., Baotic, T., Bemporad, A., Morari, T.: Efficient on-line computation of constrained optimal control. In: Proceedings of the 40th IEEE Conference on Decision and Control - CDC 2001, pp. 1187–1192. Orlando, Florida, USA (2001)
7. Doležal, J.: Existence of optimal solutions in general discrete systems. Kybernetika **11**(4), 301–312 (1975)
8. Dorato, P., Levis, A.H.: Optimal linear regulators: the discrete-time case. IEEE Trans. Automat. Control **16**, 613–620 (1971)
9. Findeisen, R.: Nonlinear Model Predictive Control: A Sampled-Data Feedback Perspective. University of Stuttgart, VDI-Verlag, Düsseldorf, PhD-thesis (2004)
10. Keerthi, S.S., Gilbert, E.G.: An existence theorem for discrete-time infinite-horizon optimal control problems. IEEE Trans. Automat. Control **30**(9), 907–909 (1985)
11. Tøndel, P., Johansen, T.A., Bemporad, A.: An algorithm for multi-parametric quadratic programming and explicit MPC solutions. Automatica **39**(3), 489–497 (2003)

Chapter 4
Infinite Horizon Optimal Control

In this chapter we give an introduction to nonlinear infinite horizon optimal control. The dynamic programming principle as well as several consequences of this principle are proved. One of the main results of this chapter is that the infinite horizon optimal feedback law asymptotically stabilizes the system and that the infinite horizon optimal value function is a Lyapunov function for the closed-loop system. Motivated by this property we formulate a relaxed version of the dynamic programming principle, which allows to prove stability and suboptimality results for nonoptimal feedback laws and without using the optimal value function. A practical version of this principle is provided, too. These results will be central in the following chapters for the stability and performance analysis of NMPC algorithms. For the special case of sampled data systems we finally show that for suitable integral costs asymptotic stability of the continuous time sampled data closed-loop system follows from the asymptotic stability of the associated discrete time system.

4.1 Definition and Well Posedness of the Problem

For the finite horizon optimal control problems from the previous chapter we can define infinite horizon counterparts by replacing the upper limits $N-1$ in the respective sums by ∞. Since for this infinite horizon formulation the terminal state $x_u(N)$ vanishes from the problem, it is not reasonable to consider terminal conditions. Furthermore, we will not consider any weights in the infinite horizon case. Hence, the most general infinite horizon problem we consider is the following.

L. Grüne and J. Pannek, *Nonlinear Model Predictive Control*,
Communications and Control Engineering, DOI 10.1007/978-3-319-46024-6_4

$$\begin{array}{|l|}\hline \text{minimize} \quad J_\infty(n, x_0, u(\cdot)) := \sum_{k=0}^{\infty} \ell(n+k, x_u(k, x_0), u(k)) \\ \text{with respect to } u(\cdot) \in \mathbb{U}^\infty(x_0), \quad \text{subject to} \\ x_u(0, x_0) = x_0, \quad x_u(k+1, x_0) = f(x_u(k, x_0), u(k)) \\ \hline \end{array} \qquad (\mathrm{OCP}_\infty^{\mathrm{n}})$$

We optimize over the set of admissible control sequences $\mathbb{U}^\infty(x_0)$ defined in Definition 3.2 and assume that this set is nonempty for all $x_0 \in \mathbb{X}$, which is equivalent to the viability of $\mathbb{X}$ according to Assumption 3.3. In order to keep the presentation self contained all subsequent statements are formulated for general time varying stage cost ℓ. For some of the results we assume that ℓ is as in (3.8), i.e., it penalizes the distance to a (possibly time varying) reference trajectory x^{ref}. In the special case of constant reference $x^{\mathrm{ref}} \equiv x_*$ the stage cost ℓ and the functional J_∞ in $(\mathrm{OCP}_\infty^{\mathrm{n}})$ do not depend on the time n. In this case, we denote the problem by (OCP_∞).

Similar to Definition 3.14 we define the optimal value function and optimal trajectories.

Definition 4.1 Consider the optimal control problem $(\mathrm{OCP}_\infty^{\mathrm{n}})$ with initial value $x_0 \in \mathbb{X}$ and time instant $n \in \mathbb{N}_0$.

(i) The function

$$V_\infty(n, x_0) := \inf_{u(\cdot) \in \mathbb{U}^\infty(x_0)} J_\infty(n, x_0, u(\cdot))$$

is called *optimal value function*.

(ii) A control sequence $u^\star(\cdot) \in \mathbb{U}^\infty(x_0)$ is called *optimal control sequence* for x_0 if

$$V_\infty(n, x_0) = J_\infty(n, x_0, u^\star(\cdot))$$

holds. The corresponding trajectory $x_{u^\star}(\cdot, x_0)$ is called *optimal trajectory*.

Since now—in contrast to the finite horizon problem—an infinite sum appears in the definition of J_∞, it is no longer straightforward that V_∞ is finite. In order to ensure that this is the case, the following definition is helpful.

Definition 4.2 Consider the control system (2.1) and a reference trajectory $x^{\mathrm{ref}} : \mathbb{N}_0 \to \mathbb{X}$ with reference control sequence $u^{\mathrm{ref}} \in \mathbb{U}^\infty(x^{\mathrm{ref}}(0))$. We say that the system is (uniformly) asymptotically controllable to x^{ref} if there exists a function $\beta \in \mathcal{KL}$ such that for each initial time $n_0 \in \mathbb{N}_0$ and for each admissible initial value $x_0 \in \mathbb{X}$ there exists an admissible control sequence $u \in \mathbb{U}^\infty(x_0)$ such that the inequality

$$|x_u(n, x_0)|_{x^{\mathrm{ref}}(n+n_0)} \leq \beta(|x_0|_{x^{\mathrm{ref}}(n_0)}, n) \tag{4.1}$$

holds for all $n \in \mathbb{N}_0$. We say that this asymptotic controllability has the small control property if $u \in \mathbb{U}^\infty(x_0)$ can be chosen such that the inequality

$$|x_u(n, x_0)|_{x^{\text{ref}}(n+n_0)} + |u(n)|_{u^{\text{ref}}(n+n_0)} \leq \beta(|x_0|_{x^{\text{ref}}(n_0)}, n) \tag{4.2}$$

holds for all $n \in \mathbb{N}_0$. Here, as in Sect. 2.3 we write $|x_1|_{x_2} = d_X(x_1, x_2)$ and $|u_1|_{u_2} = d_U(u_1, u_2)$.

Observe that uniform asymptotic controllability is a necessary condition for uniform feedback stabilization. Indeed, if we assume asymptotic stability of the closed-loop system $x^+ = g(n, x) = f(x, \mu(n, x))$, then we immediately get asymptotic controllability with control $u(n) = \mu(n, x(n, n_0, x_0))$. The small control property, however, is not satisfied in general.

In order to use Definition 4.2 for deriving bounds on the optimal value function, we need a result known as Sontag's $\mathcal{KL}$-Lemma [24, Proposition 7]. This proposition states that for each $\mathcal{KL}$-function β there exist functions $\gamma_1, \gamma_2 \in \mathcal{K}_\infty$ such that the inequality

$$\beta(r, n) \leq \gamma_1(e^{-n}\gamma_2(r))$$

holds for all $r, n \geq 0$ (in fact, the result holds for real $n \geq 0$ but we only need it for integers here). Using the functions γ_1 and γ_2 we can define stage cost functions

$$\ell(n, x, u) := \gamma_1^{-1}(|x|_{x^{\text{ref}}(n)}) + \lambda\gamma_1^{-1}(|u|_{u^{\text{ref}}(n)}) \tag{4.3}$$

for $\lambda \geq 0$. The following theorem states that under Definition 4.2 this stage cost ensures (uniformly) finite upper and positive lower bounds on V_∞.

Theorem 4.3 (Bounds on V_∞) *Consider the control system (2.1) and a reference trajectory $x^{\text{ref}} : \mathbb{N}_0 \to \mathbb{X}$ with reference control sequence $u^{\text{ref}} \in \mathbb{U}^\infty(x^{\text{ref}}(0))$. If the system is asymptotically controllable to x^{ref}, then there exist $\alpha_1, \alpha_2 \in \mathcal{K}_\infty$ such that the optimal value function V_∞ corresponding to the cost function $\ell : \mathbb{N}_0 \times X \times U \to \mathbb{R}_0^+$ from* (4.3) *with $\lambda = 0$ satisfies*

$$\alpha_1(|x_0|_{x^{\text{ref}}(n_0)}) \leq V_\infty(n_0, x_0) \leq \alpha_2(|x_0|_{x^{\text{ref}}(n_0)}) \tag{4.4}$$

for all $n_0 \in \mathbb{N}_0$ and all $x_0 \in \mathbb{X}$.
If, in addition, the asymptotic controllability has the small control property then the statement also holds for ℓ from (4.3) *with arbitrary $\lambda \geq 0$.*

Proof For each x_0, n_0, and $u \in \mathbb{U}^\infty(x_0)$ we get

$$J_\infty(n_0, x_0, u) = \sum_{k=0}^{\infty} \ell(n_0 + k, x_u(k, x_0), u(k)) \geq \ell(n, x_u(0, x_0), u(0)) \geq \gamma_1^{-1}(|x_0|_{x^{\text{ref}}(n_0)})$$

for each $\lambda \geq 0$. Hence, from the definition of V_∞ we get

$$V_\infty(n_0, x_0) = \inf_{u(\cdot) \in \mathbb{U}^\infty(x_0)} J_\infty(n_0, x_0, u(\cdot)) \geq \gamma_1^{-1}(|x_0|_{x^{\text{ref}}(n_0)}).$$

This proves the lower bound in (4.4) for $\alpha_1 = \gamma_1^{-1}$.

For proving the upper bound, we first consider the case $\lambda = 0$. For all n_0 and x_0 the control $u \in \mathbb{U}^\infty(x_0)$ from Definition 4.2 yields

$$\begin{aligned} V_\infty(n_0, x_0) &\le J_\infty(n_0, x_0, u) \\ &= \sum_{k=0}^{\infty} \ell(n_0 + k, x_u(k, x_0), u(k)) \\ &= \sum_{k=0}^{\infty} \gamma_1^{-1}(|x_u(k, x_0)|_{x^{\text{ref}}(n_0+k)}) \\ &\le \sum_{k=0}^{\infty} \gamma_1^{-1}(\beta(|x_0|_{x^{\text{ref}}(n_0)}, k)) \;\le\; \sum_{k=0}^{\infty} e^{-k}\gamma_2(|x_0|_{x^{\text{ref}}(n_0)}) \\ &= \frac{e}{e-1}\gamma_2(|x_0|_{x^{\text{ref}}(n_0)}), \end{aligned}$$

i.e., the upper inequality from (4.4) with $\alpha_2(r) = e\gamma_2(r)/(e-1)$. If the small control property holds, then the upper bound for $\lambda > 0$ follows similarly with $\alpha_2(r) = (1+\lambda)e\gamma_2(r)/(e-1)$. □

In fact, the specific form (4.3) is just one possible choice of ℓ for which this theorem holds. It is rather easy to extend the result to any ℓ which is bounded from below by some $\mathcal{K}_\infty$-function in x (uniformly for all u and n) and bounded from above by ℓ from (4.3) in balls $\mathcal{B}_\varepsilon(x^{\text{ref}}(n))$. Since, however, the choice of appropriate cost functions ℓ for infinite horizon optimal control problems is not a central topic of this book, we leave this extension to the interested reader.

4.2 The Dynamic Programming Principle

In this section we essentially restate and reprove the results from Sect. 3.4 for the infinite horizon case. We begin with the *dynamic programming principle* for the infinite horizon problem ($\text{OCP}_\infty^{\text{n}}$). Throughout this section we assume that $V_\infty(x)$ is finite for all $x \in \mathbb{X}$ as ensured, e.g., by Theorem 4.3.

Theorem 4.4 (Dynamic programming principle) *Consider the optimal control problem* ($\text{OCP}_\infty^{\text{n}}$) *with $x_0 \in \mathbb{X}$ and $n \in \mathbb{N}_0$. Then for all $K \in \mathbb{N}$ the equation*

$$V_\infty(n, x_0) = \inf_{u(\cdot)\in\mathbb{U}^K(x_0)} \left\{ \sum_{k=0}^{K-1} \ell(n+k, x_u(k, x_0), u(k)) + V_\infty(n+K, x_u(K, x_0)) \right\} \tag{4.5}$$

holds. If, in addition, an optimal control sequence $u^\star(\cdot)$ exists for x_0, then we get the equation

$$V_\infty(n, x_0) = \sum_{k=0}^{K-1} \ell(n+k, x_{u^\star}(k, x_0), u^\star(k)) + V_\infty(n+K, x_{u^\star}(K, x_0)). \tag{4.6}$$

In particular, in this case the "inf" in (4.5) is a "min".

Proof From the definition of J_∞ for $u(\cdot) \in \mathbb{U}^\infty(x_0)$ we immediately obtain

$$J_\infty(n, x_0, u(\cdot)) = \sum_{k=0}^{K-1} \ell(n+k, x_u(k, x_0), u(k)) + J_\infty(n+K, x_u(K, x_0), u(\cdot + K)), \tag{4.7}$$

where $u(\cdot + K)$ denotes the shifted control sequence defined by $u(\cdot + K)(k) = u(k+K)$, which is admissible for $x_u(K, x_0)$.

We now prove (4.5) by showing "$\geq$" and "$\leq$" separately: From (4.7) we obtain

$$\begin{aligned} J_\infty(n, x_0, u(\cdot)) &= \sum_{k=0}^{K-1} \ell(n+k, x_u(k, x_0), u(k)) + J_\infty(n+K, x_u(K, x_0), u(\cdot + K)) \\ &\geq \sum_{k=0}^{K-1} \ell(n+k, x_u(k, x_0), u(k)) + V_\infty(n+K, x_u(K, x_0)). \end{aligned}$$

Since this inequality holds for all $u(\cdot) \in \mathbb{U}^\infty$, it also holds when taking the infimum on both sides. Hence we get

$$\begin{aligned} V_\infty(n, x_0) &= \inf_{u(\cdot) \in \mathbb{U}^\infty(x_0)} J_\infty(n, x_0, u(\cdot)) \\ &\geq \inf_{u(\cdot) \in \mathbb{U}^K(x_0)} \left\{ \sum_{k=0}^{K-1} \ell(n+k, x_u(k, x_0), u(k)) + V_\infty(n+K, x_u(K, x_0)) \right\}, \end{aligned}$$

i.e., (4.5) with "$\geq$".

In order to prove "$\leq$", fix $\varepsilon > 0$ and let $u^\varepsilon(\cdot)$ be an approximately optimal control sequence for the right hand side of (4.7), i.e.,

$$\begin{aligned} &\sum_{k=0}^{K-1} \ell(n+k, x_{u^\varepsilon}(k, x_0), u^\varepsilon(k)) + J_\infty(n+K, x_{u^\varepsilon}(K, x_0), u^\varepsilon(\cdot + K)) \\ &\leq \inf_{u(\cdot) \in \mathbb{U}^\infty(x_0)} \left\{ \sum_{k=0}^{K-1} \ell(n+k, x_u(k, x_0), u(k)) + J_\infty(n+K, x_u(K, x_0), u(\cdot + K)) \right\} + \varepsilon. \end{aligned}$$

Now we decompose $u(\cdot) \in \mathbb{U}^\infty(x_0)$ analogously to Lemma 3.12(ii) and (iii) into $u_1 \in \mathbb{U}^K(x_0)$ and $u_2 \in \mathbb{U}^\infty(x_{u_1}(K, x_0))$ via

$$u(k) = \begin{cases} u_1(k), & k = 0, \ldots, K-1 \\ u_2(k-K), & k \geq K \end{cases}$$

This implies

$$\begin{aligned}
&\inf_{u(\cdot)\in\mathbb{U}^\infty(x_0)} \left\{ \sum_{k=0}^{K-1} \ell(n+k, x_u(k,x_0), u(k)) + J_\infty(n+K, x_u(K,x_0), u(\cdot+K)) \right\} \\
&= \inf_{\substack{u_1(\cdot)\in\mathbb{U}^K(x_0)\\ u_2(\cdot)\in\mathbb{U}^\infty(x_{u_1}(K,x_0))}} \left\{ \sum_{k=0}^{K-1} \ell(n+k, x_{u_1}(k,x_0), u_1(k)) + J_\infty(n+K, x_{u_1}(K,x_0), u_2(\cdot)) \right\} \\
&= \inf_{u_1(\cdot)\in\mathbb{U}^K(x_0)} \left\{ \sum_{k=0}^{K-1} \ell(n+k, x_{u_1}(k,x_0), u_1(k)) + V_\infty(n+K, x_{u_1}(K,x_0)) \right\}
\end{aligned}$$

Now (4.7) yields

$$\begin{aligned}
V_\infty(n,x_0) &\leq J_\infty(n,x_0,u^\varepsilon(\cdot)) \\
&= \sum_{k=0}^{K-1} \ell(n+k, x_{u^\varepsilon}(k,x_0), u^\varepsilon(k)) + J_\infty(n+K, x_{u^\varepsilon}(K,x_0), u^\varepsilon(\cdot+K)) \\
&\leq \inf_{u(\cdot)\in\mathbb{U}^K(x_0)} \left\{ \sum_{k=0}^{K-1} \ell(n+k, x_u(k,x_0), u(k)) + V_\infty(n+K, x_u(K,x_0)) \right\} + \varepsilon,
\end{aligned}$$

i.e.,

$$\begin{aligned}
&V_\infty(n,x_0) \\
&\leq \inf_{u(\cdot)\in\mathbb{U}^K(x_0)} \left\{ \sum_{k=0}^{K-1} \ell(n+k, x_u(k,x_0), u(k)) + V_\infty(n+K, x_u(K,x_0)) \right\} + \varepsilon.
\end{aligned}$$

Since $\varepsilon > 0$ was arbitrary and the expressions in this inequality are independent of ε, this inequality also holds for $\varepsilon = 0$, which shows (4.5) with "$\leq$" and thus (4.5).

In order to prove (4.6) we use (4.7) with $u(\cdot) = u^\star(\cdot)$. This yields

$$\begin{aligned}
V_\infty(n,x_0) &= J_\infty(n,x_0,u^\star(\cdot)) \\
&= \sum_{k=0}^{K-1} \ell(n+k, x_{u^\star}(k,x_0), u^\star(k)) + J_\infty(n+K, x_{u^\star}(K,x_0), u^\star(\cdot+K)) \\
&\geq \sum_{k=0}^{K-1} \ell(n+k, x_{u^\star}(k,x_0), u^\star(k)) + V_\infty(n+K, x_{u^\star}(K,x_0))
\end{aligned}$$

$$\begin{aligned} &\geq \inf_{u(\cdot)\in\mathbb{U}^K(x_0)} \left\{ \sum_{k=0}^{K-1} \ell(n+k, x_u(k,x_0), u(k)) + V_\infty(n+K, x_u(K,x_0)) \right\} \\ &= V_\infty(n,x_0), \end{aligned}$$

where we used the (already proved) equality (4.5) in the last step. Hence, the two "$\geq$" in this chain are actually "$=$" which implies (4.6). □

The following corollary states an immediate consequence from the dynamic programming principle. It shows that tails of optimal control sequences are again optimal control sequences for suitably adjusted initial value and time.

Corollary 4.5 *If $u^\star(\cdot)$ is an optimal control sequence for* ($\mathrm{OCP}_\infty^{\mathrm{n}}$) *with initial value x_0 and initial time n, then for each $K \in \mathbb{N}$ the sequence $u_K^\star(\cdot) = u^\star(\cdot + K)$, i.e.,*

$$u_K^\star(k) = u^\star(K+k), \quad k = 0, 1, \ldots$$

is an optimal control sequence for initial value $x_{u^\star}(K, x_0)$ and initial time $n + K$.

Proof Inserting $V_\infty(n,x_0) = J_\infty(n,x_0,u^\star(\cdot))$ and the definition of $u_K^\star(\cdot)$ into (4.7) we obtain

$$V_\infty(n,x_0) = \sum_{k=0}^{K-1} \ell(n+k, x_{u^\star}(k,x_0), u^\star(k)) + J_\infty(n+K, x_{u^\star}(K,x_0), u_K^\star(\cdot))$$

Subtracting (4.6) from this equation yields

$$0 = J_\infty(n+K, x_{u^\star}(K,x_0), u_K^\star(\cdot)) - V_\infty(n+K, x_{u^\star}(K,x_0))$$

which shows the assertion. □

The next two results are the analogues of Theorem 3.17 and Corollary 3.18 in the infinite horizon setting.

Theorem 4.6 *Consider the optimal control problem* ($\mathrm{OCP}_\infty^{\mathrm{n}}$) *with $x_0 \in \mathbb{X}$ and $n \in \mathbb{N}_0$ and assume that an optimal control sequence $u^\star(\cdot)$ exists. Then the feedback law $\mu_\infty(n,x_0) = u^*(0)$ satisfies*

$$\mu_\infty(n,x_0) = \operatorname*{argmin}_{u\in\mathbb{U}^1(x_0)} \{\ell(n,x_0,u) + V_\infty(n+1, f(x_0,u))\}. \tag{4.8}$$

and

$$V_\infty(n,x_0) = \ell(n,x_0,\mu_\infty(n,x_0)) + V_\infty(n+1, f(x_0,\mu_\infty(n,x_0))) \tag{4.9}$$

where in (4.8)*—as usual—we interpret $\mathbb{U}^1(x_0)$ as a subset of U, i.e., we identify the one element sequence $u = u(\cdot)$ with its only element $u = u(0)$.*

Proof The proof is identical to the finite horizon counterpart Theorem 3.17. □

As in the finite horizon case, the following corollary shows that the feedback law (4.8) can be used in order to construct the optimal control sequence.

Corollary 4.7 *Consider the optimal control problem* ($\mathrm{OCP}^{\mathrm{n}}_{\infty}$) *with* $x_0 \in \mathbb{X}$ *and* $n \in \mathbb{N}_0$ *and consider an admissible feedback law* $\mu_\infty : \mathbb{N}_0 \times \mathbb{X} \to U$ *in the sense of Definition 3.2(iv). Denote the solution of the closed-loop system*

$$x(0) = x_0, \quad x(k+1) = f(x(k), \mu_\infty(n+k, x(k)), \; k = 0, 1, \ldots \tag{4.10}$$

by x_{μ_∞} *and assume that* μ_∞ *satisfies* (4.8) *for initial values* $x_0 = x_{\mu_\infty}(k)$ *for all* $k = 0, 1, \ldots$. *Then*

$$u^\star(k) = \mu_\infty(n+k, x_{u^\star}(k, x_0)), \quad k = 0, 1, \ldots \tag{4.11}$$

is an optimal control sequence for initial time n and initial value x_0 *and the solution of the closed-loop system* (4.10) *is a corresponding optimal trajectory.*

Proof From (4.11) for $x(n)$ from (4.10) we immediately obtain

$$x_{u^\star}(n, x_0) = x(n), \quad n = 0, 1, \ldots.$$

Hence, we need to show that

$$V_\infty(n, x_0) = J_\infty(n, x_0, u^\star),$$

where it is enough to show "$\geq$" because the opposite inequality follows by definition of V_∞. Using (4.11) and (4.9) we get

$$V_\infty(n+k, x_0) = \ell(n+k, x(k), u^\star(k)) + V_\infty(n+k+1, x(k+1))$$

for $k = 0, 1, \ldots$. Summing these equalities for $k = 0, \ldots, K-1$ for arbitrary $K \in \mathbb{N}$ and eliminating the identical terms $V_\infty(n+k, x_0)$, $k = 1, \ldots, K-1$ on the left and on the right we obtain

$$V_\infty(n, x_0) = \sum_{k=0}^{K-1} \ell(n+k, x(k), u^\star(k)) + V_\infty(n+K, x(K)) \geq \sum_{k=0}^{K-1} \ell(n+k, x(k), u^\star(k)).$$

Since the sum is monotone increasing in K and bounded from above, for $K \to \infty$ the right hand side converges to $J_\infty(n, x_0, u^\star)$ showing the assertion. □

Corollary 4.7 implies that infinite horizon optimal control is nothing but NMPC with $N = \infty$: Formula (4.11) for $k = 0$ yields that if we replace the optimization problem ($\mathrm{OCP}^{\mathrm{n}}_{\mathrm{N}}$) in Algorithm 3.7 by ($\mathrm{OCP}^{\mathrm{n}}_{\infty}$), then the feedback law resulting from this algorithm equals μ_∞. The following theorem shows that for stage cost of the form (3.8) this infinite horizon NMPC feedback law yields an asymptotically stable closed loop and thus solves the stabilization and tracking problem.

Theorem 4.8 (Asymptotic stability) *Consider the optimal control problem* ($\mathrm{OCP}^{\mathrm{n}}_{\infty}$) *for the control system (2.1) and a reference trajectory* $x^{\mathrm{ref}} : \mathbb{N}_0 \to \mathbb{X}$ *with reference control sequence* $u^{\mathrm{ref}} \in \mathbb{U}^{\infty}(x^{\mathrm{ref}}(0))$. *Assume that there exist* $\alpha_1, \alpha_2, \alpha_3 \in \mathcal{K}_{\infty}$ *such that the inequalities*

$$\alpha_1(|x|_{x^{\mathrm{ref}}(n)}) \le V_{\infty}(n,x) \le \alpha_2(|x|_{x^{\mathrm{ref}}(n)}) \quad \text{and} \quad \ell(n,x,u) \ge \alpha_3(|x|_{x^{\mathrm{ref}}(n)}) \tag{4.12}$$

hold for all $x \in \mathbb{X}$, $n \in \mathbb{N}_0$, *and* $u \in U$. *Assume furthermore that an optimal feedback* μ_{∞} *exists, i.e., an admissible feedback law* $\mu_{\infty} : \mathbb{N}_0 \times \mathbb{X} \to U$ *satisfying* (4.8) *for all* $n \in \mathbb{N}_0$ *and all* $x \in \mathbb{X}$. *Then this optimal feedback asymptotically stabilizes the closed-loop system*

$$x^{+} = g(n,x) = f(x, \mu_{\infty}(n,x))$$

on $\mathbb{X}$ *in the sense of Definition 2.16.*

Proof For the closed-loop system, (4.9) and the last inequality in (4.12) yield

$$\begin{aligned} V_{\infty}(n,x) &= \ell(n,x,\mu_{\infty}(n,x)) + V_{\infty}(n+1, f(x,\mu_{\infty}(n,x))) \\ &\ge \alpha_3(|x|_{x^{\mathrm{ref}}(n)}) + V_{\infty}(n+1, f(x,\mu_{\infty}(n,x))). \end{aligned}$$

Together with the first two inequalities in (4.12) this shows that V_{∞} is a Lyapunov function on $\mathbb{X}$ in the sense of Definition 2.21 with $\alpha_V = \alpha_3$. Thus, Theorem 2.22 yields asymptotic stability on $\mathbb{X}$. □

By Theorem 4.3 we can replace (4.12) by the asymptotic controllability condition from Definition 4.2 if ℓ is of the form (4.3). This is used in the following corollary in order to give a stability result without explicitly assuming (4.12).

Corollary 4.9 *Consider the optimal control problem* ($\mathrm{OCP}^{\mathrm{n}}_{\infty}$) *for the control system (2.1) and a reference trajectory* $x^{\mathrm{ref}} : \mathbb{N}_0 \to \mathbb{X}$ *with reference control sequence* $u^{\mathrm{ref}} \in \mathbb{U}^{\infty}(x^{\mathrm{ref}}(0))$. *Assume that the system is asymptotically controllable to* x^{ref} *and that an optimal feedback* μ_{∞}, *i.e., a feedback satisfying* (4.8), *exists for the cost function* $\ell : \mathbb{N}_0 \times X \times U \to \mathbb{R}^{+}_{0}$ *from* (4.3) *with* $\lambda = 0$. *Then this optimal feedback asymptotically stabilizes the closed-loop system*

$$x^{+} = g(n,x) = f(x, \mu_{\infty}(n,x))$$

on $\mathbb{X}$ *in the sense of Definition 2.16.*

If, in addition, the asymptotic controllability has the small control property then the statement also holds for ℓ *from* (4.3) *with arbitrary* $\lambda \ge 0$.

Proof Theorem 4.3 yields

$$\alpha_1(|x_0|_{x^{\mathrm{ref}}(n_0)}) \le V_{\infty}(n_0, x_0) \le \alpha_2(|x_0|_{x^{\mathrm{ref}}(n_0)})$$

for suitable $\alpha_1, \alpha_2 \in \mathcal{K}_\infty$. Furthermore, by (4.3) the third inequality in (4.12) holds with $\alpha_3 = \gamma_1^{-1}$. Hence, (4.12) holds and Theorem 4.8 yields asymptotic stability on $\mathbb{X}$. □

4.3 Relaxed Dynamic Programming

The last results of the previous section show that infinite horizon optimal control can be used in order to derive a stabilizing feedback law. Unfortunately, a direct solution of infinite horizon optimal control problem is in general impossible, both analytically and numerically. Still, infinite horizon optimal control plays an important role in our analysis since we will interpret the model predictive control algorithm as an approximation of the infinite horizon optimal control problem. Here the term "approximation" is not necessarily to be understood in the sense of "being close to" (although this aspect is not excluded) but rather in the sense of "sharing the important structural properties".

Looking at the proof of Theorem 4.8 we see that the important property for stability is the inequality

$$V_\infty(n,x) \geq \ell(n,x,\mu_\infty(n,x)) + V_\infty(n+1, f(x,\mu_\infty(n,x)))$$

which follows from the feedback version (4.9) of the dynamic programming principle. Observe that although (4.9) yields equality, only this inequality is needed in the proof of Theorem 4.8.

This observation motivates a relaxed version of this dynamic programming inequality, which on the one hand yields asymptotic stability and on the other hand provides a quantitative measure of the closed-loop performance of the system. This relaxed version will be formulated in Theorem 4.11, below. In order to quantitatively measure the closed-loop performance, we use the infinite horizon cost functional evaluated along the closed-loop trajectory which we define as follows.

Definition 4.10 (*Infinite horizon cost*) Let $\mu : \mathbb{N}_0 \times \mathbb{X} \to U$ be an admissible feedback law. For the trajectories $x_\mu(n)$ of the closed-loop system $x^+ = f(x, \mu(n,x))$ with initial value $x_\mu(n_0) = x_0 \in \mathbb{X}$ we define the *infinite horizon cost* as

$$J_\infty^{cl}(n_0, x_0, \mu) := \sum_{k=0}^{\infty} \ell(n_0+k, x_\mu(n_0+k), \mu(x_\mu(n_0+k))).$$

For stage costs of the form (3.8) our stage cost ℓ is always nonnegative, hence either the infinite sum has a well defined finite value or it diverges to infinity, in which case we write $J_\infty^{cl}(n_0, x_0, \mu) = \infty$. More general stage costs will be discussed in Chap. 8.

By Corollary 4.7 for the infinite horizon optimal feedback law μ_∞ we obtain

$$J_\infty^{cl}(n_0, x_0, \mu_\infty) = V_\infty(n_0, x_0)$$

while for all other admissible feedback laws μ we get

$$J^{cl}_\infty(n_0, x_0, \mu) \geq V_\infty(n_0, x_0).$$

In other words, V_∞ is a strict lower bound for $J^{cl}_\infty(n_0, x_0, \mu)$.

The following theorem now gives a relaxed dynamic programming condition from which we can derive both asymptotic stability and an upper bound on the infinite horizon cost $J^{cl}_\infty(n_0, x_0, \mu)$ for an arbitrary admissible feedback law μ.

Theorem 4.11 (Asymptotic stability and suboptimality estimate) *Consider a stage cost $\ell : \mathbb{N}_0 \times X \times U \to \mathbb{R}^+_0$ and a function $V : \mathbb{N}_0 \times X \to \mathbb{R}^+_0$. Let $\mu : \mathbb{N}_0 \times \mathbb{X} \to U$ be an admissible feedback law and let $S(n) \subseteq \mathbb{X}$, $n \in \mathbb{N}_0$ be a family of forward invariant sets for the closed-loop system*

$$x^+ = g(n, x) = f(x, \mu(n, x)). \tag{4.13}$$

Assume there exists $\alpha \in (0, 1]$ such that the relaxed dynamic programming inequality

$$V(n, x) \geq \alpha\ell(n, x, \mu(n, x)) + V(n + 1, f(x, \mu(n, x))) \tag{4.14}$$

holds for all $n \in \mathbb{N}_0$ and all $x \in S(n)$. Then the suboptimality estimate

$$J^{cl}_\infty(n, x, \mu) \leq V(n, x)/\alpha \tag{4.15}$$

holds for all $n \in \mathbb{N}_0$ and all $x \in S(n)$.

If, in addition, there exist $\alpha_1, \alpha_2, \alpha_3 \in \mathcal{K}_\infty$ such that the inequalities

$$\alpha_1(|x|_{x^{\text{ref}}(n)}) \leq V(n, x) \leq \alpha_2(|x|_{x^{\text{ref}}(n)}) \quad \textit{and} \quad \ell(n, x, u) \geq \alpha_3(|x|_{x^{\text{ref}}(n)})$$

hold for all $x \in \mathbb{X}$, $n \in \mathbb{N}_0$, $u \in \mathbb{U}$ and a reference trajectory $x^{\text{ref}} : \mathbb{N}_0 \to \mathbb{X}$, then the closed-loop system (4.13) *is asymptotically stable on $S(n)$ in the sense of Definition 2.16.*

Proof In order to prove (4.15) consider $n \in \mathbb{N}_0$, $x \in S(n)$ and the trajectory $x_\mu(\cdot)$ of (4.13) with $x_\mu(n) = x$. By forward invariance of the sets $S(n)$ this trajectory satisfies $x_\mu(n + k) \in S(n + k)$. Hence from (4.14) for all $k \in \mathbb{N}_0$ we obtain

$$\begin{aligned} &\alpha\ell(n + k, x_\mu(n + k), \mu(n + k, x_\mu(k))) \\ &\quad\leq\; V(n + k, x_\mu(n + k)) - V(n + k + 1, x_\mu(n + k + 1)). \end{aligned}$$

Summing over k yields for all $K \in \mathbb{N}$

$$\alpha \sum_{k=0}^{K-1} \ell(n + k, x_\mu(k), \mu(x_\mu(k))) \leq V(n, x_\mu(n)) - V(n + K, x_\mu(n + K)) \leq V(n, x)$$

since $V(n+K, x_\mu(n+K)) \geq 0$ and $x_\mu(n) = x$. Since the stage cost ℓ is nonnegative, the term on the left is monotone increasing and bounded, hence for $K \to \infty$ it converges to $\alpha J_\infty^{cl}(n, x, \mu)$. Since the right hand side is independent of K, this yields (4.15).

The stability assertion now immediately follows by observing that V satisfies all assumptions of Theorem 2.22 with $\alpha_V = \alpha\,\alpha_3$. □

Remark 4.12 An inspection of the proof of Theorems 2.19 and 2.22 reveals that for fixed $\alpha_1, \alpha_2 \in \mathcal{K}_\infty$ and $\alpha_V = \alpha\,\alpha_3$ with fixed $\alpha_3 \in \mathcal{K}_\infty$ and varying $\alpha \in (0, 1]$ the attraction rate $\beta \in \mathcal{KL}$ constructed in this proof depends on α in the following way: if β_α and $\beta_{\alpha'}$ are the attraction rates from Theorem 2.22 for $\alpha_V = \alpha\,\alpha_3$ and $\alpha_V = \alpha'\alpha_3$, respectively, with $\alpha' \geq \alpha$, then $\beta_{\alpha'}(r, t) \leq \beta_\alpha(r, t)$ holds for all $r, t \geq 0$. This in particular implies that for every $\bar{\alpha} \in (0, 1)$ the attraction rate $\beta_{\bar{\alpha}}$ is also an attraction rate for all $\alpha \in [\bar{\alpha}, 1]$, i.e., we can find an attraction rate $\beta \in \mathcal{KL}$ which is independent of $\alpha \in [\bar{\alpha}, 1]$.

Remark 4.13 Theorem 4.11 proves asymptotic stability of the discrete time closed-loop system (4.13) or (2.5). For a sampled data system (2.8) with sampling period $T > 0$ this implies the discrete time stability estimate (2.47) for the sampled data closed-loop system (2.30). For sampled data systems we may define the stage cost ℓ as an integral over a running cost function L according to (3.4), i.e.,

$$\ell(x, u) := \int_0^T L(\varphi(t, 0, x, u), u(t))dt.$$

We show that for this choice of ℓ a mild condition on L ensures that the sampled data closed-loop system (2.30) is also asymptotically stable in the continuous time sense, i.e., that (2.48) holds. For simplicity, we restrict ourselves to time invariant reference $x^{\text{ref}} \equiv x_*$.

The condition we use is that there exists $\delta \in \mathcal{K}_\infty$ such that the vector field f_c in (2.6) satisfies

$$\|f_c(x, u)\| \leq \max\{\varepsilon, \delta(1/\varepsilon)L(x, u)\} \tag{4.16}$$

holds for all $x \in X$, all $u \in U$ and all $\varepsilon > 0$. For instance, in a linear–quadratic problem with $X = \mathbb{R}^d$, $U = \mathbb{R}^m$, and $x_* = 0$ we have $\|f_c(x, u)\| = \|Ax + Bu\| \leq C_1(\|x\| + \|u\|)$ and $L(x, u) = x^\top Qx + u^\top Ru \geq C_2(\|x\| + \|u\|)^2$ for suitable constants $C_1, C_2 > 0$ provided Q and R are positive definite. In this case, (4.16) holds with $\delta(r) = C_1^2/C_2 r$, since $\|f_c(x, u)\| > \varepsilon$ implies $C_1(\|x\| + \|u\|) > \varepsilon$ and thus

$$C_1(\|x\| + \|u\|) \leq \frac{C_1^2}{\varepsilon}(\|x\| + \|u\|)^2 \leq \frac{C_1^2}{C_2\varepsilon}C_2(\|x\| + \|u\|)^2 = \delta(1/\varepsilon)L(x, u).$$

In the general nonlinear case, (4.16) holds if f_c is continuous with $f_c(x_*, u_*) = 0$, $L(x, u)$ is positive definite and the inequality $\|f_c(x, u)\| \leq CL(x, u)$ holds for some constant $C > 0$ whenever $\|f_c(x, u)\|$ is sufficiently large.

We now show that (4.16) together with Theorem 4.11 implies the continuous time stability estimate (2.48). If the assumptions of Theorem 4.11 hold, then (4.15) implies $\ell(x, \mu(x)) \leq V(x)/\alpha \leq \alpha_2(|x|_{x_*})/\alpha$. Thus, for $t \in [0, T]$ inequality (4.16) yields

$$\begin{aligned}
|\varphi(t, 0, x, \mu)|_{x_*} &\leq |x|_{x_*} + \int_0^t \|f_c(\varphi(\tau, 0, x, \mu), \mu(x)(\tau))\| d\tau \\
&\leq |x|_{x_*} + \max\left\{t\varepsilon,\ \delta(1/\varepsilon) \int_0^t L(\varphi(\tau, 0, x, \mu), \mu(x)(\tau)) d\tau\right\} \\
&\leq |x|_{x_*} + \max\{T\varepsilon,\ \delta(1/\varepsilon)\ell(x, u)\} \\
&\leq |x|_{x_*} + \max\left\{T\varepsilon,\ \delta(1/\varepsilon)\alpha_2(|x|_{x_*})/\alpha\right\}
\end{aligned}$$

Setting $\varepsilon = \tilde{\gamma}(|x|_{x_*})$ with

$$\tilde{\gamma}(r) = \frac{1}{\delta^{-1}\left(\frac{1}{\sqrt{\alpha_2(r)}}\right)}$$

for $r > 0$ and $\tilde{\gamma}(0) = 0$ yields $\tilde{\gamma} \in \mathcal{K}_\infty$ and

$$\delta(1/\varepsilon)\alpha_2(|x|_{x_*}) = \sqrt{\alpha_2(|x|_{x_*})}.$$

Hence, defining

$$\gamma(r) = r + \max\{T\tilde{\gamma}(r), \sqrt{\alpha_2(r)}/\alpha\}$$

we finally obtain

$$|\varphi(t, 0, x, \mu)|_{x_*} \leq \gamma(|x|_{x_*})$$

for all $t \in [0, T]$ with $\gamma \in \mathcal{K}_\infty$.

Hence, if (4.16) and the assumptions of Theorem 4.11 hold, then the sampled data closed-loop system (2.30) fulfills the uniform boundedness over T property from Definition 2.24 and consequently by Theorem 2.27 the sampled data closed-loop system (2.30) is asymptotically stable.

We now turn to investigating practical stability. Recalling the Definitions 2.15 and 2.17 of P-practical asymptotic stability and their Lyapunov function characterizations in Theorems 2.20 and 2.23 we can formulate the following practical version of Theorem 4.11.

Theorem 4.14 (Asymptotic stability and suboptimality estimate) *Consider a stage cost* $\ell : \mathbb{N}_0 \times X \times U \to \mathbb{R}_0^+$ *and a function* $V : \mathbb{N}_0 \times X \to \mathbb{R}_0^+$. *Let* $\mu : \mathbb{N}_0 \times \mathbb{X} \to U$ *be an admissible feedback law and let* $S(n) \subseteq \mathbb{X}$, *and* $P(n) \subset S(n)$, $n \in \mathbb{N}_0$ *be families of forward invariant sets for the closed-loop system* (4.13).

Assume there exists $\alpha \in (0, 1]$ *such that the relaxed dynamic programming inequality* (4.14) *holds for all* $n \in \mathbb{N}_0$ *and all* $x \in S(n) \setminus P(n)$. *Then the suboptimality estimate*

$$J_{k^*}^{\mathrm{cl}}(n, x, \mu) \leq V(n, x)/\alpha \tag{4.17}$$

holds for all $n \in \mathbb{N}_0$ *and all* $x \in S(n)$*, where* $k^* \in \mathbb{N}_0$ *is the minimal time with* $x_\mu(k^* + n, n, x) \in P(k^* + n)$ *and*

$$J^{\mathrm{cl}}_{k^*}(n, x, \mu) := \sum_{k=0}^{k^*-1} \ell(n + k, x_\mu(n + k, n, x), \mu(x_\mu(n + k, n, x)))$$

is the truncated closed-loop performance functional from Definition 4.10.

If, in addition, there exist $\alpha_1, \alpha_2, \alpha_3 \in \mathcal{K}_\infty$ *such that the inequalities*

$$\alpha_1(|x|_{x^{\mathrm{ref}}(n)}) \leq V(n, x) \leq \alpha_2(|x|_{x^{\mathrm{ref}}(n)}) \quad \textit{and} \quad \ell(n, x, u) \geq \alpha_3(|x|_{x^{\mathrm{ref}}(n)})$$

hold for all $x \in \mathbb{X}$, $n \in \mathbb{N}_0$*, and* $u \in U$ *and a reference* $x^{\mathrm{ref}} : \mathbb{N}_0 \to \mathbb{X}$*, then the closed-loop system* (4.13) *is P-asymptotically stable on* $S(n)$ *in the sense of Definition 2.17.*

Proof The proof follows with analogous arguments as the proof of Theorem 4.11 by only considering $k < k^*$ in the first part and using Theorem 2.23 with $Y(n) = S(n)$ instead of Theorem 2.22 in the second part. □

Remark 4.15 (i) Note that Remark 4.12 holds accordingly for Theorem 4.14. Furthermore, it is easily seen that both Theorems 4.11 and 4.14 remain valid if f in (4.13) depends on n.

(ii) The suboptimality estimate (4.17) states that the closed-loop trajectories $x_\mu(\cdot, x)$ from (4.13) behave like suboptimal trajectories until they reach the sets $P(\cdot)$.

As a consequence of Theorem 4.11, we can show the existence of a stabilizing almost optimal infinite horizon optimal feedback even if no infinite horizon optimal feedback exists. The assumptions of the following Theorem 4.16 are identical with the assumptions of Theorem 4.8 except that we do not assume the existence of an infinite horizon optimal feedback law μ_∞.

Theorem 4.16 *Consider the the optimal control problem* ($\mathrm{OCP}^{\mathrm{n}}_\infty$) *with stage cost* ℓ *of the form (3.8) for the control system (2.1) and a reference trajectory* $x^{\mathrm{ref}} : \mathbb{N}_0 \to \mathbb{X}$ *with reference control sequence* $u^{\mathrm{ref}} \in \mathbb{U}^\infty(x^{\mathrm{ref}}(0))$*. Assume that there exist* $\alpha_1, \alpha_2, \alpha_3 \in \mathcal{K}_\infty$ *such that the inequalities* (4.12) *hold for all* $x \in \mathbb{X}$, $n \in \mathbb{N}_0$*, and* $u \in U$.

Then for each $\alpha \in (0, 1)$ *there exists an admissible feedback* $\mu_\alpha : \mathbb{N}_0 \times \mathbb{X} \to U$ *which asymptotically stabilizes the closed-loop system*

$$x^+ = g(n, x) = f(x, \mu_\alpha(n, x))$$

on $\mathbb{X}$ *in the sense of Definition 2.16 and satisfies*

$$J^{cl}_\infty(n, x, \mu_\alpha) \leq V_\infty(n, x)/\alpha$$

for all $x \in \mathbb{X}$ *and* $n \in \mathbb{N}_0$.

Proof Fix $\alpha \in (0, 1)$ and pick an arbitrary $x \in \mathbb{X}$. From (4.5) for $K = 1$ for each $x \in \mathbb{X}$ and each $\varepsilon > 0$ there exists $u_x^\varepsilon \in \mathbb{U}^1(x)$ with

$$V_\infty(n, x) \geq \ell(n, x, u_x^\varepsilon) + V_\infty(n + 1, f(x, u_x^\varepsilon)) - \varepsilon.$$

If $V_\infty(n, x) > 0$, then (4.12) implies $x \neq x^{\mathrm{ref}}(n)$ and thus again (4.12) yields the inequality $\inf_{u \in U} \ell(n, x, u) > 0$. Hence, choosing $\varepsilon = (1 - \alpha) \inf_{u \in U} \ell(n, x, u)$ and setting $\mu_\alpha(n, x) = u_x^\varepsilon$ yields

$$V_\infty(n, x) \geq \alpha \ell(n, x, \mu_\alpha(n, x)) + V_\infty(n + 1, f(x, \mu_\alpha(n, x))). \tag{4.18}$$

If $V_\infty(n, x) = 0$, then (4.12) implies $x = x^{\mathrm{ref}}(n)$ and thus from the definition of u^{ref} we get $f(x, u^{\mathrm{ref}}(n)) = x^{\mathrm{ref}}(n + 1)$. Using (4.12) once again gives us $V_\infty(n + 1, f(x, u^{\mathrm{ref}}(n))) = 0$ and from (3.8) we get $\ell(n, x, u^{\mathrm{ref}}(n)) = 0$. Thus, $\mu_\alpha(n, x) = u^{\mathrm{ref}}(n)$ satisfies (4.18). Hence, we obtain (4.14) with $V = V_\infty$ for all $x \in \mathbb{X}$. In conjunction with (4.12) this implies that all assumptions of Theorem 4.11 are satisfied for $V = V_\infty$ with $S(n) = \mathbb{X}$. Thus, the assertion follows. □

Again we can replace (4.12) by the asymptotic controllability condition from Definition 4.2.

Corollary 4.17 *Consider the the optimal control problem* ($\mathrm{OCP}_\infty^{\mathrm{n}}$) *for the control system (2.1) and a reference trajectory* $x^{\mathrm{ref}} : \mathbb{N}_0 \to \mathbb{X}$ *with reference control sequence* $u^{\mathrm{ref}} \in \mathbb{U}^\infty(x^{\mathrm{ref}}(0))$. *Assume that the system is asymptotically controllable to* x^{ref} *and that the cost function* $\ell : \mathbb{N}_0 \times X \times U \to \mathbb{R}_0^+$ *is of the form* (4.3) *with* $\lambda = 0$. *Then for each* $\alpha \in (0, 1)$ *there exists an admissible feedback* $\mu_\alpha : \mathbb{N}_0 \times \mathbb{X} \to U$ *which asymptotically stabilizes the closed-loop system*

$$x^+ = g(n, x) = f(x, \mu_\alpha(n, x))$$

on $\mathbb{X}$ *in the sense of Definition 2.16 and satisfies*

$$J_\infty^{cl}(n, x, \mu_\alpha) \leq V_\infty(n, x)/\alpha$$

for all $x \in \mathbb{X}$ *and* $n \in \mathbb{N}_0$.

If, in addition, the asymptotic controllability has the small control property then the statement also holds for ℓ *from* (4.3) *with arbitrary* $\lambda \geq 0$.

Proof Theorem 4.3 yields

$$\alpha_1(|x_0|_{x^{\mathrm{ref}}(n_0)}) \leq V_\infty(n_0, x_0) \leq \alpha_2(|x_0|_{x^{\mathrm{ref}}(n_0)})$$

for suitable $\alpha_1, \alpha_2 \in \mathscr{K}_\infty$. Furthermore, by (4.3) the third inequality in (4.12) holds with $\alpha_3 = \gamma_1^{-1}$. Hence, (4.12) holds and Theorem 4.16 yields the assertion. □

While Theorem 4.16 and Corollary 4.17 are already nicer than Theorem 4.8 and Corollary 4.9, respectively, in the sense that no existence of an optimal feedback law is needed, for practical applications both theorems require the (at least approximate) solution of an infinite horizon optimal control problem, which is in general a hard, often infeasible computational task, see also the discussion in Sect. 4.4, below.

Hence, in the following chapters we are going to use Theorems 4.11 and 4.14 in a different way: we will derive conditions under which (4.14) is satisfied by the finite horizon optimal value function $V = V_N$ and the corresponding NMPC feedback law $\mu = \mu_N$. The advantage of this approach lies in the fact that in order to compute $\mu_N(x_0)$ it is sufficient to know the finite horizon optimal control sequence $u^\star$ for initial value x_0. This is a much easier computing task, at least if the optimization horizon N is not too large.

4.4 Notes and Extensions

Infinite horizon optimal control is a classical topic in control theory. The version presented in Sect. 4.1 can be seen as a nonlinear generalization of the classical (discrete time) linear-quadratic regulator (LQR) problem, see, e.g., Dorato and Levis [6]. A rather general existence result for optimal control sequences and trajectories in the metric space setting considered here was given by Keerthi and Gilbert [15]. Note, however, that by Theorem 4.16 we do not need the existence of optimal controls for the existence of almost optimal stabilizing feedback controls.

Dynamic programming as introduced in Sect. 4.2 is a very common approach also for infinite horizon optimal control and we refer to the discussion in Sect. 3.5 for some background information. As in the finite horizon case, the monographs of Bertsekas [2, 3] provide a good source for more information on this method.

The connection between infinite horizon optimal control and stabilization problems for nonlinear systems has been recognized for quite a while. Indeed, the well known construction of control Lyapunov functions in continuous time by Sontag [23] is based on techniques from infinite horizon optimal control. As already observed after Corollary 4.7, discrete time infinite horizon optimal control is nothing but NMPC with $N = \infty$. This has lead to the investigation of infinite horizon NMPC algorithms, e.g., by Keerthi and Gilbert [16], Meadows and Rawlings [19], and Alamir and Bornard [1]. For linear systems, this approach was also considered in the monograph of Bitmead, Gevers, and Wertz [4].

The stability results in this chapter are easily generalized to the stability of sets when ℓ is of the form (3.24). In this case, it suffices to replace the bounds $\alpha_j(|x|_{x^{\mathrm{ref}}(n)})$, $j = 1, 2, 3$, in, e.g., Theorem 4.11 by bounds of the form

$$\alpha_j\left(\min_{y\in X^{\mathrm{ref}}(n)} |x|_y\right). \tag{4.19}$$

Alternatively, one could formulate these bounds via so called proper indicator functions as used, e.g., by Grimm et al. in [8].

By Formula (4.8) the optimal—and stabilizing—feedback law μ_∞ can be computed by solving a rather simple optimization problem once the optimal value function V_∞ is known. This has motivated a variety of approaches for solving the dynamic programming equation (4.5) (usually for $K = 1$) numerically in order to obtain an approximation of μ_∞ from a numerical approximation of V_∞. Approximation techniques like linear and multilinear approximations are proposed, e.g., in Kreisselmeier and Birkhölzer [17], Camilli, Grüne, and Wirth [5], or by Falcone [7]. A set oriented approach was developed in Junge and Osinga [14] and used for computing stabilizing feedback laws in Grüne and Junge [10] (see also [11, 12] for further improvements of this method). All such methods, however, suffer from the so called *curse of dimensionality*, which means that the numerical effort grows exponentially with the dimension of the state space X. In practice, this means that these approaches can only be applied for low-dimensional systems, typically not higher than 4–5. For homogeneous systems, Tuna [25] (see also Grüne [9]) observed that it is sufficient to compute V_∞ on a sphere, which reduces the dimension of the problem by one. Still, this only slightly reduces the computational burden. In contrast to this, a numerical approximation of the optimal control sequence $u^\star$ for finite horizon optimal control problems like (OCP$_N$) and its variants is possible also in rather high space dimensions, at least when the optimization horizon N is not too large. This makes the NMPC approach computationally attractive.

Relaxed dynamic programming in the form introduced in Sect. 4.3 was originally developed by Lincoln and Rantzer [18] and Rantzer [20] in order to lower the computational complexity of numerical dynamic programming approaches. Instead of trying to solve the dynamic programming equation (4.5) exactly, it is only solved approximately using numerical approximations for V_∞ from a suitable class of functions, e.g., polynomials. The idea of using such relaxations is classical and can be realized in various other ways, too, see, e.g., [2, Chapter 6]. Here we use relaxed dynamic programming not for solving (4.5) but rather for proving properties of closed-loop solutions, cf. Theorems 4.11 and 4.14. While the specific form of the assumptions in these theorems were first used in an NMPC context in Grüne and Rantzer [13], the conceptual idea is actually older and can be found, e.g., in Shamma and Xiong [22] or in Scokaert, Mayne and Rawlings [21]. The fact that stability of the sampled data closed-loop can be derived from the stability of the associated discrete time system for integral costs (3.4), cf. Remark 4.13, was, to the best of our knowledge, not observed before.

Problems

1. Consider the problem (OCP$^{\mathrm{n}}_\infty$) with finite optimal value function $V_\infty : \mathbb{N}_0 \times X \to \mathbb{R}_0^+$ and asymptotically stabilizing admissible optimal feedback law $\mu_\infty : \mathbb{N}_0 \times \mathbb{X} \to U$. Let $V : \mathbb{N}_0 \times X \to \mathbb{R}_0^+$ be a function which satisfies

$$V(n, x_0) = \min_{u \in \mathbb{U}^1(x_0)} \{\ell(n, x_0, u) + V_\infty(n+1, f(x_0, u))\} \tag{4.20}$$

for all $n \in \mathbb{N}_0$ and all $x_0 \in X$.

(a) Prove that $V(n, x) \geq V_\infty(n, x)$ holds for all $n \in \mathbb{N}_0$ and all $x \in \mathbb{X}$.
(b) Prove that for the optimal feedback law the inequality

$$V(n, x_0) - V_\infty(n, x_0) \leq V(n+1, f(x, \mu_\infty(n, x))) - V_\infty(n+1, f(x, \mu_\infty(n, x)))$$

holds for all $n \in \mathbb{N}_0$ and all $x \in \mathbb{X}$.
(c) Assume that in addition there exist $\alpha_2 \in \mathcal{K}_\infty$ such that the inequality

$$V(n, x) \leq \alpha_2(|x|_{x^{\text{ref}}(n)})$$

holds for all $n \in \mathbb{N}_0$, $x \in \mathbb{X}$ and a reference trajectory $x^{\text{ref}} : \mathbb{N}_0 \to \mathbb{X}$. Prove that under this condition $V(n, x) = V_\infty(n, x)$ holds for all $n \in \mathbb{N}_0$ and all $x \in \mathbb{X}$.
(d) Find a function $V : \mathbb{N}_0 \times X \to \mathbb{R}_0^+$ satisfying (4.20) but for which $V(n, x) = V_\infty(n, x)$ does not hold. Of course, for this function the additional condition on V from (c) must be violated.

Hint for (a): Define a feedback μ which assigns to each pair (n, x) a minimizer of the right hand side of (4.20), check that Theorem 4.11 is applicable for $S(n) = \mathbb{X}$ (for which $\alpha \in (0, 1]$?) and conclude the desired inequality from (4.15).
Hint for (c): Perform an induction over the inequality from (b) along the optimal closed-loop trajectory.

2. Consider the unconstrained linear control system

$$x^+ = Ax + Bu$$

with matrices $A \in \mathbb{R}^{d\times d}$, $B \in \mathbb{R}^{d\times m}$. Consider problem ($\mathrm{OCP}^{\mathrm{n}}_\infty$) with

$$\ell(x, u) = x^\top Qx + u^\top Ru$$

with symmetric positive definite matrices Q, R of appropriate dimension (this setting is called the linear-quadratic regulator (LQR) problem). If the pair (A, B) is stabilizable, then it is known that the discrete time algebraic Riccati equation

$$P = Q + A^\top(P - PB(B^\top PB + R)^{-1}B^\top P)A.$$

has a unique symmetric and positive definite solution $P \in \mathbb{R}^{d\times d}$.

(a) Show that the function $V(x) = x^\top Px$ satisfies (4.20). Note that since the problem here is time invariant we do not need the argument n.
(b) Use the results from Problem 1 to conclude that $V_\infty(x) = x^\top Px$ holds.

(c) Prove that the corresponding optimal feedback law asymptotically stabilizes the equilibrium $x_* = 0$.

Hint for (a): You may use that for symmetric matrices C, D, E of appropriate dimension with D positive definite the formula

$$\min_{u \in \mathbb{R}^m} \{x^\top Cx + u^\top Du + u^\top Ex + x^\top Eu\} = c^\top Cx - x^\top ED^{-1}x$$

holds. This formula is proved by computing the zero of the derivative of the expression in the "min" with respect to u (which is also a nice exercise).
Hint for (b) and (c): Use that for any symmetric and positive definite matrix $M \in \mathbb{R}^{d\times d}$ there exist constants $C_2 \geq C_1 > 0$ such that the inequality $C_1\|x\|^2 \leq x^\top Mx \leq C_2\|x\|^2$ holds for all $x \in \mathbb{R}^d$.

3. Consider the finite horizon counterpart ($\mathrm{OCP_N}$) of Problem 2. For this setting one can show that the optimal value function is of the form $V_N(x) = x^\top P_N x$ and that the matrix P_N converges to the matrix P from Problem 2 as $N \to \infty$. This convergence implies that for each $\varepsilon > 0$ there exists $N_\varepsilon > 0$ such that the inequality

$$|x^\top P_N x - x^\top Px| \leq \varepsilon\|x\|^2$$

holds for all $N \geq N_\varepsilon$. Use this property and Theorem 4.11 in order to prove that the NMPC feedback law from Algorithm 3.1 is asymptotically stabilizing for sufficiently large optimization horizon $N > 0$.
Hint: Look at the hint for Problem 2 (b) and (c).

4. Consider the scalar control system

$$x^+ = x + u$$

with $x \in X = \mathbb{R}$, $u \in U = \mathbb{R}$ which shall be controlled via the NMPC Algorithm 3.1 using the quadratic stage cost function

$$\ell(x, u) = x^2 + u^2$$

Compute $V_N(x_0)$ und $J_\infty^{cl}(x_0, \mu_N(\cdot))$ for $N = 2$ (cf. Chap. 3, Problem 3). Using these values, derive the degree of suboptimality α from the relaxed dynamic programming inequality (4.14) and from the suboptimality estimate (4.15).

References

1. Alamir, M., Bornard, G.: Stability of a truncated infinite constrained receding horizon scheme: the general discrete nonlinear case. Automatica **31**(9), 1353–1356 (1995)
2. Bertsekas, D.P.: Dynamic Programming and Optimal Control, vol. I, 3rd edn. Athena Scientific, Belmont (2005)

3. Bertsekas, D.P.: Dynamic Programming and Optimal Control, vol. II, 2nd edn. Athena Scientific, Belmont (2001)
4. Bitmead, R.R., Gevers, M., Wertz, V.: Adaptive optimal control. The thinking man's GPC. International Series in Systems and Control Engineering. Prentice-Hall, New York (1990)
5. Camilli, F., Grüne, L., Wirth, F.: A regularization of Zubov's equation for robust domains of attraction. In: Isidori, A., Lamnabhi-Lagarrigue, F., Respondek, W. (eds.) Nonlinear Control in the Year 2000, vol. 1. Lecture Notes in Control and Information Science, vol. 258, pp. 277–289. Springer, London (2001)
6. Dorato, P., Levis, A.H.: Optimal linear regulators: the discrete-time case. IEEE Trans. Automat. Control **16**, 613–620 (1971)
7. Falcone, M.: Numerical solution of dynamic programming equations. Appendix A in Bardi, M. and Capuzzo Dolcetta, I., Optimal control and viscosity solutions of Hamilton–Jacobi–Bellman equations, Birkhäuser, Boston (1997)
8. Grimm, G., Messina, M.J., Tuna, S.E., Teel, A.R.: Model predictive control: for want of a local control Lyapunov function, all is not lost. IEEE Trans. Automat. Control **50**(5), 546–558 (2005)
9. Grüne, L.: Homogeneous state feedback stabilization of homogeneous systems. SIAM J. Control Optim. **38**, 1288–1314 (2000)
10. Grüne, L., Junge, O.: A set oriented approach to optimal feedback stabilization. Syst. Control Lett. **54**, 169–180 (2005)
11. Grüne, L., Junge, O.: Global optimal control of perturbed systems. J. Optim. Theory Appl. **136**, 411–429 (2008)
12. Grüne, L., Junge, O.: Set oriented construction of globally optimal controllers. at -. Automatisierungstechnik **57**, 287–295 (2009)
13. Grüne, L., Rantzer, A.: On the infinite horizon performance of receding horizon controllers. IEEE Trans. Automat. Control **53**, 2100–2111 (2008)
14. Junge, O., Osinga, H.M.: A set oriented approach to global optimal control. ESAIM Control Optim. Calc. Var. **10**, 259–270 (2004)
15. Keerthi, S.S., Gilbert, E.G.: An existence theorem for discrete-time infinite-horizon optimal control problems. IEEE Trans. Automat. Control **30**(9), 907–909 (1985)
16. Keerthi, S.S., Gilbert, E.G.: Optimal infinite-horizon feedback laws for a general class of constrained discrete-time systems: stability and moving-horizon approximations. J. Optim. Theory Appl. **57**(2), 265–293 (1988)
17. Kreisselmeier, G., Birkhölzer, T.: Numerical nonlinear regulator design. IEEE Trans. Automat. Control **39**, 33–46 (1994)
18. Lincoln, B., Rantzer, A.: Relaxing dynamic programming. IEEE Trans. Automat. Control **51**(8), 1249–1260 (2006)
19. Meadows, E.S., Rawlings, J.B.: Receding horizon control with an infinite cost. In: Proceedings of the American Control Conference - ACC 1993, pp. 2926–2930. San Francisco, California, USA (1993)
20. Rantzer, A.: Relaxed dynamic programming in switching systems. IEE Proc. - Control Theory Appl. **153**(5), 567–574 (2006)
21. Scokaert, P.O.M., Mayne, D.Q., Rawlings, J.B.: Suboptimal model predictive control (feasibility implies stability). IEEE Trans. Automat. Control **44**(3), 648–654 (1999)
22. Shamma, J.S., Xiong, D.: Linear nonquadratic optimal control. IEEE Trans. Automat. Control **42**(6), 875–879 (1997)
23. Sontag, E.D.: A Lyapunov-like characterization of asymptotic controllability. SIAM J. Control Optim. **21**(3), 462–471 (1983)
24. Sontag, E.D.: Comments on integral variants of ISS. Syst. Contr. Lett. **34**, 93–100 (1998)
25. Tuna, E.S.: Optimal regulation of homogeneous systems. Automatica **41**, 1879–1890 (2005)

Chapter 5
Stability and Suboptimality Using Stabilizing Terminal Conditions

In this chapter, we present a comprehensive stability and suboptimality analysis for NMPC schemes with stabilizing terminal conditions. Both endpoint constraints as well as regional constraints plus Lyapunov function terminal cost are covered. We show that viability of the state constraint set can be replaced by viability of the terminal constraint set in order to ensure admissibility of the resulting NMPC-feedback law. The "reversing of monotonicity" of the finite time optimal value functions is proved and used in order to apply the relaxed dynamic programming framework introduced in the previous chapter. Using this framework, stability, suboptimality (i.e., estimates about the infinite horizon performance of the NMPC closed-loop system), and inverse optimality results are proved.

5.1 The Relaxed Dynamic Programming Approach

In this chapter, we investigate stability and performance of NMPC schemes with stabilizing terminal conditions for stage costs ℓ satisfying (3.8). Before we turn to the precise definition of these conditions, we outline the main arguments that we will use in our analysis. The central idea is to apply the relaxed dynamic programming result from Theorem 4.11 to $\mu = \mu_N$ and $V = V_N$ from Algorithm 3.11 and its variants and Definition 3.14, respectively.

According to the assumptions of Theorem 4.11, in order to obtain the suboptimality estimate $J_\infty^{cl}(n, x, \mu_N) \leq V_N(n, x)/\alpha$ we have to ensure the inequality

$$V_N(n, x) \geq \alpha\ell(n, x, \mu_N(n, x)) + V_N(n + 1, f(x, \mu_N(n, x))) \tag{5.1}$$

to hold for all $x \in \mathbb{X}$, $n \in \mathbb{N}_0$, and some $\alpha \in (0, 1]$, preferably as close to one as possible.

L. Grüne and J. Pannek, *Nonlinear Model Predictive Control*,
Communications and Control Engineering, DOI 10.1007/978-3-319-46024-6_5

For asymptotic stability, in addition we have to ensure the existence of α_1, α_2, $\alpha_3 \in \mathcal{K}_\infty$ such that the inequalities

$$\alpha_1(|x|_{x^{\text{ref}}(n)}) \leq V_N(n,x) \leq \alpha_2(|x|_{x^{\text{ref}}(n)}) \quad \text{and} \quad \ell(n,x,u) \geq \alpha_3(|x|_{x^{\text{ref}}(n)}) \tag{5.2}$$

hold for all $x \in \mathbb{X}$, $n \in \mathbb{N}_0$, and $u \in U$.

In order to motivate why stabilizing terminal conditions can be helpful when verifying (5.1), let us consider the problem of verifying (5.1) for the optimal control problem ($\text{OCP}_\text{N}^\text{n}$). Looking at Equality (3.20) from Theorem 3.17 and noting that $\omega_N = 1$ holds if we specialize ($\text{OCP}_\text{N,e}^\text{n}$) to ($\text{OCP}_\text{N}^\text{n}$) we see that μ_N and V_N satisfy

$$V_N(n,x_0) \geq \ell(n,x_0,\mu_N(n,x_0)) + V_{N-1}(n+1, f(x_0,\mu_N(n,x_0))). \tag{5.3}$$

This is "almost" (5.1), even with $\alpha = 1$, except that (5.3) contains the function V_{N-1} at the place where we would like to have V_N.

The trouble now is that V_N is obtained by optimizing over N steps while V_{N-1} by optimizing over only $N-1$ steps. Hence, without terminal conditions, for each admissible control sequence $u \in \mathbb{U}^N(x_0)$ we get

$$J_N(n,x_0,u) \geq J_{N-1}(n,x_0,u).$$

This inequality immediately carries over to the corresponding optimal value functions, i.e., we obtain

$$V_N(n,x_0) \geq V_{N-1}(n,x_0). \tag{5.4}$$

Unfortunately, this is exactly the opposite of what we would need in order to conclude (5.1) from (5.3).

This is the point where suitable terminal conditions provide a way out. In the following sections, we discuss terminal constrained variants ($\text{OCP}_\text{N,e}$) and ($\text{OCP}_\text{N,e}^\text{n}$) of the optimal control problems (OCP_N) and ($\text{OCP}_\text{N}^\text{n}$),respectively, under which inequality (5.4) is reversed. Throughout this chapter we will not need viability of the state constraint set $\mathbb{X}$, i.e., Assumption 3.3 will not be needed. As we will see, viability of the terminal constraint set $\mathbb{X}_0$ is sufficient in order to prove that the resulting NMPC-feedback law maintains the imposed state constraints, see also the comments after Lemma 5.2 and before Assumption 5.9, below.

5.2 Equilibrium Endpoint Constraint

A simple way of constructing stabilizing terminal conditions consists of explicitly including the desired reference solution in the optimization constraints. We introduce this variant for the case of a constant reference $x^{\text{ref}} = x_*$ and the corresponding Algorithm 3.10 and discuss the general case of a time-varying reference at the end of this section, cf. Problem (5.14). Within Algorithm 3.10, we use the optimization problem ($\mathrm{OCP_{N,e}}$) with $\mathbb{X}_0 = \{x_*\}$, $F \equiv 0$, and $\omega_k = 1$ for $k = 0, \dots, N-1$. This means that we specialize ($\mathrm{OCP_{N,e}}$) to

$$
\begin{aligned}
&\text{minimize} \quad J_N(x_0, u(\cdot)) := \sum_{k=0}^{N-1} \ell(x_u(k, x_0), u(k)) \\
&\text{with respect to } u(\cdot) \in \mathbb{U}^N_{\mathbb{X}_0}(x_0) \text{ with } \mathbb{X}_0 = \{x_*\} \\
&\text{subject to} \\
&x_u(0, x_0) = x_0, \quad x_u(k+1, x_0) = f(x_u(k, x_0), u(k))
\end{aligned}
\tag{5.5}
$$

Recall from Definition 3.9 that each trajectory $x_u(\cdot, x_0)$ with $u(\cdot) \in \mathbb{U}^N_{\mathbb{X}_0}(x_0)$ satisfies $x_u(N, x_0) \in \mathbb{X}_0$, i.e., $x_u(N, x_0) = x_*$. Thus, in (5.5) we only optimize over trajectories satisfying this equilibrium endpoint constraint. Note that (5.5) is only well defined if x_0 is an element of the feasible set $\mathbb{X}_N$ from Definition 3.9.

The idea behind the equilibrium endpoint constraint $x_u(N, x_0) = x_*$ is intuitive: since we want our closed-loop system to converge to x_* we simply add this requirement as a constraint to the optimal control problem. And since "convergence" is difficult to formalize for the finite horizon predictions, it appears reasonable to require the predictions to end exactly at the desired equilibrium.

For the analysis of this problem we will use the following assumptions.

Assumption 5.1 (*Equilibrium endpoint constraint*) (i) The point $x_* \in \mathbb{X}$ is an equilibrium for an admissible control value u_*, i.e., there exists a control value $u_* \in \mathbb{U}(x_*)$ with $f(x_*, u_*) = x_*$.

(ii) The stage cost $\ell : X \times U \to \mathbb{R}^+_0$ satisfies $\ell(x_*, u_*) = 0$ for u_* from (i).

Observe that Assumption 5.1(i) is nothing else but a viability assumption for $\mathbb{X}_0 = \{x_*\}$, cf. Assumption 3.3. In order to show that Inequality (5.4) is indeed reversed for Problem (5.5) satisfying Assumption 5.1, we first need an auxiliary result for the feasible sets $\mathbb{X}_N$ and the corresponding admissible control sequences $\mathbb{U}^N_{\mathbb{X}_0}(x_0)$ from Definition 3.9. For the specific constraint $x_u(N, x_0) = x_*$ in (5.5) the following lemma states that each admissible control sequence on the horizon $N-1$ can be extended to an admissible control sequence on the horizon N.

Lemma 5.2 *If Assumption 5.1(i) holds for the terminal constraint set $\mathbb{X}_0 = \{x_*\}$, then for each $N \geq 2$ the following properties hold.*

(i) For each $x_0 \in \mathbb{X}_{N-1}$ and each $u_{N-1}(\cdot) \in \mathbb{U}^{N-1}_{\mathbb{X}_0}(x_0)$ the control sequence

$$u_N(k) := u_{N-1}(k), \quad k = 0, \dots, N-2, \quad u_N(N-1) := u_* \tag{5.6}$$

satisfies $u_N \in \mathbb{U}^N_{\mathbb{X}_0}(x_0)$

(ii) The inclusion $\mathbb{X}_{N-1} \subseteq \mathbb{X}_N$ holds.

Proof (i) The idea of the proof is simple: since the trajectory related to $u_{N-1}(\cdot) \in \mathbb{U}^{N-1}_{\mathbb{X}_0}(x_0)$ ends up in x_*, the trajectory corresponding to the prolonged control sequence u_N from (5.6) satisfies $x_{u_N}(N, x_0) = x_*$.

In order to verify $u_N \in \mathbb{U}^N_{\mathbb{X}_0}(x_0)$ we need to show that $x_{u_N}(k, x_0) \in \mathbb{X}$ for $k = 0, \dots, N$, $u_N(k) \in \mathbb{U}(x_{u_N}(k, x_0))$ for $k = 0, \dots, N-1$ and $x_{u_N}(N, x_0) = x_*$.

From $u_{N-1} \in \mathbb{U}^{N-1}_{\mathbb{X}_0}(x_0)$ and Lemma 3.12 we obtain

$$x_{u_{N-1}}(k, x_0) \in \mathbb{X}_{N-1-k} \subseteq \mathbb{X}, \quad u_{N-1}(k) \in \mathbb{U}(x_{u_{N-1}}(k, x_0)), \quad k = 0, \dots, N-2 \tag{5.7}$$

and

$$x_{u_{N-1}}(N-1, x_0) = x_* \in \mathbb{X}_0 \subseteq \mathbb{X} \tag{5.8}$$

and from (5.8) and the definition of u_N we get

$$x_{u_N}(k, x_0) = x_{u_{N-1}}(k, x_0), \quad k = 0, \dots, N-1, \quad x_{u_N}(N, x_0) = x_*. \tag{5.9}$$

Hence (5.7) and (5.8) are also valid for x_{u_N}. In addition, we get $u_N(N-1) = u_* \in \mathbb{U}(x_*) = \mathbb{U}(x_{u_N}(N-1, x_0))$ and $x_{u_N}(N, x_0) = f(x_{u_N}(N-1, x_0), u_N(N-1)) = f(x_*, u_*) = x_*$. Thus, $u_N \in \mathbb{U}^N_{\mathbb{X}_0}(x_0)$.

(ii) Let $x_0 \in \mathbb{X}_{N-1}$. Then, there exists $u_{N-1} \in \mathbb{U}^{N-1}_{\mathbb{X}_0}(x_0)$ and by (i) we can conclude that there exists $u_N \in \mathbb{U}^N_{\mathbb{X}_0}(x_0)$. Thus, $\mathbb{U}^N_{\mathbb{X}_0}(x_0) \neq \emptyset$ from which $x_0 \in \mathbb{X}_N$ follows. □

Observe that we did not need to impose viability of the constraint set $\mathbb{X}$ in this proof. In fact, we implicitly used that under Assumption 5.1(i) the set $\mathbb{X}_N$ is forward invariant for all admissible control sequences $\mathbb{U}^N_{\mathbb{X}_0}(x)$. We explicitly formulate a consequence of this property for the NMPC closed-loop system in the following Lemma.

Lemma 5.3 *Under Assumption 5.1(i) for each $N \in \mathbb{N}$ the NMPC-feedback law μ_N obtained from Algorithm 3.10 with* ($\mathrm{OCP}_{\mathrm{N,e}}$) = (5.5) *renders the set $\mathbb{X}_N$ forward invariant, i.e., $f(x, \mu_N(x)) \in \mathbb{X}_N$ for all $x \in \mathbb{X}_N$.*

Proof Follows immediately from Corollary 3.13 and Lemma 5.2(ii). □

This lemma shows that if a state x is feasible, i.e., contained in the feasible set $\mathbb{X}_N$ then its closed-loop successor state $f(x, \mu_N(x))$ is again feasible. Thus, $\mathbb{X}_N$ is recursively feasible in the sense defined after Theorem 3.5. Using Lemma 5.2 it is now easy to establish that Inequality (5.4) is reversed for (5.5).

Lemma 5.4 *If Assumptions 5.1(i) and (ii) hold, then for each $N \geq 2$ and each $x_0 \in \mathbb{X}_{N-1}$ the optimal value functions of Problem* (5.5) *satisfy*

$$V_N(x_0) \leq V_{N-1}(x_0) \tag{5.10}$$

Proof We first show that for each $u_{N-1} \in \mathbb{U}^{N-1}_{\mathbb{X}_0}(x_0)$ the control sequence $u_N \in \mathbb{U}^N_{\mathbb{X}_0}(x_0)$ from (5.6) satisfies

$$J_N(x_0, u_N) \leq J_{N-1}(x_0, u_{N-1}). \tag{5.11}$$

To this end, recall from the proof of Lemma 5.2 that the trajectories $x_{u_N}(\cdot, x_0)$ and $x_{u_{N-1}}(\cdot, x_0)$ satisfy

$$x_{u_N}(k, x_0) = x_{u_{N-1}}(k, x_0), \quad k = 0, \dots, N-1, \quad x_{u_N}(N, x_0) = x_*.$$

Together with (5.6) this yields

$$\begin{aligned} J_N(x_0, u_N) &= \sum_{k=0}^{N-1} \ell(x_{u_N}(k, x_0), u_N(k)) \\ &= \sum_{k=0}^{N-2} \ell(x_{u_N}(k, x_0), u_N(k)) + \ell(x_{u_N}(N-1, x_0), u_N(N-1)) \\ &= \underbrace{\sum_{k=0}^{N-2} \ell(x_{u_{N-1}}(k, x_0), u_{N-1}(k))}_{=J_{N-1}(x_0, u_{N-1})} + \underbrace{\ell(x_*, u_*)}_{=0} \quad = \quad J_{N-1}(x_0, u_{N-1}). \end{aligned}$$

This shows (5.11). In fact, we even proved "=" but we will only need "≤" for proving (5.10). In order to prove (5.10), let $u^k_{N-1} \in \mathbb{U}^{N-1}_{\mathbb{X}_0}(x_0)$, $k \in \mathbb{N}$, be a sequence of control sequences such that

$$V_{N-1}(x_0) = \inf_{u \in \mathbb{U}^{N-1}_{\mathbb{X}_0}(x_0)} J_{N-1}(x_0, u) = \inf_{k \in \mathbb{N}} J_{N-1}(x_0, u^k_{N-1})$$

holds. Then, we can find $u^k_N \in \mathbb{U}^N_{\mathbb{X}_0}(x_0)$ such that (5.11) holds for $u_N = u^k_N$ and $u_{N-1} = u^k_{N-1}$. This implies

$$V_N(x_0) = \inf_{u \in \mathbb{U}^N_{\mathbb{X}_0}(x_0)} J_N(x_0, u) \leq \inf_{k \in \mathbb{N}} J_N(x_0, u^k_N) \leq \inf_{k \in \mathbb{N}} J_{N-1}(x_0, u^k_{N-1}) = V_{N-1}(x_0)$$

and thus (5.10). □

Note that for Problem (5.5) in general (5.4) does no longer hold, because the terminal constraint is more restrictive for smaller horizon than for larger ones. Thus, with the terminal constraint we do not get (5.10) on top of (5.4). Rather, we replaced (5.4) by (5.10).

Lemma 5.4 in conjunction with (5.3) enables us to conclude that the optimal value function V_N satisfies Inequality (5.1). This will be used in the proof of our following first stability theorem for an NMPC scheme in which we simply assume (5.2). Sufficient conditions for these inequalities will be discussed after the theorem.

Theorem 5.5 (Stability using endpoint constraint) *Consider the NMPC Algorithm 3.10 with* $(\mathrm{OCP}_{\mathrm{N,e}}) = (5.5)$ *and optimization horizon* $N \in \mathbb{N}$. *Let the Assumptions 5.1(i) and (ii) hold and assume that* (5.2) *holds for suitable* $\alpha_1, \alpha_2, \alpha_3 \in \mathcal{K}_\infty$. *Then the nominal NMPC closed-loop system (3.5) with NMPC-feedback law* μ_N *is asymptotically stable on* $\mathbb{X}_N$.

In addition, for $J_\infty^{cl}(x, \mu_N)$ *from Definition 4.10 the inequality*

$$J_\infty^{cl}(x, \mu_N) \le V_N(x)$$

holds for each $x \in \mathbb{X}_N$.

Proof Combining Equality (3.20) from Theorem 3.17 with Inequality (5.10) from Lemma 5.4 with $x_0 = f(x, \mu_N(x))$, for each $x \in \mathbb{X}_N$ we obtain

$$V_N(x) \ge \ell(x, \mu_N(x)) + V_{N-1}(f(x, \mu_N(x))) \ge \ell(x, \mu_N(x)) + V_N(f(x, \mu_N(x))).$$

Thus, the assumptions of Theorem 4.11 are satisfied with $V = V_N$, $\mu = \mu_N$, $S(n) = \mathbb{X}_N$ (which is forward invariant by Lemma 5.3) and $\alpha = 1$ which yields the assertion. □

The fact that each predicted trajectory x_u in (5.5) satisfies $x_u(N, x_0) = x_*$ does by no means imply that the NMPC closed-loop trajectory satisfies $x_{\mu_N}(N) = x_*$. The following example illustrates this fact.

Example 5.6 Consider again Example 2.1, i.e.,

$$x^+ = x + u =: f(x, u)$$

with $\mathbb{X} = X = \mathbb{U} = U = \mathbb{R}$ and $x_* = 0$. We use the stage cost $\ell(x, u) = x^2 + u^2$ and the terminal constraints $\mathbb{X}_0 = \{x_*\}$.

Observing that every $u(\cdot) \in \mathbb{U}^1_{\mathbb{X}_0}(x)$ must satisfy $f(x, u(0)) = 0$ we get $u(0) = -x$. Hence, (3.19) yields

$$V_1(x) = \inf_{u \in \mathbb{U}^1_{\mathbb{X}_0}(x)} \ell(x, u(0)) = x^2 + (-x)^2 = 2x^2.$$

and $\mu_1(x) = -x$. Now, using (3.19) for (5.5) with $N = 2$ we get

$$\mu_2(x) = \operatorname*{argmin}_{u \in \mathbb{R}} \{\ell(x, u) + V_1(f(x, u))\} = \operatorname*{argmin}_{u \in \mathbb{R}} \left\{x^2 + u^2 + 2(x + u)^2\right\} = -\frac{2}{3}x,$$

which is easily computed by setting the first derivative w.r.t. u of the term in braces to 0, observing that the second derivative is strictly positive. Thus, the NMPC closed loop for $N = 2$ becomes

$$x^+ = f(x, \mu_2(x)) = x - \mu_2(x) = x - \frac{2}{3}x = \frac{1}{3}x$$

with solutions

$$x_{\mu_2}(n, x_0) = \frac{1}{3^n} x_0.$$

Hence, the closed-loop solution asymptotically converges to $x_* = 0$ but never reaches 0 in finite time.

In Theorem 5.5 we have made the assumption that V_N satisfies the inequalities in (5.2). In terms of the problem data f and ℓ this is an implicit condition which may be difficult to check. For this reason, in the following proposition we give a sufficient condition on f and ℓ for these inequalities to hold true.

Proposition 5.7 *Let V_N denote the optimal value function of Problem* (5.5) *for optimization horizon $N \in \mathbb{N}$.*

(i) Assume there exists a function $\alpha_3 \in \mathcal{K}_\infty$ such that the inequality

$$\ell(x, u) \geq \alpha_3(|x|_{x_*})$$

holds for all $x \in X$ and all $u \in U$. Then

$$V_N(x) \geq \alpha_3(|x|_{x_*})$$

holds for all $x \in \mathbb{X}_N$.

(ii) Assume that f and ℓ are continuous in $X \times U$, U is compact and there exists a ball $\mathcal{B}_\nu(x_*) \subset X$, $\nu > 0$, and a function $\tilde{\alpha}_2 \in \mathcal{K}_\infty$ with the following property: For each $x \in \mathcal{B}_\nu(x_*) \cap \mathbb{X}$ there is $u_x \in \mathbb{U}(x)$ with $f(x, u_x) = x_*$ and

$$\ell(x, u_x) \leq \tilde{\alpha}_2(|x|_{x_*}). \tag{5.12}$$

Then there exists $\alpha_2 \in \mathcal{K}_\infty$ such that

$$V_N(x) \leq \alpha_2(|x|_{x_*}). \tag{5.13}$$

holds for all $x \in \mathbb{X}_N$.

Proof (i) is immediate from the definition of V_N.

In order to prove (ii), first observe that for $x \in \mathcal{B}_\nu(x_*) \cap \mathbb{X}$ the existence of u_x with $f(x, u_x) = x_*$ immediately implies $x \in \mathbb{X}_1$ and

$$V_1(x) = \inf_{u \in \mathbb{U}^1_{\mathbb{X}_0}(x)} \ell(x, u(0)) \leq \ell(x, u_x) \leq \tilde{\alpha}_2(|x|_{x_*}),$$

because the control sequence $u(\cdot) \in U^1$ defined by $u(0) = u_x$ lies in $\mathbb{U}^1_{\mathbb{X}_0}(x)$ since by assumption $f(x, u_x) = x_*$. Now by Lemma 5.4 the inequality

$$V_N(x) \leq \tilde{\alpha}_2(|x|_{x_*})$$

follows for each $N \in \mathbb{N}$ and each $x \in \mathcal{B}_\nu(x_*) \cap \mathbb{X}$.

For $x \in \mathbb{X}_N$ outside this ball consider an arbitrary closed ball $\overline{\mathcal{B}}_r(x_*)$ for $r > 0$ and an arbitrary $N \in \mathbb{N}$. Since f and ℓ are assumed to be continuous, the functional $J_N : X \times U^N \to \mathbb{R}_0^+$ is continuous, too, and since U is assumed to be compact the set $\overline{\mathcal{B}}_r(x_*) \times U^N$ is also compact (both continuity and compactness hold with the usual product topology on $X \times U^N$). Thus the value

$$\hat{\alpha}_2(r) := \max\{J_N(x, u) \mid x \in \overline{\mathcal{B}}_r(x_*),\ u \in U^N\}$$

exists and is finite and the resulting function $\hat{\alpha}_2$ is continuous and monotone increasing in r. By this definition, for each $x \in \mathbb{X}_N$ we get

$$V_N(x) \leq \hat{\alpha}_2(|x|_{x_*}).$$

Now define a function $\alpha_2 : \mathbb{R}_0^+ \to \mathbb{R}_0^+$ by

$$\alpha_2(r) := r + \begin{cases} \tilde{\alpha}_2(\nu) + \hat{\alpha}_2(r), & r \geq \nu \\ \tilde{\alpha}_2(r) + r\hat{\alpha}_2(\nu)/\nu, & r \in [0, \nu) \end{cases}$$

This function is continuous since both expressions are equal for $r = \nu$, equal to 0 at $r = 0$ and strictly increasing and unbounded due to the addition of r. Hence, $\alpha_2 \in \mathcal{K}_\infty$. Furthermore, it satisfies $\alpha_2(r) \geq \tilde{\alpha}_2(r)$ for $r \in [0, \nu]$ and $\alpha_2(r) \geq \hat{\alpha}_2(r)$ for $r \geq \nu$. Thus, it is the desired upper bound for V_N. □

The specific upper bound α_2 constructed in this proof depends on N and will in general grow unboundedly as $N \to \infty$. However, since by inequality (5.10) we know that the functions V_N are decreasing in N we can actually conclude the existence of an upper bound in $\mathcal{K}_\infty$ which is independent of N. However, since we did not (and do not want to) assume continuity of V_N, the construction of this $\mathcal{K}_\infty$-function is somewhat technically involved which is why we skip the details.

If we want to deduce local asymptotic controllability from the asymptotic stability on $\mathbb{X}_N$, we need that $\mathbb{X}_N$ contains a ball $\mathcal{B}_\nu(x_*)$ around the equilibrium x_*, cf. the comment after Definition 2.14. The following example shows that this does not necessarily mean that $\mathbb{X}_k$ for $k \leq N - 1$ must contain such a ball, too.

Example 5.8 Consider the system $x^+ = f(x, u)$ with $x \in \mathbb{X} = X = \mathbb{R}^2$, $u \in \mathbb{U} = [0, 1] \subset U = \mathbb{R}$, $x_* = 0$ and

$$f(x, u) = \begin{pmatrix} x_1(1-u) \\ \|x\|u \end{pmatrix}.$$

We use the NMPC Algorithm 3.10 with ($\text{OCP}_{\text{N,e}}$) = (5.5). For $x \neq 0$ the system is controllable to $\mathbb{X}_0 = \{0\}$ in one step if and only if $x_1 = 0$. Thus $\mathbb{X}_1 = \{x \in \mathbb{R}^2 \mid x_1 = 0\}$ which obviously does not contain a neighborhood of $x_* = 0$.

On the other hand, using the control sequence $u(0) = 1$, $u(1) = 0$ for each initial value $x \in \mathbb{R}^2$ one obtains

$$x_u(0, x) = x, \quad x_u(1, x) = (0, \|x\|)^\top, \quad x_u(2, x) = 0,$$

which implies $\mathbb{X}_2 = \mathbb{R}^2 = X$ and thus $\mathbb{X}_N = \mathbb{R}^2 = X$ for all $N \geq 2$.

Furthermore, using the stage cost $\ell(x, u) = \|x\|^2$ we obtain the upper bound

$$V_2(x) \leq \|x\|^2 + \|(0, \|x\|)^\top\|^2 = 2\|x\|^2$$

and the lower bound $\|x\|^2 \leq V_N(x)$ for all $N \geq 2$. Hence, Theorem 5.5 implies that the NMPC-feedback law μ_N stabilizes the system on $\mathbb{X}_N = \mathbb{R}^2$ for each $N \geq 2$.

The method described in this section is easily extended to time-varying reference trajectories $x^{\mathrm{ref}}(\cdot)$ replacing (5.5) by

$$\begin{aligned}
&\text{minimize} \quad J_N(n, x_0, u(\cdot)) := \sum_{k=0}^{N-1} \ell(n+k, x_u(k, x_0), u(k)) \\
&\text{with respect to } \ u(\cdot) \in \mathbb{U}^N_{\mathbb{X}_0}(n, x_0) \ \text{ with } \ \mathbb{X}_0(n) = \{x^{\mathrm{ref}}(n)\} \\
&\text{subject to} \\
&x_u(0, x_0) = x_0, \quad x_u(k+1, x_0) = f(x_u(k, x_0), u(k))
\end{aligned} \tag{5.14}$$

and choosing $(\mathrm{OCP}^{\mathrm{n}}_{\mathrm{N,e}}) = (5.14)$ in Algorithm 3.11. Since $x^{\mathrm{ref}}(\cdot)$ is known, the constraint $x_u(n + N, x_0) = x^{\mathrm{ref}}(n + N)$ is as easy to implement as its time-invariant counterpart in (5.5). All proofs in this section are easily extended to the time-varying case by including the appropriate time instants in ℓ, J_N, V_N, and μ_N.

5.3 Lyapunov Function Terminal Cost

The equilibrium terminal constraint described in the previous section provides a way to guarantee stability which is intuitive and easy to implement. Still, it has the obvious drawback that the system under consideration must be exactly controllable to x_* in finite time in order to ensure that the feasible sets $\mathbb{X}_N$ of (5.5) or (5.14) indeed contain a neighborhood of x_* or $x^{\mathrm{ref}}(n)$ in the time-varying case. Thus, it cannot be applied to systems which are merely stabilizable but not controllable to x_*. As a simple system where this is the case consider, e.g., the system with two-dimensional state $x = (x_1, x_2)^\top$ and one-dimensional control u given by

$$x^+ = f(x, u) = \begin{pmatrix} x_1 + u \\ x_2/2 \end{pmatrix}.$$

This system is obviously stabilizable at $x_* = 0$, e.g., by the feedback law $\mu(x) = -x_1/2$. However, it is not controllable to $x_* = 0$ in finite time since for any initial value $x_0 = (x_{01}, x_{02})^\top$ with $x_{02} \neq 0$ the second component of the solution $x_u(k, x_0) = (x_u(k, x_0)_1, x_u(k, x_0)_2)^\top$ satisfies $x_u(k, x_0)_2 = 2^{-k} x_{02} \neq 0$ for all $k \geq 0$ regardless of the choice of $u(\cdot)$.

Furthermore, for nonlinear and nonconvex optimal control problems the strict point constraint $x_n(N, x_0) = x_*$ may cause numerical difficulties in the optimization algorithm, such that the algorithm may not be able to find a feasible solution even if such a solution exists.

In this section, we are going to present terminal conditions in which the terminal constraint is relaxed by choosing $\mathbb{X}_0$ as a larger set containing x_*. In order to guarantee stability for this relaxed terminal constraint, we add the terminal cost function F in (OCP$_{\text{N,e}}$) to the terminal conditions. Again, we introduce the method for constant reference $x^{\text{ref}} = x_*$ and explain the necessary modifications for the general case at the end of this section.

We choose the optimal control problem (OCP$_{\text{N,e}}$) in Algorithm 3.10 as

$$\begin{aligned} &\text{minimize} \quad J_N(x_0, u(\cdot)) := \sum_{k=0}^{N-1} \ell(x_u(k, x_0), u(k)) \\ &\qquad\qquad\qquad\qquad\qquad + F(x_u(N, x_0)) \\ &\text{with respect to } u(\cdot) \in \mathbb{U}^N_{\mathbb{X}_0}(x_0) \text{ with } x_* \in \mathbb{X}_0 \\ &\text{subject to} \\ &x_u(0, x_0) = x_0, \quad x_u(k+1, x_0) = f(x_u(k, x_0), u(k)) \end{aligned} \tag{5.15}$$

Recall again that by Definition 3.9 the choice $u(\cdot) \in \mathbb{U}^N_{\mathbb{X}_0}(x_0)$ guarantees $x_u(N, x_0) \in \mathbb{X}_0$ and that (5.15) is only well defined for x_0 in the feasible set $\mathbb{X}_N$ from Definition 3.9.

We are now going to specify the properties of the terminal constraint set $\mathbb{X}_0 \subseteq \mathbb{X}$ and the terminal cost $F : \mathbb{X}_0 \to \mathbb{R}^+_0$. As in the last section we do not need to impose Assumption 3.3, i.e., viability for the state constraint set $\mathbb{X}$. However, in order to compensate for this we need viability of $\mathbb{X}_0$.

Assumption 5.9 (*Lyapunov function terminal cost*) For the closed terminal constraint set $\mathbb{X}_0 \subseteq \mathbb{X}$ defining $\mathbb{U}^N_{\mathbb{X}_0}(x_0)$ in (5.15) via Definition 3.9 and the terminal cost $F : \mathbb{X}_0 \to \mathbb{R}^+_0$ in (5.15) we assume:

(i) $\mathbb{X}_0$ is viable, i.e., for each $x \in \mathbb{X}_0$ there exists an admissible control value $u_x \in \mathbb{U}(x)$ such that

$$f(x, u_x) \in \mathbb{X}_0 \tag{5.16}$$

holds.

(ii) The terminal cost $F : \mathbb{X}_0 \to \mathbb{R}_0^+$ in (5.15) is such that for each $x \in \mathbb{X}_0$ there exists an admissible control value $u_x \in \mathbb{U}(x)$ for which (5.16) and

$$F(f(x, u_x)) + \ell(x, u_x) \leq F(x)$$

hold.

Assumption 5.9(ii) implies that F is a local control Lyapunov function of our control system. The approach of adding F is often referred to as *quasi-infinite horizon NMPC*. The reason for this denomination is that if the terminal cost F is an approximation of the infinite horizon optimal value function V_∞, then the finite horizon dynamic programming principle (3.15) is an approximation of the infinite horizon dynamic programming principle (4.5) and consequently (5.15) can be interpreted as an approximation to the infinite horizon problem (OCP_∞). While a function F satisfying Assumption 5.9(ii) exists under mild conditions on the system provided it is asymptotically controllable, cf. the discussion on control Lyapunov functions in Sect. 2.5, it is not always easy to find—we will sketch a linearization-based approach to find F and $\mathbb{X}_0$ in Remark 5.15, below. Note that the equilibrium terminal constraint problem (5.5) can be seen as a special case of (5.15) with $\mathbb{X}_0 = \{x_*\}$ and $F \equiv 0$, in which case Assumption 5.9 implies Assumption 5.1. The more interesting case, however, is obtained when $\mathbb{X}_0$ contains a whole ball around x_*, because in this case the terminal constraint is considerably weaker than in the case $\mathbb{X}_0 = \{x_*\}$ and consequently more easy to achieve for the optimization algorithm.

The subsequent analysis is analogous to the respective results in Sect. 5.2 with the goal to establish that inequality (5.4) is reversed.

Lemma 5.10 *If Assumption 5.9(i) holds for the terminal constraint set $\mathbb{X}_0 \subseteq \mathbb{X}$, then for each $N \geq 2$ the following properties hold.*

(i) For each $x_0 \in \mathbb{X}_{N-1}$ and each $u_{N-1}(\cdot) \in \mathbb{U}^{N-1}_{\mathbb{X}_0}(x_0)$ the control sequence

$$u_N(k) := u_{N-1}(k), \quad k = 0, \ldots, N-2, \quad u_N(N-1) := u_x \tag{5.17}$$

with u_x from Assumption 5.9(i) for $x = x_u(N-1, x_0)$ satisfies $u_N \in \mathbb{U}^N_{\mathbb{X}_0}(x_0)$.

(ii) The inclusion $\mathbb{X}_{N-1} \subseteq \mathbb{X}_N$ holds.

Proof The proof is completely analogous to the proof of Lemma 5.2 replacing $f(x_*, u_*) = x_*$ by $f(x, u_x) \in \mathbb{X}_0$ for $x \in \mathbb{X}_0$. □

As in the equilibrium constraint case from the previous section, the set $\mathbb{X}_N$ is invariant under the MPC feedback law μ_N as shown in the following lemma.

Lemma 5.11 *Under Assumption 5.9(i) for each $N \in \mathbb{N}$ the NMPC-feedback law μ_N obtained from Algorithm 3.10 with* ($\text{OCP}_{\text{N,e}}$) = (5.15) *renders the set $\mathbb{X}_N$ forward invariant, i.e., $f(x, \mu_N(x)) \in \mathbb{X}_N$ for all $x \in \mathbb{X}_N$.*

Proof Follows immediately from Corollary 3.13 and Lemma 5.10(ii). □

Using Lemma 5.10 we can now prove that inequality (5.4) is reversed for (5.15).

Lemma 5.12 *If Assumptions 5.9(i) and (ii) hold, then for each* $N \geq 2$ *and each* $x_0 \in \mathbb{X}_{N-1}$ *the optimal value functions of Problem* (5.15) *satisfy*

$$V_N(x_0) \leq V_{N-1}(x_0) \tag{5.18}$$

Proof We first show that for each $u_{N-1} \in \mathbb{U}^{N-1}_{\mathbb{X}_0}(x_0)$ the control sequence $u_N \in \mathbb{U}^{N}_{\mathbb{X}_0}(x_0)$ from (5.17) satisfies

$$J_N(x_0, u_N) \leq J_{N-1}(x_0, u_{N-1}). \tag{5.19}$$

To this end, observe that by construction of u_N the trajectories $x_{u_N}(\cdot, x_0)$ and $x_{u_{N-1}}(\cdot, x_0)$ satisfy

$$x_{u_N}(k, x_0) = x_{u_{N-1}}(k, x_0), \quad k = 0, \ldots, N-1, \quad x_{u_N}(N, x_0) \in \mathbb{X}_0.$$

We abbreviate $\tilde{x} = x_{u_N}(N-1, x_0) \in \mathbb{X}_0$ and $u_{\tilde{x}} = u_N(N-1)$, noting that by (5.17) $u_{\tilde{x}}$ coincides with u_x from Assumption 5.9(ii) for $x = \tilde{x}$. Thus (5.17) and Assumption 5.9(ii) yield

$$\begin{aligned}
J_N(x_0, u_N) &= \sum_{k=0}^{N-1} \ell(x_{u_N}(k, x_0), u_N(k)) + F(x_{u_N}(N, x_0))) \\
&= \sum_{k=0}^{N-2} \ell(x_{u_N}(k, x_0), u_N(k)) + \ell(x_{u_N}(N-1, x_0), u_N(N-1)) + F(x_{u_N}(N, x_0)) \\
&= \underbrace{\sum_{k=0}^{N-2} \ell(x_{u_N}(k, x_0), u_N(k))}_{=J_{N-1}(x_0, u_{N-1}) - F(\tilde{x})} + \ell(\tilde{x}, u_{\tilde{x}}) + F(f(\tilde{x}, u_{\tilde{x}})) \\
&= J_{N-1}(x_0, u_{N-1}) \underbrace{- F(\tilde{x}) + \ell(\tilde{x}, u_{\tilde{x}}) + F(f(\tilde{x}, u_{\tilde{x}}))}_{\leq 0} \\
&\leq J_{N-1}(x_0, u_{N-1}).
\end{aligned}$$

Now we can conclude (5.18) from (5.19) as in the proof of Lemma 5.4. □

As in the last section, we want to emphasize that for Problem (5.15) in general (5.4) does no longer hold.

We can obtain an similar inequality to (5.18) also for $N = 1$. Indeed, given $x \in \mathbb{X}_0$ and using the one element control sequence $u(0) = u_x$ with u_x from Assumption 5.9(ii), we obtain

$$V_1(x) \leq J_1(x, u) = \ell(x_u(0, x), u(0)) + F(x_u(1, x)) = \ell(x, u_x) + F(f(x, u_x)) \leq F(x)$$

which together with (5.18) proves

$$V_N(x) \leq F(x) \quad \text{for all } x \in \mathbb{X}_0,\ N \in \mathbb{N}. \tag{5.20}$$

Lemma 5.12 in conjunction with (5.3) enables us to conclude that the optimal value function V_N satisfies inequality (5.1). This will be used in the proof of our following second stability theorem for NMPC schemes. Again, we simply assume (5.2) and discuss sufficient conditions afterwards.

Theorem 5.13 (Stability using Lyapunov terminal costs) *Consider the NMPC Algorithm 3.10 with* ($\mathrm{OCP_{N,e}}$) = (5.15) *and optimization horizon $N \in \mathbb{N}$. Let the Assumptions 5.9(i) and (ii) hold and assume that* (5.2) *holds for suitable $\alpha_1, \alpha_2, \alpha_3 \in \mathcal{K}_\infty$. Then the nominal NMPC closed-loop system (3.5) with NMPC-feedback law μ_N is asymptotically stable on $\mathbb{X}_N$.*

In addition, for $J_\infty^{cl}(x, \mu_N)$ from Definition 4.10 the inequality

$$J_\infty^{cl}(x, \mu_N) \leq V_N(x)$$

holds for each $x \in \mathbb{X}_N$.

Proof Combining Equality (3.20) from Theorem 3.17 with Inequality (5.18) from Lemma 5.12 with $x_0 = f(x, \mu_N(x))$, for each $x \in \mathbb{X}_N$ we obtain

$$V_N(x) \geq \ell(x, \mu_N(x)) + V_{N-1}(f(x, \mu_N(x))) \geq \ell(x, \mu_N(x)) + V_N(f(x, \mu_N(x))).$$

Thus, the assumptions of Theorem 4.11 are satisfied with $V = V_N$, $\mu = \mu_N$, $S(n) = \mathbb{X}_N$ (which is forward invariant by Lemma 5.11) and $\alpha = 1$ which yields the assertion.
□

So far the results in this section were very much in parallel to the respective results for equilibrium terminal constraints in Sect. 5.2. The difference between the two approaches becomes apparent in the following proposition, where we look at sufficient conditions on the problem data under which (5.2) holds. In contrast to Proposition 5.7(ii) in which we needed a condition on f and ℓ, in Part (ii) of the following proposition we can give a sufficient condition in terms of F.

Proposition 5.14 *Let V_N denote the optimal value function of Problem* (5.15) *for some $N \in \mathbb{N}$.*

(i) Assume there exists a function $\alpha_3 \in \mathcal{K}_\infty$ such that the inequality

$$\ell(x, u) \geq \alpha_3(|x|_{x_*})$$

holds for all $x \in X$ and all $u \in U$. Then

$$V_N(x) \geq \alpha_3(|x|_{x_*})$$

holds for all $x \in \mathbb{X}_N$.

(ii) Assume that f, ℓ, and F are continuous in $X \times U$ or $\mathbb{X}_0$, respectively, that U is compact and that Assumption 5.9 is satisfied. Assume furthermore that $\mathbb{X}_0$ contains a ball $\mathscr{B}_\nu(x_*)$, $\nu > 0$, and that there exists a function $\tilde{\alpha}_2 \in \mathscr{K}_\infty$ such that

$$F(x) \leq \tilde{\alpha}_2(|x|_{x_*})$$

holds for all $x \in \mathbb{X}_0 \cap \mathscr{B}_\nu(x_*)$. Then there exists $\alpha_2 \in \mathscr{K}_\infty$ such that

$$V_N(x) \leq \alpha_2(|x|_{x_*}). \tag{5.21}$$

holds for all $x \in \mathbb{X}_N$.

Proof (i) is immediate from the definition of V_N.

In order to prove (ii), observe that for $x \in \mathscr{B}_\nu(x_*)$ the bound on F together with (5.20) implies

$$V_N(x) \leq F(x) \leq \tilde{\alpha}_2(|x|_{x_*}).$$

Now we can proceed as in the proof of Proposition 5.7 in order to construct the desired $\alpha_2 \in \mathscr{K}_\infty$. □

Remark 5.15 For nonlinear systems with $X = \mathbb{R}^d$ and $U = \mathbb{R}^m$ whose linearization at x_* is stabilizable, F and $\mathbb{X}_0$ satisfying Assumption 5.9 can be constructed by a linear-quadratic approach (LQR) via the corresponding Riccati equation, provided $\mathbb{X}$ and $\mathbb{U}$ contain neighborhoods of x_* and u_*, respectively. We briefly sketch this approach considering for simplicity of notation $x_* = 0$ and $u_* = 0$: Assume that the dynamics f satisfies

$$f(x, u) = Ax + Bu + \tilde{f}(x, u) \tag{5.22}$$

with $A \in \mathbb{R}^{d\times d}$, $B \in \mathbb{R}^{d\times m}$, and $\tilde{f} : \mathbb{R}^d \times \mathbb{R}^m \to \mathbb{R}^d$. Assume furthermore that the pair (A, B) is stabilizable and that the map $\tilde{f}$ satisfies $\|\tilde{f}(x, u)\| \leq C(\|x\|^2 + \|x\|\|u\| + \|u\|^2)$ for some constant $C > 0$ and all x, u with $\|x\|, \|u\| \leq \delta$ for some $\delta > 0$.

Under these assumptions, given symmetric and positive definite matrices $Q \in \mathbb{R}^{d\times d}$, $R \in \mathbb{R}^{m\times m}$, we can solve the infinite horizon linear-quadratic optimal control problem

$$\text{minimize } \widetilde{J}_\infty(y, u) = \sum_{k=0}^{\infty} y_u(k, y)^\top Q y_u(k, y) + u(k)^\top R u(k)$$

over $u(\cdot) \in U^\infty$, where $y_u(k, y)$ solves $y^+ = Ay + Bu$.

More precisely, the optimal value function of this problem is given by $\widetilde{V}_\infty(y) = y^\top P y$, where $P \in \mathbb{R}^{d\times d}$ is the unique symmetric and positive definite solution of the discrete time algebraic Riccati equation

$$P = A^\top PA - (A^\top PB)(R + B^\top PB)^{-1}(B^\top PA) + Q. \tag{5.23}$$

Once P is computed, the optimal control for $\widetilde{J}_\infty$ is available in feedback form

$$u^\star(y) = -(R + B^\top PB)^{-1}B^\top PAy. \tag{5.24}$$

The infinite horizon dynamic programming principle (4.6) for this problem for $K = 1$ reads

$$\widetilde{V}_\infty(y) = y^\top Qy + (u^\star(y))^\top Ru^\star(y) + \widetilde{V}_\infty(Ay + Bu^\star(y)). \tag{5.25}$$

We now show how $\widetilde{V}_\infty$, which is readily computable, e.g., by solving the algebraic Riccati equation (5.23) numerically, can be used in order to construct $\mathbb{X}_0$ and F in Assumption 5.9. To this end, we choose ℓ satisfying

$$\ell(x, u) = x^\top Qx + u^\top Ru + \tilde{\ell}(x, u) \tag{5.26}$$

with $|\tilde{\ell}(x, u)| \le D(\|x\|^3 + \|x\|^2\|u\| + \|x\|\|u\|^2 + \|u\|^3)$ for some constant $D > 0$ and all $\|x\|, \|u\| \le \delta$. For this choice, for $u = u^\star(x)$ from (5.24) with $y = x$ and using (5.25) we can compute

$$\begin{aligned}
\widetilde{V}_\infty(f(x, u)) &= (Ax + Bu + \tilde{f}(x, u))^\top P(Ax + Bu + \tilde{f}(x, u)) \\
&= (Ax + Bu)^\top P(Ax + Bu) + \underbrace{2(Ax + Bu)^\top P\tilde{f}(x, u) + \tilde{f}(x, u)^\top P\tilde{f}(x, u)}_{=:r(x,u)} \\
&= \widetilde{V}_\infty(x) - x^\top Qx - u^\top Ru + r(x, u) \\
&= \widetilde{V}_\infty(x) - \ell(x, u) + \tilde{\ell}(x, u) + r(x, u).
\end{aligned}$$

Now the structure of $r(x, u)$ and $\ell(x, u)$ together with the fact that $u = u^\star(y)$ in (5.24) depends linearly on $y = x$ implies the existence of a constant $E > 0$ with $|r(x, u)| + |\tilde{\ell}(x, u)| \le E\|x\|^3$, cf. Problem 2(a) in this chapter. Thus, for each $\sigma > 1$ we find $\delta > 0$ such that $\|x\| \le \delta$ implies

$$-\ell(x, u) + \tilde{\ell}(x, u) + r(x, u) \le -\ell(x, u)/\sigma$$

for $u = u^\star$, cf. Problem 2(b). From this inequality we obtain

$$\widetilde{V}_\infty(f(x, u)) \le \widetilde{V}_\infty(x) - \ell(x, u)/\sigma \tag{5.27}$$

whenever $\|x\| \le \delta$. Fixing some $\sigma > 1$ and the corresponding $\delta > 0$ we now pick $\nu > 0$ such that for all $x \in \mathbb{R}^d$ the inequality $\widetilde{V}_\infty(x) \le \nu$ implies

$$\|x\| \le \delta, \quad x \in \mathbb{X} \quad \text{and} \quad u^\star(x) \in \mathbb{U},$$

which exists since P is positive definite. We claim that $\mathbb{X}_0 := \{x \in \mathbb{R}^d \mid \widetilde{V}_\infty(x) \le \nu\}$ and $F(x) := \sigma\widetilde{V}_\infty(x)$ satisfy Assumption 5.9.

Indeed, picking $x \in \mathbb{X}_0$ and using the control value $u = u^\star(x)$ Inequality (5.27) implies $\widetilde{V}_\infty(f(x,u)) \le \widetilde{V}_\infty(x) \le \nu$ and thus Assumption 5.9(i) and

$$F(f(x,u)) \le \sigma \widetilde{V}_\infty(f(x,u)) \le \sigma \widetilde{V}_\infty(x) - \ell(x,u) = F(x) - \ell(x,u)$$

which is exactly Assumption 5.9(ii). Note that in this construction the set $\mathbb{X}_0$ will become smaller as $\sigma > 1$ becomes smaller.

The following example illustrates this construction.

Example 5.16 Consider the one-dimensional bilinear system

$$x^+ = x + u + xu$$

which is of the form (5.22) with $A = B = 1$ and $\tilde{f}(x,u) = xu$. For simplicity, we do not consider state and control constraints, i.e., we set $\mathbb{X} = X = \mathbb{R}$ and $\mathbb{U}(x) = U = \mathbb{R}$. We consider the stage cost

$$\ell(x,u) = x^2 + u^2 + u^4$$

which is of the form (5.26) with $Q = R = 1$ and $\tilde{\ell}(x,u) = u^4$. The Riccati equation for the linearization reads

$$P = P - P(1+P)^{-1}P + 1$$

and its solution is $P = \frac{1}{2}(1+\sqrt{5})$. Thus, the optimal value function of the linear-quadratic problem becomes

$$\widetilde{V}_\infty(y) = \frac{1}{2}(1+\sqrt{5})y^2 \approx 1.618y^2$$

and the corresponding optimal feedback control reads

$$u^\star(y) = -\frac{1+\sqrt{5}}{3+\sqrt{5}}y \approx -0.618y.$$

Numerical evaluation then yields that Assumption 5.9(ii) holds for $F(x) = \sigma \widetilde{V}_\infty(x)$ and $u_x = u^\star(x)$, e.g., for $\sigma = 1.1$ and all $x \in \mathbb{R}$ with $\widetilde{V}_\infty(x) \le \nu = 0.1$.

Remark 5.17 All results in this section can be easily generalized to tracking time-varying references x^{ref} replacing $\mathbb{X}_0$ by $\mathbb{X}_0(n)$, (OCP$_\infty$) by (OCP$_\infty^{\mathrm{n}}$) and generalizing the two conditions in Assumption 5.9 by

$$f(x,u_x) \in \mathbb{X}_0(n+1)$$

and

$$F(n+1, f(x, u_x)) + \ell(n, x, u_x) \le F(n, x),$$

both for $x \in \mathbb{X}_0(n)$.

However, constructing F in the time-varying case is considerably more complicated than in the time-invariant case, because in the time-varying case linearization and linear-quadratic control does not lead to an algebraic Riccati equation which can be easily solved.

5.4 Suboptimality and Inverse Optimality

Having established the stability of the NMPC closed loop, another important question which naturally arises in the context of NMPC schemes is the performance of the NMPC-feedback law. Here and in what follows, we again consider the simpler case of time-invariant problems noting that the extension to time-varying problems is straightforward.

While performance of general stabilizing feedback laws can be measured in many different ways, in NMPC it is natural to compare the NMPC controller with the infinite horizon optimal controller. To this end, for linear MPC controllers various characteristic values like gains and poles can be compared, see, e.g., [17, Sect. 6.5]. In the nonlinear setting considered in this book, a convenient and meaningful performance measure for the optimization-based NMPC-feedback law μ_N is the infinite horizon cost along the closed-loop trajectory as defined in Definition 4.10. In the time-invariant setting it is given by

$$J_\infty^{cl}(x_0, \mu_N) := \sum_{k=0}^{\infty} \ell(k, x_{\mu_N}(k, x_0), \mu_N(x_{\mu_N}(k, x_0))).$$

In general, it is too optimistic to assume that the NMPC-feedback μ_N yields the optimal value $J_\infty^{cl}(x_0, \mu_N) = V_\infty(x_0)$. The following two examples illustrate this point.

Example 5.18 Consider again Example 5.6, i.e.,

$$x^+ = x + u, \quad \ell(x, u) = x^2 + u^2$$

with $\mathbb{X} = X = \mathbb{U} = U = \mathbb{R}$. Using the terminal constraints $\mathbb{X}_0 = \{0\}$, in Example 5.6 we have seen that the NMPC-feedback law for $N = 2$ is $\mu_2 = -\frac{2}{3}x$.

By solving the associated discrete time algebraic Riccati equation

$$P = P - P(1+P)^{-1}P + 1$$

we can obtain the infinite horizon optimal solution. The positive solution of this equation is $P = \frac{1}{2}(1+\sqrt{5})$ and thus the infinite horizon optimal value function reads

$$V_\infty(x) = \frac{1}{2}(1+\sqrt{5})x^2 \approx 1.618x^2,$$

cf. also Example 5.16. Evaluating $J_\infty^{cl}(x,\mu_2)$ for the NMPC-feedback law $\mu_2(x) = -\frac{2}{3}x$ with trajectory $x_{\mu_2}(n,x) = \frac{1}{3^n}x$ yields

$$J_\infty^{cl}(x,\mu_2) = \sum_{k=0}^{\infty} x_{\mu_2}(k,x)^2 + \mu_2(x_{\mu_2}(k,x))^2 = \sum_{k=0}^{\infty}\left(1+\frac{4}{9}\right)\frac{1}{9^k}x^2 = \frac{13}{8}x^2 = 1.625x^2.$$

Although this value is quite close to $V_\infty(x)$, it is not optimal.

Example 5.19 Consider again Example 5.6 but now with a non quadratic cost, i.e.,

$$x^+ = x+u, \quad \ell(x,u) = x^2 + u^4$$

in which large u are penalized more heavily. As in Example 5.6 we use the equilibrium terminal constraint $\mathbb{X}_0 = \{0\}$. Repeating the computations of Example 5.6 we obtain

$$V_1(x) = x^2 + x^4$$

and

$$\mu_2(x) = \operatorname*{argmin}_{u\in\mathbb{R}} \left\{x^2 + u^4 + (x+u)^2 + (x+u)^4\right\}.$$

Solving this minimization problem (e.g., with MAPLE™, cf. Sect. A.2) yields the nonlinear feedback law

$$\begin{aligned}\mu_2(x) &= \frac{1}{12}\left(-108x + 12\sqrt{324x^6+324x^4+189x^2+12}\right)^{\frac{1}{3}} \\ &\quad - \frac{3x^2+1}{\left(-108x+12\sqrt{324x^6+324x^4+189x^2+12}\right)^{\frac{1}{3}}} - \frac{1}{2}x\end{aligned}$$

Numerical evaluation of the corresponding infinite horizon cost along the closed-loop solution for $x_0 = 20$ yields

$$J_\infty^{cl}(20,\mu_2) \approx 11240.39.$$

Since this problem is not linear-quadratic we cannot use the Riccati equation in order to compute the exact optimal value. However, the feedback law

$$\mu(x) = \frac{1}{6}(-54x + 6\sqrt{6+81x^2})^{\frac{1}{3}} - \frac{1}{(-54x+6\sqrt{6+81x^2})^{\frac{1}{3}}}$$

—whose derivation we will explain in Example 7.23 in Chap. 7—yields $J_\infty^{cl}(20, \mu) \approx 1725.33$ which was again evaluated numerically. Hence, the optimal value function satisfies $V_\infty(20) \leq 1725.33$ which shows that μ_2 is far from optimal in this example.

An alternative to the direct evaluation of $J_\infty^{cl}(x_0, \mu_N)$ as performed in these examples is readily available from the Theorems 5.5 and 5.13. These theorems provide the explicit upper bound

$$J_\infty^{cl}(x_0, \mu_N) \leq V_N(x_0) \tag{5.28}$$

for the NMPC-feedback law μ_N derived from Algorithm 3.10 with either ($\mathrm{OCP}_{\mathrm{N,e}}$) = (5.5) or ($\mathrm{OCP}_{\mathrm{N,e}}$) = (5.15). Unfortunately, in general there is no simple formula for the mismatch between V_N and the infinite horizon optimal value function V_∞. This is due to the fact that in both (5.5) and (5.15), the optimization is restricted to the controls $u(\cdot) \in \mathbb{U}^N_{\mathbb{X}_0}(x_0)$ whose corresponding trajectories enter the terminal constraint set $\mathbb{X}_0$ after at most N steps. The following proposition shows that this inevitably leads to the inequality $V_N(x) \geq V_\infty(x)$ for all $x \in \mathbb{X}_N$.

Proposition 5.20 *Consider the optimal control problem* (5.5) *or* (5.15) *and the infinite horizon optimal control problem* (OCP_∞) *with same stage cost* ℓ *and constraints* $\mathbb{X}$ *and* $\mathbb{U}(x)$. *Let the assumptions of the respective Theorem 5.5 or 5.13 be satisfied. Then the inequality*

$$V_N(x) \geq V_\infty(x)$$

holds for all $x \in \mathbb{X}_N$.

Proof Given $x \in \mathbb{X}_N$ we define a control sequence $u(\cdot) \in \mathbb{U}^N(x)$ by evaluating $\mu_N(x)$ along the NMPC closed-loop trajectory, i.e.,

$$u(k) := \mu_N(x_{\mu_N}(k, x)).$$

Then we get

$$V_\infty(x) \leq J_\infty(x, u) = J_\infty^{cl}(x, \mu_N) \leq V_N(x),$$

where the last inequality is obtained from Theorems 5.5 and 5.13, respectively. This shows the assertion. □

Unfortunately, the quantitative effect of the stabilizing constraints, i.e., the question about how much larger V_N is compared to V_∞ is in general difficult if not impossible to answer. Still, we can provide asymptotic results, i.e., conditions under which V_N converges to V_∞ on arbitrary (but fixed) balls around x_* as $N \to \infty$.

Theorem 5.21 (Convergence of V_N) *Consider the optimal control problem* (5.5) *or* (5.15) *and the infinite horizon optimal control problem* (OCP_∞) *with same stage cost* ℓ *and constraints* $\mathbb{X}$ *and* $\mathbb{U}(x)$. *Let the assumptions of Theorem 4.16 hold and assume that there is* $N_0 \in \mathbb{N}$ *such that the assumptions of the respective Theorem*

5.5 or 5.13 hold for $N = N_0$. Assume, furthermore, that $\mathbb{X}_{N_0}$ contains a ball $\mathscr{B}_\nu(x_*)$. Then for each $R > 0$ and each $\varepsilon > 0$ there exists $N_{R,\varepsilon} > 0$ such that the inequality

$$V_N(x) \le V_\infty(x) + \varepsilon$$

holds for all $N \ge N_{R,\varepsilon}$ and all $x \in \mathbb{X}_N$ with $|x|_{x_*} \le R$.

Proof For N_0 from the assumption, α_2 from the assumption of Theorem 5.5 or 5.13, respectively, and from (5.10) we get

$$V_N(x) \le V_{N_0}(x) \le \alpha_2(|x|_{x_*})$$

for all $N \ge N_0$ and all $x \in \mathscr{B}_\nu(x_*)$. Since the assumptions of Theorem 4.16 are satisfied, for any $\alpha \in (0, 1)$ the feedback law μ_α from Theorem 4.16 satisfies

$$J_\infty^{cl}(x, \mu_\alpha) \le V_\infty(x)/\alpha$$

for all $x \in \mathbb{X}$. Furthermore, μ_α asymptotically stabilizes the system, i.e., there exists $\beta \in \mathscr{KL}$ such that

$$|x_{\mu_\alpha}(n, x)|_{x_*} \le \beta(|x|_{x_*}, n)$$

holds for all $x \in \mathbb{X}$ and all $n \in \mathbb{N}_0$. Fixing some arbitrary $\bar{\alpha} \in (0, 1)$, by Remark 4.12 we may assume that β is independent of $\alpha \in [\bar{\alpha}, 1)$.

Now fix $R > 0$ and $\varepsilon > 0$ and let $n_{R,\varepsilon} \in \mathbb{N}$ be large enough such that the inequalities

$$\beta(R, n_{R,\varepsilon}) < \nu \quad \text{and} \quad \alpha_2(\beta(R, n_{R,\varepsilon})) \le \varepsilon$$

hold and set $N_{R,\varepsilon} = N_0 + n_{R,\varepsilon}$. By the monotonicity properties of α_2 and β the choice of $n_{R,\varepsilon}$ implies $x_{\mu_\alpha}(n, x) \in \mathscr{B}_\nu(x_*) \subset \mathbb{X}_{N_0}$ and

$$V_{N_0}(x_{\mu_\alpha}(n, x)) \le \alpha_2(\beta(|x|_{x_*}, n_{R,\varepsilon})) \le \varepsilon$$

for all $n \ge n_{R,\varepsilon}$ and all $x \in \mathbb{X}_N$ with $|x|_{x_*} \le R$.

For such an x and an $N \ge N_{R,\varepsilon}$, consider the control sequence $u(\cdot) \in \mathbb{U}^N_{\mathbb{X}_0}(x)$

$$u(n) = \begin{cases} \mu_\alpha(x_{\mu_\alpha}(n, x)), & n = 0, \dots, N - N_0 - 1 \\ u^*(n - N + N_0), & n = N - N_0, \dots, N - 1 \end{cases}$$

where u^* is the optimal control sequence for (5.5) or (5.15) with $N = N_0$ and $x_0 = x_{\mu_\alpha}(N - N_0, x)$. The control $u(\cdot)$ lies in $\mathbb{U}^N_{\mathbb{X}_0}(x)$ because it inherits $u(\cdot) \in \mathbb{U}^N(x)$ from μ_α and u^* and the corresponding trajectory ends in $\mathbb{X}_0$ because $u^*(\cdot) \in \mathbb{U}^{N_0}_{\mathbb{X}_0}(x_{\mu_\alpha}(N - N_0, x))$. For this control sequence we get

$$
\begin{aligned}
V_N(x) \le J_N(x,u) &= \sum_{n=0}^{N-1} \ell(x_u(n,x),u(n)) + F(x_u(N,x)) \\
&= \sum_{n=0}^{N-N_0-1} \ell(x_u(n,x),u(n)) + \sum_{n=N-N_0}^{N-1} \ell(x_u(n,x),u(n)) + F(x_u(N,x)) \\
&= \sum_{n=0}^{N-N_0-1} \ell(x_u(n,x),u(n)) + J_{N_0}(x_u(N-N_0,x),u(\cdot+N-N_0)) \\
&= \sum_{n=0}^{N-N_0-1} \ell(x_{\mu_\alpha}(n,x),\mu_\alpha(x_{\mu_\alpha}(n,x))) + J_{N_0}(x_{\mu_\alpha}(N-N_0,x),u^*(\cdot)) \\
&\le J_\infty^{cl}(x,\mu_\alpha) + V_{N_0}(x_{\mu_\alpha}(N-N_0,x)) \;\le\; V_\infty(x)/\alpha + \varepsilon
\end{aligned}
$$

where we used $N - N_0 \ge N_{R,\varepsilon} - N_0 \ge n_{R,\varepsilon}$ for estimating $V_{N_0}(x_{\mu_\alpha}(N-N_0,x))$ in the last inequality. From this inequality, we obtain the assertion since $\alpha \in (\bar{\alpha}, 1)$ was arbitrary and $N_{R,\varepsilon}$ is independent of $\alpha \in (\bar{\alpha}, 1)$. □

While in Theorem 5.21 the lower bound on the necessary optimization horizon N depends on R and ε, we can exploit the special structure of (5.15) in order to give a condition under which the bound on N merely depends on R and on properties of F.

Theorem 5.22 *Consider the optimal control problem* (5.15) *and the infinite horizon optimal control problem* (OCP_∞) *with same stage cost ℓ and constraints $\mathbb{X}$ and $\mathbb{U}(x)$. Let the assumptions of Theorem 4.16 hold and assume that there is $N_0 \in \mathbb{N}$ such that the assumptions of Theorem 5.13 hold for $N = N_0$. Assume in addition that $\mathbb{X}_0$ contains a ball $\mathcal{B}_\nu(x_*)$ and that the terminal cost F satisfies*

$$|F(x) - V_\infty(x)| \le \varepsilon$$

for all $x \in \mathcal{B}_\nu(x_*)$ and some $\varepsilon > 0$. Then for each $R > 0$ there exists $N_R > 0$ such that the inequality

$$V_N(x) \le V_\infty(x) + \varepsilon$$

holds for all $N \ge N_R$ and all $x \in \mathbb{X}_N$ with $|x|_{x_*} \le R$.

Proof Using $\bar{\alpha}$ and β as in the proof of Theorem 5.21 we choose $N_R \in \mathbb{N}$ such that the inequality

$$\beta(R, N_R) < \nu$$

holds. Given $x \in \mathbb{X}$ with $|x|_{x_*} \le R$ and $N \ge N_R$ and an arbitrary $\alpha \in [\bar{\alpha}, 1)$, we define the control sequence

$$u(n) = \mu_\alpha(x_{\mu_\alpha}(n,x)), \quad n = 0, \ldots, N-1$$

with μ_α from Theorem 4.16, i.e., satisfying (4.18). This control sequence lies in $\mathbb{U}^N_{\mathbb{X}_0}(x_0)$ since

$$|x_{\mu_\alpha}(N,x)|_{x_*} \le \beta(|x|_{x_*}, N) \le \beta(R, N_R) < \nu,$$

thus $x_u(N,x) = x_{\mu_\alpha}(N,x) \in \mathscr{B}_\nu(x_*) \subseteq \mathbb{X}_0$. For this $u(\cdot)$ we get

$$\begin{aligned} V_N(x) \le J_N(x,u) &= \sum_{n=0}^{N-1} \ell(x_u(n,x), u(n)) + F(x_u(N,x)) \\ &= \sum_{n=0}^{N-1} \ell(x_{\mu_\alpha}(n,x), \mu_\alpha(x_{\mu_\alpha}(n,x))) + F(x_{\mu_\alpha}(N,x)) \\ &\le \sum_{n=0}^{N-1} \ell(x_{\mu_\alpha}(n,x), \mu_\alpha(x_{\mu_\alpha}(n,x))) + V_\infty(x_{\mu_\alpha}(N,x)) + \varepsilon \\ &\le \sum_{n=0}^{N-1} \ell(x_{\mu_\alpha}(n,x), \mu_\alpha(x_{\mu_\alpha}(n,x))) + V_\infty(x_{\mu_\alpha}(N,x))/\alpha + \varepsilon. \end{aligned} \tag{5.29}$$

where we used $V_\infty(x_{\mu_\alpha}(N,x)) \ge 0$ and $\alpha < 1$ in the last inequality. Now (4.18) implies

$$\ell(x_{\mu_\alpha}(n,x), \mu_\alpha(x_{\mu_\alpha}(n,x))) \le V_\infty(x_{\mu_\alpha}(n,x))/\alpha - V_\infty(x_{\mu_\alpha}(n+1,x))/\alpha.$$

Inserting this inequality into (5.29) yields

$$\begin{aligned} V_N(x) &\le \sum_{n=0}^{N-1} \ell(x_{\mu_\alpha}(n,x), \mu_\alpha(x_{\mu_\alpha}(n,x))) + V_\infty(x_{\mu_\alpha}(N,x))/\alpha + \varepsilon \\ &\le \sum_{n=0}^{N-1} \big[V_\infty(x_{\mu_\alpha}(n,x))/\alpha - V_\infty(x_{\mu_\alpha}(n+1,x))/\alpha\big] + V_\infty(x_{\mu_\alpha}(N,x))/\alpha + \varepsilon \\ &\le V_\infty(x_{\mu_\alpha}(0,x))/\alpha + \varepsilon \;=\; V_\infty(x)/\alpha + \varepsilon \end{aligned}$$

and since $\alpha \in [\bar{\alpha}, 1)$ was arbitrary the assertion follows. □

Example 5.23 We illustrate Theorem 5.21 by Examples 5.6 and 5.18. Observing that $\mathbb{X}_N = \mathbb{R}$ holds for $N \ge 1$, the dynamic programming equation (3.15) for $K = 1$ and $N \ge 2$ becomes

$$V_N(x) = \inf_{u\in\mathbb{R}} \{x^2 + u^2 + V_{N-1}(x+u)\}.$$

Using this equation in order to iteratively compute V_N starting from $V_1(x) = 2x^2$, cf. Example 5.6, we obtain the (approximate) values

$$\begin{array}{llll} V_1(x) = 2x^2 & V_2(x) & = 1.666666667\,x^2 \\ V_3(x) = 1.625\,x^2 & V_4(x) & = 1.619047619\,x^2 \\ V_5(x) = 1.618181818\,x^2 & V_6(x) & = 1.618055556\,x^2 \\ V_7(x) = 1.618037135\,x^2 & V_8(x) & = 1.618034448\,x^2 \\ V_9(x) = 1.618034056\,x^2 & V_{10}(x) & = 1.618033999\,x^2, \end{array}$$

cf. Problem 4. Since, as computed in Example 5.18, the infinite horizon optimal value function is given by

$$V_\infty(x) = \frac{1}{2}(1+\sqrt{5})x^2 \approx 1.618033988\,x^2,$$

this shows that, e.g., for $R = 1$, the inequality $V_N(x) - V_\infty(x) \le \varepsilon$ holds for $\varepsilon = 2.2 \cdot 10^{-5}$ for $N = 6$, for $\varepsilon = 4.6 \cdot 10^{-7}$ for $N = 8$ and for $\varepsilon = 1.1 \cdot 10^{-8}$ for $N = 10$.

We end this section by investigating the inverse optimality of the NMPC-feedback law μ_N. While the suboptimality estimates provided so far in this section give bounds on the infinite horizon performance of μ_N, inverse optimality denotes the fact that μ_N is in fact an infinite horizon optimal feedback law—but not for the stage cost ℓ but for a suitably adjusted stage cost $\tilde{\ell}$. The motivation for such a result stems from the fact that optimal feedback laws have desirable robustness properties. This can be made precise for continuous time control affine systems

$$\dot{x}(t) = g_0(x) + g_1(x)u \tag{5.30}$$

with $x \in \mathbb{R}^d$, $u \in \mathbb{R}^m$, $g_0 : \mathbb{R}^d \to \mathbb{R}^d$, and $g_1 : \mathbb{R}^d \to \mathbb{R}^{d\times m}$. For these systems, it is known that a stabilizing (continuous time) infinite horizon optimal feedback law μ_∞ has a sector margin $(1/2, \infty)$ which means that $u = \mu_\infty(x)$ stabilizes not only (5.30) but also

$$\dot{x}(t) = g_0(x) + g_1(x)\phi(u) \tag{5.31}$$

for any $\phi : \mathbb{R}^m \to \mathbb{R}^m$ satisfying $\|u\|_2/2 \le u^\top \phi(u) \le \infty$ for all $u \in \mathbb{R}^m$, see Magni and Sepulchre [9] for details.

Although we are not aware of analogous discrete time results in the literature, it seems reasonable to expect that this robustness is inherited in an approximate way for optimal control of sampled data systems with sufficiently fast sampling. This justifies the investigation of inverse optimality also in the discrete time setting.

For the NMPC schemes presented in this chapter, we can make the following inverse optimality statement.

Theorem 5.24 (Inverse optimality) *Consider the optimal control problem* (5.5) *or* (5.15) *for some $N \in \mathbb{N}$ with the usual constraints $x \in \mathbb{X}$ and $u \in \mathbb{U}(x)$. Let the assumptions of the respective Theorem 5.5 or 5.13 hold for this N. Then on the set $\mathbb{X}_N$ the feedback μ_N equals the infinite horizon optimal feedback law for* (OCP_∞) *with stage cost*

$$\tilde{\ell}(x,u) := \ell(x,u) + V_{N-1}(f(x,u)) - V_N(f(x,u)) \tag{5.32}$$

and constraints $x \in \mathbb{X}_{N-1}$ *and* $u \in \mathbb{U}(x)$.

Proof First observe that the assumptions of Theorem 5.5 or 5.13 imply (5.10) or (5.18), respectively, and $V(x_*) = 0$. Hence, (5.32) satisfies $\tilde{\ell} \geq \ell$, is of the form (3.2), and the inequality for ℓ in (5.2) remains valid for $\tilde{\ell}$. We denote the infinite horizon optimal value function of (OCP$_\infty$) with stage cost $\tilde{\ell}$ by $\widetilde{V}_\infty$ and the corresponding optimal feedback law by $\tilde{\mu}_N$.

From the dynamic programming principle (3.15) with $K = 1$ and the definition of $\tilde{\ell}$ we get

$$\begin{aligned} V_N(x_0) &= \inf_{u \in \mathbb{U}^1_{\mathbb{X}_{N-1}}(x_0)} \{\ell(x_0,u) + V_{N-1}(f(x_0,u))\} \\ &= \inf_{u \in \mathbb{U}^1_{\mathbb{X}_{N-1}}(x_0)} \{\tilde{\ell}(x_0,u) + V_N(f(x_0,u))\}. \end{aligned}$$

Similarly, (3.19) implies

$$V_N(x_0) = \tilde{\ell}(x_0, \mu_N(x_0)) + V_N(f(x_0, \mu_N(x_0))).$$

From these two equations, by induction for each $K \in \mathbb{N}$ we get

$$V_N(x_0) \leq \sum_{k=0}^{K-1} \tilde{\ell}(x_u(k,x_0), u(k)) + V_N(x_u(K,x_0)) \tag{5.33}$$

for every $u \in \mathbb{U}^\infty(x_0)$ with $\mathbb{U}^\infty(x_0)$ defined with respect to the constraint $x \in \mathbb{X}_{N-1}$, and

$$V_N(x_0) = \sum_{k=0}^{K-1} \tilde{\ell}(x_{\mu_N}(k,x_0), \mu_N(x_{\mu_N}(k,x_0))) + V_N(x_{\mu_N}(K,x_0)). \tag{5.34}$$

Since $\tilde{\ell} \geq 0$, for an arbitrary $u \in \mathbb{U}^\infty(x_0)$ in (5.33), for $K \to \infty$ the sum

$$\sum_{k=0}^{K-1} \tilde{\ell}(x_u(k,x_0), u(k))$$

either grows unboundedly or converges to some finite value. Since $\ell(x,u) \geq \alpha_3(|x|_{x_*})$, convergence is only possible if $x_u(k,u)$ converges to x_* as $k \to \infty$, i.e., if $V_N(x_u(K,u)) \to 0$ as $k \to \infty$. Thus, in either case letting $K \to \infty$ in (5.33) we get

$$V_N(x_0) \leq \sum_{k=0}^{\infty} \tilde{\ell}(x_u(k,x_0), u(k))$$

for all $u \in \mathbb{U}^\infty(x_0)$, which implies $V_N(x_0) \le \widetilde{V}_\infty(x_0)$.

On the other hand, since μ_N asymptotically stabilizes the system, in (5.34) we get $V_N(x_{\mu_N}(K, x_0)) \to 0$ as $K \to \infty$ and thus letting $K \to \infty$ in (5.34) yields

$$V_N(x_0) = \sum_{k=0}^{\infty} \tilde{\ell}(x_{\mu_N}(k, x_0), \mu_N(x_{\mu_N}(k, x_0))) \tag{5.35}$$

which implies $V_N(x_0) \ge \widetilde{V}_\infty(x_0)$. Consequently, we get $V_N(x_0) = \widetilde{V}_\infty(x_0)$ and from (5.35) it follows that $\tilde{\mu}_\infty = \mu_N$ is the infinite horizon optimal feedback law for stage cost $\tilde{\ell}$. □

Observe that for the inverse optimality statement to hold, we need to replace the constraints $x \in \mathbb{X}$ in (OCP$_\infty$) by the in general tighter constraints $x \in \mathbb{X}_{N-1}$, where $\mathbb{X}_{N-1}$ is the feasible set for (5.5) or (5.15) with horizon $N-1$. This is because by (3.19), the feedback μ_N is obtained by minimization with respect to these constraints. Thus, it cannot in general be optimal for the infinite horizon problem with the usually weaker original constraints $x \in \mathbb{X}$.

5.5 Notes and Extensions

Most of the results in this chapter are classical and can be found in several places in the NMPC literature. In view of the huge amount of this literature, here we do not make an attempt to give a comprehensive list of references but rather restrict ourselves just to the literature from which we learned the results presented in this chapter.

While the proofs in the NMPC literature are similar to the proofs given here, the relaxed dynamic programming arguments outlined in Sect. 5.1 are usually applied in a more ad hoc manner. The reason we have put more emphasis on this approach and, in particular, used Theorem 4.11 in the stability proofs is because the analysis of NMPC schemes without stabilizing terminal conditions in the following Chap. 6 will also be based on Theorem 4.11. Hence, proceeding this way we can highlight the similarities in the analysis of these different classes of NMPC schemes.

For discrete time NMPC schemes with equilibrium terminal constraints as featured in Sect. 5.2, a version of Theorem 5.5 was published by Keerthi and Gilbert [8] in 1988, even for the more general case in which the optimization horizon may vary with time. Their approach was inspired by earlier results for linear systems, for more information on these linear results we refer to the references in [8]. Even earlier, in 1982 Chen and Shaw [1] proved stability of an NMPC scheme with equilibrium terminal constraint in continuous time, however, in their setting the whole optimal control function on the optimization horizon is applied to the plant, as opposed to only the first part. Continuous time and sampled data versions of Theorem 5.5 were given by Mayne and Michalska [10] in 1990, using, however, a differentiability assumption on the optimal value function which is quite restrictive in the presence of state constraints.

The "quasi infinite horizon" idea of imposing regional terminal constraints $\mathbb{X}_0$ plus a terminal cost satisfying Assumption 5.9 as presented in Sect. 5.3 came up in the second half of the 1990s in papers by De Nicolao, Magni and Scattolini [3, 4], Magni and Sepulchre [9] or Chen and Allgöwer [2], both in discrete and continuous time. Typically, these papers provide specific constructions of F and $\mathbb{X}_0$ satisfying Assumption 5.9 rather than imposing this assumption in an abstract way as we did here. The abstract formulation of these conditions given here was inspired by the survey article by Mayne, Rawlings, Rao, and Scokaert [11] which also contains a comparative discussion of the approaches in some of the cited papers. For a continuous time version of such abstract conditions we refer to Fontes [5]. A terminal cost meeting Assumption 5.9 was already used before by Parisini and Zoppoli [14], however, without terminal constraint; we will investigate this setting in Sect. 10.1. The construction of F and $\mathbb{X}_0$ in Remark 5.15 is similar to the construction in [2, 14]. A related NMPC variant which may have motivated some of the authors cited above was proposed by Michalska and Mayne [12]. In this so-called dual mode NMPC, the prediction horizon length is an additional optimization variable and the prediction is stopped once the set $\mathbb{X}_0$ is reached. Inside this set, the control value u_x from Assumption 5.9(ii) is used.

Establishing the existence of a suitable upper bound of V_N is essential for being able to use V_N as a Lyapunov function. The argument used here in the proofs of Propositions 5.7(ii) and 5.14(ii) was adopted from Rawlings and Mayne [16, Proposition 2.18]. Of course, this is not the only way to obtain an upper bound on V_N. Other sufficient conditions, like, e.g., the controllability condition "C" in Keerthi and Gilbert [8, Definition 3.2], may be used as well.

Regarding the suboptimality results in Sect. 5.4, for the special case of equilibrium terminal constraints $\mathbb{X}_0 = \{x_*\}$ and $F \equiv 0$, a version of the suboptimality result in Theorem 5.21 was given by Keerthi and Gilbert [8]. For the case of general $\mathbb{X}_0$ and F we are not aware of a result similar to Theorem 5.21, although we would not be surprised to learn that such a result exists in the huge body of NMPC literature. Theorem 5.22 is a variant of Grüne and Rantzer [6, Theorem 6.2] and extends Theorem 2 of Hu and Linnemann [7] in which the case $F = V_\infty$ is considered.

Inverse optimality was extensively investigated already for linear MPC leading to the famous "fake" algebraic Riccati equation introduced by Poubelle, Bitmead and Gevers [15]. For nonlinear systems in continuous time, this property was proved by Magni and Sepulchre [9]. While the discrete time nonlinear version given in Theorem 5.24 is used in an ad hoc manner in several papers (e.g., in Nešić and Grüne [13]), we were not able to find it in the literature in the general form presented here.

Problems

1. Consider the scalar control system

$$x^+ = x + u, \quad x(0) = x_0$$

with $x \in X = \mathbb{R}$, $u \in U = \mathbb{R}$ which shall be controlled via NMPC using the quadratic stage cost

$$\ell(x, u) = x^2 + u^2$$

and the stabilizing endpoint constraint $x_u(N, x_0) = x_* = 0$. For the horizon $N = 2$, compute an estimate for the closed-loop costs $J_\infty^{cl}(x, \mu_2(\cdot))$.

2. Consider the setting of Remark 5.15 and prove the following properties.

    (a) There exists a constant $E > 0$ such that $|r(x, u)| + |\tilde{\ell}(x, u)| \le E\|x\|^3$ holds for each $x \in \mathbb{R}^d$ and $u = u^\star(x)$.

    (b) For each $\sigma > 1$ there exists $\delta > 0$ such that $\|x\| \le \delta$ implies

    $$-\ell(x, u) + \tilde{\ell}(x, u) + r(x, u) \le -\ell(x, u)/\sigma$$

    for $u = u^\star(x)$.

    Hint for (b): Look at the hints for Problem 2 in Chap. 4.

3. Consider f, ℓ, $\mathbb{X}_0$ and F satisfying the assumptions of Proposition 5.14(i) and (ii). Prove the following properties.

    (a) The stage cost satisfies $\ell(x, u_x) \le \tilde{\alpha}_2(|x|_{x_*})$ for $x \in \mathbb{X}_0$, u_x from Assumption 5.9(ii) and $\tilde{\alpha}_2$ from the assumption of Proposition 5.14(ii).
    (b) For the feedback law $\mu(x) := u_x$ with u_x from Assumption 5.9(ii) the closed-loop system $x^+ = f(x, \mu(x))$ is asymptotically stable on $\mathbb{X}_0$.

4. Consider the setting from Problem 1. Prove without using Theorem 5.21 that for all $\varepsilon > 0$ and $R > 0$ there exists $N_\varepsilon \in \mathbb{N}$ such that

    $$V_N(x) \le V_\infty(x) + \varepsilon$$

    holds for all $N \ge N_\varepsilon$ and all $x \in \mathbb{R}$ with $|x| \le R$. Proceed as follows:

    (a) Use dynamic programming in order to show $V_N(x) = C_N x^2$ with $C_1 = 2$ and

    $$C_N = \frac{8C_{N-1}^2 + 12C_{N-1} + 4}{4C_{N-1}^2 + 8C_{N-1} + 4}.$$

    (b) Use the expression from (a) to conclude that $C_N \to \frac{1}{2}(1 + \sqrt{5})$ holds as $N \to \infty$.

    (c) Use the exact expression for V_∞ from Example 5.23 in order to prove the claim.

5. Consider Example 2.3, i.e.,

$$f(x,u) := \begin{pmatrix} x_1^+ \\ x_2^+ \end{pmatrix} = \begin{pmatrix} \sin(\vartheta(x)+u) \\ \cos(\vartheta(x)+u)/2 \end{pmatrix}$$

with

$$\vartheta(x) = \begin{cases} \arccos 2x_2, & x_1 \geq 0 \\ 2\pi - \arccos 2x_2, & x_1 < 0 \end{cases},$$

initial value $(0, 1/2)$ and stage cost $\ell(x,u) = \|x - x_*\|^2 + u^2$ with $x_* = (0, -1/2)$. The control values are restricted to the set $\mathbb{U} = [0, 0.2]$ which allows the car to only move clockwise on the ellipse

$$\mathbb{X} = \left\{ x \in \mathbb{R}^2 \;\middle|\; \left\| \begin{pmatrix} x_1 \\ 2x_2 \end{pmatrix} \right\| = 1 \right\}.$$

Perform the following numerical simulations for this problem.

(a) Implement the NMPC closed loop for $N = 8$ and confirm that the closed-loop trajectory does not converge toward x_*.
(b) Modify the NMPC problem by introducing the terminal constraint $\mathbb{X}_0 = \{x_*\}$. Again considering the horizon length $N = 8$, verify that now $x(n) \to x_*$.
(c) Check the control constraints for each NMPC iterate from (b) more closely, verify that they are violated at some sampling instants and explain why this happens. Determine by simulations how large N needs to be such that these violations vanish.

Hint: Instead of implementing the problem from scratch you may suitably modify the MATLAB® code for Example 6.31, cf. Appendix A.1.

References

1. Chen, C.C., Shaw, L.: On receding horizon feedback control. Automatica **18**(3), 349–352 (1982)
2. Chen, H., Allgöwer, F.: A quasi-infinite horizon nonlinear model predictive control scheme with guaranteed stability. Automatica J. IFAC **34**(10), 1205–1217 (1998)
3. De Nicolao, G., Magni, L., Scattolini, R.: Stabilizing nonlinear receding horizon control via a nonquadratic terminal state penalty. In: CESA'96 IMACS Multiconference: Computational Engineering in Systems Applications, Lille, France, pp. 185–187 (1996)
4. De Nicolao, G., Magni, L., Scattolini, R.: Stabilizing receding-horizon control of nonlinear time-varying systems. IEEE Trans. Automat. Control **43**(7), 1030–1036 (1998)
5. Fontes, F.A.C.C.: A general framework to design stabilizing nonlinear model predictive controllers. Syst. Control Lett. **42**(2), 127–143 (2001)

6. Grüne, L., Rantzer, A.: On the infinite horizon performance of receding horizon controllers. IEEE Trans. Automat. Control **53**, 2100–2111 (2008)
7. Hu, B., Linnemann, A.: Toward infinite-horizon optimality in nonlinear model predictive control. IEEE Trans. Automat. Control **47**(4), 679–682 (2002)
8. Keerthi, S.S., Gilbert, E.G.: Optimal infinite-horizon feedback laws for a general class of constrained discrete-time systems: stability and moving-horizon approximations. J. Optim. Theory Appl. **57**(2), 265–293 (1988)
9. Magni, L., Sepulchre, R.: Stability margins of nonlinear receding-horizon control via inverse optimality. Syst. Control Lett. **32**(4), 241–245 (1997)
10. Mayne, D.Q., Michalska, H.: Receding horizon control of nonlinear systems. IEEE Trans. Automat. Control **35**(7), 814–824 (1990)
11. Mayne, D.Q., Rawlings, J.B., Rao, C.V., Scokaert, P.O.M.: Constrained model predictive control: stability and optimality. Automatica **36**(6), 789–814 (2000)
12. Michalska, H., Mayne, D.Q.: Robust receding horizon control of constrained nonlinear systems. IEEE Trans. Automat. Control **38**(11), 1623–1633 (1993)
13. Nešić, D., Grüne, L.: A receding horizon control approach to sampled-data implementation of continuous-time controllers. Syst. Control Lett. **55**, 660–672 (2006)
14. Parisini, T., Zoppoli, R.: A receding-horizon regulator for nonlinear systems and a neural approximation. Automatica **31**(10), 1443–1451 (1995)
15. Poubelle, M.A., Bitmead, R.R., Gevers, M.R.: Fake algebraic Riccati techniques and stability. IEEE Trans. Automat. Control **33**(4), 379–381 (1988)
16. Rawlings, J.B., Mayne, D.Q.: Model Predictive Control: Theory and Design. Nob Hill Publishing, Madison (2009)
17. Wang, L.: Model Predictive Control System Design and Implementation using MATLAB. Advances in Industrial Control. Springer, London (2009)

Chapter 6
Stability and Suboptimality Without Stabilizing Terminal Conditions

In this chapter, we present a comprehensive stability and suboptimality analysis for NMPC schemes without stabilizing terminal conditions. After defining the setting and presenting motivating examples, we introduce a boundedness condition on the optimal value function and an asymptotic controllability assumption. Moreover, we give a detailed derivation of stability and performance estimates based on these assumptions and the relaxed dynamic programming framework introduced before. We show that our stability criterion is tight for the class of systems satisfying the controllability assumption and give conditions under which the level of suboptimality and a bound on the optimization horizon length needed for stability can be explicitly computed from the parameters in the controllability condition. As a spinoff we recover the well-known result that—under suitable conditions—stability of the NMPC closed loop can be expected if the optimization horizon is sufficiently large. We further deduce qualitative properties of the stage cost which lead to stability with small optimization horizons and illustrate by means of two examples how these criteria can be used even if the parameters in the controllability assumption cannot be evaluated precisely. Finally, we give weaker conditions under which semiglobal and semiglobal practical stability of the NMPC closed loop can be ensured.

6.1 Setting and Preliminaries

In this chapter, we consider the NMPC schemes without stabilizing terminal conditions for stage costs ℓ satisfying (3.8). Throughout this chapter we will use the basic NMPC Algorithms 3.1 and 3.7 with optimal control problems ($\mathrm{OCP_N}$) and ($\mathrm{OCP_N^n}$), respectively. Weights ω_k as in ($\mathrm{OCP_{N,e}}$) and ($\mathrm{OCP_{N,e}^n}$)—more precisely, terminal weights—will be discussed in Sect. 10.2. We consider state and control constraints $\mathbb{X}$ and $\mathbb{U}(x)$ as introduced in Sect. 3.2 and the respective sets $\mathbb{Y}$ of admissible pairs

L. Grüne and J. Pannek, *Nonlinear Model Predictive Control*,
Communications and Control Engineering, DOI 10.1007/978-3-319-46024-6_6

and $\mathbb{U}^N(x)$ of admissible control sequences from Definition 3.2. Throughout this chapter, the state constraint set $\mathbb{X} \subset X$ is supposed to be viable in the sense of Assumption 3.3. Relaxations of this viability assumption on $\mathbb{X}$ will be discussed in Sects. 7.1–7.3.

As in the previous chapter, our goal is to apply Theorem 4.11, for which we need to establish the inequalities

$$V_N(n, x) \geq \alpha\ell(n, x, \mu_N(n, x)) + V_N(n+1, f(x, \mu_N(n, x))) \qquad (5.1)$$

for all $x \in \mathbb{X}$, $n \in \mathbb{N}_0$ and some $\alpha \in (0, 1]$, and the existence of $\alpha_1, \alpha_2, \alpha_3 \in \mathcal{K}_\infty$ such that the inequalities

$$\alpha_1(|x|_{x^{\text{ref}}(n)}) \leq V_N(n, x) \leq \alpha_2(|x|_{x^{\text{ref}}(n)}) \quad \text{and} \quad \ell(n, x, u) \geq \alpha_3(|x|_{x^{\text{ref}}(n)}) \qquad (5.2)$$

hold for all $x \in \mathbb{X}$, $n \in \mathbb{N}_0$ and $u \in U$, cf. Sect. 5.1. Again, the inequality

$$V_N(n, x_0) \geq \ell(n, x_0, \mu_N(n, x_0)) + V_{N-1}(n+1, f(x_0, \mu_N(n, x_0))) \qquad (5.3)$$

which follows from (3.20) plays a vital role in our analysis. In Sect. 6.7, we will also use Theorem 4.14 in order to prove practical stability properties.

For the optimal value functions V_N of (OCP_N) or ($\text{OCP}_\text{N}^\text{n}$) and V_∞ of the corresponding infinite horizon problems ($\text{OCP}_\infty^\text{n}$) it is immediate that we get the inequalities

$$V_{N-1}(n, x_0) \leq V_N(n, x_0) \leq V_\infty(n, x_0) \qquad (6.1)$$

for all $n \in \mathbb{N}_0$, all $N \in \mathbb{N}$ and all $x_0 \in \mathbb{X}$. This inequality follows since minimization is carried out with the same constraints for all N (including ∞) and the stage cost ℓ is nonnegative. Hence, J_N is increasing in N, which carries over to V_N.

For this reason, the arguments of the last section in which we used the terminal conditions in order to reverse the inequalities in (6.1) do not work anymore.

Nevertheless, NMPC without terminal conditions works, as the following two simple examples show, in which α in (5.1) can be computed explicitly.

Example 6.1 Consider again Example 5.6, i.e.,

$$x^+ = x + u, \quad \ell(x, u) = x^2 + u^2$$

with $\mathbb{X} = X = \mathbb{U} = U = \mathbb{R}$. Here, we get

$$V_1(x_0) = \inf_{u \in \mathbb{R}} \sum_{k=0}^{0} \ell(x_u(k, x_0), u(k)) = \inf_{u \in \mathbb{R}} x_0^2 + u^2 = x_0^2.$$

and, by (3.15) for $N = 2$ and $K = 1$,

$$\begin{aligned} V_2(x_0) &= \inf_{u(\cdot)\in\mathbb{U}^1(x_0)} \{\ell(x_u(0,x_0),u(0)) + V_1(x_u(1,x_0))\} \\ &= \inf_{u\in U}\{x_0^2 + u^2 + (x_0+u)^2\} \;=\; \inf_{u\in U}\{2x_0^2 + 2u^2 + 2x_0u\} \end{aligned}$$

The minimum of this expression is attained at $u = -x_0/2$ which by (3.19) implies $\mu_2(x_0) = -x_0/2$. The resulting minimum is

$$V_2(x_0) = \frac{3}{2}x_0^2.$$

Thus, (5.1) for $N = 2$ becomes

$$\frac{3}{2}x^2 \geq \alpha(x^2 + x^2/4) + \frac{3}{2}(x^2/4)$$

which is satisfied for all x by $\alpha = 0.9$, i.e., (5.1) is satisfied for $N = 2$ and this α for all $x \in X = \mathbb{R}$.

Note that the closed-loop system for μ_2 is given by $x^+ = x/2$ for which asymptotic stability at $x_* = 0$ is also easily seen directly.

Example 6.2 We consider the previous Example 6.1 again, i.e.,

$$x^+ = x + u, \quad \ell(x,u) = x^2 + u^2$$

with $\mathbb{X} = X = U = \mathbb{R}$ but now we impose the control constraint $\mathbb{U}(x) \equiv \mathbb{U} = [-1,1]$. While the computation for V_1 again yields $V_1(x_0) = x_0^2$, the expression for V_2 now becomes

$$V_2(x_0) = \inf_{u\in[-1,1]}\{2x_0^2 + 2u^2 + 2x_0u\},$$

whose minimizer is

$$\mu_2(x_0) = \begin{cases} \max\{-x_0/2,\ -1\}, & x \geq 0 \\ \min\{x_0/2,\ 1\}, & x < 0. \end{cases}$$

Consequently, V_2 becomes

$$V_2(x_0) = \begin{cases} \frac{3}{2}x_0^2, & x_0 \in [-2,2] \\ 2(x_0^2 - |x_0| + 1), & x_0 \notin [-2,2] \end{cases}$$

For $|x| \leq 2$, inequality (5.1) for $N = 2$ is as in the previous example and is thus satisfied for $\alpha = 0.9$. For $|x| \in (2,3)$, (5.1) becomes

$$2(x_0^2 - |x_0| + 1) \geq \alpha(x^2 + 1) + \frac{3}{2}(x-1)^2$$

which is satisfied for all $|x| \in (2, 3)$ with $\alpha = 0.8$. For $|x| > 3$, we get

$$2(x_0^2 - |x_0| + 1) \geq \alpha(x^2 + 1) + 2(x_0^2 - 3|x_0| + 3)$$

which is satisfied for $\alpha = 4(|x| - 1)/(x^2 + 1)$.

Hence, if we restrict the state space $\mathbb{X} = \mathbb{R}$ to $\mathbb{X} = [-a, a]$ for some $a > 0$ (note that each such $\mathbb{X}$ is forward invariant under the closed-loop system), then we always find $\alpha \in (0, 1)$ such that (5.1) holds for all $x \in \mathbb{X} = [-a, a]$. Consequently, the feedback μ_2 asymptotically stabilizes the system at $x_* = 0$ on each set of the form $\mathbb{X} = [-a, a]$ and thus also globally.

The last example shows one of the advantages of NMPC without stabilizing terminal conditions over the constrained approach from the previous chapter: Since in this example $|f(x, u) - x| \leq 1$ holds for each $x \in \mathbb{R}$ and each $u \in \mathbb{U} = [-1, 1]$, using a terminal constraint of the form $\mathbb{X}_0 = [-\varepsilon, \varepsilon]$ implies that the feasible sets are given by $\mathbb{X}_N = [-N - \varepsilon, N + \varepsilon]$. Hence, in order to compute a controller defined on a large set $\mathbb{X}_N$, a large horizon length N is needed. In contrast to this, the NMPC-feedback μ_N without stabilizing terminal constraints globally asymptotically stabilizes the system already for $N = 2$.

Further advantages of NMPC without stabilizing terminal conditions are that no Lyapunov function terminal cost F has to be computed in advance and that no additional constraints have to be added to the optimization problem; a detailed comparative discussion of schemes with and without stabilizing terminal constraints is given in Sect. 7.4. For these reasons, stabilizing terminal conditions are often avoided in practice.

A rigorous proof for the fact that one can obtain asymptotic stability without imposing stabilizing terminal conditions was—to the best of our knowledge—first given by Alamir and Bornard in [1] for nonlinear discrete time systems whose linearization with respect to u satisfies a specific rank condition and for quadratic stage costs ℓ. Ten years later, this result was extended by Jadbabaie and Hauser in [13] (for continuous time systems) and by Grimm et al. in [9] (for discrete time systems) to systems without any rank conditions and to arbitrary positive definite costs using an exponential controllability condition in [13] and a bound on the finite horizon optimal value function V_N in [9]. The proofs in these references exploit that for $N \to \infty$ the open-loop optimal trajectories converge to a region in which ℓ is small. This fact is either used directly as in [9] or indirectly by exploiting that it implies the convergence $V_N - V_{N-1} \to 0$ for $N \to \infty$ as in [1, 13]. This convergence was also used by Grüne and Rantzer in [12] in order to estimate α in (5.1) from suitable bounds on V_N, which in turn can be guaranteed by an appropriate controllability condition.

Even though [12] showed that the convergence $V_N - V_{N-1} \to 0$ for $N \to \infty$ can be used in order to estimate α in (5.1), in this book we present a different approach in order to estimate α which we consider advantageous, because it uses the available information—either from bounds on the optimal value functions or from a controllability condition—in a more efficient way; a discussion on this fact is given in

Sect. 6.9. In particular, we will not try to establish (5.1) from the fact that $V_N - V_{N-1}$ becomes small. Rather, we will use a direct argument based on properties of optimal trajectories in order to derive an upper bound for $V_N(n+1, f(x, \mu_N(n, x)))$ in (5.1). An argument of this type was first used by Tuna, Messina and Teel [19] for NMPC with terminal cost (but without terminal constraints) and we extend this argument to the case without terminal costs. Moreover, we formulate an optimization problem in order to get the best possible estimate for α in (5.1) from this approach. As we will see, proceeding this way leads to a tight characterization of α and—under suitable conditions—to an explicit formula relating α to the bounds on the value functions or to the parameters in the controllability condition.

Following the structure from the previous chapter, we will first present our concepts and results for the time-invariant case of Algorithm 3.1 and then discuss the extensions to the time varying case of Algorithm 3.7 at the end of Sect. 6.5.

6.2 Bounds on V_N and Asymptotic Controllability with Respect to ℓ

The central assumption we will use in order to ensure asymptotic stability and performance bounds imposes upper bounds on the optimal value functions V_N. These bounds are formulated relative to the stage cost ℓ. To this end, we define

$$\ell^*(x) := \inf_{u \in U} \ell(x, u). \tag{6.2}$$

With this notation, we can formulate our central assumption.

Assumption 6.3 (*Bounds on V_N*) Consider the optimal control problem (OCP$_\text{N}$). We assume that there exist functions $B_K \in \mathcal{K}_\infty$, $K \in \mathbb{N}$ such that for each $x \in \mathbb{X}$ the inequality

$$V_K(x) \leq B_K(\ell^*(x)) \tag{6.3}$$

holds for all $K \in \mathbb{N}$.

We observe that $V_K(x) \geq \ell(x, u^\star(0)) \geq \ell^*(x)$ implies $B_K(r) \geq r$.

Before we state consequences from this assumption in the next section, we discuss a sufficient controllability condition which ensures Assumption 6.3. To this end, we first slightly enlarge the class of $\mathcal{KL}$-functions introduced in Definition 2.13.

Definition 6.4 We say that a continuous function $\beta : \mathbb{R}_{\geq 0} \times \mathbb{R}_{\geq 0} \to \mathbb{R}_{\geq 0}$ is of class $\mathcal{KL}_0$ if for each $r > 0$ we have $\lim_{t\to\infty} \beta(r, t) = 0$ and for each $t \geq 0$ we either have $\beta(\cdot, t) \in \mathcal{K}_\infty$ or $\beta(\cdot, t) \equiv 0$.

Compared to the class $\mathcal{KL}$, here we do not assume monotonicity in the second argument and we allow for $\beta(\cdot, t)$ being identically zero for some t. This allows for tighter bounds for the actual controllability behavior of the system. It is, however, easy to see that each $\beta \in \mathcal{KL}_0$ can be overbounded by a $\tilde{\beta} \in \mathcal{KL}$, e.g., by setting $\tilde{\beta}(r,t) = \max_{\tau \geq t} \beta(r,\tau) + e^{-t}r$. Using the $\mathcal{KL}_0$ functions we now formulate our controllability assumption.

Assumption 6.5 (*Asymptotic controllability wrt.* ℓ) Consider the optimal control problem ($\mathrm{OCP_N}$). We assume that the system is *asymptotically controllable with respect to* ℓ *with rate* $\beta \in \mathcal{KL}_0$, i.e., for each $x \in \mathbb{X}$ and each $N \in \mathbb{N}$ there exists an admissible control sequence $u_x \in \mathbb{U}^N(x)$ satisfying

$$\ell(x_{u_x}(n,x), u_x(n)) \leq \beta(\ell^*(x), n)$$

for all $n \in \{0, \dots, N-1\}$.

Special cases for $\beta \in \mathcal{KL}_0$ are

$$\beta(r,n) = C\sigma^n r \tag{6.4}$$

for real constants $C \geq 1$ and $\sigma \in (0,1)$, i.e., *exponential controllability*, and

$$\beta(r,n) = c_n r \tag{6.5}$$

for some real sequence $(c_n)_{n\in\mathbb{N}_0}$ with $c_n \geq 0$ and $c_n = 0$ for all $n \geq n_0$, i.e., *finite time controllability* with linear overshoot bound.

The following lemma links Assumptions 6.5 and 6.3.

Lemma 6.6 *If Assumption 6.5 holds then Assumption 6.3 holds. More precisely, for each* $K \in \mathbb{N}$ *and each* $x \in \mathbb{X}$ *the inequality*

$$V_K(x) \leq J_K(x, u_x) \leq B_K(\ell^*(x)) \tag{6.6}$$

holds for u_x *from Assumption 6.5 and*

$$B_K(r) := \sum_{n=0}^{K-1} \beta(r,n). \tag{6.7}$$

Proof The inequality follows immediately from

$$\begin{aligned} V_K(x) \leq J_K(x,u_x) &= \sum_{n=0}^{K-1} \ell(x(n,u_x), u_x(n)) \\ &\leq \sum_{n=0}^{K-1} \beta(\ell^*(x), n) = B_K(\ell^*(x)). \end{aligned}$$

□

In the special case (6.4) the values B_K, $K \in \mathbb{N}$, evaluate to

$$B_K(r) = C\frac{1-\sigma^K}{1-\sigma}r$$

while for (6.5) we obtain

$$B_K(r) = C_K r \quad \text{with} \quad C_K = \sum_{j=0}^{\min\{n_0, K-1\}} c_j.$$

It should be noted that Lemma 6.6 does not necessarily lead to the best bound on V_K in Assumption 6.3. A setting in which additional information leads to significantly better bounds can be found in Sect. 10.5.

It is easily seen that if the state trajectories itself are exponentially controllable to some equilibrium x_* then exponential controllability, i.e., Assumption 6.5 with β from (6.4), holds if ℓ has polynomial growth. In particular, this covers the usual linear-quadratic setting for stabilizable systems.

However, even if the system itself is not exponentially controllable, exponential controllability in the sense of Assumption 6.5 can be achieved by proper choice of ℓ, as the following example shows.

Example 6.7 Consider the control system

$$x^+ = x + ux^3$$

with $\mathbb{X} = [-1, 1]$ and $U = [-1, 1]$. The system is controllable to $x_* = 0$, which can be seen by choosing $u = -1$. This results in the system $x^+ = x - x^3$ whose solutions approach $x_* = 0$ monotonically for $x_0 \in \mathbb{X}$.

However, the system is not exponentially controllable to 0: exponential controllability would mean that there exist constants $C > 0$, $\sigma \in (0, 1)$ such that for each $x \in \mathbb{X}$ there is $u_x \in \mathbb{U}^\infty(x)$ with

$$|x_{u_x}(n, x)| \leq C\sigma^n |x|.$$

This implies that by choosing $n^* > 0$ so large such that $C\sigma^{n^*} \leq 1/2$ holds the inequality

$$|x_{u_x}(n^*, x)| \leq |x|/2 \tag{6.8}$$

must hold for each $x \in \mathbb{X}$. However, for each $x \geq 0$ the restriction $u \in [-1, 1]$ implies $x^+ \geq x - x^3 = (1 - x^2)x$ which by induction yields

$$x_u(n^*, x) \geq (1 - x^2)^{n^*} x$$

for all $u \in \mathbb{U}^\infty(x)$ which contradicts (6.8) for $x < 1 - 2^{-1/n^*}$.

On the other hand, since $|x| \leq 1$ we obtain $(1-x^2)^2(2x^2+1) = 1+2x^6-3x^4 \leq 1$ which implies

$$\frac{1}{(1-x^2)^2} \geq 2x^2+1 \;\Rightarrow\; -\frac{1}{2x^2(1-x^2)^2} \leq -\frac{2x^2+1}{2x^2} = -1 - \frac{1}{2x^2}.$$

Hence, choosing

$$\ell(x,u) = \ell(x) = e^{-\frac{1}{2x^2}},$$

for $u \equiv -1$ we obtain

$$\ell(x^+) = \ell(x - x^3) = e^{-\frac{1}{2(x-x^3)^2}} = e^{-\frac{1}{2x^2(1-x^2)^2}} \leq e^{-1}e^{-\frac{1}{2x^2}} = e^{-1}\ell(x).$$

By induction this implies Assumption 6.5 with β from (6.4) with $C = 1$ and $\sigma = e^{-1}$.

For certain results it will be useful that β in Assumption 6.5 has the property

$$\beta(r, n+m) \leq \beta(\beta(r,n), m) \quad \text{for all } r \geq 0, n, m \in \mathbb{N}_0. \tag{6.9}$$

Inequality (6.9), often referred to as *submultiplicativity*, ensures that any sequence of the form $b_n = \beta(r,n)$, $r > 0$, also fulfills $b_{n+m} \leq \beta(b_n, m)$. It is, for instance, always satisfied in case (6.4) and satisfied in case (6.5) if $c_{n+m} \leq c_n c_m$. If needed, this property can be assumed without loss of generality, because by Sontag's $\mathcal{KL}$-Lemma [18, Proposition 7] (cf. also the discussion before Theorem 4.3) the function β in Assumption 6.5 can be replaced by a β of the form $\beta(r,t) = \alpha_1(\alpha_2(r)e^{-t})$ for $\alpha_1, \alpha_2 \in \mathcal{K}_\infty$. Then, (6.9) is easily verified if $\alpha_2 \circ \alpha_1(r) \geq r$, which is equivalent to $\alpha_1 \circ \alpha_2(r) \geq r$, which in turn is a necessary condition for Assumption 6.5 to hold for $n = 0$ and $\beta(r,t) = \alpha_1(\alpha_2(r)e^{-t})$.

Remark 6.8 Computing β satisfying Assumption 6.5 is in general a hard task for nonlinear systems. One way to obtain such a β is via a suitable control Lyapunov function, similar to the procedure described in Khalil [14, Sect 4.4] or used in Nešić and Teel [15, Proof of Proposition 1]. However, as we will see later, the precise knowledge of β is not necessarily needed in order to apply our results, because we will be able to identify structural properties of β which guarantee good performance of the NMPC closed loop, cf. Sect. 6.6.

Remark 6.9 Note that Assumption 6.5 is an assumption in discrete time. In the case of sampled data systems with zero order hold, this implies that the discrete time system obtained from (2.8) with constant control function $u : [0,T] \to \mathbb{R}^m$ needs to satisfy this assumption. Due to the fact that this system corresponds to the continuous time system (2.6) with control functions $v : \mathbb{R} \to \mathbb{R}^m$ which are constant on the sampling intervals, this is in general a stronger assumption than requiring (2.6) to be asymptotically controllable with measurable control functions $v \in L^\infty(\mathbb{R}, \mathbb{R}^m)$, cf. also Remark 2.9.

6.3 Implications of the Bound on V_N

In this section, we will use the bound on the V_N from Assumption 6.3 in order to establish two lemmas which yield bounds for optimal value functions and functionals along pieces of optimal trajectories. In the subsequent section, these bounds will then be used for the calculation of α in (5.1).

In order to be able to calculate α in (5.1), we will need an upper bound for $V_N(f(x, \mu_N(x)))$. To this end, recall from Step (3) of Algorithm 3.1 that $\mu_N(x_0)$ is the first element of the optimal control sequence $u^\star(\cdot)$ for (OCP_N) with initial value x_0. In particular, this implies $f(x_0, \mu_N(x_0)) = x_{u^\star}(1, x_0)$. Hence, if we want to derive an upper bound for $V_N(f(x_0, \mu_N(x_0)))$ then we can alternatively derive an upper bound for $V_N(x_{u^\star}(1, x_0))$. This will be done in the following lemma.

Lemma 6.10 *Suppose Assumption 6.3 holds and consider $x_0 \in \mathbb{X}$ and an optimal control $u^\star \in \mathbb{U}^N(x_0)$ for* (OCP_N). *Then, for each $j = 0, \ldots, N-2$ the inequality*

$$V_N(x_{u^\star}(1, x_0)) \leq J_j(x_{u^\star}(1, x_0)), u^\star(1 + \cdot)) + B_{N-j}(\ell^*(x_{u^\star}(1 + j, x_0)))$$

holds for B_K from (6.3).

Proof We define the control sequence

$$\tilde{u}(n) = \begin{cases} u^\star(1 + n), \ n \in \{0, \ldots, j-1\} \\ u_x(n-j), \ n \in \{j, \ldots, N-1\}, \end{cases}$$

where u_x is an optimal control for initial value $x = x_{u^\star}(1 + j, x_0)$ and $N = N - j$. By construction, this control sequence is admissible for $x_{u^\star}(1, x_0)$ and we obtain

$$\begin{aligned} V_N(x_{u^\star}(1, x_0)) &\leq J(x_{u^\star}(1, x_0), \tilde{u}) \\ &= J_j(x_{u^\star}(1, x_0), u^\star(1 + \cdot)) + J_{N-j}(x_{u^\star}(1 + j, x_0), u_x) \\ &\leq J_j(x_{u^\star}(1, x_0), u^\star(1 + \cdot)) + B_{N-j}(\ell^*(x_{u^\star}(1 + j, x_0))) \end{aligned}$$

where we used $J_{N-j}(x_{u^\star}(1 + j, x_0), u_x) = V_{N-j}(x_{u^\star}(1 + j, x_0))$ and Assumption 6.3 in the last step. This is the desired inequality. □

In words, the idea of this proof is as follows. The upper bound for each $j \in \{0, \ldots, N-2\}$ is obtained from a specific trajectory. We follow the optimal trajectory for initial value x_0 and horizon N for j steps and for the point x reached this way we use the optimal control sequence for initial value x and horizon $N - j$ for another $N - j$ steps.

In the next lemma, we derive upper bounds for the J_k-terms along tails of the optimal trajectory $x_{u^\star}$, which will later be used in order to bound the right hand side of the inequality from Lemma 6.10. To this end we use that these tails are optimal trajectories themselves.

Lemma 6.11 *Suppose Assumption 6.3 holds and consider* $x_0 \in \mathbb{X}$ *and an optimal control* $u^\star \in \mathbb{U}^N(x)$ *for* (OCP$_\mathrm{N}$)*. Then for each* $k = 0, \ldots, N-1$ *the inequality*

$$J_{N-k}(x_{u^\star}(k, x_0), u^\star(k + \cdot)) \le B_{N-k}(\ell^*(x_{u^\star}(k, x_0)))$$

holds for B_K *from* (6.3).

Proof Corollary 3.16 implies $J_{N-k}(x_{u^\star}(k, x_0), u^\star(k + \cdot)) = V_{N-k}(x_{u^\star}(k, x_0))$. Hence, the assertion follows immediately from Assumption 6.3. □

Remark 6.12 Since $u^\star \in \mathbb{U}^N(x_0)$ we obtain $x_{u^\star}(k, x_0) \in \mathbb{X}$ for $k = 0, \ldots, N$. For $k = 0, \ldots, N-1$ this property is crucial for the proof of Lemma 6.10 because it ensures that an optimal control for initial value $x = x_{u^\star}(1 + j, x_0)$ exists. Note, however, that we do not need $x_{u^\star}(N, x_0) \in \mathbb{X}$. In fact, all results in this and the ensuing sections remain true if we remove the state constraint $x_{u^\star}(N, x_0) \in \mathbb{X}$ from the definition of $\mathbb{U}^N(x_0)$ or replace it by some weaker constraint.

6.4 Computation of α

We will now use the inequalities derived in the previous section in order to compute α for which (5.1) holds for all $x \in \mathbb{X}$. When trying to put together these inequalities in order to bound $V_N(x_{u^\star}(1, x_0))$ from above, one notices that the functionals in Lemmas 6.10 and 6.11 are not exactly the same. Hence, in order to combine these results into a closed form which is suitable for computing α we need to look at the single terms of the stage cost ℓ contained in these functionals.

To this end, let $u^\star$ be an optimal control for (OCP$_\mathrm{N}$) with initial value $x_0 = x$. Then from the definition of V_N and μ_N it follows that (5.1) is equivalent to

$$\sum_{k=0}^{N-1} \ell(x_{u^\star}(k, x), u^\star(k)) \ge \alpha \ell(x, u^\star(0)) + V_N(x_{u^\star}(1, x)). \tag{6.10}$$

Thus, in order to compute α for which (5.1) holds for all $x \in \mathbb{X}$ we can equivalently compute α for which (6.10) holds for all optimal trajectories $x_{u^\star}(\cdot, x)$ with initial values $x \in \mathbb{X}$.

For this purpose, we now consider arbitrary real values $\lambda_0, \ldots, \lambda_{N-1}, \nu \ge 0$ and start by deriving necessary conditions which hold if these values coincide with the cost along an optimal trajectory $\ell(x_{u^\star}(k, x), u^\star(k))$ and an optimal value $V_N(x_{u^\star}(1, x))$, respectively.

Proposition 6.13 *Suppose Assumption 6.3 holds and consider* $N \ge 1$*, values* $\lambda_n \ge 0$*,* $n = 0, \ldots, N-1$*, and a value* $\nu \ge 0$*. Consider* $x \in \mathbb{X}$ *and assume that there exists an optimal control sequence* $u^\star \in \mathbb{U}^N(x)$ *for* (OCP$_\mathrm{N}$) *such that*

$$\lambda_k = \ell(x_{u^\star}(k, x), u^\star(k)), \quad k = 0, \ldots, N-1$$

holds. Then,

$$\sum_{n=k}^{N-1} \lambda_n \leq B_{N-k}(\lambda_k), \quad k = 0, \ldots, N-2 \tag{6.11}$$

holds. If, furthermore,

$$\nu = V_N(x_{u^\star}(1, x))$$

holds then

$$\nu \leq \sum_{n=0}^{j-1} \lambda_{n+1} + B_{N-j}(\lambda_{j+1}), \quad j = 0, \ldots, N-2 \tag{6.12}$$

holds.

Proof If the stated conditions hold, then λ_n and ν must meet the inequalities given in Lemmas 6.10 and 6.11, which is exactly (6.12) and (6.11). □

Using this proposition, we can give a sufficient condition for (6.10) and thus for (5.1). The idea behind the following proposition is to express the terms in inequality (6.10) using the values $\lambda_0, \ldots, \lambda_{N-1}$ and ν introduced above.

Proposition 6.14 *Consider $N \geq 1$ and $B_K \in \mathcal{K}_\infty$, $K = 2, \ldots, N$ and assume that all values $\lambda_n \geq 0$, $n = 0, \ldots, N-1$ and $\nu \geq 0$ fulfilling* (6.11) *and* (6.12) *satisfy the inequality*

$$\sum_{n=0}^{N-1} \lambda_n - \nu \geq \alpha \lambda_0 \tag{6.13}$$

for some $\alpha \in (0, 1]$. Then for this α and each optimal control problem ($\mathrm{OCP_N}$) *satisfying Assumption 6.3 inequality* (5.1) *holds for μ_N from Algorithm 3.1 and all $x \in \mathbb{X}$.*

Proof Consider a control system satisfying Assumption 6.3 and an optimal control sequence $u^\star \in \mathbb{U}^N(x)$ for initial value $x \in \mathbb{X}$. Then, by Proposition 6.13 the values $\lambda_k = \ell(x_{u^\star}(k, x), u^\star(k))$ and $\nu = V_N(x_{u^\star}(1, x))$ satisfy (6.11) and (6.12), hence by assumption also (6.13). Thus, using $\ell(x, u^\star(0)) = \ell(x_{u^\star}(0, x), u^\star(0)) = \lambda_0$ we obtain

$$V_N(x_{u^\star}(1, x)) + \alpha \ell(x, u^\star(0)) = \nu + \alpha \lambda_0 \quad \leq \quad \sum_{k=0}^{N-1} \lambda_k = \sum_{k=0}^{N-1} \ell(x_{u^\star}(k, x), u^\star(k)).$$

This proves (6.10) and thus also (5.1). □

Proposition 6.14 is the basis for computing α as specified in the following theorem.

Theorem 6.15 (Abstract optimization problem) *Consider $N \geq 1$ and $B_K \in \mathcal{K}_\infty$, $K = 2, \ldots, N$ and assume that the optimization problem*

$$\begin{aligned} \alpha := \inf_{\lambda_0,\ldots,\lambda_{N-1},\nu} & \frac{\sum_{n=0}^{N-1}\lambda_n - \nu}{\lambda_0} \\ & \textit{subject to the constraints } (6.11), (6.12), \textit{ and} \\ & \lambda_0 > 0, \lambda_1, \ldots, \lambda_{N-1}, \nu \geq 0 \end{aligned} \tag{6.14}$$

has an optimal value $\alpha \in (0, 1]$. *Then for this* α *and each optimal control problem* (OCP$_\mathrm{N}$) *satisfying Assumption 6.3 inequality* (5.1) *holds for* μ_N *from Algorithm 3.1 and all* $x \in \mathbb{X}$.

Proof Consider arbitrary values $\lambda_0, \ldots, \lambda_{N-1}, \nu \geq 0$ satisfying (6.11) and (6.12).
If $\lambda_0 > 0$ then the definition of Problem (6.14) immediately implies (6.13).
If $\lambda_0 = 0$, then inequality (6.11) for $k = 0$ together with $B_K(0) = 0$ implies $\lambda_1, \ldots, \lambda_{N-1} = 0$. Thus, (6.12) for $j = 1$ yields $\nu = 0$ and again (6.13) holds.
Hence, (6.13) holds in both cases and Proposition 6.14 yields the assertion. □

Remark 6.16 (i) Theorem 6.15 shows Inequality (5.1) for all $x \in \mathbb{X}$ if Assumption 6.3 or, alternatively, Assumption 6.5 holds for all $x \in \mathbb{X}$ and $K = 2, \ldots, N$.

If we want to establish Inequality (5.1) only for states $x_0 \in Y$ for a subset $Y \subset \mathbb{X}$, then from the proofs of the Lemmas 6.10 and 6.11 it follows that Proposition 6.13 holds for all $x_0 \in Y$ (instead of for all $x \in \mathbb{X}$) under the following condition:

$$\begin{aligned} & (6.3) \text{ holds for } x = x_{u^\star}(k, x_0) \text{ for all } k = 0, \ldots, N-1, \text{ all } x_0 \in Y \\ & \text{and all } K = 2, \ldots, N, \text{ where } u^\star \text{ is the optimal control for } J_N(x_0, u). \end{aligned} \tag{6.15}$$

This implies that under condition (6.15) Theorem 6.15 holds for all $x_0 \in Y$ and consequently (5.1) holds for all $x_0 \in Y$.

(ii) A further relaxation of the assumptions of Theorem 6.15 can be obtained by observing that if we are interested in establishing Inequality (5.1) only for states $x_0 \in Y$, then in (6.14) we only need to optimize over those λ_i which correspond to optimal trajectories starting in Y. In particular, if we know that $\inf_{x_0 \in Y} \ell^*(x_0) \geq \zeta$ for some $\zeta > 0$, then the constraint $\lambda_0 > 0$ can be tightened to $\lambda_0 \geq \zeta$.

The following lemma shows that the optimization problem (6.14) specializes to a linear program if the functions $B_K(r)$ are linear in r.

Lemma 6.17 *If the functions* $B_K(r)$ *from* (6.3) *in the constraints* (6.11), (6.12) *are linear in* r, *then* α *from Problem* (6.14) *coincides with*

$$\begin{aligned} \alpha := \min_{\lambda_0,\ldots,\lambda_{N-1},\nu} & \sum_{n=0}^{N-1} \lambda_n - \nu \\ & \textit{subject to the (now linear) constraints} (6.11), (6.12), \textit{ and} \\ & \lambda_0 = 1, \lambda_1, \ldots, \lambda_{N-1}, \nu \geq 0. \end{aligned} \tag{6.16}$$

In particular, this holds if Assumption 6.5 holds with functions $\beta(r, t)$ *being linear in* r.

Proof Due to the linearity, all sequences $\bar{\lambda}_0, \ldots, \bar{\lambda}_{N-1}, \bar{\nu}$ satisfying the constraints in (6.14) can be written as $\gamma\lambda_0, \ldots, \gamma\lambda_{N-1}, \gamma\nu$ for some $\lambda_0, \ldots, \lambda_{N-1}, \nu$ satisfying the constraints in (6.16), where $\gamma = 1/\bar{\lambda}_0$. Since

$$\frac{\sum_{n=0}^{N-1} \bar{\lambda}_n - \bar{\nu}}{\bar{\lambda}_0} = \frac{\sum_{n=0}^{N-1} \gamma\lambda_n - \gamma\nu}{\gamma\lambda_0} = \frac{\sum_{n=0}^{N-1} \lambda_n - \nu}{\lambda_0} = \sum_{n=0}^{N-1} \lambda_n - \nu,$$

the values α in Problems (6.14) and (6.16) coincide. □

The next result gives an explicit solution of Problem (6.16) and thus also (6.14) if the functions B_K are linear.

Proposition 6.18 *If the functions $B_K(r)$ from* (6.3) *in the constraints* (6.11), (6.12) *are linear in r, then the solution of Problems* (6.14) *and* (6.16) *satisfies the inequality*

$$\alpha \geq \underline{\alpha}_N \tag{6.17}$$

for

$$\underline{\alpha}_N := 1 - \frac{(\gamma_N - 1)\prod_{k=2}^{N}(\gamma_k - 1)}{\prod_{k=2}^{N}\gamma_k - \prod_{k=2}^{N}(\gamma_k - 1)} \quad \textit{with } \gamma_k = B_k(r)/r, \tag{6.18}$$

where γ_k is well defined by linearity of B_k.

If, in addition, the functions B_K are of the form (6.7) *with $\beta \in \mathcal{KL}_0$ being linear in its second argument and satisfying* (6.9), *then equality holds in* (6.17), *i.e.,*

$$\alpha = \underline{\alpha}_N. \tag{6.19}$$

Proof The rather technical proof of this proposition can be found in Sect. 6.8.

In the special cases of exponential controllability (6.4) and finite time controllability (6.5) we get

$$\gamma_k = C\frac{1-\sigma^k}{1-\sigma} \quad \text{and} \quad \gamma_k = C_k = \sum_{j=0}^{\min\{n_0, k-1\}} c_j,$$

respectively. We will further investigate Formula (6.18) in Sect. 6.6.

Due to the optimization-based approach leading to Proposition 6.18, Eq. (6.18) yields a very good (and under suitable conditions even tight, cf. Theorem 6.27) estimate for α in (5.1). The price for this is a very involved proof, cf. Sect. 6.8. The following, conceptually similar version of Proposition 6.18, yields a weaker estimate but allows for a much more simple proof.

Proposition 6.19 *If the functions $B_K(r)$ from (6.3) in the constraints (6.11), (6.12) are linear in r, then the solution of Problems (6.14) and (6.16) satisfies the inequality*

$$\alpha \geq \tilde{\alpha}_N \tag{6.20}$$

for

$$\tilde{\alpha}_N := 1 - (\gamma_2 - 1)(\gamma_N - 1) \prod_{k=2}^{N-1} \left(\frac{\gamma_k - 1}{\gamma_k} \right) \quad \text{with } \gamma_k = B_k(r)/r. \tag{6.21}$$

Proof We prove the theorem by showing the inequality

$$\lambda_{N-1} \leq (\gamma_N - 1) \prod_{k=2}^{N-1} \left(\frac{\gamma_k - 1}{\gamma_k} \right) \lambda_0 \tag{6.22}$$

for all feasible $\lambda_0, \ldots, \lambda_{N-1}$. From this (6.20) follows since (6.12) with $j = N - 2$ implies

$$\nu \leq \sum_{n=1}^{N-2} \lambda_n + \gamma_2 \lambda_{N-1}$$

and thus (6.22), $\gamma_2 \geq 1$ and $\lambda_0 = 1$ yield

$$\sum_{n=0}^{N-1} \lambda_n - \nu \geq \lambda_0 + (1 - \gamma_2)\lambda_{N-1} \geq \lambda_0 - (\gamma_2 - 1)(\gamma_N - 1) \prod_{k=2}^{N-1} \left(\frac{\gamma_k - 1}{\gamma_k} \right) \lambda_0 = \tilde{\alpha}_N$$

for all feasible $\lambda_1, \ldots, \lambda_{N-1}$ and ν, which yields $\alpha \geq \tilde{\alpha}_N$.

In order to prove (6.22), we start by observing that (6.11) with $j = p$ implies

$$\sum_{k=p+1}^{N-1} \lambda_k \leq (\gamma_{N-p} - 1)\lambda_p \tag{6.23}$$

for $p = 0, \ldots, N - 2$. From this we can conclude

$$\lambda_p + \sum_{k=p+1}^{N-1} \lambda_k \geq \frac{\sum_{k=p+1}^{N-1} \lambda_k}{\gamma_{N-p} - 1} + \sum_{k=p+1}^{N-1} \lambda_k = \frac{\gamma_{N-p}}{\gamma_{N-p} - 1} \sum_{k=p+1}^{N-1} \lambda_k.$$

Using this inequality inductively for $p = 1, \ldots, N - 2$ yields

$$\sum_{k=1}^{N-1} \lambda_k \geq \prod_{k=1}^{N-2} \left(\frac{\gamma_{N-k}}{\gamma_{N-k} - 1} \right) \lambda_{N-1} = \prod_{k=2}^{N-1} \left(\frac{\gamma_k}{\gamma_k - 1} \right) \lambda_{N-1}.$$

Using (6.23) for $p = 0$ we then obtain

$$(\gamma_N - 1)\lambda_0 \geq \sum_{k=1}^{N-1} \lambda_k \geq \prod_{k=2}^{N-1} \left(\frac{\gamma_k}{\gamma_k - 1}\right) \lambda_{N-1}$$

which implies (6.22). □

A comparison of the two formulas (6.19) and (6.20) can be found in Remark 6.22, below.

6.5 Main Stability and Performance Results

We are now ready to state our main result on stability and performance of the basic NMPC Algorithm 3.1 without stabilizing terminal conditions. In this section, we deal with global asymptotic stability, i.e., asymptotic stability on the whole state constraint set $\mathbb{X}$. Further results on semiglobal and practical asymptotic stability will be provided in Sect. 6.7.

Theorem 6.20 (Stability without terminal conditions) *Consider the NMPC Algorithm 3.1 with optimization horizon $N \in \mathbb{N}$ and stage cost ℓ satisfying $\alpha_3(|x|_{x_*}) \leq \ell^*(x) \leq \alpha_4(|x|_{x_*})$ for suitable $\alpha_3, \alpha_4 \in \mathcal{K}_\infty$. Suppose that Assumption 6.3 holds and that α from Theorem 6.15 satisfies $\alpha \in (0, 1]$. Then the nominal NMPC closed-loop system* (3.5) *with NMPC-feedback law μ_N is asymptotically stable on $\mathbb{X}$.*

In addition, the inequality

$$J_\infty^{cl}(x, \mu_N) \leq V_N(x)/\alpha \leq V_\infty(x)/\alpha$$

holds for each $x \in \mathbb{X}$.

Proof The assertion follows readily from Theorem 4.11 and Inequality (6.1) if we prove the inequalities (5.1) and (5.2). Inequality (5.1) follows directly from Theorem 6.15.

Regarding (5.2), observe that the inequality for ℓ follows immediately from our assumptions. From the definition of V_N, we get

$$V_N(x) = \inf_{u \in \mathbb{U}^N(x)} J_N(x, u) \geq \inf_{u \in \mathbb{U}^N(x)} \ell(x, u(0)) = \ell^*(x) \geq \alpha_3(|x|_{x_*}),$$

thus the lower inequality for V_N follows with $\alpha_1 = \alpha_3$. The upper inequality in (5.2) follows from Assumption 6.3 and the upper bound on ℓ^* via

$$V_N(x) \leq B_N(\ell^*(x)) \leq B_N(\alpha_4(|x|_{x_*})),$$

i.e., for $\alpha_2 = B_N \circ \alpha_4$. □

The next corollary is an immediate consequence of Theorem 6.20.

Corollary 6.21 *Consider the NMPC Algorithm 3.1 with optimization horizon $N \in \mathbb{N}$ and stage cost ℓ satisfying $\alpha_3(|x|_{x_*}) \le \ell^*(x) \le \alpha_4(|x|_{x_*})$ for suitable $\alpha_3, \alpha_4 \in \mathcal{K}_\infty$. Suppose that Assumption 6.3 holds for linear functions $B_K \in \mathcal{K}_\infty$ and that $\alpha = \underline{\alpha}_N$ from Formula* (6.18) *satisfies $\alpha \in (0, 1]$. Then the nominal NMPC closed-loop system* (3.5) *with NMPC-feedback law μ_N is asymptotically stable on $\mathbb{X}$.*

In addition, the inequality

$$J_\infty^{cl}(x, \mu_N) \le V_N(x)/\alpha \le V_\infty(x)/\alpha$$

holds for each $x \in \mathbb{X}$.

Proof The assertion follows from Theorem 6.20 and Proposition 6.18. □

The main advantage of Corollary 6.21 over Theorem 6.20 lies in the fact that α is given explicitly by Formula (6.18) rather than implicitly by the Optimization Problem (6.14).

Remark 6.22 The essential condition for stability in Corollary 6.21 is the inequality $\underline{\alpha}_N > 0$. This could alternatively be formulated using $\tilde{\alpha}_N$ from (6.21) instead of $\underline{\alpha}_N$ from (6.18), leading to a more conservative criterion. In order to illustrate that the criterion $\tilde{\alpha}_N > 0$ is more conservative than the criterion $\underline{\alpha}_N > 0$, we consider the case where $\gamma_k = \gamma$ for all k, i.e., the γ_k are independent of k, and compute the minimal N for which $\tilde{\alpha}_N > 0$ and $\underline{\alpha}_N > 0$, respectively, hold. For $\gamma_k = \gamma$ the expressions simplify to

$$\tilde{\alpha}_N = 1 - \frac{(\gamma - 1)^N}{\gamma^{N-2}} \quad \text{and} \quad \underline{\alpha}_N = 1 - \frac{(\gamma - 1)^N}{\gamma^{N-1} - (\gamma - 1)^{N-1}}.$$

Thus, an optimization horizon N for which $\tilde{\alpha}_N > 0$ must satisfy

$$N > 2 + 2\frac{\ln(\gamma - 1)}{\ln \gamma - \ln(\gamma - 1)}$$

while the same condition for $\underline{\alpha}_N > 0$ is given by

$$N > 2 + \frac{\ln(\gamma - 1)}{\ln \gamma - \ln(\gamma - 1)}.$$

This means that the estimate for the minimal stabilizing horizon based on $\tilde{\alpha}_N$ is about twice as large as the estimate based on $\underline{\alpha}_N$.

In this context, it is interesting to look at the asymptotic behavior of the bounds on N for $\gamma \to \infty$. For large γ the denominator is approximately $1/\gamma$. This implies that asymptotically for $\gamma \to \infty$ the first estimate for N behaves like $2\gamma \ln \gamma$ while the second behaves like $\gamma \ln \gamma$.

The class of systems which is covered by Corollary 6.21 is quite large, since, e.g., exponential controllability holds on compact sets $\mathbb{X}$ whenever the linearization of f in x_* is stabilizable and ℓ is quadratic.

The following simple example illustrates the use of Corollary 6.21 for the case of a nonexponentially controllable system.

Example 6.23 We reconsider Example 6.7, i.e.,

$$x^+ = x + ux^3 \quad \text{with} \quad \ell(x, u) = e^{-\frac{1}{2x^2}}.$$

As shown in Example 6.7, Assumption 6.5 holds with $\beta(r, k) = C\sigma^k r$ with $C = 1$ and $\sigma = e^{-1}$. The bounds in Assumption 6.3 resulting from this β according to (6.7) are

$$B_K(r) = C\frac{1-\sigma^K}{1-\sigma}r = C\frac{1-e^{-K}}{1-e^{-1}}r,$$

thus Corollary 6.21 is applicable and we obtain $\alpha \geq \underline{\alpha}_N$ with $\underline{\alpha}_N$ from Formula (6.18). The γ_k in Formula (6.18) are given by

$$\gamma_k = C\frac{1-e^{-k}}{1-e^{-1}}.$$

A straightforward computation reveals that for these values, Formula (6.18) simplifies to

$$1 - \frac{(\gamma_N - 1)\prod_{k=2}^{N}(\gamma_k - 1)}{\prod_{k=2}^{N}\gamma_k - \prod_{k=2}^{N}(\gamma_k - 1)} = 1 - e^{-N}.$$

Hence, for $N = 2$ we obtain $\alpha = 1 - e^{-2} \approx 0.865$ and for $N = 3$ we get $\alpha \geq 1 - e^{-3} \approx 0.95$. Hence, Corollary 6.21 ensures asymptotic stability for all $N \geq 2$ and—since $1/0.95 \approx 1.053$—for $N = 3$ the performance of the NMPC controller is at most about 5.3 % worse than the infinite horizon controller.

While in this simple example the computation of α via Formula (6.18) is possible, in many practical examples this will not be the case. However, Formula (6.18) can still be used to obtain valuable information for the design of NMPC schemes. This aspect will be discussed in detail in Sect. 6.6.

Although the main benefit of the approach developed in this chapter compared to other approaches is that we can get rather precise quantitative estimates, it is nevertheless good to know that our approach also guarantees asymptotic stability for sufficiently large optimization horizons N under suitable assumptions. This is the statement of our final stability result.

Theorem 6.24 (Stability for sufficiently large N) *Consider the NMPC Algorithm 3.1 with optimization horizon $N \in \mathbb{N}$ and stage cost ℓ satisfying $\alpha_3(|x|_{x_*}) \leq \ell^*(x) \leq \alpha_4(|x|_{x_*})$ for suitable $\alpha_3, \alpha_4 \in \mathcal{K}_\infty$. Suppose that Assumption 6.3 holds for linear $B_K \in \mathcal{K}_\infty$ of the form $B_K(r) = \gamma_K r$ with $\gamma_\infty := \sup_{k\in\mathbb{N}} \gamma_k < \infty$.*

Then the nominal NMPC closed-loop system (3.5) with NMPC-feedback law μ_N is asymptotically stable on $\mathbb{X}$ provided N is sufficiently large.

Furthermore, for each $C > 1$ there exists $N_C > 0$ such that

$$J_\infty^{cl}(x, \mu_N) \leq CV_N(x) \leq CV_\infty(x)$$

holds for each $x \in \mathbb{X}$ and each $N \geq N_C$.

Proof The assertion follows immediately from Corollary 6.21 if we show that $\underline{\alpha}_N \to 1$ holds in (6.18) as $N \to \infty$. This property holds if and only if

$$\lim_{N\to\infty} \frac{(\gamma_N - 1)\prod_{k=2}^{N}(\gamma_k - 1)}{\prod_{k=2}^{N}\gamma_k - \prod_{k=2}^{N}(\gamma_k - 1)} = 0 \tag{6.24}$$

with γ_k defined in (6.18). Note that $\gamma_k \geq 1$ holds for all $k \in \mathbb{N}$ and that we may assume without loss of generality $\gamma_{k'} \geq \gamma_k$ for $k' > k \geq 1$, because otherwise we may replace $\gamma_{k'}$ by $\max_{k=1,\ldots,k'} \gamma_k$. This implies the convergence $\gamma_k \to \gamma_\infty$ for $k \to \infty$.

If $\gamma_k = 1$ holds for some $k \geq 2$, then we immediately get the assertion since then the expression in (6.24) equals 0 for all $N \geq j$. Thus, we may assume $\gamma_k > 1$ for all $k \geq 2$.

The inequality $\gamma_k \leq \gamma_\infty$ shows that the factor $(\gamma_N - 1)$ in (6.24) is uniformly bounded by $\gamma_\infty - 1$ for all $N \in \mathbb{N}$. The remaining factor in (6.24) can be written as

$$\frac{\prod_{k=2}^{N}(\gamma_k - 1)}{\prod_{k=2}^{N}\gamma_k - \prod_{k=2}^{N}(\gamma_k - 1)} = \frac{1}{\frac{\prod_{k=2}^{N}\gamma_k - \prod_{k=2}^{N}(\gamma_k - 1)}{\prod_{k=2}^{N}(\gamma_k - 1)}} = \frac{1}{\prod_{k=2}^{N}\frac{\gamma_k}{\gamma_k - 1} - 1}$$

Now, we pick $K > 0$ so large that $|\gamma_k - \gamma_\infty| \leq 1/2$ for all $k \geq K$. Such a K exists since $\gamma_k \to \gamma_\infty$ as $k \to \infty$. Then the estimates $\gamma_K \leq \gamma_k \leq \gamma_\infty$ for all $k \geq K \in \mathbb{N}$ imply for $N > K$

$$\prod_{k=2}^{N}\frac{\gamma_k}{\gamma_k - 1} \geq \prod_{k=2}^{K}\frac{\gamma_k}{\gamma_k - 1}\left(\frac{\gamma_K}{\gamma_\infty - 1}\right)^{N-K} \to \infty$$

as $N \to \infty$, since $\gamma_\infty - 1 < \gamma_K$. Thus,

$$\lim_{N\to\infty} \frac{\prod_{k=2}^{N}(\gamma_k - 1)}{\prod_{k=2}^{N}\gamma_k - \prod_{k=2}^{N}(\gamma_k - 1)} = \lim_{N\to\infty} \frac{1}{\prod_{k=2}^{N}\frac{\gamma_k}{\gamma_k - 1} - 1} = 0$$

which shows the claim. □

Remark 6.25 For B_K of the form (6.7), a sufficient condition for the γ_k being bounded by γ_∞ is that Assumption 6.5 holds for a $\beta \in \mathcal{KL}_0$ which is linear in its first argument and is summable, i.e.,

$$\sum_{k=0}^{\infty} \beta(r, k) < \infty \quad \text{for all } r > 0.$$

Theorem 6.24 justifies what is often done in practice: we set up an NMPC scheme using a reasonable stage cost ℓ and increase N until the closed-loop system becomes stable. While this procedure may work in many applications, it is certainly not the most sophisticated way to proceed and a clever design of ℓ may significantly improve the performance. Examples in which this is the case can be found in Sect. 6.6.

Remark 6.26 Recall from Sect. 6.1 that throughout this chapter, we use our standing assumption that $\mathbb{X}$ is viable. This property is needed in order to ensure recursive feasibility of $\mathbb{X}$, cf. the discussion after Theorem 3.5. Approaches which allow to relax these assumptions are discussed in Sects. 7.1–7.3, cf. also the discussion in Sect. 7.4(iv).

Theorem 6.20 and Corollary 6.21 give a sufficient condition for asymptotic stability for the nominal NMPC closed-loop system (3.5) in terms of the value α. Particularly, if α from (6.14) or $\underline{\alpha}_N$ from (6.18), respectively, is positive for the B_K induced by a $\beta \in \mathcal{KL}_0$ via (6.7), then we can conclude asymptotic stability whenever Assumption 6.5 is satisfied with this β for the optimization problem ($\mathrm{OCP_N}$) in Algorithm 3.1.

The following theorem shows that for $\beta \in \mathcal{KL}_0$ satisfying (6.9)—which implies $\alpha = \underline{\alpha}_N$ if β is linear in its first argument—this condition is tight for the class of systems satisfying Assumption 6.5 in the following sense: if α from (6.14) is negative, then there exists a control system (2.1) and a stage cost ℓ such that Assumption 6.5 holds but the nominal NMPC closed-loop system (3.5) is *not* asymptotically stable.

Theorem 6.27 *Consider $\beta \in \mathcal{KL}_0$ satisfying* (6.9), *let $N \geq 1$ and assume that the optimization problem* (6.14) *for B_K from* (6.7) *has an optimal value $\alpha < 0$.*

Then there exists a control system (2.1) and a stage cost ℓ satisfying Assumption 6.5 and $\alpha_3(|x|_{x_*}) \leq \ell^*(x) \leq \alpha_4(|x|_{x_*})$ for suitable $\alpha_3, \alpha_4 \in \mathcal{K}_\infty$, such that the nominal NMPC closed-loop system (3.5) *is not asymptotically stable.*

Proof We first show that $\alpha < 0$ in (6.14) implies the following property:

$$\begin{array}{c}\text{there exists } \lambda_0, \ldots, \lambda_{N-1}, \nu > 0 \text{ satisfying} \\ \text{(6.11)with strict inequalities, (6.12)and} \\ \sum_{n=0}^{N-1} \lambda_n - \nu < 0.\end{array} \tag{6.25}$$

In order to prove (6.25) we use that $\alpha < 0$ in (6.14) yields the existence of $\bar{\lambda}_0 > 0$, $\bar{\lambda}_1, \ldots, \bar{\lambda}_{N-1}, \nu \geq 0$ satisfying (6.11), (6.12) and

$$\sum_{n=0}^{N-1} \bar{\lambda}_n - \bar{\nu} < 0. \tag{6.26}$$

These properties imply $\bar{\lambda}_1 > 0, \ldots, \bar{\lambda}_{N-1} > 0$ and $\bar{\nu} > 0$: the inequality $\bar{\nu} > 0$ immediately follows from (6.26) and $\bar{\lambda}_k \geq 0$. Assuming $\bar{\lambda}_k = 0$ for some $k \in \{1, \ldots, N-1\}$, the respective inequality from (6.11) together with $B_{N-k}(0) = 0$ implies $\bar{\lambda}_{k+1} = \ldots = \bar{\lambda}_{N-1} = 0$. Thus, in particular $\bar{\lambda}_{N-1} = 0$ which using (6.12) for $j = N-2$ implies $\bar{\nu} \leq \sum_{n=0}^{N-3} \bar{\lambda}_{n+1}$ contradicting (6.26).

Now we pick an arbitrary $\varepsilon > 0$ and set

$$\lambda_0 := \bar{\lambda}_0, \ \ldots, \ \lambda_{N-2} := \bar{\lambda}_{N-2}, \ \lambda_{N-1} := \bar{\lambda}_{N-1} - \varepsilon,$$

and

$$\nu := \min\left\{\bar{\nu}, \sum_{n=0}^{N-2} \lambda_{n+1} + B_2(\lambda_{N-1})\right\}.$$

We claim that for these values (6.25) holds for all sufficiently small $\varepsilon > 0$. In order to see this, first one easily checks that (6.11) and (6.12) hold. Furthermore, since $\lambda_{N-1} < \bar{\lambda}_{N-1}$ appears on the left hand side but not on the right hand side of each inequality in (6.11), it follows that the inequalities in (6.11) are indeed strict. Furthermore, for $\varepsilon > 0$ sufficiently small the inequality $\lambda_{N-1} > 0$ holds. In order to complete the proof of (6.25) it remains to show that for $\varepsilon > 0$ sufficiently small the inequality $\nu > 0$ and the inequality in the last line of (6.25) holds.

To this end, we use that the second term in the "min" is exactly (6.12) for $j = N-2$. Thus, by continuity of B_2 the value ν converges to $\bar{\nu} > 0$ as $\varepsilon \to 0$. Hence, for $\varepsilon > 0$ sufficiently small $\bar{\nu} > 0$ implies $\nu > 0$ and (6.26) implies the inequality in the last line of (6.25) which completes the proof of (6.25).

Now, we construct a control system (2.1) on the state-space

$$X = \left\{(q, p) \in \mathbb{R}^2 \,|\, q \in \{0\} \cup \{2^{-k} | k \in \mathbb{N}_0\}, \, p \in \{(-N+1)q, \ldots, Nq\}\right\}.$$

For the control values $U = \{-1, 0, 1\}$ we define the dynamics

$$\begin{aligned}
f((1,p),-1) &= (1,\max\{-N+1,p-1\}), && p\in\{(-N+1)q,\dots,Nq\}\\
f((1,p),0) &= (1/2,p/2), && p\in\{(-N+1)q,\dots,Nq\}\\
f((1,p),1) &= (1,\min\{N,p+1\}), && p\in\{(-N+1)q,\dots,Nq\}\\
f((q,p),u) &= (q/2,p/2), && (q,p)\in X,\ q\le 1/2,\ u\in U
\end{aligned}$$

for which $x_* = (0,0)$ is an equilibrium for all $u \in U$. On X we use the metric induced by the usual Euclidean norm $\|(q,p)\| = \sqrt{q^2+p^2}$ implying $\|x\|_{x_*} = \|x\|$. We do not impose any constraints, i.e., we set $\mathbb{X} = X$ and $\mathbb{U}(x) = U$ for all $x \in X$.

Using the values $\lambda_1,\dots,\lambda_{N-1}$ and ν from (6.25) we define the stage cost ℓ in ($\mathrm{OCP_N}$) as

$$\begin{aligned}
\ell((1,p),1) &= \lambda_p, && p\in\{0,N-1\}\\
\ell((1,p),1) &= \nu, && p\notin\{0,N-1\}\\
\ell((1,p),-1) &= \ell((1,-p+1),1)\\
\ell((1,p),0) &= \beta(\min\{\ell((1,p),1),\ell((1,p),-1)\},0)\\
\ell((2^{-k},p),u) &= \beta(\min\{\ell((1,2^kp),1),\ell((1,2^kp),-1)\},k), && k\ge 1,\ u\in U.
\end{aligned}$$

We first verify that f and ℓ satisfy the stated assumptions.

The stage cost ℓ satisfies the inequalities from the assumption for $\alpha_3'(r) = \inf_{x\in X,\|x\|\ge r}\ell^*(x)$ and $\tilde{\alpha}_4'(r) = \sup_{x\in X,\|x\|\le r}\ell^*(x)$. Due to the discrete nature of the state space α_3' and α_4' are discontinuous but they are easily under- and overbounded by continuous $\mathscr{K}_\infty$ functions α_3 and α_4, respectively, for which the assumed inequalities $\alpha_3(|x|_{x_*}) \le \ell^*(x) \le \alpha_4(|x|_{x_*})$ hold.

In order to see that Assumption 6.5 is satisfied for the given β, first observe that Assumption 6.5 for $n = 0$ implies $\beta(r,0) \ge r$. From this inequality and the definition of ℓ, we obtain

$$\ell((1,p),0) \ge \min\{\ell((1,p),1),\ell((1,p),-1)\}$$

and thus

$$\ell^*((1,p)) = \min\{\ell((1,p),1),\ell((1,p),-1)\}.$$

Furthermore, for $k \ge 1$ we have that $\ell((2^{-k},p),u)$ is independent of u which yields

$$\ell^*((2^{-k},p)) = \ell((2^{-k},p),0).$$

Now for $u_x \equiv 0$ and initial value $x = (q,p) \in X$ with $q = 2^{-k_0}$ the trajectory becomes

$$x_{u_x}(k,x) = (2^{-k-k_0},2^{-k}p).$$

Thus, by construction of ℓ and (6.9) we obtain

$$\begin{aligned}\ell(x_{u_x}(k,x),u_x(k)) &= \beta(\min\{\ell((1,2^{k_0}p),1),\ell((1,2^{k_0}p),-1)\},k+k_0)\\ &= \beta(\ell^*(1,2^{k_0}p),k+k_0)\\ &\le \beta(\beta(\ell^*(1,2^{k_0}p),k_0),k) \;=\; \beta(\ell((2^{-k_0},p),0),k)\\ &= \beta(\ell^*((2^{-k_0},p)),k) \;=\; \beta(\ell^*(x),k)\end{aligned}$$

which yields Assumption 6.5.

Now, we prove the existence of a closed-loop trajectory which does not converge to x_*, which shows that asymptotic stability does not hold. To this end we abbreviate $\Lambda = \sum_{n=0}^{N-1}\lambda_n$ (note that (6.25) implies $\nu > \Lambda$) and investigate the values $J_N((1,0),u)$ for different choices of u:

Case 1: $u(0) = 0$. In this case, regardless of the values $u(n)$, $n \ge 1$, we obtain $x(n,u) = (2^{-n},0)$ and thus

$$\begin{aligned}J_N((1,0),u) &= \sum_{n=0}^{N-1}\beta(\min\{\ell((1,0),1),\ell((1,0),-1)\},n)\\ &= B_N(\min\{\ell((1,0),1),\ell((1,0),-1)\}) \;=\; B_N(\min\{\lambda_0,\lambda_1\}).\end{aligned}$$

In case that the minimum is attained in λ_0, by the (strict) inequality (6.11) for $k = 0$ we obtain $J_N((1,0),u) > \Lambda$. If the minimum is attained in λ_1 then by (6.12) for $j = 0$ and (6.25) we obtain $J_N((1,0),u) \ge \nu > \Lambda$. Thus, in both cases the inequality $J_N((1,0),u) > \Lambda$ holds.

Case 2: $u(n) = -1, n = 0,\dots,N-2$. This choice yields $x(n,u) = (1,-n)$ for $n = 0,\dots,N-1$ and thus

$$\begin{aligned}J_N((1,0),u) &= \sum_{n=0}^{N-2}\lambda_{n+1} + \ell((1,-(N-1)),-1) \ge \ell((1,-(N-1)),-1)\\ &= \ell((1,N),1) \;=\; \nu \;>\; \Lambda.\end{aligned}$$

Case 3: $u(n) = -1, n = 0,\dots,k-1$, and $u(k) = 1$ for a $k \in \{1,\dots,N-2\}$. In this case we obtain $x(n,u) = (1,-n)$ for $n = 0,\dots,k$ implying

$$J_N((1,0),u) = \sum_{n=0}^{k-1}\lambda_{n+1} + \ell((1,-k),1) \ge \ell((1,-k),1) = \nu > \Lambda.$$

Case 4: $u(n) = -1, n = 0,\dots,k-1$, and $u(k) = 0$ for a $k \in \{1,\dots,N-2\}$. This control sequence yields $x(n,u) = (1,-n)$ for $n = 0,\dots,k$ while for $n = k+1,\dots,N-1$ we get $x(n,u) = (2^{-(n-k)}, -2^{-(n-k)}k)$. Thus

$$
\begin{aligned}
J_N((1,0),u) &= \sum_{n=0}^{k-1} \lambda_{n+1} + \sum_{n=k}^{N-1} \beta(\min\{\ell((1,-k),1), \ell((1,-k),-1)\}, n-k) \\
&= \sum_{n=0}^{k-1} \lambda_{n+1} + B_{N-k}(\lambda_{k+1}) \geq \nu > \Lambda,
\end{aligned}
$$

where we have used (6.12) for $j = k$ in the second last inequality.

Case 5: $u(n) = 1, n = 0, \ldots, N-1$. This yields $x(n,u) = (1,n)$ and thus

$$
J_N((1,0),u) = \sum_{n=0}^{N-1} \lambda_n = \Lambda.
$$

Summarizing, we obtain that any optimal control $u_x^\star$ for $x = (1,0)$ must satisfy $u_x^\star(0) = 1$ because for $u(0) = 1$ we can realize a value $\leq \Lambda$ while for $u(0) \neq 1$ we inevitably obtain a value $> \Lambda$. Consequently, the NMPC-feedback law $\mu_N(x) = u_x^\star(0)$ will steer the system from $x = (1,0)$ to $x^+ := (1,1)$.

Now, we use that by construction f and ℓ have the symmetry properties

$$
\begin{aligned}
f((q,p),u) - (0,p) &= -f((q,-p+q),-u) + (0,-p+q) \\
\ell((q,p),u) &= \ell((q,-p+q),-u)
\end{aligned}
$$

for all $(q,p) \in X$ which implies $J((q,p),u) = J(q,-p+q),-u)$. Observe that $x^+ = (1,1)$ is exactly the symmetric counterpart of $x = (1,0)$. Thus, any optimal control $u_{x^+}^\star$ for x^+ must satisfy $u_{x^+}^\star(n) = -u_x^\star(n)$ for some optimal control $u_x^\star$ for initial value x. Hence, we obtain $u_{x^+}^\star(0) = -1$, which means that the NMPC-feedback $\mu_N(x^+) = u_{x^+}^\star(0)$ steers x^+ back to x. Thus, under the NMPC-feedback law, we obtain the closed-loop trajectory $(x, x^+, x, x^+, \ldots)$, which clearly does not converge to $x_* = (0,0)$. This shows that the closed-loop system is not asymptotically stable. $\square$

Remark 6.28 If we weaken the assumptions of Theorem 6.27 to $\alpha = 0$ instead of $\alpha < 0$, then the inequalities in (6.25) will not be strict. Under this weaker assumption, in Cases 1–4 in the proof of Theorem 6.27 we get $J_N((1,0),u) \geq \Lambda$ instead of $J_N((1,0),u) > \Lambda$. This means that the control sequence $u(n) \equiv 1$ from Case 5 is still optimal but it is no longer the *unique* optimal control sequence. Consequently, the value of $\mu_N((1,0))$ depends on the optimization algorithm. The algorithm may select $\mu_N((1,0)) = 1$—leading to a closed-loop system which is not asymptotically stable—or it may select a control value which yields asymptotic stability. Thus, instability may occur for $\alpha = 0$ but it does not necessarily need to occur.

All results developed so far in this chapter remain valid for the time varying case when the bounds on the V_N are uniform in time according to the following definition.

Assumption 6.29 (*Bounds on V_N*) Consider the optimal control problem (OCP_N^n). We assume that there exist functions $B_K \in \mathcal{K}_\infty$, $K \in \mathbb{N}$, such that for each $x \in \mathbb{X}$ and $n \in \mathbb{N}$ the inequality

$$V_K(n, x) \leq B_K(\ell^*(x)) \tag{6.27}$$

holds for all $K \in \mathbb{N}$.

A straightforward extension of Lemma 6.6 shows that Assumption 6.29 is implied by the following assumption.

Assumption 6.30 (*Asymptotic controllability wrt. ℓ*) Consider the optimal control problem (OCP_N^n). We assume that the system is uniformly asymptotically controllable with respect to ℓ with rate $\beta \in \mathcal{KL}_0$, i.e., for each $x \in \mathbb{X}$, each $N \in \mathbb{N}$ and each $n_0 \in \mathbb{N}_0$ there exists an admissible control sequence $u_x \in \mathbb{U}^N(x)$ satisfying

$$\ell(n_0 + n, x_{u_x}(n, x), u_x(n)) \leq \beta(\ell^*(n_0, x), n)$$

for all $n \in \{0, \ldots, N-1\}$.

Under these assumptions, all results in this chapter carry over to the time-dependent setting when we replace λ_k and ν in Proposition 6.13 by

$$\lambda_k = \ell(n_0 + k, x_{u^\star}(k, x), u^\star(k)) \quad \text{and} \quad \nu = V_N(n_0 + 1, x_{u^\star}(1, x)).$$

Proceeding this way, one easily sees that all results remain valid.

6.6 Design of Good Stage Costs ℓ

In this section, we illustrate by means of several examples how our theoretical findings and in particular Theorem 6.15 in conjunction with Lemma 6.6 and Proposition 6.18 can be used in order to identify and design stage costs ℓ such that the NMPC feedback law μ_N exhibits stability and good performance with small optimization horizons N. To this end we first visualize Formula (6.18) for linear B_K induced by different $\beta \in \mathcal{KL}_0$ via (6.7), starting with the case of exponential controllability (6.4). Note that in this case (6.9) and thus (6.19) always holds.

Given a desired suboptimality level $\alpha^\star \geq 0$, we use Formula (6.18) in order to determine the regions in the (σ, C)-plane for which $\underline{\alpha}_N \geq \alpha^\star$ holds for different optimization horizons N. Figure 6.1 shows the resulting regions for $\alpha^\star = 0$ (i.e., "plain" stability) and $\alpha^\star = 0.5$.

Looking at Fig. 6.1 one sees that the parameters C and σ play a very different role. While for both parameters the necessary optimization horizon N becomes the smaller the smaller these parameters are, small overshoot C (i.e., values of C close to 1) have a much stronger effect than small decay rates σ (i.e., values of σ close to 0). Indeed, Fig. 6.1(left) shows that for sufficiently small C we can always achieve stability for $N = 2$ while for $C \geq 8$ even values of σ very close to 0 will not yield stability for $N \leq 16$. For the required higher suboptimality level $\alpha \geq 0.5$, Fig. 6.1(right) indicates a qualitatively similar behavior.

For finite time control, i.e., controllability with $\mathcal{KL}_0$-functions satisfying (6.5), the situation is very similar. For instance, consider functions of the form $\beta(r, 0) = c_0 r$, $\beta(r, 1) = c_1 r$, $c_0 \geq c_1$, and $\beta(r, n) = 0$ for $n \geq 2$, i.e., $n_0 = 2$. This function again satisfies (6.9), hence (6.19) holds. For this β, Fig. 6.2 shows the analogous graphs as in Fig. 6.1.

One immediately sees that the qualitative behavior depicted in Fig. 6.2 is very similar to the analogous graphs in Fig. 6.1: again, reducing the overshoot c_0 we can

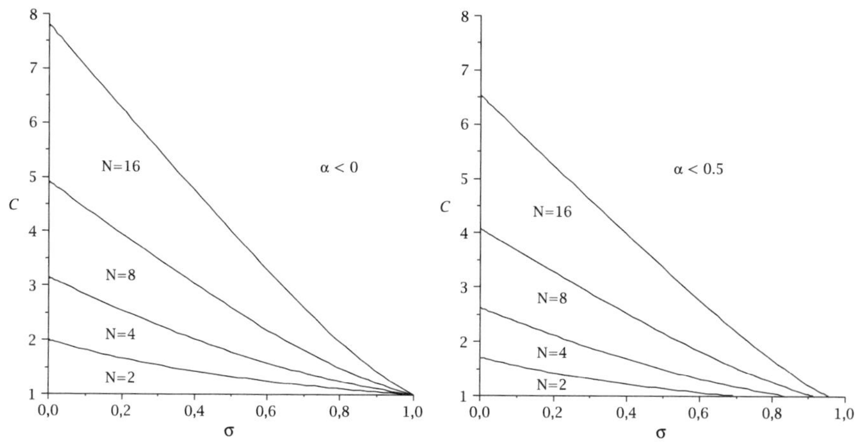

Fig. 6.1 Suboptimality regions for different optimization horizons N depending on C and σ in (6.4) for $\alpha^\star = 0$ (*left*) and $\alpha^\star = 0.5$ (*right*)

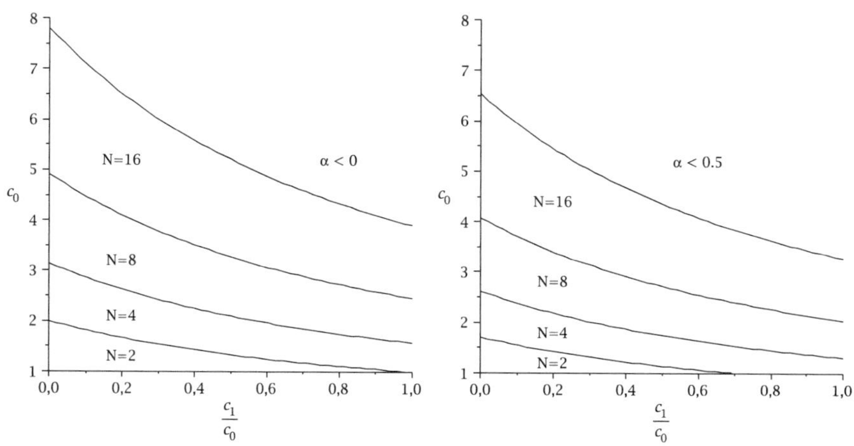

Fig. 6.2 Suboptimality regions for different optimization horizons N depending on c_0 and c_1/c_0 in (6.5) with $n_0 = 2$ for $\alpha^\star = 0$ (*left*) and $\alpha^\star = 0.5$ (*right*)

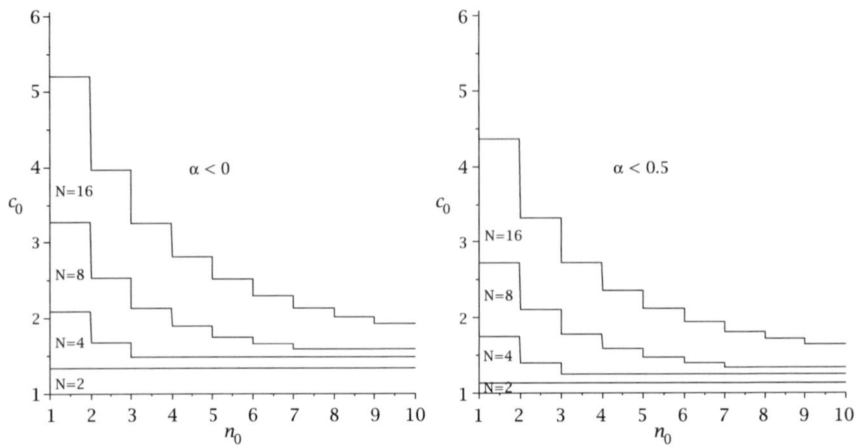

Fig. 6.3 Suboptimality regions for different optimization horizons N depending on c_0 and n_0 in (6.5) with $c_n = c_0/2$ for $n = 1, \ldots, n_0 - 1$ for $\alpha^\star = 0$ (*left*) and $\alpha^\star = 0.5$ (*right*)

always achieve stability with $N = 2$ regardless of the ratio c_1/c_0 while reducing c_1 and keeping c_0 fixed, in general we need $N > 2$ in order to guarantee stability.

Finally, in Fig. 6.3 we compare the effect of the overshoot c_0 and the time n_0 in (6.5) by using $\beta(r, 0) = c_0 r$, $\beta(r, n) = c_0 r/2$ for $n = 1, \ldots, n_0$ and $\beta(r, n) = 0$ for $n \geq n_0$. Again, it turns out that the time n_0 needed to control the system to x_* is less important than the overshoot: for all times $n_0 \geq 1$ we can always achieve stability for c_0 sufficiently close to 1 while for fixed c_0 this can in general not be achieved even for $n_0 = 1$, i.e., for controllability in one step. Note that for $c_0 < 2$ this function β does not satisfy (6.9), thus for these values of c_0 Formula (6.18) only provides a lower bound for α, cf. (6.17). Consequently, for $c_0 < 2$ the regions depicted in Fig. 6.3 may underestimate the true regions. Still, for all n_0 the lower bounds obtained from (6.17) ensure both asymptotic stability and the desired performance bound $\alpha \geq 0.5$ for $N = 2$ whenever c_0 is sufficiently close to 1.

Together, these examples lead to a conclusion which is as intuitive as simple: an NMPC controller without stabilizing terminal conditions will yield stability and good performance for small horizons N if ℓ can be chosen such that Assumption 6.5 is satisfied with a $\beta \in \mathcal{KL}_0$ with small overshoot. Thus, the criterion "small overshoot" can be used as a design guideline for selecting a good stage cost ℓ.

For some systems, it is possible to rigorously compute β in Assumption 6.5 which leads to a precise determination of, e.g., C and σ in (6.4). Examples where this is possible also include infinite dimensional systems, like the linear wave equation or certain classes of semilinear parabolic equations, cf. [7]. However, more often than not precise estimates for β cannot be obtained due to the complexity of the dynamics. Still, using heuristic arguments it may be possible to determine stage costs ℓ for which the overshoot is reduced. In the remainder of this section, we will illustrate this procedure for two examples.

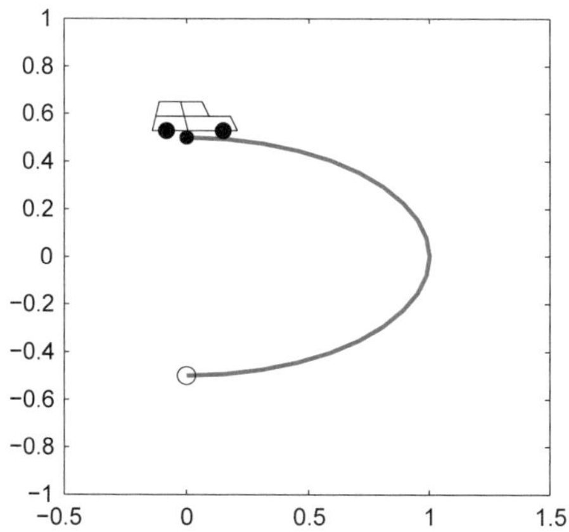

Fig. 6.4 Illustration of the stabilization problem

Example 6.31 We consider Example 2.3, i.e.,

$$f(x,u) := \begin{pmatrix} x_1^+ \\ x_2^+ \end{pmatrix} = \begin{pmatrix} \sin(\vartheta(x)+u) \\ \cos(\vartheta(x)+u)/2 \end{pmatrix}$$

with

$$\vartheta(x) = \begin{cases} \arccos 2x_2, & x_1 \geq 0 \\ 2\pi - \arccos 2x_2, & x_1 < 0 \end{cases}$$

using the control values $\mathbb{U} = [0, 0.2]$, i.e., the car can only move clockwise on the ellipse

$$\mathbb{X} = \left\{ x \in \mathbb{R}^2 \;\middle|\; \left\| \begin{pmatrix} x_1 \\ 2x_2 \end{pmatrix} \right\| = 1 \right\}.$$

As in illustrated in Fig. 6.4, we want to stabilize the system at the equilibrium $x_* = (0, -1/2)^\top$ starting from the initial value $x_0 = (0, 1/2)^\top$.

Interpreting $\mathbb{X}$ as a subset of $\mathbb{R}^2$, we can try to achieve this goal by using NMPC without terminal conditions with the stage cost

$$\ell(x,u) = \|x - x_*\|^2 + u^2. \tag{6.28}$$

As the simulations in Fig. 6.5 show, asymptotic stability of $x_* = (0, -1/2)$ is achieved for $N = 11$ but not for $N = 10$.

The reason for the closed loop not being asymptotically stable for $N = 10$ (and, in fact, for all $N \leq 10$) is the overshoot in the stage cost ℓ when moving along the ellipse; see Fig. 6.6.

The fact that this overshoot of ℓ appears along the NMPC closed-loop trajectory does in general not imply that the overshoot is present for all possible control

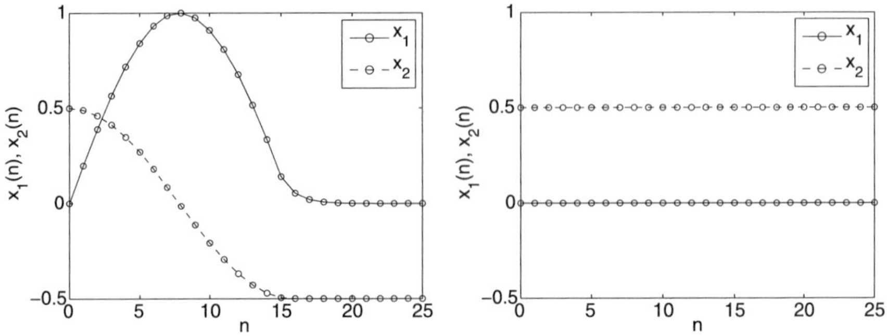

Fig. 6.5 NMPC closed-loop trajectories for Example 2.3 with stage cost (6.28) and optimization horizons $N = 11$ (*left*), $N = 10$ (*right*)

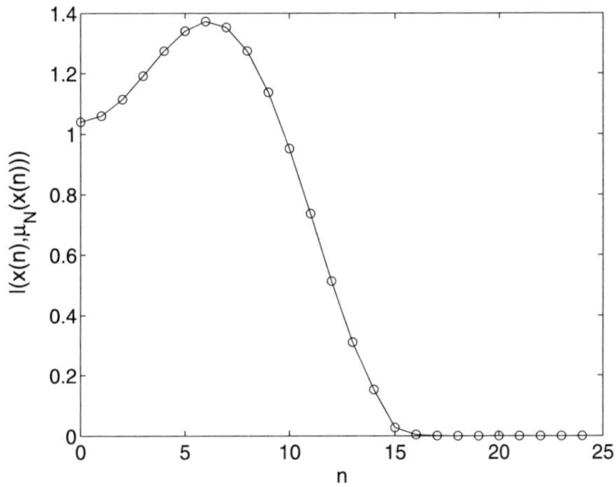

Fig. 6.6 Stage cost (6.28) along the NMPC closed-loop trajectory for $N = 11$

sequences u controlling the system to x_*. However, in this example a look at the geometry reveals that for ℓ from (6.28) the overshoot is in fact not avoidable: no matter how we control the system to x_*, before we can eventually reduce ℓ to 0, we need to increase ℓ when moving along the ellipse around the curve. Thus, loosely speaking, the loss of asymptotic stability for $N \leq 10$ is caused by the fact that the optimizer does not “see” that in the long run it is beneficial to move around the curve and thus stays at the initial value x_0 for all future times.

Looking closer at the geometry of the example, one easily sees that the overshoot is entirely due to the x_1-component of the solution: while x_2 converges monotonically to the desired position $x_{*2} = -0.5$, x_1 first needs to move from 0 to 1 before we can eventually control it to $x_{*1} = 0$, again. From this observation it follows that the overshoot in ℓ can be avoided by putting more weight on the x_2–component. Indeed, if we replace $\ell(x, u) = \|x - x_*\|^2 + u^2 = (x_1 - x_{*1})^2 + (x_2 - x_{*2})^2 + u^2$ from (6.28) by

$$\ell(x, u) = (x_1 - x_{*1})^2 + 5(x_2 - x_{*2})^2 + u^2, \tag{6.29}$$

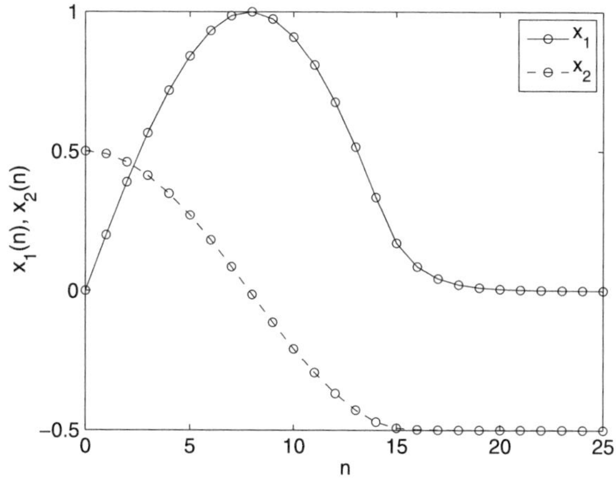

Fig. 6.7 NMPC closed-loop trajectories for Example 2.3 with stage cost (6.29) and optimization horizon $N = 2$

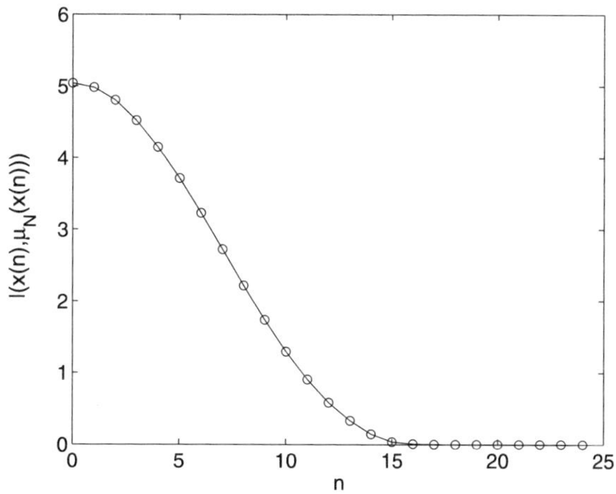

Fig. 6.8 Stage cost (6.29) along the NMPC closed loop for $N = 2$

then we obtain asymptotic stability even for $N = 2$, cf. Fig. 6.7.

Figure 6.8 shows the stage cost along the closed loop trajectory for this example. The figure clearly shows that the overshoot has been removed completely which explains why the NMPC closed-loop is stable for $N = 2$.

We would like to emphasize that for removing the overshoot we did not use any quantitative information, i.e., we did not attempt to estimate the function β in Assumption 6.5. For selecting a good cost function ℓ it was sufficient to observe that putting a larger weight on x_2 will reduce the overshoot. On the basis of this observation, the fact that the weight "5" used in (6.29) is sufficient to achieve asymptotic stability with $N = 2$ was then determined by a simple try-and-error procedure using numerical simulations.

Example 6.32 As a second example we consider the infinite dimensional PDE models introduced in Example 2.12. We first consider the system with distributed control, i.e.,

$$y_t(t,x) = \theta y_{xx}(t,x) - y_x(t,x) + \rho\left(y(t,x) - y(t,x)^3\right) + u(t,x) \tag{6.30}$$

with control function $u \in L^\infty(\mathbb{R} \times \Omega, \mathbb{R})$, domain $\Omega = (0, 1)$ and real parameters $\theta = 0.1$, $\rho = 10$. Here y_t and y_x denote the partial derivatives with respect to t and x, respectively, and y_{xx} denotes the second partial derivative with respect to x.

The solution y of (6.30) is supposed to be continuous in $\overline{\Omega}$ and to satisfy the boundary and initial conditions

$$y(t,0) = 0,\ y(t,1) = 0 \text{ for all } t \geq 0 \quad \text{and} \quad y(0,x) = y_0(x) \text{ for all } x \in \Omega \tag{6.31}$$

for some given continuous function $y_0 : \overline{\Omega} \to \mathbb{R}$ with $y_0(0) = y_0(1) = 0$.

Observe that we have changed notation here in order to be consistent with the usual PDE notation: $x \in \Omega$ is the independent space variable while the unknown function $y(t,\cdot) : \Omega \to \mathbb{R}$ in (6.30) is the state now. Hence, the state is now denoted by y (instead of x) and the state space of this PDE control system is a function space, more precisely the Sobolev space $H_0^1(\Omega)$, although the specific form of this space is not crucial for the subsequent reasoning.

Figure 6.9 shows the solution of the uncontrolled system (6.30), (6.31), i.e., with $u \equiv 0$. For growing t the solution approaches an asymptotically stable steady state $y_{**} \neq 0$. The figure (as well as all other figures in this section) was computed numerically using a finite difference scheme with 50 equidistant nodes on (0, 1) (finer resolutions did not yield significantly different results) and initial value y_0 with $y_0(0) = y_0(1) = 0$, $y_0|_{[0.02,0.3]} \equiv -0.1$, $y_0|_{[0.32,0.98]} \equiv 0.1$ and linear interpolation in between.

By symmetry of (6.30) the function $-y_{**}$ must be an asymptotically stable steady state, too. Furthermore, from (6.30) it is obvious that $y_* \equiv 0$ is another steady state,

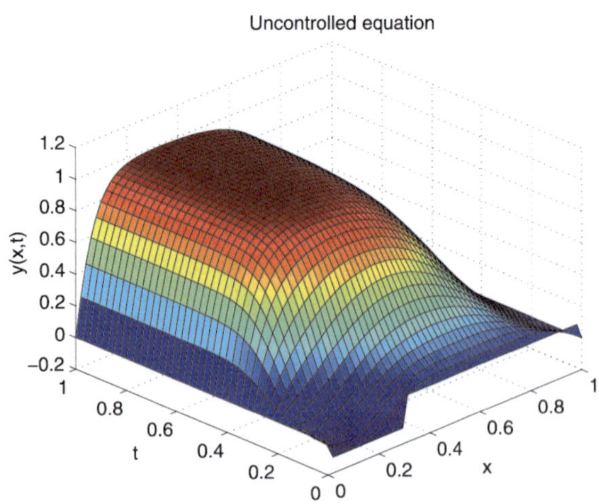

Fig. 6.9 Solution $y(t,x)$ of (6.30), (6.31) with $u \equiv 0$

which is, however, unstable. Our goal is now to use NMPC in order to stabilize the unstable equilibrium $y_* \equiv 0$.

To this end we consider the sampled data system corresponding to (6.30) with sampling period $T = 0.025$. In order to obtain a more intuitive notation for the solution of the sampled data system, instead of introducing the abstract variable z as in Example 2.12 here we denote the state of the sampled data system at the n-th sampling instant, i.e., at time nT by $y(n, \cdot)$. For penalizing the distance of the state $y(n, \cdot)$ to $y_* \equiv 0$ a popular choice in the literature is the L^2 functional

$$\ell(y(n, \cdot), u(n, \cdot)) = \|y(n, \cdot)\|^2_{L^2(\Omega)} + \lambda \|u(n, \cdot)\|^2_{L^2(\Omega)} \tag{6.32}$$

which penalizes the mean squared distance from $y(n, \cdot)$ to $y_* \equiv 0$ and the control effort with weighting parameter $\lambda > 0$. Here we choose $\lambda = 0.1$.

Another possible choice of measuring the distance to $y_* \equiv 0$ is obtained by using the H^1 norm

$$\|y(n, \cdot)\|_{H^1(\Omega)} = \|y(n, \cdot)\|^2_{L^2(\Omega)} + \|y_x(n, \cdot)\|^2_{L^2(\Omega)}.$$

This leads to defining

$$\ell(y(n, \cdot), u(n, \cdot)) = \|y(n, \cdot)\|^2_{L^2(\Omega)} + \|y_x(n, \cdot)\|^2_{L^2(\Omega)} + \lambda \|u(n, \cdot)\|^2_{L^2(\Omega)}, \tag{6.33}$$

which in addition to the L^2 distance and the control effort as in (6.32) also penalizes the mean squared distance from $y_x(n, \cdot)$ to $y_{*,x} \equiv 0$. Figures 6.10 and 6.11 show the respective NMPC closed-loop solutions with optimization horizons $N = 3$ and $N = 11$.

Figure 6.10 indicates that for $N = 3$ the NMPC scheme with ℓ from (6.32) does not stabilize the system at $y_* \equiv 0$, while for ℓ from (6.33) it does. For (6.32) we need an optimization horizon of at least $N = 11$ in order to obtain a stable closed-loop solution, cf. Fig. 6.11. For ℓ from (6.33) the right images in Figs. 6.10 and 6.11 show that enlarging the horizon does not improve the closed-loop behavior any further.

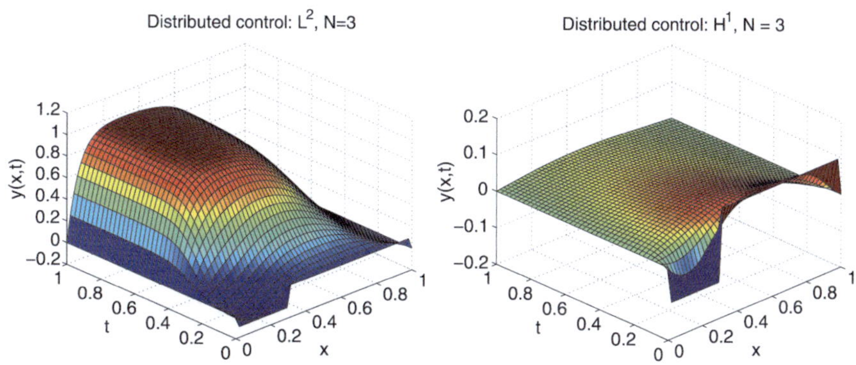

Fig. 6.10 NMPC closed loop for (6.30) with $N = 3$ and ℓ from (6.32) (*left*) and (6.33) (*right*)

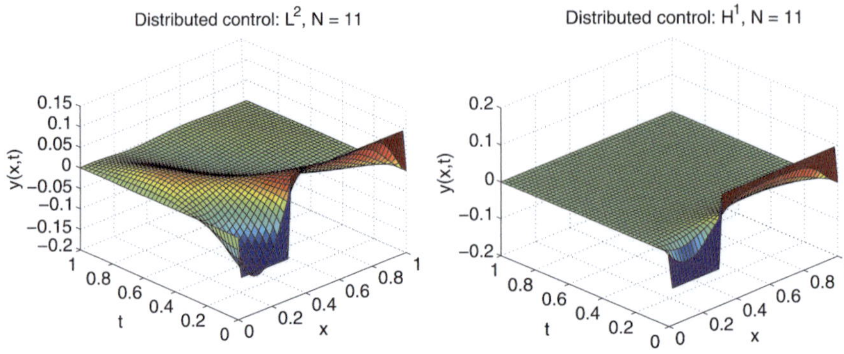

Fig. 6.11 NMPC closed loop for (6.30) with $N = 11$ and ℓ from (6.32) (*left*) and (6.33) (*right*)

Using our theoretical results we can explain why ℓ from (6.33) performs much better for small horizons N. For this example, our controllability condition Assumption 6.5 reads

$$\ell(y(n,\cdot), u(n,\cdot)) \le C\sigma^n \ell^*(y(0,\cdot)). \tag{6.34}$$

For ℓ from (6.32) this becomes

$$\|y(n,\cdot)\|^2_{L^2(\Omega)} + \lambda \|u(n,\cdot)\|^2_{L^2(\Omega)} \le C\sigma^n \|y(0,\cdot)\|^2_{L^2(\Omega)}. \tag{6.35}$$

Now in order to control the system to $y^* \equiv 0$, in (6.30) the control needs to compensate for y_x and $\rho\left(y(t,x) - y(t,x)^3\right)$, i.e., any control steering $y(n,\cdot)$ to 0 must satisfy

$$\|u(n,\cdot)\|^2_{L^2(\Omega)} \approx \|y_x(n,\cdot)\|^2_{L^2(\Omega)} + \|\rho\left(y(n,\cdot) - y(n,\cdot)^3\right)\|^2_{L^2(\Omega)}. \tag{6.36}$$

Inserting this approximate equality into (6.35) implies—regardless of the value of σ—that the overshoot bound C in (6.35) is large if $\|y_x(n,\cdot)\|^2_{L^2(\Omega)} \gg \|y(0,\cdot)\|^2_{L^2(\Omega)}$ holds, which is the case in our example.

For ℓ from (6.33) inequality (6.34) becomes

$$\begin{aligned} &\|y(n,\cdot)\|^2_{L^2(\Omega)} + \|y_x(n,\cdot)\|^2_{L^2(\Omega)} + \lambda \|u(n,\cdot)\|^2_{L^2(\Omega)} \\ &\le \; C\sigma^n \left(\|y(0,\cdot)\|^2_{L^2(\Omega)} + \|y_x(0,\cdot)\|^2_{L^2(\Omega)}\right). \end{aligned} \tag{6.37}$$

Due to the fact that $\|y_x(0,\cdot)\|^2_{L^2(\Omega)} \gg \|y(0,\cdot)\|^2_{L^2(\Omega)}$ holds in our example, inserting the approximate Eq. (6.36) into (6.37) does not imply large C, which explains the considerable better performance for ℓ from (6.33).

The fact that the H^1-norm penalizes the distance to $y_* \equiv 0$ in a "stronger" way than the L^2-norm may lead to the conjecture that the better performance for this norm is intuitive. Our second example shows that this is not the case. This example is similar to Eqs. (6.30), (6.31), except that the distributed control is changed to Dirichlet boundary control. Thus, (6.30) becomes

$$y_t(t,x) = \theta y_{xx}(t,x) - y_x(t,x) + \rho\left(y(t,x) - y(t,x)^3\right), \tag{6.38}$$

again with $\theta = 0.1$ and $\rho = 10$, and (6.31) changes to

$$y(t,0) = u_0(t),\ \ y(t,1) = u_1(t) \text{ for all } t \geq 0,\ \ y(0,x) = y_0(x) \text{ for all } x \in \Omega$$

with $u_0, u_1 \in L^\infty(\mathbb{R}, \mathbb{R})$. The cost functions (6.32) and (6.33) change to

$$\ell(y(n,\cdot), u(n,\cdot)) = \|y(n,\cdot)\|^2_{L^2(\Omega)} + \lambda(u_0(n)^2 + u_1(n)^2) \tag{6.39}$$

and

$$\ell(y(n,\cdot), u(n,\cdot)) = \|y(n,\cdot)\|^2_{L^2(\Omega)} + \|y_x(n,\cdot)\|^2_{L^2(\Omega)} + \lambda(u_0(n)^2 + u_1(n)^2), \tag{6.40}$$

respectively, again with $\lambda = 0.1$.

Due to the more limited possibilities to control the equation, the problem obviously becomes more difficult, hence we expect to need larger optimization horizons for stability of the NMPC closed loop. However, what is surprising at the first glance is that ℓ from (6.39) stabilizes the system for smaller horizons than ℓ from (6.40), as the numerical results in Fig. 6.12 confirm.

A closer look at the dynamics reveals that we can again explain this behavior with our theoretical results. In fact, steering the chosen initial solution to $y_* = 0$ requires u_1 to be such that a rather large gradient appears close to $x = 1$. Thus, during the transient phase $\|y_x(n,\cdot)\|^2_{L^2(\Omega)}$ becomes large which in turn causes ℓ from (6.40) to become large and thus causes a large overshoot bound C in (6.34). In ℓ from (6.39), on the other hand, these large gradients are not "visible" which is why the overshoot in (6.34) is smaller and thus allows for stabilization with smaller N.

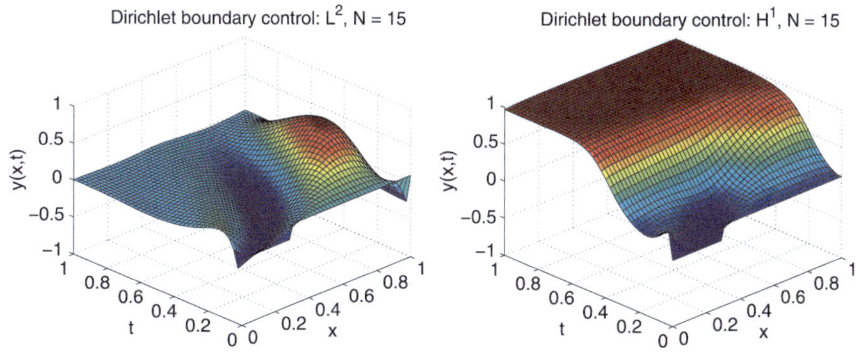

Fig. 6.12 NMPC closed loop for (6.38) with $N = 15$ and ℓ from (6.32) (*left*) and (6.33) (*right*)

6.7 Semiglobal and Practical Asymptotic Stability

We have seen in Theorem 6.24 that linearity of the B_K in Assumption 6.3—together with boundedness of the corresponding γ_K—guarantees that for sufficiently large optimization horizon N the nominal NMPC closed-loop system (3.5) will be asymptotically stable on the whole set $\mathbb{X}$. Even though the examples in the last section show that the linearity condition can be fulfilled, it is easy to come up with examples in which linearity is not satisfied or at least difficult or almost impossible to check. In this section we show that also in this case one can guarantee that NMPC without stabilizing terminal conditions has reasonable stability properties. However, to this end we have to weaken the stability notion according to the following definition.

Definition 6.33 Consider the NMPC Algorithm 3.1 and the resulting nominal closed-loop system (3.5) with feedback law μ_N and solutions $x_{\mu_N}(k, x)$.

(i) We call the closed-loop system (3.5) *semiglobally asymptotically stable with respect to the optimization horizon* N if there exists $\beta \in \mathcal{KL}$ such that the following property holds: for each $\Delta > 0$ there exists $N_\Delta \in \mathbb{N}$ such that for all $N \geq N_\Delta$ and all $x \in \mathbb{X}$ with $|x|_{x_*} \leq \Delta$ the inequality

$$|x_{\mu_N}(k, x)|_{x_*} \leq \beta(|x|_{x_*}, k)$$

holds for all $k \in \mathbb{N}_0$.

(ii) We call the closed-loop system (3.5) *semiglobally practically asymptotically stable with respect to the optimization horizon* N if there exists $\beta \in \mathcal{KL}$ such that the following property holds: for each $\delta > 0$ and $\Delta > \delta$ there exists $N_{\delta,\Delta} \in \mathbb{N}$ such that for all $N \geq N_{\delta,\Delta}$ and all $x \in \mathbb{X}$ with $|x|_{x_*} \leq \Delta$ the inequality

$$|x_{\mu_N}(k, x)|_{x_*} \leq \max\{\beta(|x|_{x_*}, k), \delta\}$$

holds for all $k \in \mathbb{N}_0$.

Semiglobal asymptotic stability relaxes the asymptotic stability condition by requiring asymptotic stability only for the set of initial values $x \in \mathbb{X}$ with $|x|_{x_*} \leq \Delta$. Although Δ can be chosen arbitrarily large by suitably adjusting the optimization horizon N, for each finite N it will in general be a finite value.

Semiglobal practical asymptotic stability additionally relaxes the requirement that the solution exactly tends to the equilibrium x_* by only requiring that the solution behaves like an asymptotically stable solution until it reaches a δ-neighborhood of x_*. Similar to the value of Δ, the size δ of this neighborhood can be arbitrarily tuned by adjusting the optimization horizon N, but for each finite N it will in general be a positive value.

Of course, both definitions can be straightforwardly extended to the time varying case with reference $x^{\mathrm{ref}}(n)$ instead of x_*.

Semiglobal and semiglobal practical asymptotic stability can be expressed via the stability properties already introduced in Chap. 2. This is made precise in the following lemma.

Lemma 6.34 *(i) The NMPC closed-loop is semiglobally asymptotically stable with respect to the optimization horizon N if for each $\Delta > 0$ there exists $N_\Delta > 0$ such that for all $N \geq N_\Delta$ there exists a forward invariant set Y with $\overline{\mathscr{B}}_\Delta(x_*) \subset Y$ such that the system is asymptotically stable on Y in the sense of Definition 2.14.*

(ii) The NMPC closed loop is semiglobally practically asymptotically stable with respect to the optimization horizon N if for each $\delta > 0$ and $\Delta > \delta$ there exists $N_{\delta,\Delta} > 0$ such that for all $N \geq N_{\delta,\Delta}$ there exist forward invariant sets Y and P with $\overline{\mathscr{B}}_\Delta(x_*) \subset Y$ and $P \subseteq \overline{\mathscr{B}}_\delta(x_*)$ such that the system is P-practically asymptotically stable on Y in the sense of Definition 2.15.

Proof (i) follows immediately from the definition. (ii) follows from the fact that according to Definition 2.15 for each $k \in \mathbb{N}_0$ either $|x_{\mu_N}(k,x)|_{x_*} \leq \beta(|x|_{x_*},k)$ or $x_{\mu_N}(k,x) \in P$ holds. Since the latter implies $|x_{\mu_N}(k,x)|_{x_*} \leq \delta$ we obtain the assertion. □

For obtaining semiglobal stability, it turns out that a "semiglobal" linearity assumption on the B_K is sufficient. This is the statement of the following theorem.

Theorem 6.35 *Consider the NMPC Algorithm 3.1 with optimization horizon $N \in \mathbb{N}$ and stage cost ℓ satisfying $\alpha_3(|x|_{x_*}) \leq \ell^*(x) \leq \alpha_4(|x|_{x_*})$ for suitable $\alpha_3, \alpha_4 \in \mathscr{K}_\infty$. Assume that Assumption 6.3 holds for functions $B_K \in \mathscr{K}_\infty$ which for each $R > 0$ and all $r \in [0, R]$ satisfy the inequality*

$$B_k(r) \leq \gamma_k^R r \quad \text{for all } k \in \mathbb{N}, \text{ with constants } \gamma_k^R \text{ satisfying } \sup_{k\in\mathbb{N}} \gamma_k^R < \infty.$$

Then the nominal NMPC closed-loop system (3.1) with NMPC-feedback law μ_N is semiglobally asymptotically stable on $\mathbb{X}$ with respect to the optimization horizon N.

Furthermore, for each $C > 1$ and each $\Delta > 0$ there exists $N_{C,\Delta} > 0$ such that

$$J_\infty^{cl}(x, \mu_N) \leq C V_N(x) \leq C V_\infty(x)$$

holds for each $x \in \overline{\mathscr{B}}_\Delta(x_*) \cap \mathbb{X}$ and each $N \geq N_{C,\Delta}$.

Proof We first show the existence of $\overline{\alpha} \in \mathscr{K}_\infty$ such that the inequality $B_k(r) \leq \overline{\alpha}(r)$ holds for all $r \geq 0$ and all $k \in \mathbb{N}$. To this end, we define $\gamma_\infty^R := \sup_{k\in\mathbb{N}} \gamma_k^R$ for each $R > 0$. Then the inequality $B_k(r) \leq \gamma_\infty^R r$ holds for all $k \in \mathbb{N}$ and all $r \in [0, R]$. Now for $R = 1, 2, \ldots$ we inductively define $\overline{\gamma}^1 = \gamma_\infty^1$ and

$$\overline{\gamma}^{R+1} = \max\{\overline{\gamma}^R, \gamma_\infty^{R+1}\}.$$

This definition implies $\overline{\gamma}^{R+1} \geq \overline{\gamma}^R$ and $B_k(r) \leq \overline{\gamma}^R r$ for all $r \in [0, R]$, $R \in \mathbb{N}$. Setting

$$\overline{\alpha}(r) := (R-r)\overline{\gamma}^R r + (r-R+1)\overline{\gamma}^{R+1} r, \quad r \in [R-1, R],\ R \in \mathbb{N}$$

we obtain a continuous, strictly increasing and unbounded function with $\overline{\alpha}(0) = 0$, hence $\overline{\alpha} \in \mathcal{K}_\infty$. For $r \in [R-1, R]$ and $R \in \mathbb{N}$ we obtain

$$B_k(r) \leq \overline{\gamma}^R r = (R-r)\overline{\gamma}^R r + (r-R+1)\overline{\gamma}^R r \leq (R-r)\overline{\gamma}^R r + (r-R+1)\overline{\gamma}^{R+1} r$$

which shows $B_k(r) \leq \overline{\alpha}(r)$ for $r \in [R-1, R]$. Since this holds for each $R \in \mathbb{N}$, we get the desired inequality $B_k(r) \leq \overline{\alpha}(r)$ for all $r \geq 0$.

Now fix $\Delta > 0$ and set $L := \overline{\alpha}(\tilde{\alpha}(\Delta))$. Since for each $N \in \mathbb{N}$ we have the inequality $V_N(x) \leq \overline{\alpha}(\ell^*(x)) \leq \overline{\alpha}(\alpha_4(|x|_{x_*}))$, for $x \in \overline{\mathcal{B}}_\Delta(x_*)$ we obtain $V_N(x) \leq L$ and thus the inclusion

$$\overline{\mathcal{B}}_\Delta(x_*) \subseteq V_N^{-1}([0, L]) =: S_N, \tag{6.41}$$

where V_N^{-1} denotes the sublevel set

$$V_N^{-1}([0, L]) := \{x \in \mathbb{X} \mid V_N(x) \in [0, L]\}.$$

Defining further $L' := \alpha_3^{-1}(L)$, for all $x \in S_N$ and all $y \notin \overline{\mathcal{B}}_{L'}(x_*)$ we obtain

$$V_N(x) \leq L = \alpha_3(L') < \alpha_3(|y|_{x_*}) \leq \ell^*(y).$$

This implies that for all $N \in \mathbb{N}$ and all $x \in S_N$ each optimal trajectory $x_{u^\star}(\cdot, x)$ of length N will remain in $\overline{\mathcal{B}}_{L'}(x_*)$. This holds because if there exists $k' \in \{0, \ldots, N-1\}$ with $y = x_{u^\star}(k', x) \notin \overline{\mathcal{B}}_{L'}(x_*)$ we obtain

$$J_N(x, u^\star) = \sum_{k=0}^{N-1} \ell(x_{u^\star}(k, x), u^\star(k)) \geq \ell(x_{u^\star}(k', x), u^\star(k')) \geq \ell^*(y) > V_N(x)$$

contradicting the optimality of $u^\star$. Setting $R := \alpha_4(L')$ then implies $\ell^*(x_{u^\star}(k, x)) \leq R$ for all $k = 0, \ldots, N-1$ and each optimal trajectory for $V_N(x)$ with $x \in S_N$ and arbitrary $N \in \mathbb{N}$.

Now we fix an arbitrary $\alpha_0 \in (0, 1)$ and note that the values γ_k^R can without loss of generality be assumed to be increasing in k; otherwise we may replace γ_k^R by $\max_{k' \leq k} \gamma_{k'}^R$. Then by the same arguments as in the proof of Theorem 6.24 we find $N_\Delta > 0$ such that for all $N \geq N_\Delta$ the inequality $\underline{\alpha}_N \geq \alpha_0$ holds in (6.18) with $\gamma_k = \gamma_k^R$. Now for each $x \in S_N$ we have shown above that the optimal trajectory for $V_N(x)$ satisfies $\ell^*(x_{u^\star}(k, x)) \leq R$ for all $k = 0, \ldots, N-1$ and thus $B_k(\ell^*(x_{u^\star}(k, x))) \leq \gamma_k^R \ell^*(x_{u^\star}(k, x))$ holds. Hence, by Remark 6.16(ii) Inequality (5.1) holds for all $x \in S_N$. In particular, this implies $V_N(f(x, \mu_N(x))) \leq V_N(x)$ for all $x \in S_N$ and thus by definition of S_N as a sublevel set of V_N this set is forward invariant. Hence, Theorem 4.11 can be applied with $S(n) = S_N$. Together with Lemma 6.34(i) and (6.41) this proves semiglobal asymptotic stability and with $\alpha_0 = 1/C$ we obtain the estimate for $J_\infty^{cl}(x, \mu_N)$. □

Let us now turn to practical (and semiglobal) stability. We have seen so far that a global linearity assumption on the B_k implies global stability while a "semiglobal" linearity assumption, i.e., the existence of a linear upper bound for B_k on each interval of the form $[0, R]$, implies semiglobal stability. This observation naturally leads to the conjecture that a "semiglobal practical" linearity assumption, i.e., a linear bound on the B_k on each interval $[\rho, R]$ with $R > \rho > 0$ should be sufficient for semiglobal practical stability. As we will see, this is indeed the case, however, we can formulate this condition in an even weaker way by simply assuming the existence of $\overline{\alpha} \in \mathcal{K}_\infty$ with $B_k(r) \le \overline{\alpha}(r)$ for all $k \in \mathbb{N}$ and all $r \ge 0$. This is because on each interval $[\rho, R]$ for $R > \rho > 0$ any $\mathcal{K}_\infty$-function $\overline{\alpha}$ can be bounded from above by the linear function $r \mapsto \gamma r$ for $\gamma = \max_{r\in[\rho,R]} \overline{\alpha}(r)/r$. Hence, any $\mathcal{K}_\infty$ function automatically satisfies a "semiglobal practical" linearity assumption.

Before we can formulate the respective Theorem 6.37, we have to provide a technical lemma which we will need in its proof. Without the linearity assumption the functions B_k appearing in the constraints (6.11), (6.12) in (6.14) become nonlinear functions. Hence, (6.14) does no longer reduce to the linear problem (6.16) for which our formula (6.18) is valid. In the semiglobal case in Theorem 6.35 we could circumvent this problem in the proof by ensuring that all finite time optimal trajectories starting in S_N stay in the region where B_k is linear. In the following semiglobal practical case, we have to cope with nonlinearities in the $B_k(r)$ not only for large r but also for small r which correspond to small neighborhoods of x_*. Since there is no way to exclude that the finite time optimal trajectories enter small neighborhoods of x_*—after all, this is precisely what we want them to do when we minimize the distance from x_*—we cannot use the same trick as in the proof of Theorem 6.35. Instead, we show in the following lemma that changing the B_k in a region where B_k is small does only slightly change the optimal value of (6.14), at least for trajectories starting sufficiently far away from 0, i.e., for values λ_0 in (6.14) which are bounded from below by some sufficiently large constant ζ. This statement is made precise in the following lemma.

Lemma 6.36 *Consider increasing functions $B_k^i : \mathbb{R}_0^+ \to \mathbb{R}_0^+$ for $k \in \mathbb{N}$ and $i = 1, 2$. Assume that these functions satisfy $B_k^i(r) \ge r$ for all $k \in \mathbb{N}$, $r \ge 0$ and that there exist constants $\sigma, \rho > 0$ with*

$$|B_k^i(r)| \le \sigma \quad \text{for all } r \le \rho \text{ and } k \in \mathbb{N}$$

for $i = 1, 2$ and

$$B_k^1(r) = B_k^2(r) \quad \text{for all } r \ge \rho \text{ and } k \in \mathbb{N}.$$

For $i = 1, 2$ and a constant $\zeta \ge \rho$ consider the optimization problems

$$\begin{aligned} \alpha^i := \inf_{\lambda_0,\dots,\lambda_{N-1},\nu} \ & \frac{\sum_{n=0}^{N-1} \lambda_n - \nu}{\lambda_0} \\ & \text{subject to the constraints (6.11), (6.12) with } B_k = B_k^i, \text{ and} \\ & \lambda_0 \ge \zeta, \lambda_1, \dots, \lambda_{N-1}, \nu \ge 0. \end{aligned} \tag{6.42}$$

Then the inequality $|\alpha^1 - \alpha^2| \le \sigma/\zeta$ *holds.*

Proof We show the inequality $\alpha^1 \le \alpha^2 + \sigma/\zeta$. Then the assertion follows by symmetry of the two problems.

In order to show the assertion, fix $\varepsilon > 0$ and pick ε-optimal values $\lambda_0^2, \ldots, \lambda_{N-1}^2$, ν^2, i.e., values which satisfy the constraints in (6.42) for $i = 2$ and

$$\frac{\sum_{n=0}^{N-1} \lambda_n^2 - \nu^2}{\lambda_0^2} \le \alpha^2 + \varepsilon.$$

The proof now consists in constructing λ_k^1, ν^1 satisfying the constraints in (6.42) for $i = 1$ and

$$\frac{\sum_{n=0}^{N-1} \lambda_n^1 - \nu^1}{\lambda_0^1} \le \alpha^2 + \varepsilon + \sigma/\zeta.$$

To this end, we distinguish two cases:

Case 1: $\lambda_n^2 \ge \rho$ for all $n \in \{0, \ldots, N-1\}$. In this case $B_k^1(\lambda_n^2)$ and $B_k^2(\lambda_n^2)$ coincide, hence $\lambda_n^1 := \lambda_n^2$, $n = 0, \ldots, N-1$, and $\nu^1 := \nu^2$ satisfy the constraints (6.11), (6.12) for $B_k = B_k^1$. This implies

$$\alpha^1 \le \frac{\sum_{n=0}^{N-1} \lambda_n^1 - \nu^1}{\lambda_0^1} = \frac{\sum_{n=0}^{N-1} \lambda_n^2 - \nu^2}{\lambda_0^2} = \alpha^2 + \varepsilon$$

Case 2: $\lambda_n^2 < \rho$ for some $n \in \{0, \ldots, N-1\}$. In this case, let $n^* \in \{0, \ldots, N-1\}$ be minimal with $\lambda_{n^*}^2 < \rho$, which implies $B_{N-n^*+1}^2(\lambda_{n^*}^2) \le \sigma$. Since $\lambda_0^2 \ge \zeta \ge \rho$ we obtain $n^* \ge 1$. From (6.12) with $j = n^* - 1$ it follows that

$$\nu^2 \le \sum_{n=1}^{n^*-1} \lambda_n^2 + B_{N-n^*+1}(\lambda_{n^*}^2) \le \sum_{n=1}^{n^*-1} \lambda_n^2 + \sigma. \tag{6.43}$$

We now set $\lambda_n^1 := \lambda_n^2$ for $n = 0, \ldots, n^* - 1$, $\lambda_n^1 := 0$, $n = n^*, \ldots, N-1$ and $\nu^1 := \max\{\nu^2 - \sigma, 0\}$. This definition implies

$$B_{N-k}^1(\lambda_k^1) = B_{N-k}^2(\lambda_k^2) \quad \text{for } k = 0, \ldots, n^* - 1,$$

$$\sum_{n=k}^{N-1} \lambda_k^1 \le \sum_{n=k}^{N-1} \lambda_k^2 \quad \text{for } k = 0, \ldots, N-1$$

and

$$\sum_{n=k}^{N-1} \lambda_k^1 = 0 \quad \text{for } k = n^*, \ldots, N-1,$$

which implies (6.11) for $B_k = B_k^1$. Since ν_2 satisfies (6.12), for ν_1 we get the inequality

$$\nu^1 \le \nu^2 \le \sum_{n=0}^{j-1} \lambda_{n+1}^2 + B_{N-j}^2(\lambda_{j+1}^2) = \sum_{n=0}^{j-1} \lambda_{n+1}^1 + B_{N-j}^1(\lambda_{j+1}^1) \quad \text{for } j \le n^* - 2.$$

In case $\nu^1 = 0$ we furthermore get

$$\nu^1 = 0 \le \sum_{n=0}^{j-1} \lambda_{n+1}^1 + B_{N-j}^1(\lambda_{j+1}^2) \quad \text{for } j \ge n^* - 1$$

and in case $\nu^1 = \nu^2 - \sigma$ from (6.43) and the definition of the λ_n^1 we obtain

$$\nu^1 = \nu^2 - \sigma \le \sum_{n=0}^{n^*-2} \lambda_{n+1}^2 \le \sum_{n=0}^{j-1} \lambda_{n+1}^1 + B_{N-j}^1(\lambda_{j+1}^1) \quad \text{for } j \ge n^* - 1.$$

This shows (6.12) for $B^1 = B_k^1$. Thus, since $\lambda_0^1 = \lambda_0^2 \ge \zeta$, the values $\lambda_0^1, \dots, \lambda_{N-1}^1, \nu^1$ satisfy all constraints in (6.42) and we can conclude

$$\alpha^1 \le \frac{\sum_{n=0}^{N-1} \lambda_n^1 - \nu^1}{\lambda_0^1} \le \frac{\sum_{n=0}^{N-1} \lambda_n^2 - \nu^1}{\lambda_0^2} \le \frac{\sum_{n=0}^{N-1} \lambda_n^2 - \nu^2 + \sigma}{\lambda_0^2} \le \alpha^2 + \varepsilon + \frac{\sigma}{\zeta}$$

where we used $\lambda_0^1 = \lambda_0^2$ and $\lambda_0^2 \ge \zeta$.

Thus, in both cases we obtain

$$\alpha^1 \le \alpha^2 + \varepsilon + \frac{\sigma}{\zeta}$$

which shows the assertion since $\varepsilon > 0$ was arbitrary. □

Now we are able to prove our main result on semiglobal practical asymptotic stability of the NMPC closed loop. Here we define the finite horizon closed-loop performance $J_{k^*}^{cl}$ analogous to (4.10) with k^* instead of ∞ as upper summation limit.

Theorem 6.37 (Practical stability without terminal conditions) *Consider the NMPC Algorithm 3.1 with optimization horizon $N \in \mathbb{N}$ and stage cost ℓ satisfying $\alpha_3(|x|_{x_*}) \le \ell^*(x) \le \alpha_4(|x|_{x_*})$ for suitable $\alpha_3, \alpha_4 \in \mathcal{K}_\infty$. Assume that Assumption 6.3 holds for functions $B_K \in \mathcal{K}_\infty$ which satisfy*

$$B_K(r) \le \overline{\alpha}(r)$$

for some $\overline{\alpha} \in \mathcal{K}_\infty$, all $K \in \mathbb{N}$ and all $r \ge 0$. Then the nominal NMPC closed-loop system (3.5) with NMPC-feedback law μ_N is semiglobally practically asymptotically stable on $\mathbb{X}$ with respect to the optimization horizon N.

Furthermore, for each $C > 1$ and each $\Delta > \delta > 0$ there exists $N_{C,\delta,\Delta} > 0$ such that

$$J^{cl}_{k^*}(x, \mu_N) \le C V_N(x) \le C V_\infty(x)$$

for all $x \in \mathscr{B}_\Delta(x_*) \cap \mathbb{X}$ and all $N \ge N_{C,\delta,\Delta}$ where $k^* \in \mathbb{N}_0$ is the minimal time instant with $|x_{\mu_N}(k^*, x)|_{x_*} \le \delta$.

Proof We first show the a priori estimate

$$V_N(f(x, \mu_N(x))) \le \overline{\alpha}(V_N(x)) \tag{6.44}$$

for all $N \ge 2$ and all $x \in \mathbb{X}$. Indeed, we have

$$V_N(x) = \sum_{k=0}^{N-1} \ell(x_{u_*}(k, x), u_*(k)) \ge \ell^*(x_{u_*}(1, x)) = \ell^*(f(x, \mu_N(x))).$$

By Assumption 6.3 this implies

$$V_N(f(x, \mu_N(x))) \le B_N(\ell^*(f(x, \mu_N(x)))) \le \overline{\alpha}(\ell^*(f(x, \mu_N(x)))) \le \overline{\alpha}(V_N(x)),$$

i.e., (6.44). Furthermore, we observe that with $\alpha_1 = \alpha_3$ and $\alpha_2 = \overline{\alpha} \circ \alpha_4$ the inequalities

$$\alpha_1(|x|_{x_*}) \le V_N(x) \le \alpha_2(|x|_{x_*})$$

hold for all $N \ge 2$ and all $x \in \mathbb{X}$.

Now we fix arbitrary $\Delta > \delta > 0$. We pick $R > 0$ as in the proof of Theorem 6.35 and define the values

$$r_0 := \alpha_1(\delta), \quad r_1 := \overline{\alpha}^{-1}(r_0) \quad \text{and} \quad \zeta := \overline{\alpha}^{-1}(r_1).$$

These definitions yield the implications

$$V_N(x) \le r_0 \quad \Rightarrow \quad |x|_{x_*} \le \delta, \tag{6.45}$$

$$V_N(x) \le r_1 \quad \Rightarrow \quad V_N(f(x, \mu_N(x))) \le r_0 \tag{6.46}$$

and

$$V_N(x) \ge r_1 \quad \Rightarrow \quad \overline{\alpha}(\ell^*(x)) \ge r_1 \quad \Rightarrow \quad \ell^*(x) \ge \zeta, \tag{6.47}$$

where we used (6.44) for (6.46) and Assumption 6.3 together with the bound $\overline{\alpha}$ on the B_k for (6.47).

Now, we pick $\alpha_0 \in (0, 1)$, set $\sigma = (1 - \alpha_0)\zeta/2$ and $\rho = \overline{\alpha}^{-1}(\sigma)$. Defining $\gamma := \max_{r \in [\rho, R]} \overline{\alpha}(r)/r$ we obtain

$$B_k(r) \le \overline{\alpha}(r) \le \gamma r \quad \text{for all } r \in [\rho, R].$$

Defining further

$$B_k^1(r) = \gamma r \quad \text{and} \quad B_k^2(r) = \begin{cases} \max\{\gamma r, B_k(r)\}, & r \in [0, R] \\ \gamma r, & r \geq R \end{cases}$$

we get $B_k^1(r) = B_k^2(r)$ for $r \geq \rho$ and $B_k^1(r) \leq B_k^2(r) \leq \overline{\alpha}(r) \leq \sigma$ for $r \in [0, \rho]$. Hence, B_k^1 and B_k^2 satisfy the assumptions of Lemma 6.36. Since $B_k^1(r)$ is linear in r, by the same arguments as in the proof of Theorem 6.24 we find $N_{\delta,\Delta} > 0$ such that (6.18) for $B_k = B_k^1$ yields $\underline{\alpha}_N \geq \alpha_0/2 + 1/2$ for all $N \geq N_{\delta,\Delta}$. This implies $\alpha^1 \geq \alpha_0/2 + 1/2$ in Lemma 6.36 and consequently

$$\alpha^2 \geq \alpha_0/2 + 1/2 - \sigma/\zeta = \alpha_0/2 + 1/2 - (1-\alpha_0)/2 = \alpha_0.$$

Now using the set $S_N = V_N^{-1}([0, L]) \supseteq \mathscr{B}_\Delta(x_*)$ defined in the proof of Theorem 6.35, as in this proof we obtain that each optimal trajectory starting in $x \in S_N$ satisfies $\ell^*(x_{u^*}(k, x)) \leq R$. Setting $Y = S_N \setminus V_N^{-1}([0, r_1])$, by (6.47) we furthermore obtain $\ell^*(x) \geq \zeta$ for all $x \in Y$. Hence, for this set Y the variant of the optimization problem (6.14) obtained from Remark 6.16(i) and (ii) coincides with the optimization problem from Lemma 6.36 with $i = 2$, yielding $\alpha = \alpha^2 \geq \alpha_0$ in Theorem 6.15. Consequently, we obtain (5.1) with $\alpha = \alpha_0$ for all $x \in Y$.

We claim that this implies that the sublevel set $V_N^{-1}([0, r_0])$ is forward invariant for the closed-loop system, i.e., that $V_N(f(x, \mu_N(x))) \leq r_0$ holds whenever $V_N(x) \leq r_0$ holds. Indeed, if $V_N(x) \in [r_1, r_0]$, then we have $x \in Y$ and thus (5.1) holds with $\alpha = \alpha_0 > 0$, which implies $V_N(f(x, \mu_N(x))) \leq V_N(x) \leq r_0$. On the other hand, if $V_N(x) \leq r_1$ then (6.46) yields $V_N(f(x, \mu_N(x))) \leq r_0$.

Thus, defining $P = V_N^{-1}([0, r_0])$ and $S = S_N$, all assumptions of Theorem 4.14 are satisfied and furthermore the inclusions $\overline{\mathscr{B}}_\Delta(x_*) \subseteq S$ and $P \subseteq \overline{\mathscr{B}}_\delta(x_*)$ hold by (6.41) and (6.45). Hence, Theorem 4.14 yields semiglobal practical stability using Lemma 6.34(ii) and the estimate for $J_{k^*}(x, \mu_N)$ by choosing $\alpha_0 = 1/C$. □

We end this section by a simple example which illustrates the practical stability.

Example 6.38 Consider the control system (2.1), i.e.,

$$x^+ = x + u,$$

with equilibrium $x_* = 0$ and stage cost

$$\ell(x, u) = x^2 + |u|.$$

The system is controllable to 0 in finite time, which is easily seen if for initial value x we choose $u_x(0) = -x$ and $u_x(n) = 0$ for $n \geq 1$. The resulting trajectories satisfy $x_{u_x}(n) = 0$ for $n \geq 1$. In particular, we obtain

$$V_N(x) \leq \sum_{n=0}^{N-1} \ell(x_{u_x}(n), u_x(n)) = \ell(x, u_x(0)) = x^2 + |x|$$

and since $\ell^*(x) = x^2$ this shows that Assumption 6.3 holds with $B_K(r) = r + \sqrt{r}$. In particular, the assumptions of Theorem 6.37 hold with $\overline{\alpha}(r) = r + \sqrt{r}$ which ensures semiglobal practical asymptotic stability of the NMPC closed loop.

On the other hand, for the given stage cost Assumption 6.3 does not hold with uniformly linearly bounded functions B_K. We show this fact by contradiction: Suppose there exists $\gamma \in \mathbb{R}$ auch that Assumption 6.3 holds with $B_K(r) \leq \gamma r$ for all $K \in \mathbb{N}$ and all $r \geq 0$. Then Assumption 6.3 implies for any $x \in \mathbb{R}$ and the corresponding optimal control sequence $u_N^\star$ with horizon N

$$\sum_{n=0}^{N-1} |u_N^\star(n)| \leq \sum_{n=0}^{N-1} \ell(x_{u_N^\star}(n), u_N^\star(n)) = V_N(x) \leq B_N(\ell^*(x)) \leq \gamma x^2. \qquad (6.48)$$

This implies

$$|x_{u_N^\star}(n, x)| \geq |x| - \gamma n x^2.$$

Together with (6.48) we obtain

$$\gamma x^2 \geq \sum_{n=0}^{N-1} \ell(x_{u_N^\star}(n), u_N^\star(n)) \geq \sum_{n=0}^{N-1} |x_{u_N^\star}(n, x)|^2 \geq \sum_{n=0}^{N-1} (|x| - \gamma n x^2)^2.$$

Fixing $0 < \eta < 1$ and choosing $|x|$ so small that $\gamma(N-1)x^2 < \eta|x|$, this inequality yields

$$\gamma x^2 \geq N(1-\eta)x^2.$$

This, however, is not possible for $N > \gamma/(1-\eta)$, hence Assumption 6.3 cannot hold with $B_K(r) \leq \gamma r$.

As a consequence, Theorem 6.24 is not applicable and we cannot expect asymptotic stability of the closed loop. The numerical simulations shown in Fig. 6.13 confirm this behavior. From left to right the closed-loop trajectory for $N = 2, 3, 4$ with

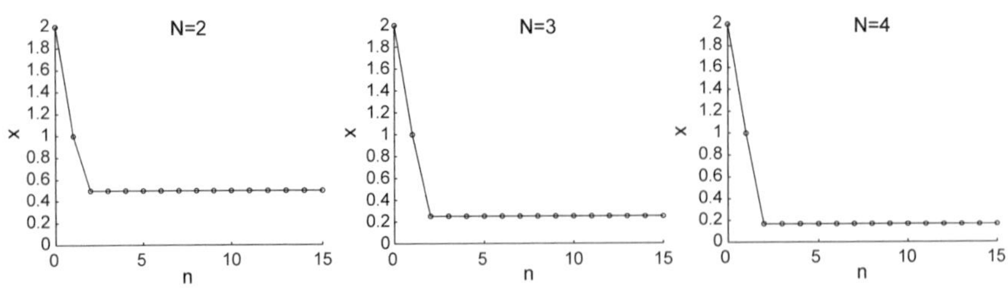

Fig. 6.13 Closed-loop behavior for optimization horizons $N = 2, 3, 4$

$x_0 = 2$ is shown. As Theorem 6.37 predicts, the solutions converge to smaller and smaller neighborhoods of $x_* = 0$ as N increases, but they do not converge to $x_* = 0$ for fixed N.

6.8 Proof of Proposition 6.18

In this section we provide the proof of Proposition 6.18. We start by recalling that Assumption 6.3 and $V_N(x) \geq \ell^*(x)$ imply $B_k(r) \geq r$ and thus $\gamma_k \geq 1$.

Now, the main part of the proof consists of three steps. In the first step, we transform (6.16) into an equivalent form more suitable for our analysis. In the second step we show that $\underline{\alpha}_N$ from (6.18) is the explicit solution of this equivalent problem if we remove some of the constraints. Since the solution of the minimization problem with fewer constraints is always less or equal than the solution of the problem with all constraints, this proves (6.17). Finally, in the third step we show that under condition (6.9) the removed constraints are always satisfied for the optimal solution of the problem from Step 2, which shows (6.19). Some technical equalities we need throughout the proof are collected in Lemma 6.40 at the end of this section.

Step 1: The optimal value α of (6.16) equals the optimal value of the following optimization problem:

$$\min_{\lambda} \; 1 - (\gamma_2 - 1)\lambda_{N-1} \tag{6.49}$$

subject to the (componentwise) constraints $\lambda = (\lambda_1, \ldots, \lambda_{N-1})^\top \geq 0$ and

$$\sum_{n=1}^{N-2} \lambda_n + \lambda_{N-1} \leq \gamma_N - 1 \tag{6.50}$$

$$\sum_{n=j}^{N-2} \lambda_n - \gamma_{N-j}\lambda_j + \lambda_{N-1} \leq 0, \qquad j = 1, \ldots, N-2 \tag{6.51}$$

$$\sum_{n=j}^{N-2} \lambda_n - \gamma_{N-j+1}\lambda_j + \gamma_2\lambda_{N-1} \leq 0, \qquad j = 1, \ldots, N-2 \tag{6.52}$$

Proof of Step 1: We first show that for the optimal values $\lambda_0, \ldots, \lambda_{N-1}$ and ν in (6.16) the inequality (6.12) for $j = N-2$ is an equality. To this end, assume that inequality (6.12) for $j = N-2$ is strict, i.e., that it holds with $<$.

If $\lambda_{N-1} > 0$, then we can—at least slightly—reduce λ_{N-1} without violating the inequalities (6.11) and (6.12). Since this reduces the value under the minimum in (6.16), this contradicts the optimality of $\lambda_0, \ldots, \lambda_{N-1}$ and ν.

If $\lambda_{N-1} = 0$, then the strict inequality (6.12) for $j = N-2$ and the inequality $\lambda_{N-2} \leq B_3(\lambda_{N-2})$ implies that (6.12) for $j = N-3$ must be strict, too. Thus, assuming $\lambda_{N-2} > 0$ leads to a contradiction similarly to the case $\lambda_{N-1} > 0$, above. Proceeding inductively yields $\lambda_1 = \lambda_2 = \ldots = \lambda_{N-1} = 0$ and consequently the right hand side of (6.12) for $j = N-2$ equals zero. Since $\nu \geq 0$ this contradicts this inequality being strict.

Since we have shown that equality holds in (6.12) for $j = N-2$, we obtain the expression

$$\nu = \sum_{n=1}^{N-2} \lambda_n + B_2(\lambda_{N-1}) = \sum_{n=1}^{N-2} \lambda_n + \gamma_2 \lambda_{N-1}. \tag{6.53}$$

Inserting this into the expression under the minimum in (6.16) and using $\lambda_0 = 1$ yields

$$\sum_{n=0}^{N-1} \lambda_n - \sum_{n=1}^{N-2} \lambda_n - \gamma_2 \lambda_{N-1} = \lambda_0 + \lambda_{N-1} - \gamma_2 \lambda_{N-2} = 1 - (\gamma_2 - 1)\, \lambda_{N-1}.$$

This shows that the optimization objectives in (6.16) and (6.49) coincide. In order to show that the optimal values coincide, it hence remains to show that the constraints (6.11), (6.12) with $\lambda_0 = 1$ are equivalent to (6.50)–(6.52).

To this end, we first note that (6.11) for $k = 0$ becomes

$$\sum_{n=0}^{N-1} \lambda_n \leq \gamma_N \lambda_0 = \gamma_N$$

which is equivalent to (6.50) since $\lambda_0 = 1$.

The remaining inequalities (6.11) for $k = 1, \ldots, N-2$ can be rewritten as

$$\lambda_{N-1} \leq \gamma_{N-k} \lambda_k - \sum_{n=k}^{N-2} \lambda_n, \quad k = 1, \ldots, N-2$$

which is exactly (6.51) if we change the index from k to j.

Inserting the expression (6.53) into (6.12) and shifting the summation index by 1, the inequalities (6.12) can be equivalently rewritten as

$$\gamma_2 \lambda_{N-1} \leq \gamma_{N-j+1} \lambda_j - \sum_{n=j}^{N-2} \lambda_n, \quad j = 1, \ldots, N-2.$$

which is (6.52). This shows the claim in Step 1.

In the following second step, we remove the constraints (6.51) from Problem (6.49) and provide an explicit solution for this relaxed problem.

Step 2: The optimization problem (6.49) with constraints (6.50), (6.52) and $\lambda \geq 0$ has the solution

$$\min_{\lambda} \; 1 - (\gamma_2 - 1)\lambda_{N-1} \;\; = \;\; \underline{\alpha}_N$$

with $\underline{\alpha}_N$ from (6.18). Furthermore, $\lambda^\star = (\lambda_1^\star, \ldots, \lambda_{N-1}^\star)$ with

$$\lambda_{N-1-i}^\star = -\left(\prod_{j=1}^{i-1} \frac{d_{N-1-j} - 1}{d_{N-1-j}}\right) \frac{\gamma_2}{d_{N-1-i}} \lambda_{N-1}^\star, \quad i = 1, \ldots, N-2 \tag{6.54}$$

with $d_j = 1 - \gamma_{N-j+1}$ is a corresponding minimizer.

Proof of Step 2: First observe that $\gamma_2 = 1$ implies $\min_\lambda \ 1 - (\gamma_2 - 1)\lambda_{N-1} = 1$ and that $\lambda_1 = \ldots = \lambda_{N-1} = 0$ is a minimizer. In this case one easily verifies the assertion of Step 2, hence in what follows we will assume $\gamma_2 > 1$ which implies $\gamma_j > 1$ for all $j \geq 2$.

We can equivalently rewrite the constraints (6.50), (6.52) compactly as $A\lambda \leq b$ (interpreted componentwise), where

$$A := \begin{pmatrix} 1 & 1 & \ldots & 1 & 1 \\ d_1 & 1 & \ldots & 1 & \gamma_2 \\ 0 & d_2 & \ddots & \vdots & \vdots \\ \vdots & \ddots & \ddots & 1 & \gamma_2 \\ 0 & \ldots & 0 & d_{N-2} & \gamma_2 \end{pmatrix} \quad \text{and} \quad b := \begin{pmatrix} \gamma_N - 1 \\ 0 \\ \vdots \\ 0 \\ 0 \end{pmatrix}$$

with d_j defined as above. This equivalence follows since the first inequality in $A\lambda \leq b$ is equivalent to (6.50) while the remaining $N-2$ inequalities are equivalent to (6.52), $j = 1, \ldots N-2$.

Now denote by $\lambda^\star \geq 0$ a minimizer of (6.49) satisfying the constraints $A\lambda^\star \leq b$. We show by contradiction that $A\lambda^\star = b$ holds. To this end, assume that $A\lambda^\star \neq b$ holds, i.e., that there exists $k \in \{1, \ldots, N-1\}$ such that

$$\sum_{n=1}^{N-1} A_{kn}\lambda_n^\star < b_k \tag{6.55}$$

holds. In order to obtain the contradiction, note that $\gamma_2 > 1$ implies that minimizing (6.49) is equivalent to maximizing λ_{N-1}.

If (6.55) holds for $k = 1$, then we define the constants

$$\varepsilon := b_1 - \sum_{n=1}^{N-1} A_{1n}\lambda_n^\star > 0, \quad \delta := -\max_{i=1,\ldots,N} d_i > 0$$

and choose $\tilde{\varepsilon} > 0$ such that

$$\tilde{\varepsilon}\left(1 + \gamma_2 \sum_{i=1}^{N-2} \frac{(1+\delta)^{N-2-i}}{\delta^{N-1-i}}\right) \leq \varepsilon.$$

We set $\lambda_{N-1} = \lambda^\star_{N-1} + \tilde{\varepsilon}$ and

$$\lambda_i = \lambda_i^\star + \tilde{\varepsilon}\gamma_2 \frac{(1+\delta)^{N-2-i}}{\delta^{N-1-i}}, \quad i = 1, \ldots, N-2.$$

This implies

$$\sum_{n=1}^{N-1} A_{1n}\lambda_n = \sum_{n=1}^{N-1} A_{1n}\lambda_n^\star + \tilde{\varepsilon} + \sum_{n=2}^{N-1} \tilde{\varepsilon}\gamma_2 \frac{(1+\delta)^{N-2-i}}{\delta^{N-1-i}} \le A_{1n}\lambda_n^\star + \varepsilon = b_1$$

and, for $k = 2, \ldots, N-1$,

$$\begin{aligned}\sum_{n=1}^{N-1} A_{kn}\lambda_n &= \sum_{n=1}^{N-2} A_{kn}\left(\lambda_n^\star + \tilde{\varepsilon}\gamma_2 \frac{(1+\delta)^{N-2-n}}{\delta^{N-1-n}}\right) + A_{kN-1}(\lambda^\star_{N-1} + \tilde{\varepsilon}) \\ &= \sum_{n=1}^{N-1} A_{kn}\lambda_n^\star + d_{k-1}\tilde{\varepsilon}\gamma_2 \frac{(1+\delta)^{N-1-k}}{\delta^{N-k}} + \sum_{n=k}^{N-2} \tilde{\varepsilon}\gamma_2 \frac{(1+\delta)^{N-2-n}}{\delta^{N-1-n}} + \gamma_2\tilde{\varepsilon}\end{aligned}$$

Now we can estimate

$$\begin{aligned}& d_{k-1}\tilde{\varepsilon}\gamma_2 \frac{(1+\delta)^{N-1-k}}{\delta^{N-k}} + \sum_{n=k}^{N-2} \tilde{\varepsilon}\gamma_2 \frac{(1+\delta)^{N-2-n}}{\delta^{N-1-n}} + \gamma_2\tilde{\varepsilon} \\ =\ & \tilde{\varepsilon}\left(d_{k-1}\gamma_2 \frac{(1+\delta)^{N-1-k}}{\delta^{N-k}} + \sum_{n=k}^{N-2} \gamma_2 \frac{(1+\delta)^{N-2-n}}{\delta^{N-1-n}} + \gamma_2\right) \\ \le\ & \tilde{\varepsilon}\left(-\delta\gamma_2 \frac{(1+\delta)^{N-1-k}}{\delta^{N-k}} + \sum_{n=k}^{N-2} \gamma_2 \frac{(1+\delta)^{N-2-n}}{\delta^{N-1-n}} + \gamma_2\right) \\ =\ & \frac{\tilde{\varepsilon}\gamma_2}{\delta^{N-1-k}}\left(-(1+\delta)^{N-1-k} + \sum_{n=0}^{N-2-k} (1+\delta)^{N-2-k-n} + \delta^{N-1-k}\right) \ = \ 0\end{aligned}$$

where we used (6.62) in the last step. This shows

$$\sum_{n=1}^{N-1} A_{kn}\lambda_n \le \sum_{n=1}^{N-1} A_{kn}\lambda_n^\star \le b_k.$$

Thus, we have constructed a vector $\lambda \ge 0$ satisfying the constraints $A\lambda \le b$ and $\lambda_{N-1} > \lambda^\star_{N-1}$. Since λ_{N-1} must be maximal for the optimal solution, this contradicts the optimality of $\lambda^\star$. Hence, (6.55) cannot hold for $k = 1$.

Now assume (6.55) for some $k \ge 2$. Let $k^\star$ be maximal such that (6.55) holds for $k = k^\star$. Then, since $d_j < 0$, $\lambda^\star_{k^\star-1}$ is the only entry with negative sign in this inequality and thus it must be strictly positive since $b_{k^*} = 0$. On the other hand, $\lambda^\star_{k^\star-1}$

appears with positive sign in all inequalities for $k \leq k^\star - 1$ and it does not appear at all in all inequalities for $k \geq k^\star + 1$. Thus, for $\varepsilon > 0$ sufficiently small the sequence

$$\lambda = (\lambda_1, \ldots, \lambda_{N-1}) = (\lambda_1^\star, \ldots, \lambda_{k^\star-2}^\star, \lambda_{k^\star-1}^\star - \varepsilon, \lambda_{k^\star}^\star, \ldots, \lambda_{N-1}^\star)$$

satisfies the constraints $A\lambda \leq b$ and yields the same optimal value in (6.49) as $\lambda^\star$. Thus, λ is optimal, too. However, the inequality $\lambda_{k^\star-1} < \lambda_{k^\star-1}^\star$ implies that (6.55) holds for λ and $k = 1$. By the first part of the proof, this contradicts the optimality of λ. Hence, (6.55) cannot hold for $k \geq 2$.

The considerations made so far show that the optimal $\lambda^\star$ satisfies $A\lambda^\star = b$. We use this linear system of equations in order to prove (6.54) by induction over i.

For $i = 1$, (6.54) follows immediately from the last equation in $A\lambda^\star = b$. For the induction step $i - 1 \to i$ we use the $(N - i)$th equation in $A\lambda^\star = b$ in order to obtain

$$\lambda_{N-1-i}^\star = \frac{1}{-d_{N-1-i}} \left(\gamma_2 \lambda_{N-1}^\star + \sum_{k=1}^{i-1} \lambda_{N-1-k}^\star \right).$$

Using the induction assumption, i.e., (6.54) for $i - 1$ instead of i we can continue

$$\begin{aligned}
&\frac{1}{-d_{N-1-i}} \left(\gamma_2 \lambda_{N-1}^\star + \sum_{k=1}^{i-1} \lambda_{N-1-k}^\star \right) \\
&= \frac{\gamma_2 \lambda_{N-1}^\star}{-d_{N-1-i}} \left(1 - \sum_{k=1}^{i-1} \prod_{j=1}^{k-1} \frac{d_{N-1-j} - 1}{d_{N-1-j}} \frac{1}{d_{N-1-k}} \right) \\
&= \frac{\gamma_2 \lambda_{N-1}^\star}{-d_{N-1-i}} \left(1 + \sum_{k=1}^{i-1} \prod_{j=1}^{k-1} \frac{1 - d_{N-1-j}}{-d_{N-1-j}} \frac{1}{-d_{N-1-k}} \right) \\
&= \frac{\gamma_2 \lambda_{N-1}^\star}{\prod_{j=1}^{i} (-d_{N-1-j})} \sum_{k=0}^{i-1} \left(\prod_{j=1}^{k-1} (1 - d_{N-1-j}) \prod_{j=k+1}^{i-1} (-d_{N-1-j}) \right) \\
&= \frac{\gamma_2 \lambda_{N-1}^\star}{\prod_{j=1}^{i} (-d_{N-1-j})} \sum_{k=0}^{i-1} \prod_{j=1}^{i-1} (1 - d_{N-1-j}) \\
&= - \left(\prod_{j=1}^{i-1} \frac{d_{N-1-j} - 1}{d_{N-1-j}} \right) \frac{\gamma_2}{d_{N-1-i}} \lambda_{N-1}^\star
\end{aligned}$$

where in the second last step we have used Lemma 6.40(i) with $\delta_j = -d_j$. This shows (6.54) for i.

Finally, we use this formula in order to show that $\underline{\alpha}_N$ from (6.18) is the optimal value for the problem defined in Step 2. To this end we rewrite the first equation of $A\lambda^\star = b$ as $\gamma_N - 1 - \lambda_{N-1}^\star = \sum_{k=1}^{N-2} \lambda_k$. Inserting (6.54) into this equation and using the definition $d_j = 1 - \gamma_{N-j+1}$ we obtain

$$
\begin{aligned}
\gamma_N - 1 - \lambda^\star_{N-1} = \sum_{k=1}^{N-2} \lambda_k &= \left(-\sum_{k=1}^{N-2} \left(\prod_{j=1}^{k-1} \frac{d_{N-1-j} - 1}{d_{N-1-j}} \right) \frac{\gamma_2}{d_{N-1-k}} \right) \lambda^\star_{N-1} \\
&= \left(\sum_{k=1}^{N-2} \left(\prod_{j=1}^{k-1} \frac{\gamma_{j+2}}{\gamma_{j+2} - 1} \right) \frac{\gamma_2}{\gamma_{k+2} - 1} \right) \lambda^\star_{N-1} \\
&= \prod_{j=1}^{N-2} \frac{1}{\gamma_{j+2} - 1} \left(\sum_{k=1}^{N-2} \left(\prod_{j=1}^{k-1} \gamma_{j+2} \prod_{j=k+1}^{N-2} (\gamma_{j+2} - 1) \right) \right) \gamma_2 \lambda^\star_{N-1} \\
&= \prod_{j=3}^{N} \frac{1}{\gamma_j - 1} \left(\sum_{k=3}^{N} \left(\prod_{j=3}^{k-1} \gamma_j \prod_{j=k+1}^{N} (\gamma_j - 1) \right) \right) \gamma_2 \lambda^\star_{N-1} \\
&= \prod_{j=3}^{N} \frac{1}{\gamma_j - 1} \left(\prod_{j=3}^{N} \gamma_j - \prod_{j=3}^{N} (\gamma_j - 1) \right) \gamma_2 \lambda^\star_{N-1} \\
&= \underbrace{\left(\prod_{j=3}^{N} \frac{\gamma_j}{\gamma_j - 1} - 1 \right)}_{=:\rho} \gamma_2 \lambda^\star_{N-1}
\end{aligned}
$$

where we used (6.63) in the second last equality. Solving for $\lambda^\star_{N-1}$ yields

$$
\lambda^\star_{N-1} = \frac{\gamma_N - 1}{\rho\gamma_2 + 1}
$$

and inserting this into (6.49) we obtain

$$
\min\ 1 - (\gamma_2 - 1)\lambda_{N-1} = 1 - (\gamma_2 - 1)\lambda^\star_{N-1} = 1 - \frac{(\gamma_2 - 1)(\gamma_N - 1)}{\rho\gamma_2 + 1}. \tag{6.56}
$$

The denominator of this fraction can be written as

$$
\begin{aligned}
\rho\gamma_2 + 1 = \left(\prod_{j=3}^{N} \frac{\gamma_j}{\gamma_j - 1} - 1 \right) \gamma_2 + 1 &= \frac{\prod_{j=3}^{N} \gamma_j}{\prod_{j=3}^{N} (\gamma_j - 1)} \gamma_2 - (\gamma_2 - 1) \\
&= \frac{\prod_{j=2}^{N} \gamma_j - \prod_{j=2}^{N} (\gamma_j - 1)}{\prod_{j=3}^{N} (\gamma_j - 1)}.
\end{aligned}
$$

Inserting this into (6.56) we finally obtain

$$\min\ 1-(\gamma_2-1)\lambda_{N-1}=1-\frac{(\gamma_2-1)(\gamma_N-1)\prod_{j=3}^{N}(\gamma_j-1)}{\prod_{j=2}^{N}\gamma_j-\prod_{j=2}^{N}(\gamma_j-1)}$$

which is exactly $\underline{\alpha}_N$ from (6.18). This finishes the proof of Step 2.

Let us summarize what we have proved so far: In Step 1, we have shown that Problem (6.16) can be equivalently reformulated as (6.49) subject to the constraints $\lambda \geq 0$, (6.50), (6.51) and (6.52). In Step 2, we have shown that the optimal value of the Problem (6.49) subject to the constraints $\lambda \geq 0$, (6.50) and (6.52) is exactly $\underline{\alpha}_N$ from (6.18). Since this is the optimal value of a minimization problem which is equivalent to (6.16) but with fewer constraints, $\underline{\alpha}_n$ must be less or equal than the optimal value of (6.16). Hence, we have shown (6.17).

The proof of the remaining Eq. (6.19) provided (6.9) holds is an immediate consequence of our final Step 3.

Step 3: If (6.9) holds and the B_k are defined by (6.7), then the optimal solution $\lambda^\star$ of Problem (6.49) subject to the constraints (6.50) and (6.52) satisfies the constraints (6.51).

Proof of Step 3: We prove the assertion by showing that for $\lambda = \lambda^\star$ the inequalities (6.51) for $j = 2, \ldots, N-2$ are implied by the respective inequalities (6.52). Since (6.52) holds for $\lambda^\star$ by definition of the constraints in Step 2, this shows (6.51).

To this end, it is sufficient to show that

$$-\gamma_{N-j}\lambda_j^\star+\lambda_{N-1}^\star\leq-\gamma_{N-j+1}\lambda_j^\star+\gamma_2\lambda_{N-1}^\star$$

or, equivalently,

$$(\gamma_2-1)\lambda_{N-1}^\star\geq(\gamma_{N-j+1}-\gamma_{N-j})\lambda_j^\star \tag{6.57}$$

holds for $j = 2, \ldots, N-2$. Inserting (6.54) one sees that (6.57) is equivalent to

$$\prod_{i=2}^{N-j+1}(\gamma_i-1)\geq(\gamma_{N-j+1}-\gamma_{N-j})\prod_{i=2}^{N-j}\gamma_i. \tag{6.58}$$

Now we define $c_n := \beta(r, n)/r$. Note that the c_n are well defined since β is linear in r and that (6.7) yields the identity $\gamma_k = \sum_{n=0}^{k-1} c_n$. Property (6.9) then implies[1]

$$c_{n+m}\leq c_n c_m \quad \text{for all} \quad n,m\in\mathbb{N}_0. \tag{6.59}$$

In order to prove (6.58) we prove the auxiliary inequality

$$\prod_{i=2}^{N-j}(\gamma_i-1)\sum_{n=k}^{N-j+k-1}c_n-c_{N-j+k-1}\prod_{i=2}^{N-j}\gamma_i\geq 0 \tag{6.60}$$

[1] In fact, (6.59) is the reason for calling (6.9) "submultiplicativity".

for arbitrary $k \in \mathbb{N}$ and $j = 1, \ldots, N-2$ by induction over j, starting with $j = N-2$. In this case for arbitrary $k \in \mathbb{N}$ we get

$$\begin{aligned}\prod_{i=2}^{N-j}(\gamma_i - 1) \sum_{n=k}^{N-j+k-1} c_n - c_{N-j+k-1}\prod_{i=2}^{N-j}\gamma_i &= (\gamma_2 - 1)(c_k + c_{k+1}) - c_{k+1}\gamma_2 \\ &= \gamma_2 c_k - c_k - c_{k+1} \\ &= c_0 c_k + c_1 c_k - c_k - c_{k+1} \\ &\geq c_k + c_{k+1} - c_k - c_{k+1} \;=\; 0\end{aligned}$$

using (6.59) for $m = k$ and $n = 0$ and 1 in the $\geq$-estimate. For the induction step $j+1 \to j$ and arbitrary $k \in \mathbb{N}$ we can write the right hand side of (6.60)

$$\begin{aligned}&\prod_{i=2}^{N-j}(\gamma_i - 1) \sum_{n=k}^{N-j+k-1} c_n - c_{N-j+k-1}\prod_{i=2}^{N-j}\gamma_i \\ &= (-1)\prod_{i=2}^{N-j-1}(\gamma_i - 1)\sum_{n=k}^{N-j+k-1} c_n \\ &\quad + \gamma_{N-j}\left[\prod_{i=2}^{N-j-1}(\gamma_i - 1)\sum_{n=k}^{N-j+k-1} c_n - c_{N-j+k-1}\prod_{i=2}^{N-j-1}\gamma_i\right] \\ &= \prod_{i=2}^{N-j-1}(\gamma_i - 1)\left(c_k\gamma_{N-j} - \sum_{n=k}^{N-j+k-1} c_n\right) \\ &\quad + \gamma_{N-j}\left[\prod_{i=2}^{N-j-1}(\gamma_i - 1)\sum_{n=k+1}^{N-j+k-1} c_n - c_{N-j+k-1}\prod_{i=2}^{N-j-1}\gamma_i\right].\end{aligned}$$

Now the induction assumption implies that the second summand is ≥ 0, since the term in square brackets is the right hand side of (6.60) with $j+1$ and $k+1$ instead of j and k. For the first summand, using (6.59) we can estimate

$$c_k\gamma_{N-j} = c_k\sum_{n=0}^{N-j-1} c_n = \sum_{n=0}^{N-j-1} c_k c_n \geq \sum_{n=0}^{N-j-1} c_{k+n} = \sum_{n=k}^{N-j+k-1} c_n.$$

This shows that the term in brackets and thus the whole first summand is ≥ 0, which proves (6.60).

Using the equality $\gamma_k = \sum_{n=0}^{k-1} c_k$ and $c_0 \geq 1$, the left hand side of the desired inequality (6.58) can be estimated by

$$\prod_{i=2}^{N-j+1}(\gamma_i - 1) = \prod_{i=2}^{N-j}(\gamma_i - 1)\left(\sum_{n=0}^{N-j} c_n - 1\right) \geq \prod_{i=2}^{N-j}(\gamma_i - 1)\sum_{n=1}^{N-j} c_n.$$

For the right hand side we obtain

$$(\gamma_{N-j+1} - \gamma_{N-j}) \prod_{i=2}^{N-j} \gamma_i = c_{N-j} \prod_{i=2}^{N-j} \gamma_i.$$

Hence, (6.60) for $k = 1$ implies (6.58) and thus (6.57). This proves Step 3.

Step 3 implies that the optimal solution $\underline{\alpha}_N$ from (6.18) of (6.49) subject to the constraints (6.50) and (6.52) obtained in Step 2 equals the optimal solution of (6.49) subject to the constraints (6.50)–(6.52). Since by Step 1 the latter problem is equivalent to (6.16), this proves (6.19) and thus finishes the proof of Proposition 6.18.

Remark 6.39 In Step 3, we have shown that under the condition (6.9) the conditions (6.51) are redundant in (6.49). Since in Step 1 we have shown that the conditions (6.51) are equivalent to (6.11) for $k = 1, \dots, N-2$, this shows that under condition (6.9) the optimal value of Problem (6.16) does not change if we remove the constraints (6.11) for $k = 1, \dots, N-2$. While this has no consequences for the results in this book—since we get these constraints for free from the optimality of the trajectory $x_{u^\star}(\cdot, x_0)$ via Lemma 6.11—this observation may be useful in other settings, e.g., when analyzing NMPC with nonoptimal trajectories.

We end this section with a technical lemma we needed in the preceding proof.

Lemma 6.40 *(i) For all $\delta_1, \dots, \delta_{N-2} \in \mathbb{R}$ and all $i \in \{1, \dots, N-1\}$ the equation*

$$\prod_{j=1}^{i-1} (1 + \delta_{N-1-j}) = \sum_{k=0}^{i-1} \left(\prod_{j=1}^{k-1} (1 + \delta_{N-1-j}) \prod_{j=k+1}^{i-1} \delta_{N-1-j} \right). \tag{6.61}$$

holds (with the usual convention $\prod_{j=j_1}^{j_2} = 1$ if $j_2 < j_1$).

(ii) For all $\delta \in \mathbb{R}$ and $k \in \{1, \dots, N-1\}$ the equation

$$(1+\delta)^{N-1-k} = \sum_{n=0}^{N-2-k} (1+\delta)^{N-2-k-n} \delta^n + \delta^{N-1-k} \tag{6.62}$$

holds.

(iii) For all $\gamma_3, \dots, \gamma_N \in \mathbb{R}$ the equation

$$\prod_{j=3}^{N} \gamma_j = \prod_{j=3}^{N} (\gamma_j - 1) + \sum_{k=3}^{N} \left(\prod_{j=3}^{k-1} \gamma_j \prod_{j=k+1}^{N} (\gamma_j - 1) \right) \tag{6.63}$$

holds.

Proof (i) We prove (6.61) by induction over i. For $i = 1$ the equality is obvious.

Under the induction assumption that (6.61) holds for $i-1$ instead of i we obtain

$$\begin{aligned}
\prod_{j=1}^{i-1}(1+\delta_{N-1-j}) &= \prod_{j=1}^{i-2}(1+\delta_{N-1-j}) + \delta_{N-1-(i-1)} \prod_{j=1}^{(i-1)-1} (1+\delta_{N-1-j}) \\
&= \prod_{j=1}^{i-2}(1+\delta_{N-1-j}) \\
&\quad + \ \delta_{N-1-(i-1)} \sum_{k=1}^{(i-1)-1} \left(\prod_{j=1}^{k-1}(1+\delta_{N-1-j}) \prod_{j=k+1}^{(i-1)-1} \delta_{N-1-j} \right) \\
&= \sum_{k=0}^{i-1} \left(\prod_{j=1}^{k-1}(1+\delta_{N-1-j}) \prod_{j=k+1}^{i-1} \delta_{N-1-j} \right),
\end{aligned}$$

i.e., (6.61) for i.

(ii) Formula (6.62) follows immediately from (6.61) by setting $\delta_1 = \ldots, \delta_{N-1} = \delta$ and $k = N - i$.

(iii) Using (6.61) with $\delta_{N-1-j} = \gamma_{j+2} - 1$ and $i = N - 1$ yields

$$\prod_{j=1}^{N-2} \gamma_{j+2} = \sum_{k=0}^{N-2} \left(\prod_{j=1}^{k-1} \gamma_{j+2} \prod_{j=k+1}^{N-2} (\gamma_{j+2} - 1) \right).$$

Writing the summand for $k = 0$ on the right hand side separately and using the summation indices $j + 2$ instead of j and $k + 2$ instead of k yields (6.63).

6.9 Notes and Extensions

The conceptual idea of establishing stability and performance of NMPC schemes via relaxed dynamic programming ideas as outlined in Sect. 6.1 was to our knowledge first used by Shamma and Xiong [17]. However, in this reference different inequalities than (5.1) were used and the inequalities were verified by numerical evaluation. In the form presented here, relaxed dynamic programming was introduced for the analysis of NMPC schemes in Grüne and Rantzer [12], where also a first controllability condition for verifying this inequality was given.

The asymptotic controllability condition with respect to ℓ from Sect. 6.2 was introduced in Grüne [10] and most of the results in Sects. 6.2–6.5 were taken from this reference with minor modifications and extensions. Exceptions are Proposition 6.18 and Theorem 6.24 and their respective proofs which were taken from Grüne, Pannek, Seehafer and Worthmann [11] and Proposition 6.19, which can be seen as a variant of Tuna, Messina and Teel [19] without terminal cost. Note that in [11]

Proposition 6.18 is proved in a more general setting (and with an even more involved proof); we will sketch this setting in Sect. 10.4.

Section 6.6 summarizes and extends discussions from [10, 11]. Example 6.31 in this section has not been published before while Example 6.32 was taken from Altmüller, Grüne and Worthmann [6]. Further examples of the analysis and design of stage costs can be found, e.g., in Worthmann et al. [20] for a nonholonomic ODE model and in Altmüller [3] and Altmüller and Grüne [4, 5] for controlled PDEs. Section 6.7 consists of previously unpublished material, however, the semiglobal practical stability result in Theorem 6.37 was proved before by Grimm, Messina, Tuna, and Teel in [9, Theorem 1] using a different proof technique and slightly different technical assumptions. Corollaries 2 and 3 in [9] provide counterparts of Theorems 6.35 and 6.24 proving semiglobal and "real" asymptotic stability, respectively, under similar conditions as in our theorems. Since the results in this reference are quite similar to our approach presented in this chapter, we will briefly discuss the main differences.

The decisive difference to our results is that in [9] bounds of the type $V_N(x) \leq \alpha_V(|x|_{x_*})$ for $\alpha_V \in \mathcal{K}_\infty$ independent of N together with suitable bounds on α_V and the other $\mathcal{K}_\infty$ functions involved are used while our analysis relies on the Assumptions 6.3 or 6.5. The main advantage of using these assumptions instead of the bound $V_N(x) \leq \alpha_V(|x|_{x_*})$ lies in the fact that the fine structure of $\beta(r,k)$ or B_k—in particular the dependence of these functions on k—can be used. The benefit of using this structure is nicely illustrated in the exponential controllability case (6.4): for $\beta(r,k) = C\sigma^k r$ and $\ell^*(x) \leq \alpha_4(|x|_{x_*})$ we obtain the bound $V_N(x) \leq \alpha_V(|x|_{x_*})$ for $\alpha_V(r) = C\alpha_4(r)/(1-\sigma)$. Given $C_1 < C_2$ and $\sigma_1 > \sigma_2$ with

$$\frac{C_1}{1-\sigma_1} = \frac{C_2}{1-\sigma_2}$$

this will thus yield the same α_V. Hence, using the upper bound α_V in the stability analysis, we cannot distinguish between large overshoot C_2 and fast decay σ_2 and small overshoot C_1 and slow decay σ_1. However, our analysis in Sect. 6.6 based on Assumption 6.5 shows that the pair C_1, σ_1 provides a much better stability behavior than C_2, σ_2. Particularly, Fig. 6.1 shows that for C_1 sufficiently close to 1 we always enter the region where stability holds for $N = 2$, a fact which remains invisible when looking only at α_V.

Another advantage of our approach is that it automatically leads to suboptimality estimates which are not provided in [9]. On the other hand, a major advantage of the approach in [9] is that it allows to handle nonpositive definite stage costs ℓ via a suitable detectability condition. We will discuss this aspect in Sect. 10.3. It should be noted that the advantages of our approach are maintained if we replace Proposition 6.18 by Proposition 6.19, which is an adaptation of a result for NMPC with terminal costs (but without terminal constraints) in Tuna, Messina and Teel [19]. Proposition 6.19 is easier to prove than Proposition 6.18, but we also arrive at more conservative estimates, as Remark 6.22 shows.

Since the first edition of this book, continuous time counterparts of the most important results presented in this chapter have been developed by Reble and Allgöwer [16] and the relation between the discrete time analysis presented here and the continuous time analysis in this reference has been investigated for sampled data systems by Worthmann, Reble, Grüne and Allgöwer [21]. The continuous time approach was subsequently extended and applied to the Burgers PDE by Azmi and Kunisch [8] and an analysis of the value of α in the context of model reduction of semilinear parabolic PDEs was presented by Alla and Volkwein [2].

Besides the approach presented in this chapter and [9] there are various other approaches which ensure stability of NMPC schemes without stabilizing terminal conditions. Probably the earliest reference is Alamir and Bornard [1], another well-known reference is Jadbabaie and Hauser [13]. Both were already briefly discussed at the end of Sect. 6.1; here we only remark that these references do not provide bounds on the optimization horizon N.

We finally mention that inverse optimality can be obtained in a similar way as for terminal conditioned schemes in Theorem 5.24, see Problem 6, below.

Problems

General remark: all NMPC algorithms in the following problems are meant without stabilizing terminal conditions.

1. Consider a control system (2.1) on $X = \mathbb{R}^d$ and $\mathbb{U} = \mathbb{R}^m$ which is exponentially controllable to $x_* = 0$. This means that there exist constants $K > 0$ and $\eta \in (0, 1)$ such that for each $x \in \mathbb{R}^d$ there is a control sequence $u_x \in \mathbb{U}^N(x)$ such that

$$\|x_{u_x}(k, x)\| \leq K\eta^n \|x_0\|$$

holds for all $k \in \mathbb{N}_0$.

   (a) Show that the system satisfies the controllability Assumption 6.5 with $\beta \in \mathcal{KL}_0$ of type (6.4) for any stage cost of the form $\ell(x, u) = x^\top Qx$ with positive definite matrix $Q \in \mathbb{R}^{d\times d}$.
   (b) Does the assertion from (a) also hold for a stage cost of the form $\ell(x, u) = x^\top Qx + u^\top Ru$ for positive definite matrices $Q \in \mathbb{R}^{d\times d}$ and $R \in \mathbb{R}^{m\times m}$? If not, which additional property must be satisfied?

   Hint: Look at the hints for Problem 2 in Chap. 4.
2. Consider a function $\beta \in \mathcal{KL}_0$ of the form (6.5).

   (a) Prove that there exists $C \geq 1$ and $\sigma \in (0, 1)$ such that for the function $\tilde{\beta} \in \mathcal{KL}_0$ of type (6.4) with $\tilde{\beta}(r, n) = C\sigma^n r$ the inequality

$$\beta(r, n) \leq \tilde{\beta}(r, n)$$

   holds for all $n \in \mathbb{N}_0$ and all $r \geq 0$.

(b) Determine $C \geq 1$ and $\sigma \in (0, 1)$ (with C as small as possible) such that the inequality from (a) holds for the values $n_0 = 2$, $c_0 = 2$ and $c_1 = 1$ in (6.5).
(c) Compute α for $\beta(r, n)$ and $\tilde{\beta}(r, n)$ for the values from (b) and $N = 3, 4, 5, 6$ using (6.19) (MAPLE™ or MATLAB® may be helpful here). Which function provides better values?

3. Consider the control system

$$x(n+1) = \begin{pmatrix} 1 & 1.1 \\ -1.1 & 1 \end{pmatrix} x(n) + \begin{pmatrix} 0 \\ 1 \end{pmatrix} u(n)$$

with $x(n) \in X = \mathbb{R}$, $u(n) \in U = \mathbb{R}$ and stage cost $\ell(x, u) = \max\{\|x\|_\infty, u\}$.

(a) Given the initial value $x(0) = (x_1, x_2)$, use the control sequence $u(0) = \frac{21}{110}x_1 - 2x_2$, $u(1) = \frac{221}{110}x_1 + \frac{221}{100}x_2$, $u(n) = 0$ for all $n \geq 2$ to compute the minimal horizon length N for which stability can be guaranteed.
(b) Try to reduce the minimal required horizon length N from (a). To this end, change the term u in the stage cost ℓ to ηu for some $\eta \in \mathbb{R}_0^+$ and analyze the impact of this change.

Hint for (a): Construct β such that Assumption 6.5 holds and Corollary 6.21 is applicable.

4. Consider a control system (2.1), an admissible feedback law $\mu : \mathbb{X} \to U$ and a Lyapunov function $V : \mathbb{X} \to \mathbb{R}_0^+$ for the closed-loop system $x^+ = g(x) := f(x, \mu(x))$. Show that the NMPC-feedback law μ_N for stage cost $\ell(x, u) = V(x)$ stabilizes the system for $N = 2$.
Hint: A direct argument may be easier than trying to apply one of the theorems from this chapter.
5. Consider the control system

$$x(n+1) = 2x(n) + u(n)$$

with $x(n) \in X = \mathbb{R}$, $u(n) \in U = \mathbb{R}$.

(a) Show that for stage cost $\ell(x, u) = x^2$ the NMPC control law $\mu_N : X \to U$ is optimal for the infinite horizon problem for arbitrary $N \geq 2$, .
(b) For the stage cost $\ell(x, u) = x^2 + u^2$, compute minimal horizon lengths N such that $\alpha \geq \overline{\alpha}$ holds for $\overline{\alpha} \in \{0.5, 0.9, 0.99\}$.

6. Consider an NMPC problem satisfying the assumptions of Theorem 6.20. Prove that the function

$$\tilde{\ell}(x, u) := \ell(x, u) + V_{N-1}(f(x, u)) - V_N(f(x, u))$$

(cf. (5.32)) is of the form (3.2) and that μ_N is the infinite horizon optimal feedback law for (OCP$_\infty$) with stage cost $\tilde{\ell}$.
Hint: Follow the proof of Theorem 5.24.

References

1. Alamir, M., Bornard, G.: Stability of a truncated infinite constrained receding horizon scheme: the general discrete nonlinear case. Automatica **31**(9), 1353–1356 (1995)
2. Alla, A., Volkwein, S.: Asymptotic stability of POD based model predictive control for a semilinear parabolic PDE. Adv. Comput. Math. **41**(5), 1073–1102 (2015)
3. Altmüller, N.: Model predictive control for partial differential equations. PhD Thesis, Universität Bayreuth (2014)
4. Altmüller, N., Grüne, L.: Distributed and boundary model predictive control for the heat equation. GAMM-Mitt. **35**, 131–145 (2012)
5. Altmüller, N., Grüne, L.: A comparative stability analysis of Neumann and Dirichlet boundary MPC for the heat equation. In: Proceedings of the 1st IFAC Workshop on Control of Systems Modeled by Partial Differential Equations – CPDE 2013, pp. 161–166 (2013)
6. Altmüller, N., Grüne, L., Worthmann, K.: Performance of NMPC schemes without stabilizing terminal constraints. In: Diehl, M., Glineur, F., Jarlebring, E., Michiels, W. (eds.) Recent Advances in Optimization and its Applications in Engineering, pp. 289–298. Springer (2010)
7. Altmüller, N., Grüne, L., Worthmann, K.: Receding horizon optimal control for the wave equation. In: Proceedings of the 49th IEEE Conference on Decision and Control - CDC 2010, pp. 3427–3432. Atlanta, Georgia (2010)
8. Azmi, B., Kunisch, K.: On the stabilizability of the Burgers equation by receding horizon control. SIAM J. Control Optim. **54**(3), 1378–1405 (2016)
9. Grimm, G., Messina, M.J., Tuna, S.E., Teel, A.R.: Model predictive control: for want of a local control Lyapunov function, all is not lost. IEEE Trans. Autom. Control **50**(5), 546–558 (2005)
10. Grüne, L.: Analysis and design of unconstrained nonlinear MPC schemes for finite and infinite dimensional systems. SIAM J. Control Optim. **48**, 1206–1228 (2009)
11. Grüne, L., Pannek, J., Seehafer, M., Worthmann, K.: Analysis of unconstrained nonlinear MPC schemes with varying control horizon. SIAM J. Control Optim. **48**, 4938–4962 (2010)
12. Grüne, L., Rantzer, A.: On the infinite horizon performance of receding horizon controllers. IEEE Trans. Autom. Control **53**, 2100–2111 (2008)
13. Jadbabaie, A., Hauser, J.: On the stability of receding horizon control with a general terminal cost. IEEE Trans. Autom. Control **50**(5), 674–678 (2005)
14. Khalil, H.K.: Nonlinear systems, 3rd edn. Prentice Hall, Upper Saddle River, NJ (2002)
15. Nešić, D., Teel, A.R.: A framework for stabilization of nonlinear sampled-data systems based on their approximate discrete-time models. IEEE Trans. Autom. Control **49**(7), 1103–1122 (2004)
16. Reble, M., Allgöwer, F.: Unconstrained model predictive control and suboptimality estimates for nonlinear continuous-time systems. Automatica **48**, 1812–1817 (2011)
17. Shamma, J.S., Xiong, D.: Linear nonquadratic optimal control. IEEE Trans. Autom. Control **42**(6), 875–879 (1997)
18. Sontag, E.D.: Comments on integral variants of ISS. Syst. Control Lett. **34**, 93–100 (1998)
19. Tuna, S.E., Messina, M.J., Teel, A.R.: Shorter horizons for model predictive control. In: Proceedings of the 2006 American Control Conference, pp. 863–868. Minneapolis, Minnesota, USA (2006)
20. Worthmann, K., Mehrez, M.W., Zanon, M., Mann, G.K.I., Gosine, R.G., Diehl, M.: Model predictive control of nonholonomic mobile robots without stabilizing constraints and costs. IEEE Trans. Control Syst. Tech. **1**, 1 (2016). doi:10.1109/TCST.2015.2488589. Printed version to appear
21. Worthmann, K., Reble, M., Grüne, L., Allgöwer, F.: The role of sampling for stability and performance in unconstrained nonlinear model predictive control. SIAM J. Cont. Optim. **52**, 581–605 (2014)

Chapter 7
Feasibility and Robustness

In this chapter we consider two different but related issues. In the first part we discuss the feasibility problem, i.e., that the nominal NMPC closed-loop solutions remain inside a set on which the finite horizon optimal control problems defining the NMPC feedback law are feasible. We formally define the property of recursive feasibility and explain why the assumptions of the previous chapters, i.e., viability of the state constraint set or of the terminal constraint set ensure this property. Then we present two ways to relax the viability assumption on the state constraint set in the case that no terminal constraints are used. After a comparative discussion on NMPC schemes with and without stabilizing terminal conditions, we start with the second part of the chapter in which robustness of the closed loop under additive perturbations and measurement errors is investigated. Here robustness concerns both the feasibility and admissibility as well as the stability of the closed loop. We provide different assumptions and resulting NMPC schemes for which we can rigorously prove such robustness results and also discuss examples which show that in general robustness may fail to hold.

7.1 The Feasibility Problem

We start by introducing the feasibility problem for the NMPC Algorithm 3.1, i.e., for the NMPC formulation without terminal constraints.

Recall from Definition 3.2 that for each $x \in \mathbb{X}$ the set of admissible control sequences $\mathbb{U}^N(x_0)$ is nonempty if and only if there exists a control sequence $u \in \mathbb{U}^N$ for which the conditions

$$(x_u(k, x_0), u(k)) \in \mathbb{Y} \text{ for all } k = 0, \ldots, N-1 \quad \text{and} \quad x_u(N, x_0) \in \mathbb{X}$$

are satisfied. Moreover, recall from the discussion after Assumption 3.3 that the optimization problem ($\mathrm{OCP_N}$) in the NMPC Algorithm 3.1 is called *feasible* for the

L. Grüne and J. Pannek, *Nonlinear Model Predictive Control*,
Communications and Control Engineering, DOI 10.1007/978-3-319-46024-6_7

initial value x_0 if $\mathbb{U}^N(x_0) \neq \emptyset$ holds. Since only feasible optimal control problems allow for an admissible solution, the points $x_0 \in \mathbb{X}$ satisfying $\mathbb{U}^N(x_0) \neq \emptyset$ are exactly the points for which the NMPC feedback law μ_N is well defined. The feasibility problem in NMPC now lies in the fact that even though the constraints $x_u(n, x_0) \in \mathbb{X}$ imply $x_{u^\star}(1, x_0) = x_{\mu_N}(1, x_0) = f(x_0, \mu_N(x_0)) \in \mathbb{X}$, it may happen that

$$\mathbb{U}^N(f(x_0, \mu_N(x_0))) = \emptyset,$$

i.e., that the optimization problem ($\mathrm{OCP_N}$) for initial value $f(x_0, \mu_N(x_0))$ to be solved at next time instant is infeasible. This means that μ_N and thus also the closed loop system (2.5) is not defined for $x = f(x_0, \mu_N(x_0))$ and the NMPC closed loop runs into a "dead end".

Example 7.1 We illustrate this fact by Example 3.4, i.e.,

$$x^+ = f(x, u) = \begin{pmatrix} x_1 + x_2 + u/2 \\ x_2 + u \end{pmatrix}.$$

We use the state constraints $\mathbb{X} = [-1, 1]^2$ and the control constraints $\mathbb{U}(x) = \mathbb{U} = [-1/4, 1/4]$. With the same computation as in Example 3.4 one sees that $\mathbb{X}$ is not viable, since, for instance, for the point $x = (1, 1)^\top \in \mathbb{X}$ we obtain $f(x, u) \notin \mathbb{X}$ for all $u \in \mathbb{U}$.

As we have seen in Example 10.2, the system can be stabilized respecting the state and control constraints starting from the initial value $x = (-1, 1)^\top$. Running the NMPC Algorithm 3.1 with $N = 2$ and $\ell(x, u) = \|x\|^2 + 5u^2$ with this initial value, however, results in the trajectory shown in Fig. 7.1. Here we have not stopped the simulation upon infeasibility but rather continued the computation with the infeasible solution returned by the optimization algorithm.

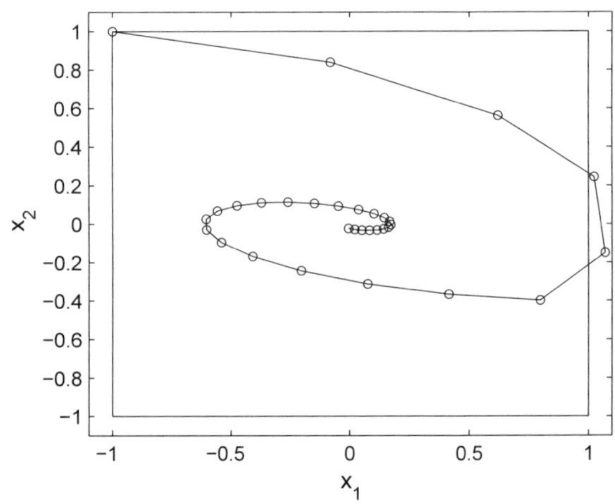

Fig. 7.1 Infeasible trajectory for Example 7.1 with initial value $x_0 = (-1, 1)^\top$ and optimization horizon $N = 2$

Although asymptotically stable, at time $n = 3$ and 4 this trajectory violates the state constraints which are indicated by the black box. Moreover, while the optimization algorithm reported infeasibility for $n = 1, 2, 3, 4$, it terminated successfully at time $n = 0$, i.e., the infeasibility at later time instants was not detected upon initialization and is not due to a failure of the optimization algorithm at time $n = 0$.

In order to formally analyze this problem we introduce the following notions.

Definition 7.2 Let a constraint set $\mathbb{X}$ and an optimization horizon $N \in \mathbb{N}_\infty$ for the NMPC Algorithm 3.1 be given.
(i) A point $x \in \mathbb{X}$ is called *feasible* for $\mathbb{X}$ and N if $\mathbb{U}^N(x) \neq \emptyset$.
(ii) The *feasible set* for $\mathbb{X}$ and N is defined as

$$\mathscr{F}_N := \{x \in \mathbb{X} \mid x \text{ is feasible for } \mathbb{X} \text{ and } N\}.$$

The set $\mathscr{F}_\infty$ is also called *viability kernel*.

(iii) A set $A \subseteq \mathbb{X}$ is called *recursively feasible* for optimization horizon $N \in \mathbb{N}$ if $A \subseteq \mathscr{F}_N$ and it is forward invariant for the NMPC feedback law μ_N, i.e., if $f(x, \mu_N(x)) \in A$ holds for all $x \in A$.

The recursive feasibility property from Definition 7.2(iii) guarantees that for any initial value $x \in A$ the NMPC closed loop will generate a solution which is admissible for all future times. Formally, this is stated in the following lemma.

Lemma 7.3 *Let $A \subseteq \mathbb{X}$ be recursively feasible for the NMPC Algorithm 3.1 with optimization horizon $N \in \mathbb{N}$. Then for each $x \in A$ the closed-loop solution $x_{\mu_N}(n, x)$ generated by (2.5) is well defined for all $n \in \mathbb{N}_0$ and satisfies $x_{\mu_N}(n, x) \in A$ and thus also $x_{\mu_N}(n, x) \in \mathbb{X}$ for all $n \in \mathbb{N}_0$.*

Proof The result follows by a straightforward induction using (2.5) and the relation $f(x, \mu_N(x)) \in A$ for all $x \in A$. □

Thus, in addition to stability, for proper operation of the NMPC scheme we also need to ensure that the desired operating range of our controller lies in a recursively feasible set. Note that $x_{\mu_N}(n, x) \in \mathbb{X}$ for all $n \in \mathbb{N}$ immediately implies the inclusion $A \subseteq \mathscr{F}_\infty$. For this reason, the viability kernel $\mathscr{F}_\infty$ is the maximal possible recursively feasible subset of $\mathbb{X}$. In particular, it is the maximal set on which an admissible feedback can be defined, independent of how this feedback is constructed.

The reason why we did not address the feasibility problem in the previous chapters lies in the fact that the assumptions imposed so far always implied feasibility. In fact, in Chap. 6 we always assumed that the constraint set $\mathbb{X}$ is viable in the sense of Assumption 3.3. Under this assumption Theorem 3.5 ensures that $A = \mathbb{X}$ is recursively feasible, as already remarked after this theorem. However, arbitrary constraint sets are in general not viable. Moreover, while in simple examples the construction of a viable subset of $\mathbb{X}$ may be possible, cf. Example 3.4, for complicated dynamics this can be a difficult if not impossible task.

The terminal constrained scheme discussed in Chap. 5 provides a remedy to this problem. For this scheme, recursive feasibility of $\mathbb{X}_N$ is always ensured by Lemma 5.3. However, the price that we pay for this nice property is that the operating range is a priori restricted to $\mathbb{X}_N$, which may be considerably smaller than the operating range of the unconstrained scheme, cf. Example 6.2. Furthermore, we either need to impose equilibrium constraints $\mathbb{X}_0 = \{x_*\}$—which may be too restrictive for some systems and may cause problems in the numerical optimization routine—or we need to find a viable terminal constraint set $\mathbb{X}_0$ along with a terminal cost F defined on $\mathbb{X}_0$ which satisfies Assumption 5.9. While the design of F can be avoided by using the mixed scheme from the first part of Sect. 10.1, in any case we need to find a viable terminal constraint set $\mathbb{X}_0$. Note that finding this "small" set is in general easier than finding a "big" viable state constraint set $\mathbb{X}$. However, both in terms of the operating range of the scheme and in terms of implementational simplicity, it would be desirable if we could use the unconstrained scheme without having to worry about the feasibility problem and without having to construct a viable terminal constraint set $\mathbb{X}_0$. In the following sections we will show two approaches in this direction.

7.2 Feasibility of Unconstrained NMPC Using Exit Sets

In this and in the subsequent section we present two results which ensure feasibility for the unconstrained NMPC Algorithm 3.1 under two different assumptions. While the first result uses properties of the interplay between the dynamics f and the constraint set $\mathbb{X}$ and is independent of any stability properties, the second approach uses asymptotic stability of the closed loop in order to ensure feasibility of subsets of the state space.

In order to introduce our first approach we need the following objects.

Definition 7.4 Consider a control system (2.1) with state constraint set $\mathbb{X} \subset X$ and control constraint set $\mathbb{U}(x) \subseteq U$, $x \in \mathbb{X}$. We recursively define the *exit sets* $E_k \subseteq X$, $k \in \mathbb{N}_0$ as

$$E_0 := X \setminus \mathbb{X}, \quad E_k := \{x \in \mathbb{X} \mid f(x,u) \in \bigcup_{i=0}^{k-1} E_i \text{ for all } u \in \mathbb{U}(x)\},\ k = 1, 2, \ldots.$$

Remark 7.5 (i) This definition immediately implies $E_{k'} \subseteq E_k$ for all $k \geq k' \geq 1$.

(ii) If $\mathbb{X}$ is viable then the definition of E_0 implies $E_1 = \emptyset$ and thus by induction $E_k = \emptyset$ for all $k \geq 1$.

In words, for $k \geq 1$ the exit set E_k consists of all points $x_0 \in \mathbb{X}$ for which it is unavoidable that the trajectory $x_u(k, x_0)$ leaves $\mathbb{X}$ after at most k steps regardless of how $u \in \mathbb{U}^k(x_0)$ is chosen. This is made precise in the following lemma.

Lemma 7.6 *A point $x \in X$ satisfies $x \in E_k$ if and only if the following property holds:*

$$\begin{array}{l} \text{for each } u \in U^k \text{ there exists } k_u \in \{0, \dots, k\} \text{ such that either} \\ u(n) \notin \mathbb{U}(x_u(n, x)) \text{ for some } n \in \{0, \dots, k_u - 1\} \text{ or } x_u(k_u, x) \notin \mathbb{X}. \end{array} \tag{7.1}$$

Proof We show the property by induction over k. For $k = 0$ the assertion follows immediately from $E_0 = X \setminus \mathbb{X}$ and $x_u(0, x) = x$. For the induction step $k \to k + 1$ assume that the assertion holds for k. We then need to show that $x \in E_{k+1}$ holds if and only if (7.1) holds for $k + 1$.

We first show that (7.1) holds for $x \in E_{k+1}$. Let $x \in E_{k+1}$ and pick $u \in U^{k+1}$. If $u(0) \notin \mathbb{U}(x)$, then (7.1) holds with $k_u = 1$. Hence, assume $u(0) \in \mathbb{U}(x)$. Then we get $x' = x_u(1, x) \in E_i$ for some $i \in \{0, \dots, k\}$. Now, by induction assumption for the shifted control $u' = u(\cdot + 1) \in U^k$ there exists $k_{u'} \in \{0, \dots, k\}$ such that either $u'(n) \notin \mathbb{U}(x_{u'}(n, x'))$ for some $n \in \{0, \dots, k_{u'} - 1\}$ or $x_{u'}(k_{u'}, x') \notin \mathbb{X}$. Since $u'(n) = u(n + 1)$ and $x_{u'}(k_{u'}, x') = x_u(k_u, x)$ for $k_u = k_{u'} + 1 \le k + 1$, this shows that (7.1) holds for $k_u = k_{u'} + 1$.

Conversely, given $x \in X$ for which (7.1) holds for $k + 1$, we need to show $x \in E_{k+1}$. Let $u_x \in \mathbb{U}(x)$ be given and denote $x' = f(x, u_x)$. We have to show that $x' \in E_i$ for some $i \le k$. If $x' \notin \mathbb{X}$ then $x' \in E_0$ and we are done. Otherwise, we pick an arbitrary control sequence $u' \in U^k$ for x' and define a control sequence $u \in U^{k+1}$ by setting $u(0) = u_x$ and $u(j) = u'(j - 1)$ for $j \in \{1, \dots, k\}$. Then by (7.1) there exists $k_u \le k + 1$ such that either $u(n) \notin \mathbb{U}(x_u(n, x))$ for some $n \in \{0, \dots, k_u - 1\}$ or $x_u(k_u, x) \notin \mathbb{X}$ and since $x_u(1, x) = x' \in \mathbb{X}$ we know that $k_u \ge 1$. By construction of u this implies that (7.1) holds for x' and u' with $k_{u'} = k_u - 1 \le k$. Hence, by the induction assumption $x' \in E_k$ and consequently by definition of the E_k we obtain $x \in E_{k+1}$. □

Our next lemma shows the relation between the exit sets E_k and the feasible sets $\mathscr{F}_N$ from Definition 7.2(ii).

Lemma 7.7 *Consider a control system (2.1) with state constraint set $\mathbb{X} \subset X$ and control constraint set $\mathbb{U}(x)$, $x \in \mathbb{X}$. Then the identity*

$$\mathscr{F}_N = \mathbb{X} \setminus E_N = \mathbb{X} \setminus \left(\bigcup_{k=1}^{N} E_k \right) \tag{7.2}$$

holds for all $N \in \mathbb{N}$.

Proof The second equality follows immediately from Remark 7.5(i). It remains to show $\mathscr{F}_N = \mathbb{X} \setminus E_N$ which we will do by proving "$\subseteq$" and "$\supseteq$".

"$\subseteq$": Consider $x \in \mathscr{F}_N$. Then $\mathbb{U}^N(x)$ is nonempty, hence we can pick $u \in \mathbb{U}^N(x)$. By definition of $\mathbb{U}^N(x)$ this implies $u(k) \in \mathbb{U}(x_u(k, x))$ for all $k = 0, \dots, N - 1$ and $x_u(k, x) \in \mathbb{X}$ for $k = 0, \dots, N$. Thus, (7.1) does not hold and Lemma 7.6 implies $x \notin E_N$ and thus $x \in \mathbb{X} \setminus E_N$.

"$\supseteq$": Let $x \in \mathbb{X} \setminus E_N$, i.e., $x \in \mathbb{X}$ and $x \notin E_N$. Then Lemma 7.6 implies that (7.1) does not hold, i.e., there exists $u \in U^N$ with $u(n) \in \mathbb{U}(x_u(n,x))$ for $n = 0, \ldots, N-1$ and $x_u(n,x) \in \mathbb{X}$ for $n = 0, \ldots, N$. This means that $u \in \mathbb{U}^N(x)$, hence $\mathbb{U}^N(x) \neq \emptyset$ and consequently $x \in \mathcal{F}_N$. □

Remark 7.8 If f is continuous then one can also show $\mathcal{F}_\infty = \mathbb{X} \setminus \left(\bigcup_{k=1}^\infty E_k\right)$.

Now we introduce the assumption we will use in order to guarantee feasibility.

Assumption 7.9 There exists $N_0 \in \mathbb{N}_0$ such that $E_k \subseteq E_{N_0}$ for all $k \geq N_0$.

By Remark 7.5(ii) this assumption is satisfied for $N_0 = 0$ if $\mathbb{X}$ is viable. Thus, Assumption 7.9 can be seen as a relaxation of Assumption 3.3. In Example 7.12, below, we will see that this condition is satisfied for the system from Example 7.1. However, before we look at this example we show that under this assumption the feasible set $\mathcal{F}_{N_0}$ becomes recursively feasible for optimization horizon $N \geq N_0 + 1$. To this end we need another preparatory lemma.

Lemma 7.10 *Under Assumption 7.9 the identity*

$$\mathcal{F}_\infty = \mathcal{F}_N$$

holds for all $N \geq N_0$.

Proof First observe that Assumption 7.9 together with Lemma 7.7 immediately implies $\mathcal{F}_N = \mathcal{F}_{N_0}$ for all $N \geq N_0$. Thus, it is sufficient to show the assertion for $N = N_0$.

Since the inclusion $\mathcal{F}_\infty \subseteq \mathcal{F}_N$ follows directly from the definition, it remains to show the converse inclusion. Thus, we need to prove that $\mathbb{U}^\infty(x) \neq \emptyset$ for all $x \in \mathcal{F}_{N_0}$. We do this by constructing $u \in \mathbb{U}^\infty(x)$. To this end, since $\mathcal{F}_{N_0} = \mathcal{F}_{N_0+1}$, we can pick $u_0 \in \mathbb{U}^{N_0+1}(x)$ and set $u(0) := u_0(0)$. Then the definition of $\mathbb{U}^N(x)$ implies $u_0(\cdot+1) \in \mathbb{U}^{N_0}(x_1)$ for $x_1 = x_{u_0}(1,x) = x_u(1,x)$. Thus $x_1 \in \mathcal{F}_{N_0} = \mathcal{F}_{N_0+1}$ and we can find $u_1 \in \mathbb{U}^{N_0+1}(x_1)$. Setting $u(1) := u_1(0)$ with the same arguments we obtain $u_1(\cdot+1) \in \mathbb{U}^{N_0}(x_2)$ for $x_2 = x_{u_1}(1,x_1) = x_u(2,x)$. Proceeding iteratively we obtain a control sequence $u \in \mathbb{U}^\infty$ which satisfies $x_k = x_u(k,x) = x_{u_{k-1}}(1,x_{k-1}) \in \mathcal{F}_{N_0+1} \subseteq \mathbb{X}$ and $u(k) = u_k(0) \in \mathbb{U}(x_k) = \mathbb{U}(x_u(k,x))$ for all $k \in \mathbb{N}_0$. Thus $u \in \mathbb{U}^\infty(x)$. □

Using Lemma 7.10 we can now prove our first recursive feasibility result.

Theorem 7.11 (Recursive feasibility for exit sets) *Consider the NMPC Algorithm 3.1 and let Assumption 7.9 hold. Then the feasible set $\mathcal{F}_{N_0} = \mathcal{F}_\infty$ is recursively feasible for all optimization horizons $N \geq N_0 + 1$.*

Proof Consider $x \in \mathscr{F}_{N_0}$. Since by Lemma 7.10 the identity $\mathscr{F}_{N_0} = \mathscr{F}_N$ holds, in particular we obtain $\mathscr{F}_{N_0} \subseteq \mathscr{F}_N$. Hence, problem (OCP$_\mathrm{N}$) in Algorithm (3.1) is feasible. Let $u^\star$ be the corresponding optimal control which implies $\mu_N(x) = u^\star(0)$. Since $u^\star \in \mathbb{U}^N(x)$ the definition of $\mathbb{U}^N(x)$ implies $u^\star(\cdot + 1) \in \mathbb{U}^{N-1}(x_{u^\star}(1, x)) = \mathbb{U}^{N-1}(f(x, \mu_N(x)))$. Thus, $\mathbb{U}^{N-1}(f(x, \mu_N(x))) \neq \emptyset$ and hence $f(x, \mu_N(x)) \in \mathscr{F}_{N-1}$. Since $N - 1 \geq N_0$, Lemma 7.10 yields $\mathscr{F}_{N-1} = \mathscr{F}_{N_0}$. This shows $f(x, \mu_N(x)) \in \mathscr{F}_{N_0}$, i.e., $\mathscr{F}_{N_0}$ is recursively feasible. □

Example 7.12 We illustrate Assumption 7.9 and Theorem 7.11 by means of Examples 3.4 and 7.1, i.e.,

$$x^+ = f(x, u) = \begin{pmatrix} x_1 + x_2 + u/2 \\ x_2 + u \end{pmatrix}.$$

As in the previous examples we use the state constraints $\mathbb{X} = [-1, 1]^2$. The control constraints are chosen more generally as $\mathbb{U}(x) = \mathbb{U} = [-\bar{u}, \bar{u}]$ with $\bar{u} > 0$.

A straightforward but tedious computation shows that the exit sets E_k are given by

$$E_k = \bigcup_{j=1}^{k} \{x \in [-1, 1]^2 \mid x_1 > -jx_2 + 1 + j^2\bar{u}/2 \text{ or } x_1 < -jx_2 - 1 - j^2\bar{u}/2\}.$$

In Example 3.4 we chose $\bar{u} = 1$. With this parameter one sees that $E_k = E_1$ for all $k \geq 1$ because the inequalities for $j \geq 2$ are never satisfied for $x \in [-1, 1]^2$. Hence, Assumption 7.9 is satisfied with $N_0 = 1$ and Theorem 7.11 yields that the set $\mathscr{F}_1 = \mathscr{F}_\infty = \mathbb{X} \setminus E_1$ from Lemma 7.7 is recursively feasible for all $N \geq 2$. This set is exactly the set defined in (3.6).

In Example 7.1 we have $\bar{u} = 1/4$. In this case one sees that $E_4 \neq E_3$ because $(-0.99, 1)^\top \in E_4$ but $(-0.99, 1)^\top \notin E_3$. On the other hand, $E_k = E_4$ for all $k \geq 4$ because for $x_2 \in [-1, 1]$ and $j \geq 5$ the inequality

$$-jx_2 + 1 + j^2\bar{u}/2 > -4x_2 + 1 + 4^2\bar{u}/2$$

holds. Hence, the inequality for $j = 4$ is always satisfied whenever the inequality for some $j \geq 5$ is satisfied, thus $x \in E_j$ for $j \geq 5$ implies $x \in E_4$. Consequently, Assumption 7.9 holds with $N_0 = 4$ and according to Theorem 7.11 for Example 7.1 the closed-loop solution satisfies the state constraints for $N \geq 5$ and all $x \in \mathbb{X} \setminus E_4$. In particular, since the point $(-1, 1)^\top$ is not contained in E_4, the infeasibility from Fig. 7.1 should disappear. Figure 7.2 shows that this is exactly what happens.[1]

[1] In fact, the infeasibility already disappears for $N = 3$ and $N = 4$ but this is not covered by our theorem.

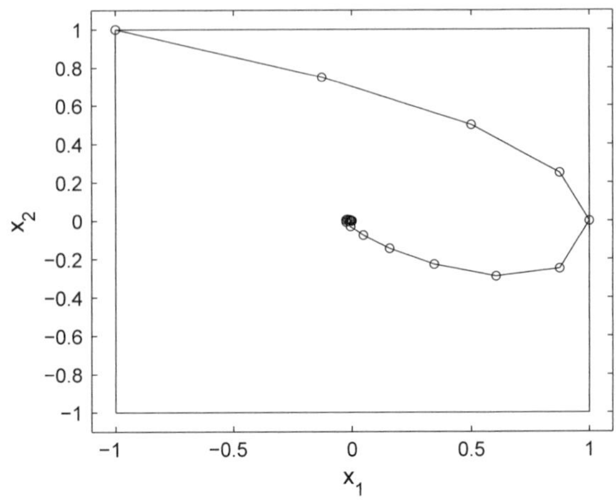

Fig. 7.2 Feasible trajectory for Example 7.1 with initial value $x_0 = (-1, 1)^\top$ and optimization horizon $N = 5$

7.3 Feasibility of Unconstrained NMPC Using Stability

A main advantage of the feasibility analysis in the previous section is that it is completely independent of any stability properties of the closed loop. Thus, feasibility and stability can be analyzed independently of each other. Unfortunately, Theorem 7.11 crucially relies on Assumption 7.9 which may not be satisfied for many practical problems and, even if it is satisfied, may be difficult to verify for complex dynamics.

Thus, in this section we present an alternative result which shows that feasibility may be inherited from the optimality properties of the solution and from the stability of the closed loop. In order to derive this result we first need three basic assumptions and a couple of preparatory lemmas.

Our stability results rely on Assumptions 6.3 or 6.5 which are both only satisfied for all $x \in \mathbb{X}$ if the state constraint set $\mathbb{X}$ is viable. We now show how to generalize Assumption 6.5 if $\mathbb{X}$ is not viable. In this case, asymptotic controllability for arbitrary horizons N only makes sense for initial values $x \in \mathscr{F}_\infty$. This is what our first assumption demands.

Assumption 7.13 *(Asymptotic controllability)* Consider the optimal control problem ($\mathrm{OCP_N}$) with a not necessarily viable state constraint set $\mathbb{X}$. We assume that on the viability kernel $\mathscr{F}_\infty$ the system is asymptotically controllable with respect to ℓ with rate $\beta \in \mathscr{KL}_0$, i.e., for each $x \in \mathscr{F}_\infty$ and each $N \in \mathbb{N}$ there exists an admissible control sequence $u_x \in \mathbb{U}^N(x)$ satisfying $x_{u_x}(n, x) \in \mathscr{F}_\infty$ for all $n = 1, \ldots, N$ and

$$\ell(x_{u_x}(n, x), u_x(n)) \leq \beta(\ell^*(x), n)$$

for all $n \in \{0, \ldots, N-1\}$ and ℓ^* from (6.2).

Under this assumption the results from the stability analysis in Chap. 6 remain valid if we replace the state constraints $x_u(k,x) \in \mathbb{X}$ for $k = 0, \dots, N$ (which in (OCP$_\text{N}$) is implicitly expressed by the requirement $u \in \mathbb{U}^N(x)$) by the state constraints

$$x_u(k,x) \in \mathscr{F}_\infty \quad \text{for } k = 0, \dots, N. \tag{7.3}$$

Indeed, since $\mathscr{F}_\infty$ is viable the standing assumption from Chap. 6, cf. Remark 6.26, is satisfied for these stricter state constraints, i.e., if we replace $\mathbb{X}$ by $\mathscr{F}_\infty$. Furthermore, from the observation in Remark 6.12 it follows that the results from Chap. 6 remain valid if we replace the constraint for $k = N$ in (7.3) by a weaker constraint. In particular, Theorem 6.15 remains valid for β from Assumption 7.13 and $\mathscr{F}_\infty$ in place of $\mathbb{X}$ if we weaken (7.3) to

$$x_u(k,x) \in \mathscr{F}_\infty \quad \text{for } k = 0, \dots, N-1, \quad x_u(N,x) \in \mathbb{X}. \tag{7.4}$$

In the remainder of this section, the reference (OCP$_\text{N}$) will always refer to the problem with the original state constraints $x_u(k,x) \in \mathbb{X}$, $k = 0, \dots, N$, while we will always explicitly refer to (7.4) if we consider (OCP$_\text{N}$) with the additional constraints (7.4).

In order to apply our stability results from Chap. 6, we need further assumptions on ℓ and V_N. For this assumption recall once again the definition $\ell^*(x) := \inf_{u \in U} \ell(x,u)$ from (6.2).

Assumption 7.14 There exist $\alpha_1, \alpha_2, \alpha_3, \alpha_4 \in \mathscr{K}_\infty$ and $N_0 \geq 2$ such that the inequalities

$$\alpha_1(|x|_{x_*}) \leq V_N(x) \leq \alpha_2(|x|_{x_*})$$

and

$$\alpha_3(|x|_{x_*}) \leq \ell^*(x) \leq \alpha_4(|x|_{x_*})$$

hold for all $N \geq N_0$.

Note that the assumptions of Theorem 6.24 imply Assumption 7.14 with $\alpha_1 = \alpha_3$ and $\alpha_2(r) = \sum_{k=0}^{\infty} \beta(\alpha_4(r), k)$. Here the linearity and summability of β ensure that α_2 is indeed a $\mathscr{K}_\infty$-function.

Finally, we want to ensure that the feedback stabilization problem under the given state constraints $\mathbb{X}$ is solvable locally around x_*. A prerequisite for this is that there exists a neighborhood of x_* whose intersection with $\mathbb{X}$ consists of points which are feasible for $N = \infty$. This is our last assumption.

Assumption 7.15 There exists a ball $\mathscr{B}_\delta(x_*)$ such that $\mathscr{B}_\delta(x_*) \cap \mathbb{X} \subseteq \mathscr{F}_\infty$.

Observe that we only require the inclusion $\mathscr{B}_\delta(x_*) \cap \mathbb{X} \subseteq \mathscr{F}_\infty$ (as opposed to $\mathscr{B}_\delta(x_*) \subseteq \mathscr{F}_\infty$), which allows for the situation that x_* is on the boundary of $\mathbb{X}$.

The following two lemmas show properties of optimal trajectories which are crucial for our feasibility analysis.

Lemma 7.16 *Assume that Assumptions 7.13–7.15 hold. Let* $\varepsilon = \alpha_2^{-1} \circ \alpha_3(\delta) > 0$ *with* $\alpha_2, \alpha_3 \in \mathcal{K}_\infty$ *from Assumption 7.14 and* $\delta > 0$ *from Assumption 7.15. Then for each* $N \geq 2$ *and each* $x \in \mathcal{B}_\varepsilon(x_*) \cap \mathbb{X}$ *the optimal trajectory for* (OCP$_\mathrm{N}$) *satisfies* $x_{u^\star}(n, x) \in \mathcal{B}_\delta(x_*) \cap \mathbb{X}$ *for all* $n \in \{0, \ldots, N-1\}$.

Proof The relation $x_{u^\star}(n, x) \in \mathbb{X}$ follows immediately from $u^\star \in \mathbb{U}^N(x)$. It remains to show $x_{u^\star}(n, x) \in \mathcal{B}_\delta(x_*)$. From the inequality for ℓ^* in Assumption 7.14 we obtain

$$\ell(y, u) \geq \alpha_3(\delta) \quad \text{for all } y \notin \mathcal{B}_\delta(x_*). \tag{7.5}$$

On the other hand, the inequality for V_N in Assumption 7.14 and the definition of ε imply

$$V_N(x) < \alpha_2(\varepsilon) = \alpha_2(\alpha_2^{-1} \circ \alpha_3(\delta)) = \alpha_3(\delta) \quad \text{for all } x \in \mathcal{B}_\varepsilon(x_*). \tag{7.6}$$

Now assuming $x_{u^\star}(n, x) \notin \mathcal{B}_\delta(x_*)$ for some $x \in \mathcal{B}_\varepsilon(x_*)$ and some $n \in \{0, \ldots, N-1\}$, and using (7.5) with $y = x_{u^\star}(n, x)$ implies

$$V_N(x) = \sum_{k=0}^{N-1} \ell(x_{u^\star}(k, x), u^\star(k)) \geq \ell(x_{u^\star}(n, x), u^\star(n)) \geq \alpha_3(\delta)$$

which contradicts (7.6). □

Lemma 7.17 *Assume that Assumptions 7.13–7.15 hold and consider some* $\varepsilon > 0$. *Let* $N \geq 2$ *and* $x \in \mathcal{F}_N$ *with* $V_N(x) < N\alpha_3(\varepsilon)$. *Then the optimal trajectory* $x_{u^\star}(n, x)$ *for* (OCP$_\mathrm{N}$) *satisfies* $x_{u^\star}(n, x) \in \mathcal{B}_\varepsilon(x_*) \cap \mathbb{X}$ *for some* $n \in \{0, \ldots, N-1\}$.

Proof Again, $x_{u^\star}(n, x) \in \mathbb{X}$ follows immediately from $u^\star \in \mathbb{U}^N(x)$ and it remains to show $x_{u^\star}(n, x) \in \mathcal{B}_\varepsilon(x_*)$. To this end, assume $x_{u^\star}(n, x) \notin \mathcal{B}_\varepsilon(x_*)$ for all $n \in \{0, \ldots, N-1\}$. Then Assumption 7.14 implies

$$V_N(x) = \sum_{k=0}^{N-1} \ell(x_{u^\star}(k, x), u^\star(k)) \geq \sum_{k=0}^{N-1} \alpha_3(|x_{u^\star}(k, x)|_{x_*}) \geq N\alpha_3(\varepsilon)$$

which contradicts $V_N(x) < N\alpha_3(\varepsilon)$. □

The next lemma shows a property of arbitrary admissible trajectories.

Lemma 7.18 *Let* $N \in \mathbb{N}$, $x \in \mathcal{F}_N$ *and* $u \in \mathbb{U}^N(x)$ *be such that the corresponding trajectory satisfies* $x_u(N-1, x) \in \mathcal{F}_\infty$. *Then* $x_u(k, x) \in \mathcal{F}_\infty$ *for all* $k = 0, \ldots, N-1$.

Proof Fix an arbitrary $k \in \{0, \ldots, N-2\}$ and abbreviate $x_k = x_u(k, x)$. Since $y = x_u(N-1, x) \in \mathcal{F}_\infty$ there exists a control sequence $u_y \in \mathbb{U}^\infty(y)$, i.e.,

$$u_y(n) \in \mathbb{U}(x_{u_y}(n, y)) \quad \text{and} \quad x_{u_y}(n, y) \in \mathbb{X}$$

for all $n \in \mathbb{N}_0$. Then the concatenated control sequence

$$\bar{u}(n) = \begin{cases} u(n+k), & n = 0, \ldots, N-k-2 \\ u_y(n-N+k+1), & n \geq N-k-1 \end{cases}$$

and the initial value x_k yield a trajectory satisfying

$$x_{\bar{u}}(n, x_k) = \begin{cases} x_u(n+k, x), & n = 0, \ldots, N-k-1 \\ x_{u_y}(n-N+k+1, y), & n \geq N-k-1. \end{cases}$$

This trajectory remains in $\mathbb{X}$ for all $n \geq 0$ and the corresponding control sequence $\bar{u}$ is admissible for all times. Thus $u \in \mathbb{U}^\infty(x_k)$, hence $\mathbb{U}^\infty(x_k) \neq \emptyset$ and consequently $x_k \in \mathscr{F}_\infty$. □

Combining the three previous lemmas we arrive at the following proposition.

Proposition 7.19 *Assume that Assumptions 7.13–7.15 hold and let $N \geq 2$ and $x \in \mathscr{F}_N$ with $V_N(x) < N\alpha_3 \circ \alpha_2^{-1} \circ \alpha_3(\delta)$ with $\alpha_2, \alpha_3 \in \mathscr{K}_\infty$ from Assumption 7.14 and $\delta > 0$ from Assumption 7.15. Then the optimal trajectory $x_{u^\star}(n, x)$ for* (OCP$_\mathrm{N}$) *with horizon N satisfies $x_{u^\star}(n, x) \in \mathscr{F}_\infty$ for all $n \in \{0, \ldots, N-1\}$.*

In particular, for these x the optimal values and the optimal trajectories do not change if we add the constraints (7.4) *to the optimal control problem* (OCP$_\mathrm{N}$).

Proof Applying Lemma 7.17 with $\varepsilon = \alpha_2^{-1} \circ \alpha_3(\delta)$ yields $x_{u^\star}(n, x) \in \mathscr{B}_\varepsilon(x_*)$ for some $n \in \{0, \ldots, N-1\}$. Since by Corollary 3.16 the trajectory $x_{u^\star}(n + \cdot, x)$ is optimal for horizon $N - n$ and initial value $x_{u^\star}(n, x)$, Lemma 7.16 yields $x_{u^\star}(k, x) \in \mathscr{B}_\delta(x_*) \cap \mathbb{X}$ for $k = n, \ldots, N-1$. In particular, this implies $x_{u^\star}(N-1, x) \in \mathscr{B}_\delta(x_*) \cap \mathbb{X} \subseteq \mathscr{F}_\infty$. Now Lemma 7.18 yields the assertion. □

Now we are ready to formulate our feasibility theorem.

Theorem 7.20 (Recursive feasibility via controllability condition) *Let Assumptions 7.13–7.15 hold, let $N \geq 2$ and assume that α from Theorem 6.15 with β from Assumption 7.13 satisfies $\alpha \in (0, 1]$. Then the set*

$$A = \{x \in \mathscr{F}_N \mid V_N(x) < N\alpha_3 \circ \alpha_2^{-1} \circ \alpha_3(\delta)\}$$

with $\alpha_2, \alpha_3 \in \mathscr{K}_\infty$ from Assumption 7.14 and $\delta > 0$ from Assumption 7.15 is recursively feasible for the NMPC feedback μ_N from Algorithm 3.1. Furthermore, the NMPC closed loop (2.5) is asymptotically stable on A.

Proof From the discussion after Assumption 7.13 it follows that Theorem 6.15 is applicable for Algorithm 3.1 with the additional constraints (7.4) in (OCP$_\mathrm{N}$) with β from Assumption 7.13. Thus, (5.1) holds for μ_N from Algorithm 3.1 and the corresponding optimal value function V_N for all $x \in \mathscr{F}_\infty$ if we add the constraints (7.4).

By Proposition 7.19, for $x \in A$ the optimal trajectories do not change if we add the constraints (7.4) to the optimal control ($\mathrm{OCP_N}$) in Algorithm 3.1. In particular, this implies that the resulting NMPC feedback μ_N does not change if we add the state constraints (7.4). Since, furthermore, for $x \in A$ the optimal trajectories of ($\mathrm{OCP_N}$) lie in $\mathscr{F}_\infty$, we get $x \in \mathscr{F}_\infty$ and $f(x, \mu_N(x)) \in \mathscr{F}_\infty$, thus V_N is defined in x and $f(x, \mu_N(x))$. Hence, for each $x \in A$ (5.1) also holds for μ_N from Algorithm 3.1 and the corresponding optimal value function V_N without the constraints (7.4). For $x \in A$ this implies

$$V_N(f(x, \mu_N(x))) \leq V_N(x) - \alpha\ell(x, \mu_N(x)) \leq V_N(x) < N\alpha_3 \circ \alpha_2^{-1} \circ \alpha_3(\delta)$$

and thus $f(x, \mu_N(x)) \in A$. This shows the recursive feasibility of A. □

Corollary 7.21 *Let Assumptions 7.13–7.15 hold, let $N_0 \geq 2$ and assume that α from Theorem 6.15 with β from Assumption 7.13 satisfies $\alpha \in (0, 1]$ for all $N \geq N_0$. Then for each bounded set $K \subseteq \mathscr{F}_\infty$ there exists $N_K \geq N_0$ such that for each $N \geq N_K$ there exists a recursively feasible set A_N for the NMPC feedback μ_N from Algorithm 3.1 with $K \subseteq A_N$. Furthermore, the NMPC closed loop* (2.5) *is asymptotically stable on A_N. In particular, if $\mathbb{X}$ is bounded, then $\mathscr{F}_\infty$ is recursively feasible for all sufficiently large optimization horizons N.*

Proof Using the upper bound α_2 on V_N from Assumption 7.14 and the inclusion $\mathscr{F}_\infty \subseteq \mathscr{F}_N$ it follows that the set A from Theorem 7.20 contains the set $\mathscr{B}_\nu(x_*) \cap \mathscr{F}_\infty$ with

$$\nu = \alpha_2^{-1}\left(N\alpha_3 \circ \alpha_2^{-1} \circ \alpha_3(\delta)\right).$$

Since $\nu \nearrow \infty$ for $N \to \infty$, for each bounded set $K \subseteq \mathscr{F}_\infty$ we can choose $N_K \geq N_0$ such that $K \subseteq \mathscr{B}_\nu(x_*) \cap \mathscr{F}_\infty$ holds for all $N \geq N_K$. This shows the claim for $A_N = A$. □

7.4 Comparing NMPC with and Without Terminal Conditions

Now that we have developed the main stability and feasibility results we will discuss the main advantages and disadvantages of NMPC schemes with and without terminal conditions. More precisely we distinguish between the following schemes.

(a) NMPC with equilibrium (or time-varying reference) endpoint constraint from Sect. 5.2
(b) NMPC with Lyapunov function terminal cost from Sect. 5.3
(c) NMPC without terminal conditions from Chap. 6

We compare the main features of these NMPC variants in terms of

(i) design, i.e., the choice of the necessary ingredients of the respective algorithms

(ii) stability, i.e., the asymptotic stability properties of the closed loop and the assumptions needed in order to guarantee them
(iii) performance, i.e., the suboptimality compared with the infinite horizon optimal value
(iv) feasibility, i.e., the guarantee that the optimal control problem in the NMPC closed loop is solvable for the given constraints
(v) numerical effort, i.e., the time needed for the online optimization

(i) Regarding the design, clearly the schemes (a) and (c) are preferable. In both cases all that needs to be designed is a desired equilibrium (or reference in the time-varying case) and a stage cost ℓ which is positive definite with respect to this reference which in the simplest case could be of the form (3.3).

In contrast to this, the additional construction of a viable terminal constraint set $\mathbb{X}_0$ and a terminal cost F meeting Assumptions 5.1(i) and (ii) necessary for (b) poses a considerable additional difficulty in the design of the scheme, notably (but not exclusively) for time-varying references.

This is probably the main reason for the fact that in our discussion with practitioners the formulations (a) and (c) turned out to be the by far preferred NMPC variants in industrial applications.

(ii) The main difference of the stability properties for the different schemes lies in the fact that for (a) and (b) the operating range, i.e., the region on which the stabilizing feedback is defined or, equivalently, the domain of attraction of the reference solution for the closed-loop solution, is a priori confined to the feasible set $\mathbb{X}_N$. In contrast to this, the unconstrained scheme (c) can yield larger and even unbounded stability regions for fixed N, cf. Example 6.2. On the other hand, for small optimization horizons N in (a) and (b) only the domain of attraction shrinks while for (c) asymptotic stability may be lost completely. Which of the two advantages is dominant can only be assessed on a case by case basis for each particular system to be controlled, usually performed with the support of numerical simulations and/or experimental results.

Regarding the conditions for stability, (a) requires the system to be controllable to the desired reference point or trajectory in finite time, while (b) requires the viability of the terminal constraint set $\mathbb{X}_0$ and the compatibility of F and ℓ in the sense of Assumption 5.1(ii). The unconstrained scheme, in turn, requires the bound from Assumption 6.3 or the asymptotic controllability from Assumption 6.5 and a positive value α in Theorem 6.15 or a sufficiently large optimization horizon N, cf. Theorem 6.37. The conditions for the existence of suitable F and ℓ for (b) and ℓ for (c), respectively, can roughly be regarded as comparably strong as they both essentially require asymptotic controllability with suitable uniformity. In contrast to this, the finite time controllability condition for (a) is stronger.

Concerning the verification of these conditions, the assumptions for (a) and (b) are typically considerably easier to check than the asymptotic controllability condition for (c). However, a considerable difference between the conditions for (b) and (c) is that F in (b) must be constructed in order to run the scheme while B_K or β in Assumption 6.3 or 6.5, respectively, are only needed for the analysis of the scheme but not at runtime.

(iii) Regarding performance, the respective Theorems 5.21 and 6.24 show that for all schemes (a)–(c) the infinite horizon performance $J_\infty^{cl}(x, \mu_N)$ from Definition (4.10) approaches the optimal value $V_\infty(x)$ if the optimization horizon N tends to infinity. The conditions for these theorems to hold are essentially equivalent to conditions needed for asymptotic stability. For scheme (b), Theorem 5.22 gives an alternative estimate under an assumption on the terminal cost F.

For fixed N, however, not only the operating range (cf. Example 6.2 and the discussion in (ii), above) but also the performance may differ, at times considerably, even if $\mathbb{X}_N = \mathbb{X}$ holds. We illustrate this effect by two examples.

Example 7.22 We reconsider the Examples 5.18 and 6.1, i.e.,

$$x^+ = x + u, \quad \ell(x, u) = x^2 + u^2$$

with $\mathbb{X} = X = \mathbb{U} = U = \mathbb{R}$. In Example 5.18 we computed that scheme (a) for $N = 2$ yields the controller $\mu_2(x) = 2x/3$ satisfying $J_\infty^{cl}(x, \mu_2) = 1.625x^2$.

In Example 6.1 it turned out that for the same example from scheme (c) we obtain the controller $\mu_2(x) = -x/2$. This yields the closed-loop solution $x_{\mu_2}(k, x) = x/2^k$ and $\ell(x, \mu_2(x)) = x^2 + (x/2)^2 = 5x^2/4$. This implies

$$J_\infty^{cl}(x, \mu_2) = \sum_{k=0}^{\infty} \ell(x_{\mu_2}(k, x), \mu_2(x_{\mu_2}(k, x))) = \sum_{k=0}^{\infty} \frac{5}{4}\frac{x^2}{2^{2k}} = \frac{5}{3}x^2 \approx 1.666x^2$$

(note that this value coincides with the upper bound $V_2(x)/\alpha$ from Theorem 6.20 since in Example 6.1 we computed $V_2(x) = 3x^2/2$ and $\alpha = 0.9$, hence $V_2(x)/\alpha = (3x^2/2)(10/9) = 5x^2/3$). Hence, for this example scheme (a) yields a better performance than scheme (c).

Example 7.23 Consider again Example 5.19, i.e.,

$$x^+ = x + u, \quad \ell(x, u) = x^2 + u^4$$

with $\mathbb{X} = X = \mathbb{U} = U = \mathbb{R}$. In this example we showed that scheme (a) yields

$$J_\infty^{cl}(20, \mu_2) \approx 11240.39.$$

On the other hand, the controller μ given in Example 5.19 is nothing but the controller μ_2 for scheme (c), which we again computed by MAPLE™. This controller yields

$$J_\infty^{cl}(20, \mu_2) \approx 1725.33,$$

i.e., a considerably better value.

Roughly speaking, the terminal conditions employed in (a) and (b) cause the NMPC feedback law to steer the system to the equilibrium or reference more rapidly at the cost of larger control effort, while the unconditioned scheme (c) typically acts more cautiously. This is why in Example 7.22, in which the control is only moderately penalized, scheme (a) performs better while in Example 7.23, in which large control values are penalized much more heavily, scheme (c) yields the better result. In general, it appears that for a stronger penalization of the control effort the unconstrained scheme (c) provides better performance. It should, however, also be mentioned that a stronger penalization of u typically yields larger B_K and β in Assumptions 6.3 and 6.5, respectively, which in turn may affect the stability of scheme (c).

(iv) Our discussion on feasibility from the last sections shows that for the terminal conditioned schemes (a) and (b) the sets $\mathbb{X}_N$ are "automatically" recursively feasible. This property is inherited from the viability of the terminal constraint set $\mathbb{X}_0$. For the unconstrained scheme recursive feasibility can be expected on (a subset of) the viability kernel, as Theorems 7.11 and 7.20 as well as Corollary 7.21 show. However, in contrast to (a) and (b) here we need additional assumptions and a sufficiently large optimization horizon N if the state constraint set $\mathbb{X}$ itself is not viable.

Regarding the detection of infeasibility, the schemes (a) and (b) have the advantage that feasibility of the nominal closed loop is guaranteed once the optimization algorithm reports that ($\mathrm{OCP}_{\mathrm{N,e}}$) has a feasible solution for the initial value x_0. In contrast to this, in scheme (c) infeasibility may occur even if ($\mathrm{OCP}_{\mathrm{N}}$) has a feasible solution for the initial value, cf. Example 7.1. Thus, in schemes (a) and (b) the infeasibility is usually detected earlier than in scheme (c).

(v) The numerical effort depends on many different parameters, most notably on the dimension of the problem, on the optimization horizon, on the structure of the dynamics f and the stage cost ℓ and on the number and type of constraints. Generally, one has to take into account that in a nonlinear and nonconvex setting it can often not be expected that the optimization algorithm is able to find a global optimum. The reason for this will become apparent in the discussion of nonlinear optimization algorithms in Chap. 12. Hence, in general it is difficult to assess which of the schemes is preferable from the numerical point of view.

However, regarding the constraints it is clear that the schemes (a) and (b) are more demanding than (c). In particular, the endpoint constraint of scheme (a) may cause severe problems in the numerical optimization routine for nonlinear and nonconvex problems. From this point of view the regional constraint in scheme (b) is typically preferable to (a) and scheme (c) without terminal conditions is certainly the best of all. However, if the terminal conditions help to significantly reduce the optimization horizon N in scheme (a) or (b) compared to scheme (c), e.g., when no good stage cost in the sense of Sect. 6.6 can be found for (c), then this effect may easily override the advantage of having fewer constraints in scheme (c). Hence, similar to what was said in the discussion in the first paragraph of (ii), again an assessment on a case by case basis must be made in order to decide which scheme is more appropriate for a given system and control task.

Summarizing the discussion in this section, one sees that both schemes with and without stabilizing terminal conditions have their specific advantages and disadvantages. In practice, it is presumably a good choice to start with an unconditioned scheme which is easier to design and assess its performance via numerical simulations and practical experiments. If the desired performance—be it in terms of stability, suboptimality or feasibility—is not achieved for reasonable choices of N and simple modifications like, e.g., the terminal weights from Sect. 10.2 do not yield a solution, then one of the more sophisticated methods like adding appropriate terminal constraints and costs, a redesign of the stage cost functions in the spirit of Sect. 6.6 or one of the mixed schemes from Sect. 10.1 should be considered.

7.5 Robustness: Basic Definition and Concepts

Real systems never exactly coincide with their mathematical models. This means that in practice the behavior of the real system will deviate from the mathematically idealized model (2.1). In this and in the following sections we will analyze the impact of these deviations on the NMPC closed loop and discuss NMPC variants which provide robustness against such errors. In order to simplify the setting we will consider the case of time invariant reference $x^{\mathrm{ref}} = x_*$. All results do, however, carry over to the time-varying case provided the necessary assumptions hold uniformly with respect to time; we will comment on this in remarks after our main results.

Sources for errors are, for instance, modeling errors, uncertain parameters, external disturbances acting on the system and measurement errors. A further source are numerical errors, which are almost inevitable in NMPC schemes because we need a numerical integration scheme for the solution of ($\mathrm{OCP}^{\mathrm{n}}_{\mathrm{N,e}}$) or its variants. This topic will be treated in detail in Chap. 11, see in particular Sect. 11.5.

As a consequence, the predicted trajectories $x_u(k, x)$ used in ($\mathrm{OCP}^{\mathrm{n}}_{\mathrm{N,e}}$) and its variants do not exactly coincide with the future behavior of the real system.

Formally, we have already taken this fact into account by referring to the closed-loop systems (2.5) and (3.5) as *nominal* closed-loop system. Recall that the nominal NMPC closed loop (3.5) whose behavior we analyzed in the preceding chapters and sections is given by

$$x^+ = f(x, \mu_N(x)).$$

Here f exactly coincides with f in (2.1) which is used in ($\mathrm{OCP}^{\mathrm{n}}_{\mathrm{N,e}}$) or its variants to compute the NMPC controller μ_N.

In order to analyze the influence of the various error sources, for simplicity of exposition we assume that our state space X is a normed vector space such that we can add elements of X and measure the size of elements $x \in X$ by their norm $\|x\|$. Then we can introduce the perturbed closed-loop model

$$\tilde{x}^+ = f(\tilde{x}, \mu_N(\tilde{x} + e)) + d \tag{7.7}$$

Here $d \in X$ is an additive perturbation which covers all kinds of errors causing f to deviate from the evolution of the real system, like modeling and numerical errors, external disturbances, uncertain parameters etc. In addition, we consider the error term $e \in X$ which models measurement errors. Note that when both f and μ_N are continuous then one could express the effects of d and e on the system via one additive perturbation $\bar{d}$. However, while all of our robustness results will rely on the continuity of f we will not assume continuity of μ_N because optimal feedback controls and thus NMPC-feedback laws are, in general, discontinuous.

In order to distinguish between the nominal and the perturbed system, we denote the states of the perturbed system by $\tilde{x}$ and the states of the nominal model by x. For initial value $x_0 \in X$ and sequences of perturbation values $d(\cdot),\ e(\cdot) \in X^{\mathbb{N}}$ we obtain solutions $\tilde{x}_{\mu_N}(k, x_0)$ of (7.7) from the iteration

$$\begin{aligned} \tilde{x}_{\mu_N}(0, x_0) &= x_0 \\ \tilde{x}_{\mu_N}(k+1, x_0) &= f(\tilde{x}_{\mu_N}(k, x_0), \mu_N(\tilde{x}_{\mu_N}(k, x_0) + e(k))) + d(k), \quad k = 0, 1, \ldots \end{aligned}$$

Although this solution depends on the particular sequences $d(\cdot)$ and $e(\cdot)$, we will not explicitly include this dependence in our notation. Instead, given a tuple of bounds $(\overline{d}, \overline{e}) \in \mathbb{R}_0^+ \times \mathbb{R}_0^+$ and an initial value x_0 we will define the following set $S_{(\overline{d},\overline{e})}(x_0)$ of solutions

$$S_{(\overline{d},\overline{e})}(x_0) := \{\tilde{x}_{\mu_N}(\cdot, x_0) \mid \|d(k)\| \le \overline{d},\ \|e(k)\| \le \overline{e} \text{ for all } k \in \mathbb{N}_0\}.$$

The desired robust stability property is now given by the following definition.

Definition 7.24 *(Practical stability with respect to perturbations)* Given a set $A \subseteq \mathbb{X}$ such that the optimal control problem defining μ_N is feasible for all $x_0 \in A$, we say that x_* is *semiglobally practically asymptotically stable on A with respect to the perturbations d and e* if there exists $\beta \in \mathcal{KL}$ such that the following property holds: for each $\delta > 0$ and $\Delta > \delta$ there exists $\overline{d}, \overline{e} > 0$, such that each solution $\tilde{x}_{\mu_N}(\cdot, x_0) \in S_{(\overline{d},\overline{e})}(x_0)$ with $x_0 \in A$ and $|x_0|_{x_*} \le \Delta$ satisfies $\tilde{x}_{\mu_N}(k, x_0) \in A$ and

$$|\tilde{x}_{\mu_N}(k, x_0)|_{x_*} \le \max\{\beta(|x_0|_{x_*}, k),\ \delta\}$$

for all $k \in \mathbb{N}_0$, provided the initial measurement error $e(0)$ satisfies $x_0 + e(0) \in A$.

Observe that this definition resembles Definition 6.33(ii) except that now the size of the perturbation plays the role of the optimization horizon N. Furthermore, we have explicitly included admissibility and feasibility into the definition in order to exclude the case that the perturbations drive the closed-loop trajectory out of the feasible or admissible set. The precise meaning of "the optimal control problem defining μ_N is feasible for all $x_0 \in A$" depends on the NMPC setting under consideration: if terminal constraints $\mathbb{X}_0$ are used we require $A \subseteq \mathbb{X}_N$, otherwise $A \subseteq \mathcal{F}_N$. The additional set $A \subseteq \mathbb{X}$ is needed if the feasible set is strictly smaller than $\mathbb{X}$. Observe that the definition in particular implies recursive feasibility of A.

In words, this definition requires that for all initial values x_0 which are both in the ball $\overline{\mathcal{B}}_\Delta(x_*)$ and in A the perturbed solutions of (7.7) stay within the state constraint set $\mathbb{X}$ and behave like asymptotically stable solutions until they reach the ball $\overline{\mathcal{B}}_\delta(x_*)$.

The condition $x_0 + e(0) \in A$ is a technical requirement needed to ensure that the optimization problem for obtaining μ_N—which we do not assume to be feasible outside A—is feasible at initial time $n = 0$. In what follows, we will often use the simpler term *robust stability* instead of semiglobal practical asymptotic stability.

It should be noted that this robust stability property is closely related to a regional version of the input-to-state stability (ISS) property. Indeed, the assumption in Definition 7.24 implies that for fixed $\Delta > 0$ we can find a $\mathscr{K}_\infty$-function ρ such that for each $\delta \in (0, \Delta]$ the stability property in Definition 7.24 is satisfied whenever $\overline{d} \leq \rho(\delta)$ and $\overline{e} \leq \rho(\delta)$ holds, i.e., the term "sufficiently small" in Definition 7.24 may be quantified by a function $\rho \in \mathscr{K}_\infty$. Then, defining $\gamma = \rho^{-1}$, for all $x_0 \in \mathscr{B}_\Delta(x_*)$ and all perturbation sequences d, e with $\|d\|_\infty := \sup_{k\in\mathbb{N}_0} \|d(k)\| \leq \rho(\Delta)$ and $\|e\|_\infty := \sup_{k\in\mathbb{N}_0} \|e(k)\| \leq \rho(\Delta)$, Definition 7.24 and the definition of γ imply

$$|\tilde{x}_{\mu_N}(k, x_0)|_{x_*} \leq \max\{\beta(|x_0|_{x_*}, k),\ \gamma(\|d\|_\infty),\ \gamma(\|e\|_\infty)\},$$

i.e., the system is input-to-state stable for inputs d and e.

7.6 Robustness Without State Constraints

In this section we will show that the robust stability property from Definition 7.24 is always satisfied under mild conditions if we do not impose state constraints. State constraints affect the robustness analysis of the stability of (7.7) in two ways: on the one hand, even if the nominal closed loop is admissible on the whole state constraint set $\mathbb{X}$, i.e., if $f(x, \mu_N(x)) \in \mathbb{X}$ holds for all $x \in \mathbb{X}$, arbitrary small perturbations d, e may lead to a violation of the state constraints, i.e., to $f(x, \mu_N(x + e)) + d \notin \mathbb{X}$, if $f(x, \mu_N(x))$ is near the boundary of $\mathbb{X}$. Likewise, the perturbations may destroy the recursive feasibility of a nominally recursively feasible set. On the other hand, state constraints may introduce instability even if we only consider perturbations satisfying $f(x, \mu_N(x + e)) + d \in \mathbb{X}$. The latter is a more subtle issue, which we will illustrate in the next section. Solutions to both problems will be discussed in Sects. 7.8 and 7.9.

Without state constraints, i.e., with $\mathbb{X} = X$, the problem considerably simplifies. As introduced in the last section we assume that X is a normed vector space with norm $\|x\|$. This implies $x + d \in X$ and $x + e \in X$ for all $x, d, e \in X$. We allow for input constraints but we assume $\mathbb{U}(x) = \mathbb{U}$, i.e., that the input constraint set is independent of x. This ensures $\mu_N(x + e) \in \mathbb{U}$ for all $x, e \in X$ while for state dependent input constraints and measurement errors we will never be able to exactly satisfy state dependent input constraints, because the control value $\mu_N(x+e)$ will be selected from $\mathbb{U}(x + e)$ instead of $\mathbb{U}(x)$. One could, however, extend the subsequent proofs to input constraint sets $\mathbb{U}(x)$ which depend continuously on the state x in a suitable set theoretic sense. Still, in order not to overload the presentation with technicalities we decided not to include this extension.

For our analysis we need the following definition.

Definition 7.25 *(Uniform continuity)* Consider vector spaces X and Y, a set $A \subset X$ and an arbitrary set $\mathbb{U}$.

(i) A function $W : X \to Y$ is called *uniformly continuous* on A if there exists a function $\omega \in \mathcal{K}$ such that for all $x, y \in A$ the inequality

$$\|W(x) - W(y)\| \leq \omega(\|x - y\|) \tag{7.8}$$

holds.

(ii) A function $W : X \times \mathbb{U} \to Y$ is called *uniformly continuous* on A *uniformly* in $u \in \mathbb{U}$ if there exists a function $\omega \in \mathcal{K}$ such that for all $x, y \in A$ and all $u \in \mathbb{U}$ the inequality

$$\|W(x, u) - W(y, u)\| \leq \omega(\|x - y\|) \tag{7.9}$$

holds. In both cases, the function ω is called *modulus of continuity.*

Note that continuity of $W : X \to Y$ implies uniform continuity on any compact set $A \subset X$. This observation will be used, e.g., in Corollary 7.29, below, exploiting the fact that closed balls in the state space $X = \mathbb{R}^d$ are always compact.

Before, however, we formulate our main result for arbitrary vector spaces X. The following theorem is formulated for Algorithm 3.1 without terminal conditions and we will comment on the case with terminal conditions afterwards. For simplicity, we will directly work with the assumptions of Theorem 4.11, which are ensured, e.g., by Theorems 6.20 and 6.24 or by Corollary 6.21. Alternatively, one could work with the weaker assumptions of Theorem 4.14 as in Theorem 6.37 but since this would cause further technicalities in the statement and the proof of the following theorem, we prefer to use the simpler setting of Theorem 4.11.

Theorem 7.26 (Practical stability using uniform continuity) *Consider the NMPC Algorithm 3.1 without state constraints, i.e., $\mathbb{X} = X$ for some vector space X and with input constraints satisfying $\mathbb{U}(x) = \mathbb{U}$ for all $x \in X$. Assume that $V = V_N$ satisfies the assumptions of Theorem 4.11 with constant reference $x^{\text{ref}} \equiv x_*$ on $S = X$. Assume furthermore that V_N and f are uniformly continuous, uniformly in u in case of f, on the closed balls $\overline{\mathcal{B}}_\rho(x_*)$ for all $\rho > 0$, with functions ω_V and ω_f in* (7.8) *and* (7.9)*, respectively.*

Then the perturbed closed-loop system (7.7) *is semiglobally practically asymptotically stable in the sense of Definition 7.24 on $A = X$.*

Proof Fix $\Delta > \delta > 0$. For all $\nu > 0$ the bounds $\alpha_1, \alpha_2 \in \mathcal{K}_\infty$ on $V = V_N$ from Theorem 4.11 imply

$$\overline{\mathcal{B}}_{\alpha_2^{-1}(\nu)}(x_*) \subseteq V_N^{-1}([0, \nu]) \subseteq \overline{\mathcal{B}}_{\alpha_1^{-1}(\nu)}(x_*).$$

Thus, defining $\sigma = \alpha_2^{-1}(\alpha_1(\delta)/2)$, $\gamma = \alpha_1^{-1}(\alpha_2(\Delta))$ and $\rho = \alpha_1^{-1}(\alpha_2(\gamma + \sigma))$ yields the inclusions

$$\overline{\mathcal{B}}_\sigma(x_*) \subseteq V_N^{-1}([0, \alpha_1(\delta)/2]), \quad V_N^{-1}([0, \alpha_1(\delta)]) \subseteq \overline{\mathcal{B}}_\delta(x_*),$$

$$\overline{\mathscr{B}}_\Delta(x_*) \subseteq V_N^{-1}([0, \alpha_2(\Delta)]) \subseteq \overline{\mathscr{B}}_\gamma(x_*)$$

and

$$\overline{\mathscr{B}}_\sigma(V_N^{-1}([0, \alpha_2(\Delta)])) \subseteq \overline{\mathscr{B}}_{\gamma+\sigma}(x_*) \subseteq V_N^{-1}([0, \alpha_2(\gamma+\sigma)]) \subseteq \overline{\mathscr{B}}_\rho(x_*).$$

Note that (4.14) yields the implication

$$x \in V_N^{-1}([0, \alpha_2(\gamma+\sigma)]) \quad \Rightarrow \quad f(x, \mu_N(x)) \in V_N^{-1}([0, \alpha_2(\gamma+\sigma)]). \tag{7.10}$$

Let ω_V and ω_f be the functions from Definition 7.25 for V_N and $f(\cdot, u)$, respectively, for $A = \overline{\mathscr{B}}_\rho(x_*)$. This implies that (7.8) and (7.9), respectively, hold for V_N and $f(\cdot, u)$ for all $x, y \in V_N^{-1}([0, \alpha_2(\gamma+\sigma)])$. Furthermore, since the lower bound $\alpha_3 \in \mathscr{K}_\infty$ on ℓ from Theorem (4.14) is continuous, it is uniformly continuous on the compact set $[0, \Delta + \sigma]$. We denote the respective function ω from (7.8) by ω_α.

Now we define the function

$$\widetilde{V}(x) := \begin{cases} V_N(x), & x \in V_N^{-1}([0, \alpha_2(\gamma+\sigma)]) \\ \alpha_2(\gamma+\sigma), & \text{otherwise.} \end{cases}$$

By construction, this function is continuous, coincides with V_N on $V_N^{-1}([0, \alpha_2(\gamma + \sigma)])$ and is constant outside this set. Hence, (7.8) holds for $W = \widetilde{V}$ with $\omega = \omega_V$ for all $x, y \in X$. Furthermore, (7.10) implies that (4.14) holds for $V = \widetilde{V}$ for all $x \in V_N^{-1}([0, \alpha_2(\gamma+\sigma)])$.

Now consider arbitrary $d, e \in X$ with $\|d\| \leq \sigma$ and $\|e\| \leq \sigma$ and a point $x \in V_N^{-1}([0, \alpha_2(\Delta)])$. This choice implies

$$x + e \in \overline{\mathscr{B}}_\sigma(V_N^{-1}([0, \alpha_2(\Delta)])) \subseteq V_N^{-1}([0, \alpha_2(\gamma+\sigma)])$$

and thus (4.14) holds for $\widetilde{V}$ in the point $x + e$. Using (7.8) and (4.14) we obtain

$$\begin{aligned}
&\widetilde{V}(f(x, \mu_N(x+e)) + d) \\
&\leq \widetilde{V}(f(x, \mu_N(x+e)) + \omega_V(\|d\|) \\
&\leq \widetilde{V}(f(x+e, \mu_N(x+e)) + \omega_V(\|d\|) + \omega_V(\omega_f(\|e\|)) \\
&\leq \widetilde{V}(x+e) - \alpha\ell(x+e, \mu(x+e)) + \omega_V(\|d\|) + \omega_V(\omega_f(\|e\|)) \\
&\leq \widetilde{V}(x) - \alpha\alpha_3(|x+e|_{x_*}) + \omega_V(\|d\|) + \omega_V(\omega_f(\|e\|)) + \omega_V(\|e\|) \\
&\leq \widetilde{V}(x) - \alpha\alpha_3(|x|_{x_*}) + \omega_V(\|d\|) + \omega_V(\omega_f(\|e\|)) + \omega_V(\|e\|) + \alpha\omega_\alpha(\|e\|).
\end{aligned}$$

Now choose $\overline{d}, \overline{e} \in (0, \sigma]$ so small that

$$\omega_V(\overline{d}) + \omega_V(\omega_f(\overline{e})) + \omega_V(\overline{e}) + \alpha\omega_\alpha(\overline{e}) \leq \min\{\alpha\alpha_3(\sigma)/2, \alpha_1(\delta)/2\}$$

holds. For $x \in X$ with $\widetilde{V}(x) \in [\alpha_1(\delta)/2, \alpha_2(\Delta)]$ and $\|d\| \leq \overline{d}$, $\|e\| \leq \overline{e}$ this implies

$$\begin{aligned}
&\widetilde{V}(f(x, \mu_N(x+e)) + d) \\
&\leq \widetilde{V}(x) - \alpha\alpha_3(|x|_{x_*}) + \omega_V(\|d\|) + \omega_V(\omega_f(\|e\|)) + \omega_V(\|e\|) + \alpha\omega_\alpha(\|e\|) \\
&\leq \widetilde{V}(x) - \alpha\alpha_3(|x|_{x_*}) + \alpha\alpha_3(\sigma)/2 \\
&\leq \widetilde{V}(x) - \alpha\alpha_3(|x|_{x_*})/2
\end{aligned}$$

and for $\widetilde{V}(x) \leq \alpha_1(\delta)/2$ we obtain

$$\begin{aligned}
&\widetilde{V}(f(x, \mu_N(x+e)) + d) \\
&\leq \widetilde{V}(x) - \alpha\alpha_3(|x|_{x_*}) + \omega_V(\|d\|) + \omega_V(\omega_f(\|e\|)) + \omega_V(\|e\|) + \alpha\omega_\alpha(\|e\|) \\
&\leq \widetilde{V}(x) + \alpha_1(\delta)/2 \ \leq \ \alpha_1(\delta)
\end{aligned}$$

In both cases we obtain $\widetilde{V}(f(x, \mu_N(x+e)) + d) \leq \alpha_2(\Delta) \leq \alpha_2(\gamma + \sigma)$, hence x and $f(x, \mu_N(x+e)) + d$ lie in the region where $\widetilde{V}$ and V_N coincide and thus we can replace every occurrence of $\widetilde{V}$ by V_N in both chains of inequalities. The leads to

$$V_N(f(x, \mu_N(x+e)) + d) \leq V_N(x) - \alpha\alpha_3(|x|_{x_*})/2 \tag{7.11}$$

if $V_N(x) \in [\alpha_1(\delta)/2, \alpha_2(\Delta)]$ and

$$V_N(f(x, \mu_N(x+e)) + d) \leq \alpha_1(\delta) \tag{7.12}$$

if $V_N(x) \leq \alpha_1(\delta)/2$. Observing that (7.11) implies (7.12) if $V_N(x) \in [\alpha_1(\delta)/2, \alpha_1(\delta)]$ we can conclude that (7.12) holds for all $x \in X$ with $V_N(x) \leq \alpha_1(\delta)$.

Defining $S = V_N^{-1}([0, \alpha_2(\Delta)])$ and $P = V_N^{-1}([0, \alpha_1(\delta)])$, Inequalities (7.11) and (7.12) imply that both sets are forward invariant and that (4.14) is satisfied for $\ell(x, u) = \alpha_3(|x|_{x_*})/2$ for all $x \in S \setminus P$. Indeed, forward invariance of S follows immediately from (7.11) while forward invariance of P follows since (7.12) holds for all x with $V_N(x) \in [0, \alpha_1(\delta)]$.

Thus, all assumptions of Theorem 4.14 are satisfied (observe that the theorem remains valid in presence of the additional n-dependence of the perturbed system, cf. Remark 4.15). Hence, we obtain the assertion from Lemma 6.34 using that by construction we have $\overline{\mathscr{B}}_R(x_*) \cap A \subseteq S$ and $P \subseteq \overline{\mathscr{B}}_\varepsilon(x_*)$. □

Remark 7.27 (i) Note that we did not use continuity of μ_N in this proof, which is important since optimal feedback laws are in general not continuous.

(ii) For the NMPC Algorithm 3.10 with terminal conditions the result also holds if we can ensure $f(x, \mu_N(x+e)) + d \in \mathbb{X}_N$ for all $x \in V_N^{-1}([0, \alpha_2(\Delta)])$. This is, for instance, guaranteed if the sublevel set $V_N^{-1}([0, \alpha_2(\gamma + \sigma)])$ used in the proof does not intersect the boundary of $\mathbb{X}_N$, i.e., if it is contained in the interior of $\mathbb{X}_N$. However, stabilizing terminal conditions may prevent V_N from being continuous, cf. Example 7.30, below.

(iii) The result can be straightforwardly generalized to time-varying references provided $W = V_N(n, \cdot)$ satisfies (7.8) for all $n \in \mathbb{N}_0$ with ω independent of n.

(iv) The function ω_V in (7.8) measures how sensitive $V_N(x)$ depends on changes in x. The proof shows that $\overline{d}$ can be chosen the larger the smaller ω_V is. Typically, the solutions $x_u(k, x)$ appearing in the definition of $V_N(x)$ depend more sensitive on x the larger k is. Thus, one can expect that ω_V grows with N, i.e., the stability becomes less robust for larger optimization horizons. This is rather intuitive since the longer the prediction horizon the more the perturbed solutions $\tilde{x}_{\mu_N}(k, x)$ deviate from the nominal predictions $x_u(k, x)$.

(v) The condition that the uniform continuity of f is uniform in u is quite strong if $\mathbb{U}$ is unbounded. However, this can be circumvented by penalizing large u in ℓ sufficiently strong such that the optimal solution will never use large u. This technique has, for instance, been used in a sampled data context in [9] and for continuous time systems in [3].

The following corollaries show that robustness can be expected under suitable continuity conditions on the problem data. The first corollary is formulated for state spaces which are arbitrary vector spaces.

Corollary 7.28 *Consider the NMPC Algorithm 3.1 without state constraints, i.e., $\mathbb{X} = X$ for some vector space X, and with input constraints satisfying $\mathbb{U}(x) = \mathbb{U}$ for all $x \in X$. Assume that $V = V_N$ satisfies the assumptions of Theorem 4.11 with constant reference $x^{\text{ref}} \equiv x_*$ on $S = X$, that f is bounded and uniformly continuous in x on each closed ball $\overline{\mathscr{B}}_\rho(x_*)$ and that ℓ is uniformly continuous in x on each such ball, both uniformly in $u \in \mathbb{U}$.*

Then the perturbed closed-loop system (7.7) *is semiglobally practically asymptotically stable in the sense of Definition 7.24 on $A = X$.*

Proof We show that under the given conditions V_N is uniformly continuous on each closed ball $\overline{\mathscr{B}}_R(x_*)$, $R > 0$. Then the assertion follows from Theorem 7.26.

To this end, observe that the boundedness assumption on f implies that there exists $\rho > 0$ such that $x_u(k, x) \in \overline{\mathscr{B}}_\rho(x_*)$ holds for all $k = 0, \ldots, N-1$, all $x \in \overline{\mathscr{B}}_R(x_*)$ and all $u \in \mathbb{U}$.

This implies that $x_u(k, x)$ is uniformly continuous in $x \in \overline{\mathscr{B}}_R(x_*)$, hence the stage cost $\ell(x_u(k, x), u(k))$ is uniformly continuous in $x \in \overline{\mathscr{B}}_R(x_*)$ and consequently $J_N(x, u)$ is uniformly continuous in $x \in \overline{\mathscr{B}}_R(x_*)$, too. This uniform continuity carries over to V_N which proves the claim. □

The second corollary shows that in finite-dimensional state space we can drop the uniform continuity and the boundedness assumptions on the problem data.

Corollary 7.29 *Consider the NMPC Algorithm 3.1 with $X = \mathbb{R}^d$ without state constraints, i.e., $\mathbb{X} = X$ and with input constraints satisfying $\mathbb{U}(x) = \mathbb{U}$ for all $x \in X$. Assume that $V = V_N$ satisfies the assumptions of Theorem 4.11 with constant reference $x^{\text{ref}} \equiv x_*$ on $S = X$. Assume furthermore that ℓ and f are continuous and that $\mathbb{U}$ is compact.*

Then the perturbed closed-loop system (7.7) *is semiglobally practically asymptotically stable in the sense of Definition 7.24 on* $A = X$.

Proof The proof follows when we show continuity of V_N, because then in $X = \mathbb{R}^d$ uniform continuity of V_N and f on each closed ball $\overline{\mathscr{B}}_\rho(x_*)$ and on $\overline{\mathscr{B}}_\rho(x_*) \times \mathbb{U}$, respectively, follows from the compactness of these sets. Note that uniform continuity of f in (x, u) in the sense of Definition 7.25(i) implies uniform continuity of f in x uniformly in u in the sense of Definition 7.25(ii).

In order to prove continuity of V_N, observe that continuity of f and ℓ implies continuity of $J_N(x, u)$ on $\mathbb{R}^d \times \mathbb{U}^N$. This continuity carries over to V_N because minima of continuous functions are again continuous. □

For infinite-dimensional systems, Corollary 7.29 does not apply since closed balls are not compact. Thus, even though Theorem 7.26 and Corollary 7.28 formally apply to infinite-dimensional systems, their practical usefulness is somewhat limited because the required uniform continuity properties may not be satisfied for most practically relevant systems. In fact, it appears doubtful whether robust stability for infinite-dimensional systems can be expected, at all, for the general class of perturbations considered here. Rather, we conjecture that suitable structural properties of the perturbations need to be imposed, as, e.g., in the robust stability results for linear infinite-dimensional systems by Curtain and Zwart in [4, Chap. 9].

However, as we will see in the next section, even for finite-dimensional systems Corollary 7.29 does in general neither extend to NMPC schemes with general state constraints nor to schemes with stabilizing terminal conditions.

7.7 Examples for Nonrobustness Under State Constraints

In this section we provide two examples, taken from Grimm, Messina, Tuna and Teel [6], which show that both general state constraints as well as terminal constraints which are part of stabilizing terminal conditions can render the stability of the NMPC closed loop non-robust.

Our first example shows that even without additional state constraints a stabilizing terminal constraint may result in a non-robust NMPC feedback law.

Example 7.30 Consider again Example 5.8, i.e., $x^+ = f(x, u)$ with $x \in \mathbb{X} = X = \mathbb{R}^2$, $u \in \mathbb{U} = [0, 1] \subset U = \mathbb{R}$, $x_* = 0$ and

$$f(x, u) = \begin{pmatrix} x_1(1-u) \\ \|x\|u \end{pmatrix}.$$

As we have seen in Example 5.8, using the NMPC Algorithm 3.1 with $(\mathrm{OCP}_{\mathrm{N,e}})$ = (5.5) and $\mathbb{X}_0 = \{0\}$ we obtain $\mathbb{X}_1 = \{x \in \mathbb{R}^2 \mid x_1 = 0\}$. Hence, for $x \in \mathbb{R}^2$ with

$x_1 \neq 0$ each admissible control sequence $u \in \mathbb{U}^2_{\mathbb{X}_0}(x)$ must satisfy $u(0) = 1$ in order to ensure $x_u(1, x) \in \mathbb{X}_1$. Thus, the NMPC-feedback law for $N = 2$ also satisfies

$$\mu_2(x) = 1 \quad \text{for all } x \in \mathbb{R}^2 \text{ with } x_1 \neq 0.$$

Now consider the perturbed closed loop (7.7) with perturbation sequences $d(\cdot) \equiv d_0 = (\varepsilon, 0)^\top$ and $e(\cdot) \equiv e_0 = 0$ for some arbitrarily small $\varepsilon > 0$. Then for any $x \in \mathbb{R}^2$ with $x \notin \mathbb{X}_1$ we obtain

$$f(x, \mu_2(x+e_0)) + d_0 = \begin{pmatrix} x_1(1-\mu_2(x)) \\ \|x\|\mu_2(x) \end{pmatrix} + \begin{pmatrix} \varepsilon \\ 0 \end{pmatrix} = \begin{pmatrix} 0 \\ \|x\| \end{pmatrix} + \begin{pmatrix} \varepsilon \\ 0 \end{pmatrix} = \begin{pmatrix} \varepsilon \\ \|x\| \end{pmatrix}$$

which implies $\tilde{x}_{\mu_2}(1, x) = f(x, \mu_2(x + e(0))) + d(0) \notin \mathbb{X}_1$ and

$$\|\tilde{x}_{\mu_2}(1, x)\| = \|f(x, \mu_2(x + e(0))) + d(0)\| = \|(\varepsilon, \|x\|)^\top\| > \|x\|.$$

Since $\tilde{x}_{\mu_2}(1, x) \notin \mathbb{X}_1$ we can go on inductively and obtain

$$\|\tilde{x}_{\mu_2}(k, x)\| > \|x\|$$

for all $k \in \mathbb{N}$. Thus, despite the fact that μ_2 globally asymptotically stabilizes the nominal closed-loop system as shown in Example 5.8, for arbitrary small perturbations d the perturbed closed loop (7.7) is not asymptotically stable.

Note that by Remark 7.27(ii), Theorem 7.26 would in principle be applicable since $\mathbb{X}_2 = \mathbb{R}^2$ and thus no sublevel set of V_2 intersects the boundary of $\mathbb{X}_2$. However, V_2 is discontinuous at $\mathbb{X}_1$, because on $\mathbb{X}_1$ we get $V_2(x) \leq V_1(x) = \|x\|^2$ while outside $\mathbb{X}_1$ the only admissible control sequence is $u(0) = 1$, $u(1) = 0$, which implies $V_2(x) = 2\|x\|^2$.

For this example the NMPC Algorithm 3.1 without terminal conditions provides an alternative which resolves the robustness problem. Indeed, for $u = 1/2$ we obtain

$$\|f(x,u)\|^2 = \left\| \begin{pmatrix} x_1/2 \\ \|x\|/2 \end{pmatrix} \right\|^2 = (x_1/2)^2 + (\|x\|/2)^2 = \frac{x_1^2}{4} + \frac{x_1^2}{4} + \frac{x_2^2}{4} \leq \frac{\|x\|^2}{2}.$$

Thus, for $\ell(x) = \|x\|^2$ and $u_x \equiv 1/2$ Assumption 6.5 is satisfied with $\beta(r, n) = C\sigma^n r$ with $C = 1$ and $\sigma = 1/2$. For $N = 2$, Proposition 6.18 yields $\alpha = \underline{\alpha}_N = 1 - (\gamma_2 - 1)^2$ and since $\gamma_2 = B_2(r)/r = C + C\sigma = 3/2$ by Lemma 6.6, we obtain $\alpha = 1 - (1/2)^2 = 3/4$. Thus, by Corollary 6.21 the NMPC-feedback μ_2 without terminal conditions stabilizes the system and since all assumptions of Corollary 7.29 are satisfied, the asymptotic stability is robust.

Example 7.31 Our second example shows that state constraints may also render NMPC without stabilizing terminal conditions to be nonrobust. The system is a discrete time version of a system known as Artstein's circles. For $x \in X = \mathbb{R}^2$ and $u \in U = [-1, 1]$ it is given by

$$x^+ = f(x,u) = \begin{pmatrix} \dfrac{-(x_1^2+x_2^2)u + x_1}{1+(x_1^2+x_2^2)u^2 - 2x_1u} \\ \dfrac{x_2}{1+(x_1^2+x_2^2)u^2 - 2x_1u} \end{pmatrix}.$$

This is the exact zero order hold sampled data system for sampling period $T = 1$ of the continuous time system $\dot{x}_1 = (x_1^2 - x_2^2)u$, $\dot{x}_2 = 2x_1x_2u$ introduced by Artstein in [1].

The peculiarity of the system is that a solution which starts on the circle

$$S_r = \{x \in \mathbb{R}^2 \mid x_1^2 + (x_2 - r)^2 = r^2\}$$

for some $r \in \mathbb{R}$ can never leave this circle regardless of how the control u is chosen. Figure 7.3 illustrates these circles for $r = -5, -4, \ldots, 5$.

For $x \neq 0$, the control can only be used in order to change the direction of rotation on each circle S_r which for $x_2 > 0$ is clockwise for $u < 0$ and counterclockwise for $u > 0$. For $x_2 < 0$ this orientation changes, for $x_2 = 0$ and $x_1 \neq 0$ the system moves left or right on the x_1-axis and $x_* = 0$ is an equilibrium for all $u \in \mathbb{U}$. For the cost function $\ell(x) = \|x\|_\infty = \max\{|x_1|, |x_2|\}$, on each circle with parameter r we have $\ell(x) = |x_2|$ if $|x_2| \geq r$ and $\ell(x) = |x_1|$ otherwise. Using this fact one can conclude that there exists $\sigma \in (0, 1)$ such that

$$\ell(f(x,u)) \leq \sigma \ell(x)$$

holds when we choose $u = -1$ for $x_1 \geq 0$ and $u = 1$ for $x_1 \leq 0$. Thus, Assumption 6.5 is satisfied for $\beta(r, n) = C\sigma^n r$ with $C = 1$ which implies $\underline{\alpha}_N > 0$ in (6.18). Hence, by Theorem 6.20 the NMPC closed loop for Algorithm 3.1 is asymptotically stable for all horizons $N \geq 2$.

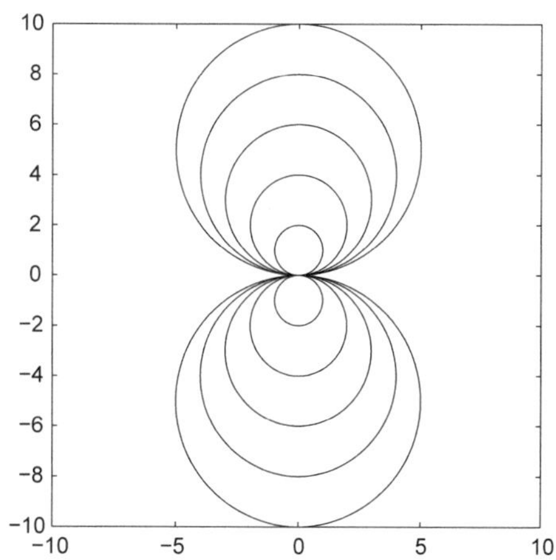

Fig. 7.3 Sketch of Artstein's circles

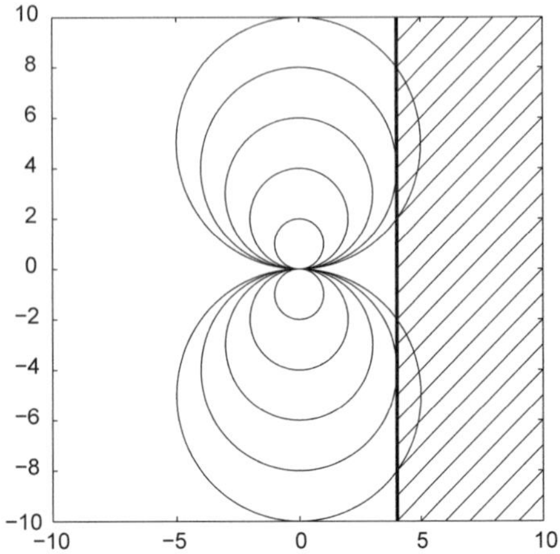

Fig. 7.4 Sketch of state constraint set for Artstein's circles. Admissible states are to the *left* of the *vertical line*

The state constraints we consider now are given by the set

$$\mathbb{X} = \{x \in \mathbb{R}^2 \mid x_1 \leq c\}$$

for some $c > 0$. For $c = 4$ the complement of this set is given by the hatched region in Fig. 7.4.

A little computation shows that for all $c \in (0, 1)$, all circles S_r with $r > r_c = c/\sqrt{1-c^2}$, i.e., circles in the upper half plane with radius $r > r_c$, and all initial values $x \in S_r \cap \mathbb{X}$ with $x_2 > r$ it is not possible to move clockwise toward 0 without violating the state constraints at some point. Hence, we have to take the counterclockwise "detour" in order to control the system to $x_* = 0$. In the function β in the Controllability Assumption 6.5 this detour shows up as an overshoot parameter $C > 1$. However, since the stage cost along the detour is at most $2\ell(x)$ and the time until ℓ decreases exponentially again is bounded from above by a number of steps which is independent of $\ell(x)$, the function $\beta(r, n)$ is still of the form $C\sigma^n r$ and hence we can conclude asymptotic stability for sufficiently large N by Lemma 6.6 and Corollary 6.21.

This asymptotic stability is, however, not robust in the sense of Definition 7.24. In order to see this, fix an arbitrary and sufficiently small $\varepsilon > 0$, consider the circle S_{r_c} and the unique point y on this circle with $y_1 = c$ and $y_2 > r_c$, i.e., the "upper" intersection of S_{r_c} with the boundary of $\mathbb{X}$. Using the control value $u = -1$ this point is mapped onto the point $z = f(x, -1)$ on S_{r_c} with $z_1 = c$ and $z_2 < r$, i.e., on the "lower" intersection of S_{r_c} with the boundary of $\mathbb{X}$. On the one hand this implies that the control sequence $u \equiv -1$ is admissible and that it controls the trajectory counterclockwise in the shortest and thus also cheapest way—in the sense of (OCP$_\mathrm{N}$)—to the origin. Hence, $\mu_N(y) = -1$.

On the other hand, the fact that $z = f(x, -1)$ is on the boundary of $\mathbb{X}$ implies that for $\varepsilon > 0$ sufficiently small all points

$$y' \in Y' := \mathscr{B}_\varepsilon(y) \cap \mathbb{X} \cap S_{r_c} \setminus \{y\},$$

i.e., all points that lie above y and close to y on S_{r_c}, the image $z' = f(y', -1)$ will not be contained in $\mathbb{X}$. Hence, in order to construct a trajectory $x_u(\cdot)$ with $x_u(0) = y'$ which converges to the origin we have two possibilities: either we pick the (unique) control value $u_{y'} < 0$ with $x_u(1) = f(y', u_{y'}) = y$, set $u(0) = u_{y'}$ and continue with $u(1) = u(2) = \cdots = -1$. This will result in a clockwise movement. Alternatively, we can choose the control sequence $u \equiv 1$ which will cause the trajectory to approach the origin counterclockwise. The stage cost ℓ decreases along the first trajectory while for $c < 1/2$ one can compute that it increases along the second—at least for one step before it starts to decrease, again. For this reason, the first trajectory will be optimal for (OCP$_\mathrm{N}$) for all horizons $N \geq 2$. Consequently, we get $\mu_N(y') = u_{y'}$ and $f(y', \mu_N(y')) = y$.

Now we can use a similar construction as in the previous example: pick an initial value $x \in Y'$ and perturbation sequences $e \equiv 0$ and $\|d(n)\| < \varepsilon$ where each $d(n)$ is chosen such that $y + d(n) \in Y'$. By induction over n this implies

$$f(\tilde{x}_{\mu_N}(n), \mu_N(\tilde{x}_{\mu_N}(n))) + d(n) = y + d(n) \in Y'.$$

Thus, for arbitrarily small perturbations the solution gets stuck in Y' and will never reach a neighborhood of the origin.

It is interesting to look at the Lyapunov function V_N of the system on the set $Y' \cup \{y\}$. Indeed, since the first move of the optimal trajectory will map $y' \in Y'$ to y, by the dynamic programming principle and since $\ell(y') \geq y'_2 > r_c$ we can conclude that $V_N(y') > V_{N-1}(y) + r_c$. Since, on the other hand, the optimal trajectory for y approaches the origin clockwise with maximal speed for all N, the additional term in $V_N(y)$ compared to $V_{N-1}(y)$ will be strictly smaller than $r_c - \varepsilon$ for some $\varepsilon > 0$ for all $N \geq 2$. Hence we get $V_N(y) \leq V_{N-1}(y) + r_c - \varepsilon \leq V_N(y') - \varepsilon$ with $\varepsilon > 0$ independent of y. Since y lies at the (lower) boundary of the set Y' this immediately implies that V_N is discontinuous at y.

It should be noted that this discontinuity is not caused by the specific choice of ℓ or the horizon N. Instead, it is solely due to the fact that for topological reasons the given state constraints separate the initial values into different sets whose optimal trajectories approach the origin in different ways: for y it moves clockwise with maximal speed and for each $y' \in Y$ it moves clockwise with passing through y. Furthermore, as pointed out after Fig. 7.4, for all $x \in S_r \cap \mathbb{X}$ with $r > r_c$ and $x_2 > r$ it is not possible at all to approach the origin clockwise, hence the optimal trajectory will move counterclockwise with maximal speed, which defines yet another different behavior. Since the counterclockwise movement is more expensive than the clockwise movement and since passing through y is more expensive than approaching the origin with maximal speed, V_N will be discontinuous at all boundaries where these different sets of initial values touch—unless ℓ is particularly tuned in order to penalize the different trajectories in exactly the same way, which is a difficult if not impossible task and in any case an exceptional situation.

Thus, in general we will have to take discontinuities of V_N into account and cannot conclude robustness from continuity as in Theorem 7.26. Consequently, in the next section we will look at a technique which allows us to conclude robustness without assuming continuity of V_N.

7.8 Robustness with State Constraints via Robust-Optimal Feasibility

In order to ensure robustness in the presence of state constraints we present two approaches. Again we assume $\mathbb{U}(x) = \mathbb{U}$ in order to avoid violation of the input constraints due to measurement errors. Like in Sect. 7.6, however, the subsequent definitions, statements and proofs could be extended to the case of input constraint sets $\mathbb{U}(x)$ depending continuously on x.

In this section we will present a first approach, which is inspired by Grimm, Messina, Tuna and Teel [8], but we simplify the setting and use a different proof based on the techniques from Chap. 6. This first approach has the advantage that we do not need continuity of the optimal value function V_N. The second approach, presented in the following Sect. 7.9, uses this continuity and presents a setting in which continuity can be proved in the presence of state constraints.

Both approaches rely on the fact that we use state constraint sets which depend on time. To this end, we denote the state constraint sets by $\mathbb{X}^k$, $k = N, N-1, \ldots, 0$ with $\mathbb{X} = \mathbb{X}^N$ and extend Definition 3.2 as follows.

Definition 7.32 For $K \in \mathbb{N}$ and $N \in \mathbb{N}$ with $K \leq N$ and an initial value $x_0 \in \mathbb{X}^K$ we call a control sequence $u \in U^K$ and the corresponding trajectory $x_u(k, x_0)$ *admissible for x_0 and K*, if

$$u(k) \in \mathbb{U}(x_u(k, x_0)) \quad \text{and} \quad x_u(k+1, x_0) \in \mathbb{X}^{K-(k+1)}$$

holds for all $k = 0, \ldots, K-1$. The respective set of admissible control sequences for x_0 is denoted by $\mathbb{U}^K(x_0)$.

Note that this definition reduces to Definition 3.2 if $\mathbb{X}^k = \mathbb{X}$ for all $k = 0, \ldots, N$. The optimization problem (OCP$_\mathrm{N}$) can be used with this setting without any changes since we use the same notation for the admissible control sequences as before in Definition 3.2. Similarly, Assumptions 6.3 and 6.5 and the definition of the feasible set $\mathscr{F}_N$ in Definition 7.2(ii) immediately extend to this time-varying state constraint concept. As always, we will assume that an optimal control $u^\star$ exists for (OCP$_\mathrm{N}$) for each initial value.

Observe that the terminal constraints from Definition 3.9 form a special case of this time-varying constraint setting by defining $\mathbb{X}^N = \mathbb{X}^{N-1} = \cdots = \mathbb{X}^1 = \mathbb{X}$ and $\mathbb{X}^0 = \mathbb{X}_0$. In this case $\mathbb{U}^K(x)$ from Definition 7.32 coincides with $\mathbb{U}^K_{\mathbb{X}_0}(x)$ from Definition 3.9 and $\mathscr{F}_K$ from Definition 7.2(ii) equals $\mathbb{X}_K$ from Definition 3.9.

For our first robustness result, we use the following property.

Definition 7.33 *(Robust-optimal feasibility)* For given optimization horizon $N \in \mathbb{N}$ the NMPC Algorithm 3.1 is said to be *robust-optimal feasible*, if for each closed ball $\overline{\mathscr{B}}_\rho(x_*)$ there exists $\eta > 0$ such that the following holds:

For each $x \in \overline{\mathscr{B}}_\rho(x_*) \cap \mathscr{F}_N$, each $z \in \mathscr{B}_\eta(f(x, \mu_N(x)))$ and the optimal control $u^\star$ for ($\mathrm{OCP_N}$) with $x_0 = x$ we have

$$x_{u^\star(\cdot+1)}(k, z) \in \mathscr{F}_{N-k} \text{ for all } k \in \{0, \dots, N-1\} \tag{7.13}$$

This condition in particular demands $z = x_{u^\star(\cdot+1)}(0, z) \in \mathscr{F}_N$ for $z = f(x, \mu_N(x))$ which implies that $\mathscr{F}_N$ is recursively feasible for the NMPC closed loop. It can be satisfied by the following construction of tightening state constraints sets $\mathbb{X}^k$ whose idea goes back to Limón, Alamo, and Camacho [14].

Assume that f is uniformly continuous and bounded in x on each closed ball uniformly in u. Then for each arbitrary but fixed horizon N and admissible control input $u \in \mathbb{U}^N$ the trajectory $x_u(k, x)$ is uniformly continuous in $x \in \overline{\mathscr{B}}_\rho(x_*) \cap \mathscr{F}_N$ uniformly in $k \in \{1, \dots, N-2\}$ and u. Let ω denote the modulus of continuity from (7.9). Now we pick $\delta > 0$ and assume that we can choose the constraint sets $\mathbb{X}^k$ such that

$$\mathbb{X}^N = \mathbb{X}, \quad \overline{\mathscr{B}}_\delta(\mathbb{X}^{k-1}) \subset \mathbb{X}^k \quad \text{and} \quad \mathscr{F}_k = \mathbb{X}^k \tag{7.14}$$

holds for all $k = 1, \dots, N$. Let $x \in \overline{\mathscr{B}}_\rho(x_*) \cap \mathscr{F}_N$ and $u^\star \in \mathbb{U}^N(x)$ be a corresponding optimal control. Then for $z = f(x, \mu_N(x))$ we obtain

$$x_{u^\star(\cdot+1)}(k, z) = x_{u^\star(\cdot)}(k+1, x) \in \mathbb{X}^{N-k-1}$$

for $k = 0, \dots, N-1$. For arbitrary $z \in \mathscr{B}_\varepsilon(f(x, \mu_N(x)))$ this implies

$$x_{u^\star(\cdot+1)}(k, z) \in \mathscr{B}_{\omega(\varepsilon)}(\mathbb{X}^{N-k-1})$$

and if we choose $\varepsilon > 0$ so small that $\omega(\varepsilon) \leq \delta$ holds then we get

$$x_{u^\star(\cdot+1)}(k, z) \in \mathscr{B}_{\omega(\varepsilon)}(\mathbb{X}^{N-k-1}) \subseteq \mathscr{B}_\delta(\mathbb{X}^{N-k-1}) \subset \mathbb{X}^{N-k}$$

for $k = 0, \dots, N-1$. Hence, we obtain (7.13).

The main limitation of this construction is the condition $\mathscr{F}_k = \mathbb{X}^k$ in (7.14), since in general the feasible sets $\mathscr{F}_k$ for the tightening constraint sets $\mathbb{X}^k$ may be very small or even empty. However, for some systems it is possible to prove rigorously that this construction is possible, as the following example shows.

Example 7.34 We reconsider Artstein's circles with the constraints from Example 7.31 and show that $\mathbb{X}^k$ with (7.14) can be constructed for this example. From the form of the dynamics it follows that for each $a > 0$ there exists $\nu \in (0, 1)$ such that for each $x \in \mathbb{R}^2$ with $x_1 > a$ there exists $u \in \mathbb{U}$ with $f(x, u)_1 \leq \nu x_1$. Hence, setting

$$\mathbb{X}^k = \{x \in \mathbb{R}^2 \mid x_1 \le c - (N-k)\varepsilon\},$$

for $\varepsilon > 0$ sufficiently small (depending on c and N) we obtain $\mathscr{F}_k = \mathbb{X}^k$. Since the first two conditions in (7.14) are obviously satisfied for this choice of the $\mathbb{X}^k$, with these state constraint sets the NMPC-feedback law satisfies (7.13).

Effectively, the tightening constraint sets cause the NMPC controller to choose the more expensive counterclockwise trajectories for a larger set of initial values and thus prevent the feedback from selecting the small steps from y' to y which caused the nonrobustness in Example 7.31.

In order to show robustness of the asymptotic stability, again we need a suitable controllability condition. The following condition is slightly different from Assumption 7.13 in the sense that we assume controllability on the sets $\mathscr{F}_{N-k}$ rather than on $\mathscr{F}_\infty$, since for a finite number of state constraint sets $\mathbb{X}^N, \mathbb{X}^{N-1}, \ldots, \mathbb{X}^0$ it is not possible to define $\mathscr{F}_\infty$.

Assumption 7.35 Consider the optimal control problem (OCP$_\text{N}$) with a not necessarily viable state constraint set $\mathbb{X}$. We assume that on the feasible sets $\mathscr{F}_{N-k}$, $k = 0, \ldots, N-1$ the system is asymptotically controllable with respect to ℓ with rate $\beta \in \mathscr{KL}_0$, i.e., for each $x \in \mathscr{F}_{N-k}$ there exists an admissible control sequence $u_x \in \mathbb{U}^{N-k}(x)$ satisfying

$$\ell(x_{u_x}(n, x), u_x(n)) \le \beta(\ell^*(x), n)$$

for all $n \in \{0, \ldots, N-k-1\}$.

The following theorem now shows that Definition 7.33 and Assumption 7.35 indeed imply robustness of the asymptotic stability.

Theorem 7.36 (Practical stability using robust-optimal feasibility) *Consider the NMPC Algorithm 3.1 with state constraint sets $\mathbb{X}$ and $\mathbb{X}^N, \ldots, \mathbb{X}^0 \subseteq X$ with $\mathbb{X}^N = \mathbb{X}$ for some vector space X and with input constraints satisfying $\mathbb{U}(x) = \mathbb{U}$ for all $x \in X$. Assume that f is bounded and uniformly continuous in x on each closed ball $\overline{\mathscr{B}}_\rho(x_*)$ and that ℓ is uniformly continuous in x on each such ball, both uniformly in $u \in \mathbb{U}$. Let the assumptions of Theorem 6.20 hold with Assumption 6.3 replaced by Assumption 7.35 and assume that Definition 7.33 is satisfied.*

Then the perturbed closed-loop system (7.7) *is semiglobally practically asymptotically stable in the sense of Definition 7.24 on $A = \mathscr{F}_N$.*

Proof We first show the assertion for $e \equiv 0$, i.e., for the case without measurement error. To this end, fix $\Delta > \delta > 0$ in Definition 7.24. Let $R > 0$ be such that $\overline{\mathscr{B}}_\Delta(x_*) \cap \mathscr{F}_N \subseteq V_N^{-1}([0, R])$, let $\rho > 0$ be such that $V_N^{-1}([0, R]) \subseteq \overline{\mathscr{B}}_\rho(x_*)$ and pick $\eta > 0$ from Definition 7.33.

We show that there exists a function $\sigma(\overline{d})$ with $\sigma(\overline{d}) \to 0$ as $\overline{d} \to 0$ such that the inequality

$$V_N(f(x, \mu_N(x)) + d) \le V_N(x) - \alpha\ell(x, \mu_N(x)) + \sigma(\overline{d}) \tag{7.15}$$

holds for all $x \in V_N^{-1}([0, R])$ and all $\|d\| \le \overline{d}$ with $\alpha \in (0, 1)$ from the assumptions of Theorem 6.20. For $\overline{d}$ sufficiently small this implies the Inequalities (7.11) and (7.12) from the proof of Theorem 7.26 from which we can conclude practical asymptotic stability as in this proof.

In order to prove (7.15), pick η corresponding to $\overline{\mathscr{B}}_\rho(x_*)$ from Definition 7.33, $x \in \overline{\mathscr{B}}_\rho(x_*) \cap \mathscr{F}_N$ and consider the trajectories $x_{u^\star(\cdot+1)}(k, z)$ for $z \in \mathscr{B}_\eta(f(x, \mu_N(x)))$. For $z = f(x, \mu_N(x))$ we obtain $x_{u^\star(\cdot+1)}(k, f(x, \mu_N(x))) = x_{u^\star}(k+1, x)$ and thus the values

$$\lambda_k := \ell(x_{u^\star(\cdot+1)}(k-1, f(x, \mu_N(x))), u^\star(k))$$

satisfy (6.11). Defining

$$\tilde{\lambda}_k(z) := \ell(x_{u^\star(\cdot+1)}(k-1, z), u^\star(k))$$

for $z \in \mathscr{B}_\eta(f(x, \mu_N(x)))$, from the uniform continuity we obtain the existence of $\tilde{\sigma}(\eta)$ with $\tilde{\sigma}(\eta) \to 0$ as $\eta \to 0$ such that

$$|\tilde{\lambda}_k(z) - \lambda_k| \le \tilde{\sigma}(\eta)$$

holds for all $z \in \mathscr{B}_\eta(f(x, \mu_N(x)))$. Using the same arguments as in the proof of Lemma 6.10, from Assumption 7.35 and (7.13) we obtain

$$V_N(z) \le \sum_{n=0}^{j-1} \tilde{\lambda}_{n+1}(z) + B_{N-j}(\tilde{\lambda}_{j+1}(z)), \quad j = 0, \ldots, N-2$$

with B_K from (6.7). Now continuity of the B_K implies the existence of $\sigma(\eta)$ with $\sigma(\eta) \to 0$ as $\eta \to 0$ such that

$$V_N(z) \le \sum_{n=0}^{j-1} \lambda_{n+1} + B_{N-j}(\lambda_{j+1}) + \sigma(\eta), \quad j = 0, \ldots, N-2. \tag{7.16}$$

Observe that this is exactly (6.12) with $V_N(z)$ in place of ν and the additional term $\sigma(\eta)$. Now the assumptions of Theorem 6.20 imply that Theorem 6.15 applies which in turn implies that (6.13) holds for some $\alpha > 0$. Hence we get that each ν satisfying (6.12) also satisfies

$$\nu \le V_N(x) - \alpha\ell(x, \mu_N(x)).$$

Hence, (7.16) implies

$$V_N(z) \le V_N(x) - \alpha\ell(x, \mu_N(x)) + \sigma(\eta)$$

which is exactly (7.15) by setting $\eta = \overline{d}$ and $z = f(x, \mu_N(x)) + d$.

It remains to show the assertion for $e \not\equiv 0$. To this end, fix $\Delta > \delta > 0$ and denote by $\tilde{d} > 0$ the bound on $\|d\|$ from the first part of the proof which ensures practical asymptotic stability for $\tilde{\delta} = \delta/2$, $\widetilde{\Delta} = \Delta + \tilde{\delta} > 0$ and $\tilde{e} \equiv 0$. Let $\rho > 0$ be as in the first part of the proof for these values. Now take a trajectory $\tilde{x}_{\mu_N}(k, x_0)$ of (7.7) with $|x|_{x_*} \leq \Delta$ and define $\hat{x}(k) = \tilde{x}_{\mu_N}(k, x) + e(k)$. Then $\hat{x}$ is a solution of the system

$$\hat{x}(n+1) = f(\hat{x}(n), \mu_N(\hat{x}(n))) + \tilde{d}(n)$$

with

$$\tilde{d}(n) = d(n) + f(\tilde{x}_{\mu_N}(k, x), \mu_N(\hat{x}(n))) - f(\hat{x}(n), \mu_N(\hat{x}(n))) + e(n+1).$$

As long as the solution stays in $\overline{\mathscr{B}}_\rho(x_*)$, by uniform continuity of f in x we find $\overline{e} > 0$ such that $\|e(n)\| \leq \overline{e}$ and $\|e(n+1)\| \leq \overline{e}$ imply

$$\|\tilde{d}(n)\| \leq \|d(n)\| + \tilde{d}/2.$$

Without loss of generality we can set $\overline{e} \leq \tilde{\delta}$ implying $\hat{x}(0) = x_0 + e(0) \in \mathscr{B}_{\widetilde{\Delta}}(x_*)$. Hence, setting $\overline{d} = \tilde{d}/2$ the trajectory $\hat{x}(n)$ can be interpreted as a solution of (7.7) with $e \equiv 0$ and $\overline{d} = \tilde{d}$ starting in $\hat{x}(0) = x_0 + e(0) \in \mathscr{B}_{\widetilde{\Delta}}(x_*) \cap \mathscr{F}_N$. Hence, the first part of the proof applies to $\hat{x}$, in particular, the trajectory stays in $\overline{\mathscr{B}}_\rho(x_*)$ for all n and thus we get the practical asymptotic stability estimate

$$|\hat{x}(k)|_{x_*} \leq \max\{\tilde{\beta}(|\hat{x}(0)|_{x_*}, k), \tilde{\delta}\}$$

which implies

$$|\tilde{x}_{\mu_N}(k, x_0)|_{x_*} \leq \max\{\tilde{\beta}(|x|_{x_*} + \overline{e}, k), \tilde{\delta}\} + \overline{e}.$$

It remains to convert this upper bound into the form from Definition 7.24. To this end, set $\beta(r, n) = 2\tilde{\beta}(2r, n)$ which is again in $\mathscr{KL}$. We reduce $\overline{e}$ further, if necessary, in order to ensure $\overline{e} \leq \tilde{\delta}$ and $\tilde{\beta}(2\overline{e}, 0) \leq \tilde{\delta}$. Then, for $|x|_{x_*} \leq \overline{e}$ we obtain

$$\tilde{\beta}(|x|_{x_*} + \overline{e}, k) \leq \tilde{\delta}$$

for all $k \geq 0$ which implies

$$|\tilde{x}_{\mu_N}(k, x_0)|_{x_*} \leq \tilde{\delta} + \overline{e} \leq \delta.$$

For $|x|_{x_*} \geq \overline{e}$ we get

$$\beta(|x|_{x_*}, k) \geq 2\tilde{\beta}(|x|_{x_*} + \overline{e}, k)$$

Hence, for all $k \geq 0$ with $\tilde{\beta}(|x|_{x_*} + \overline{e}, k) \geq \tilde{\delta}$ we get $\beta(|x|_{x_*}, k) \geq \tilde{\beta}(|x|_{x_*} + \overline{e}, k) + \tilde{\delta}$ and thus

$$|\tilde{x}_{\mu_N}(k, x_0)|_{x_*} \leq \tilde{\beta}(|x|_{x_*} + \overline{e}, k) + \overline{e} \leq \beta(|x|_{x_*}, k).$$

Finally, for $k \geq 0$ with $\tilde{\beta}(|x|_{x_*} + \overline{e}, k) \leq \tilde{\delta}$ we get

$$|\tilde{x}_{\mu_N}(k, x_0)|_{x_*} \leq \tilde{\delta} + \overline{e} \leq \delta.$$

Thus, for all $k \geq 0$ we have

$$|\tilde{x}_{\mu_N}(k, x_0)|_{x_*} \leq \max\{\beta(|x|_{x_*}, k), \delta\},$$

i.e., the desired property from Definition 7.24. □

Remark 7.37 (i) As in the proof of Theorem 7.26 we did not use continuity of μ_N.

(ii) The proof heavily relies on the controllability based analysis for NMPC schemes without terminal conditions. Hence, it does not apply to the case with terminal conditions. However, analogous statements can be made also for NMPC schemes with terminal conditions, see, e.g., [8, 14, Sect. IV].

(iii) Like Theorem 7.26 the result can be straightforwardly generalized to time-varying references provided the assumed continuity and controllability properties for ℓ are uniform with respect to the initial time.

(iv) In finite-dimensional state space, i.e., $X = \mathbb{R}^d$, the assumed uniform continuity properties on the balls $\overline{\mathscr{B}}_\rho(x_*)$ follow from mere continuity by an argument similar to the one used in the proof of Corollary 7.29. Regarding the uniformity of the assumed properties with respect to u, Remark 7.27(v) applies accordingly.

(v) As outlined in the discussion after Corollary 7.29, for infinite-dimensional systems the required uniform continuity is a rather restrictive condition.

7.9 Robustness with State Constraints via Continuity of V_N

The second approach we want to present for ensuring Definition 7.24 under state constraints uses a modification of Theorem 7.26 which exploits the continuity of V_N. The main problem here is not to extend Theorem 7.26—this will turn out to be rather straightforward—but rather to design the state constraint sets $\mathbb{X}^k$ such that continuity of V_N can be expected under a checkable condition. The following assumption defines a sufficient condition for this purpose.

Assumption 7.38 The state constraint sets $\mathbb{X}^k$, $k = N, N-1, \ldots, 0$ satisfy the following conditions.

(i) $\mathbb{X}^{N-1} = \mathbb{X}^{N-2} = \cdots = \mathbb{X}^0$ and $\overline{\mathscr{B}}_\delta(\mathbb{X}^0) \subseteq \mathbb{X}^N = \mathbb{X}$ for some $\delta > 0$.

(ii) For each $x \in \mathbb{X}^N$ there exists $u \in \mathbb{U}(x)$ with $f(x, u) \in \mathbb{X}^0$.

(iii) For each $\rho > 0$ there exist $\gamma \in \mathscr{K}$ and $\varepsilon' > 0$ such that for each $\varepsilon \in (0, \varepsilon']$, each $x \in \mathbb{X}^N \cap \overline{\mathscr{B}}_\rho(x_*)$, each $u \in \mathbb{U}(x)$ with $f(x, u) \in \mathbb{X}^0$ and each $x' \in \mathbb{X}^N \cap \overline{\mathscr{B}}_\varepsilon(x)$ there is $u' \in U(x')$ with $f(x', u') \in \mathbb{X}^0$, $\|f(x, u) - f(x', u')\| \leq \gamma(\varepsilon)$ and $|\ell(x, u) - \ell(x', u')| \leq \gamma(\varepsilon)$.

The condition again requires a tightening structure of the state constraint sets, however, in contrast to (7.14) only for $\mathbb{X}^N$ and $\mathbb{X}^{N-1}$ while all other state constraint sets equal $\mathbb{X}^{N-1}$. In words, it demands that whenever $x \in \mathbb{X}^N$ and $f(x,u) \in \mathbb{X}^0$, then for all nearby points $x' \approx x$ we find a control u' with $f(x',u') \in \mathbb{X}^0$ which is "nearby" u in the sense that $f(x',u') \approx f(x,u)$ and $\ell(x',u') \approx \ell(x,u)$. Before showing that this condition ensures continuity of V_N, let us consider an example where this condition is satisfied.

Example 7.39 We reconsider Example 3.4 (see also Examples 2.2, 7.1 and 7.12), i.e.,

$$x^+ = f(x,u) = \begin{pmatrix} x_1 + x_2 + u/2 \\ x_2 + u \end{pmatrix}$$

with state constraints $\mathbb{X} = [-1,1]^2$. Even for the largest possible set $\mathbb{U} = \mathbb{R}$ of control values it is not possible to satisfy Assumption 7.38 for $\mathbb{X}^N = \mathbb{X}$. This is because the only control which maps the point $x = (1,1)^\top$ into $\mathbb{X} = [-1,1]^2$ is $u = -2$: for larger values $u' > -2$ we get $f(x,u')_1 = 1+1+u'/2 > 2-1 = 1$ and for smaller values $u' < -2$ we get $f(x,u')_2 = 1 + u' < 1 - 2 = -1$. For $u = -2$, however, we get $f(x,u) = (1,-1)^\top \in \partial\mathbb{X}^N$ and thus we cannot reach any set $\mathbb{X}^0$ with $\overline{\mathscr{B}}_\delta(\mathbb{X}^0) \subseteq \mathbb{X}^N$, regardless of how small $\delta > 0$ is chosen.

Hence, we need to restrict $\mathbb{X}$ in order to satisfy Assumption 7.38. For simplicity, we again pick the set

$$\mathbb{X}^N = \mathbb{X} = \{(x_1,x_2)^\top \in \mathbb{R}^2 \mid x_1 \in [-1,1], x_2 \in [-1,1] \cap [-3/2 - x_1, 3/2 - x_1]\}$$

derived in Example 3.4. We claim that for $\mathbb{U} = [-1-2\eta, 2+2\eta]$ and $\mathbb{X}^{N-1} = \mathbb{X}^{N-2} = \ldots = \mathbb{X}^0$ with

$$\begin{aligned}\mathbb{X}^0 = \{(x_1,x_2)^\top \in \mathbb{R}^2 \mid\, & x_1 \in [-1+\eta, 1-\eta], \\ & x_2 \in [-1+\eta, 1-\eta] \cap [-3/2 - x_1 + \eta, 3/2 - x_1 - \eta]\}\end{aligned}$$

Assumption 7.38 holds if ℓ is continuous and $\eta > 0$ is sufficiently small.

First observe that this choice implies Assumption 7.38(i) for $\delta > 0$ sufficiently small compared to η (for instance, $\delta = \eta/4$ will work). Assumption 7.38(ii) follows since straightforward computations show that for each $x \in \mathbb{X}$ the relation $f(x,u) \in \mathbb{X}^0$ holds if and only if $u \in \mathbb{U}$ satisfies the inequalities

$$u \geq \max\left\{\frac{2}{3}\left(-x_1 - 2x_2 - \frac{3}{2} + \eta\right), -1 - x_2 + \eta, 2(-1 - x_1 - x_2 + \eta)\right\} \tag{7.17}$$

and

$$u \le \min\left\{\frac{2}{3}\left(-x_1 - 2x_2 + \frac{3}{2} - \eta\right), \quad 1 - x_2 - \eta,\ 2(\ 1 - x_1 - x_2 - \eta)\right\}. \tag{7.18}$$

Comparison of the single terms on the right hand sides of (7.17) and (7.18) yields that for all $x \in \mathbb{X}^N$ there exists $u \in \mathbb{U}$ satisfying (7.17) and (7.18) whenever $\eta \le 3/8$.

Finally, Assumption 7.38(iii) follows because the bounds on u in (7.17) and (7.18) change continuously with x. Thus, given an admissible u for some $x \in \mathbb{X}$ which hence must satisfy the bounds (7.17) and (7.18), for $x' \in \mathbb{X}$ with $x' \approx x$ the bounds change only slightly and thus we find an admissible $u' \approx u$ for x'. Since f is continuous and ℓ was assumed to be continuous, the existence of the desired $\gamma \in \mathscr{K}$ follows.

In contrast to this example, for Artstein's circles under the state constraints $\mathbb{X}$ from Example 7.31, Assumption 7.38 does not hold: no matter how the constraint sets $\mathbb{X}^k$ are chosen there will always be points x and $x' \in \mathbb{X}$ arbitrarily close to each other such that from x a clockwise movement is possible while from x' only a counterclockwise movement is admissible, or vice versa.

The following Proposition shows that under Assumption 7.38 and under additional continuity conditions on f and ℓ, V_N is uniformly continuous on closed balls.

Proposition 7.40 *Consider the optimal control problem* (OCP$_\mathrm{N}$) *with state constraint sets* $\mathbb{X}^N, \ldots, \mathbb{X}^0$ *satisfying Assumption 7.38 and assume that* f *is bounded and uniformly continuous in* x *on each closed ball* $\overline{\mathscr{B}}_\rho(x_*)$ *and that* ℓ *is uniformly continuous in* x *on each such ball, both uniformly in* $u \in \mathbb{U}$. *Then* V_N *is uniformly continuous on* $\overline{\mathscr{B}}_R(x_*) \cap \mathbb{X}$ *for each* $R > 0$.

Proof For each $\eta > 0$ and initial value $x \in \overline{\mathscr{B}}_R(x_*) \cap \mathbb{X}$ we pick a control sequence $u_{x,\eta} \in \mathbb{U}^N(x)$ with

$$J_N(x, u_{x,\eta}) \le V_N(x) + \eta.$$

We show that there exists $\omega \in \mathscr{K}$ such that for all sufficiently small $\varepsilon_0 > 0$, all $x \in \overline{\mathscr{B}}_R(x_*) \cap \mathbb{X}$, all $\eta > 0$ and all $\hat{x} \in \mathbb{X} \cap \mathscr{B}_{\varepsilon_0}(x_*)$ there exists $\hat{u} \in \mathbb{U}^N(x')$ such that the inequality

$$J_N(\hat{x}, \hat{u}) \le J_N(x, u_{x,\eta}) + \omega(\varepsilon_0) \tag{7.19}$$

holds. This implies $V_N(\hat{x}) \le V_N(x) + \eta + \omega(\varepsilon_0)$ and since $\eta > 0$ was arbitrary we obtain

$$V_N(\hat{x}) \le V_N(x) + \omega(\varepsilon_0).$$

Since this inequality holds for all $x, \hat{x} \in \overline{\mathscr{B}}_R(x_*) \cap \mathbb{X}$ with $\|x - \hat{x}\| \le \varepsilon_0$, we get

$$|V_N(\hat{x}) - V_N(x)| \le \omega(\varepsilon_0)$$

for all such x and $\hat{x}$ and thus the desired uniform continuity.

It remains to show (7.19). Since by assumption f is uniformly bounded, each trajectory $x_u(k,x)$ with $x \in \overline{\mathscr{B}}_R(x_*) \cap \mathbb{X}$ satisfies $x_u(k,x_0) \in \overline{\mathscr{B}}_\rho(x_*)$ for some sufficiently large $\rho > 0$, all $k = 0,\dots,N$ and all $u \in \mathbb{U}^N(x_0)$. We pick $\gamma \in \mathscr{K}$ from Assumption 7.38(iii) for this $\rho > 0$. Without loss of generality we may assume $\gamma(r) \geq r$ for all $r \geq 0$.

Now fix $x \in \overline{\mathscr{B}}_R(x_*) \cap \mathbb{X}$ and $\eta > 0$ and abbreviate $u = u_{x,\eta}$. Given $\hat{x} \in \mathbb{X} \cap \mathscr{B}_{\varepsilon_0}(x)$, we inductively construct a control sequence $\hat{u}$ as follows.

$$\begin{aligned}&\text{For } k = 0,\dots,N-1 \text{ we set } \hat{u}(k) := u' \text{ with } u' \text{ from}\\ &\text{Assumption 7.38(iii) for } x = x_u(k,x),\ u = u(k) \text{ and } x' = x_{\hat{u}}(k,\hat{x}).\end{aligned} \tag{7.20}$$

Note that we only need $\hat{u}(0),\dots,\hat{u}(k-1)$ in order to compute $x' = x_{\hat{u}}(k,\hat{x})$. Hence, (7.20) is well defined provided $\|x_{\hat{u}}(k,\hat{x}) - x_u(k,x)\| \leq \varepsilon'$ holds for all $k = 0,\dots,N-1$ for $\varepsilon' > 0$ from Assumption 7.38(iii). In this case, the construction of $\hat{u}$ implies $\hat{u} \in \mathbb{U}^N(\hat{x})$ and in particular $x_{\hat{u}}(k,\hat{x}) \in \mathbb{X}^0$.

We will now show by induction that we can ensure this inequality for $k = 0,\dots,N-1$ if we choose ε_0 so small that $\gamma^{N-1}(\varepsilon_0) \leq \varepsilon'$ holds, where γ^k is defined inductively by $\gamma^0(r) = r$ and $\gamma^{k+1}(r) = \gamma \circ \gamma^k(r)$.

To this end, by induction over $k = 0,\dots,N-1$ we prove the inequality

$$\|x_{\hat{u}}(k,\hat{x}) - x_u(k,x)\| \leq \gamma^k(\varepsilon_0) \leq \gamma^{N-1}(\varepsilon_0) \leq \varepsilon' \tag{7.21}$$

For $k = 0$, (7.21) immediately follows. For the induction step $k \to k+1$, using the abbreviations $u' = \hat{u}(k)$, $x = x_u(k,x)$, $u = u(k)$ and $x' = x_{\hat{u}}(k,\hat{x})$ from (7.20), we get

$$x_{\hat{u}}(k+1,\hat{x}) = f(x',u') \quad \text{and} \quad x_u(k+1,x) = f(x,u).$$

Now the induction assumption yields $x' \in \overline{\mathscr{B}}_{\gamma^k(\varepsilon_0)}(x) \subseteq \overline{\mathscr{B}}_{\varepsilon'}(x)$, hence u' from Assumption 7.38(iii) exists and we obtain

$$\|f(x',u') - f(x',u')\| \leq \gamma(\gamma^k(\varepsilon_0)) = \gamma^{k+1}(\varepsilon_0).$$

This proves (7.21) for $k+1$. In order to prove (7.19) we now use the inequality

$$|\ell(x,u) - \ell(x',u')| \leq \gamma(\|x - x'\|)$$

which follows from Assumption 7.38(iii). Combining this with (7.21) yields

$$|\ell(x_u(k,x),u(k)) - \ell(x_{\hat{u}}(k,\hat{x}),\hat{u}(k))| \leq \gamma^{k+1}(\varepsilon_0).$$

Thus,

$$J_N(\hat{x},\hat{u}) = \sum_{k=0}^{N-1} \ell(x_{\hat{u}}(k,\hat{x}),\hat{u}(k)) \le \sum_{k=0}^{N-1} \big(\ell(x_{\hat{u}}(k,\hat{x}),\hat{u}(k)) + \gamma^{k+1}(\varepsilon_0)\big)$$
$$= \sum_{k=0}^{N-1} \ell(x_{\hat{u}}(k,\hat{x}),\hat{u}(k)) + \omega(\varepsilon_0)$$

for

$$\omega(r) = \sum_{k=0}^{N-1} \gamma^k(r).$$

Since $\gamma \in \mathcal{K}$ implies $\gamma^k \in \mathcal{K}$ and consequently $\omega \in \mathcal{K}$ and since ω is independent of x and η, this shows (7.19). □

With the help of Proposition 7.40 we can now prove our second robustness theorem under state constraints. As in Sect. 7.6 we directly use the assumptions of Theorem 4.11 for ensuring stability, which can be guaranteed by, e.g., Theorems 6.20, 6.24 or Corollary 6.21 in combination with Assumption 6.5 and Lemma 6.6, using, of course, the admissible control sequences related to the state constraints from Assumption 7.38 in the Controllability Assumption 6.5.

Theorem 7.41 (Practical stability using continuity of V_N) *Consider the NMPC Algorithm 3.1 with state constraint sets $\mathbb{X} = \mathbb{X}^N, \dots, \mathbb{X}^0 \subseteq X$ for some vector space X and with input constraints satisfying $\mathbb{U}(x) = \mathbb{U}$ for all $x \in X$. Assume that f is bounded and uniformly continuous in x on each closed ball $\overline{\mathcal{B}}_\rho(x_*)$ and that ℓ is uniformly continuous in x on each such ball, both uniformly in $u \in \mathbb{U}$. Assume furthermore that $V = V_N$ satisfies the assumptions of Theorem 4.11 with constant reference $x^{\text{ref}} \equiv x_*$ on $S = \mathbb{X}$ and that Assumption 7.38 holds.*

Then the perturbed closed-loop system (7.7) is semiglobally practically asymptotically stable in the sense of Definition 7.24 on $A = \mathbb{X}$.

Proof The state constraints ensure $f(x,\mu_N(x)) \in \mathbb{X}^{N-1}$ for all $x \in \mathbb{X}$ and thus $\overline{\mathcal{B}}_\delta(f(x,\mu_N(x))) \subseteq \mathbb{X}$ by Assumption 7.38(i). Hence, in the absence of measurement errors, i.e., for $\bar{e} = 0$, and for $\bar{d} \le \delta$ we obtain $\tilde{x}_{\mu_N}(k,x) \in \mathbb{X}$ for all $\tilde{x}_{\mu_N}(\cdot,x) \in S_{(\bar{d},\bar{e})}(x)$ and all $x \in \mathbb{X}$.

Using this property and the uniform continuity of V_N guaranteed by Proposition 7.40, for $\bar{e} = 0$ the proof is analogous to the proof of Theorem 7.26 when all sets in this proof are intersected by $\mathbb{X}$ and we use the a priori restriction $\bar{d} \le \delta$. Practical asymptotic stability for $\bar{e} > 0$ then follows as in the second part of the proof of Theorem 7.36. □

Remark 7.42 (i) As in the previous robustness results again we did not use continuity of μ_N.

(ii) The continuity proof of Proposition 7.40 does not immediately extend to NMPC schemes with stabilizing terminal constraint set $\mathbb{X}_0$. If these are to be used, then additional conditions on $\mathbb{X}_0$ need to be imposed.

(iii) Remarks 7.27(v) and 7.37(iii)–(v) apply accordingly to Theorem 7.41.

We end this chapter with a brief discussion of the robustness conditions introduced in the last sections.

Without state constraints, Corollaries 7.28 and 7.29 show that robustness can be expected under reasonable continuity conditions on the problem data.

In the presence of state constraints, however, things become quite more restrictive: Looking at the conditions (7.13), (7.14) and Assumption 7.38(ii) one sees that in all cases the dynamics f must be able to map each point from the feasible set (i.e., $\mathcal{F}_N$ in the case of (7.13) and (7.14) and $\mathbb{X}$ in the case of Assumption 7.38) into the interior of the feasible set with some positive distance δ to its boundary. This requirement is considerably stronger than the viability of the feasible set, which in turn is already a rather restrictive assumption.

Comparing (7.13) with Assumption 7.38, (7.13) is less demanding in terms of regularity of the value function V_N, however, Assumption 7.38 allows for a simpler and less restrictive choice of the constraint sets $\mathbb{X}^k$ compared to (7.14), which have to shrink only from $\mathbb{X}^N$ to $\mathbb{X}^{N-1}$ instead of for each pair $\mathbb{X}^k$ and $\mathbb{X}^{k-1}$. This makes the scheme easier to design and to implement; however, as the discussion of Artstein's circles after Example 7.39 shows, there are systems for which state constraint sets satisfying (7.13) can be constructed but for which Assumption 7.38 cannot be satisfied. It is an open question whether there are examples for which the converse is true.

In practice, for systems with complex dynamics it seems doubtful whether one will ever be able to systematically construct constraint sets $\mathbb{X}^k$ for which either condition can be satisfied. Instead, in order to cover more realistic settings it seems desirable to relax these conditions. This, however, may lead to complicated situations. For instance, instead of requiring to be able to steer the system away from the boundary of the state constraint set in one step, it appears more natural and less demanding to be able to do so only after a larger number of steps. In the presence of perturbations, however, this would mean that the system would leave the admissible set for a couple of steps before it is able to enter this set again and it is not clear what kind of conditions one would have to impose on the optimization problem ($\mathrm{OCP_N}$) in order to guarantee that the NMPC closed loop will actually show this behavior. A straightforward solution to this problem is to wait with the re-optimization until the solution enters the admissible set again; this is, e.g., proposed by Michalska and Mayne [18, Sect. 5] along with suitable conditions which guarantee that this re-entering actually happens. However, during this time the system runs in open loop and hence the controller cannot react to further unforeseen disturbances. Currently, we are not aware of NMPC formulations which resolve this problem.

Finally, we note that in all approaches presented here the perturbations and errors are not explicitly taken into account in the design, i.e., in the optimization problem (OCP_N) and its variants. Alternative approaches are briefly discussed at the end of Sect. 7.10, below.

7.10 Notes and Extensions

While recursive feasibility has long been recognized as an inherent property of the feasible set $\mathbb{X}_N$ of terminal constrained NMPC schemes—at least in the nominal case—there are only few known approaches to deal with the feasibility problem for schemes without stabilizing terminal conditions. The approach via exit sets in Sect. 7.2 is a reformulation of results which appeared in Chap. 5 of the PhD thesis of Eric Kerrigan [11].[2] It appears to be closely related to the study of target problems for differential or difference inclusions when the closure of $X \setminus \mathbb{X}$ is considered as a target. More precisely, we conjecture that a discrete time version of the decomposition of the dynamical behavior at the boundary of $\mathbb{X}$ analogous to Quincampoix [21] can be used in order to decide whether Assumption 7.9 is satisfied. Still, one can only speculate whether this relation can be useful for designing appropriate state constraint sets $\mathbb{X}$.

The approach of proving recursive feasibility for nonlinear MPC via stability in Sect. 7.3 is an original contribution and has—to the best of our knowledge—not been considered before. It does, however, bear similarities with the feasibility result for linear MPC by Primbs and Nevistić [20, Theorem 3] and its proof. The essential difference of the proof in [20] and our proof is the fact that we avoid the use of V_∞ and that we do not require compactness of sublevel sets of V_∞. For this reason our proof also works for infinite-dimensional state spaces X. A variant of this result for the case where Assumption 7.13 fails to hold near the boundary of $\mathcal{F}_\infty$ can be found in Boccia, Grüne and Worthmann [2].

The discussion in Sect. 7.4 reflects our own personal experiences based on the results developed in this book and numerous numerical simulations. Of course, other researchers have made different experiences and we highly recommend reading the complementary discussion by Mayne [17].

Regarding robustness, the fact that regularity properties of Lyapunov functions—like the continuity used in Sects. 7.6 and 7.9—imply robustness is well known in the control literature, see, for instance, Kellett, Shim and Teel [10] for an analysis in a sampled data context. For NMPC schemes, this has been used before, e.g., in De Nicolao, Magni and Scattolini [5] under the assumption that V_N is C^2. The proof idea of Theorem 7.26 in Sect. 7.6 using mere continuity of V_N was borrowed from the stability analysis of nonlinear sampled data systems, cf., e.g., Nešić, Teel and Kokotović [19].

Regularity properties of optimal value functions under state constraints are frequently studied in optimal control. Assumption 7.38 and the proof of Proposition

[2] We only learned about this after the first edition of this book was published.

7.40 in Sect. 7.6 were inspired by a continuous time construction by Soner [25]; we are not aware of similar discrete time constructions in the literature. It should be noted that the condition in discrete time is more demanding than the condition in continuous time because in discrete time it is not possible to apply a control value for an arbitrary short time interval which is crucial in the proof in [25].

The idea of ensuring robustness not via continuity of the optimal value function but via tighter constraint sets as in Sect. 7.8 appears to be used for the first time by Michalska and Mayne [18] in a continuous time setting. The definition of tightening constraint sets satisfying (7.14) given here for our discrete time setting was inspired by Limón, Alamo, and Camacho [14]. A refined version of this construction allowing for suboptimal open-loop control sequences and discontinuous dynamics was recently studied in Lazar and Heemels [13]. A closely related variant is the so-called tube based MPC, cf. e.g., Langson, Chryssochoos, Raković and Mayne [12] or Sect. 5.1 in the survey article by Limón, Alamo, Raimondo, Muñoz de la Peña, Bravo, Ferramosca and Camacho [15]. In this area, also computational methods for actually computing appropriate tightening state constraint sets have been investigated. All these references typically use stabilizing terminal conditions and consider only additive disturbances.

Definition 7.33 was introduced by Grimm, Messina, Tuna and Teel [8] who also considered measurement errors and schemes without stabilizing terminal conditions similar to the setting we used here. The same authors also constructed the examples in Sect. 7.7, see [6]. The condition in [8] is more general than Definition 7.33 in the sense that—roughly speaking—(7.13) is only required for k from a subset of $\{0, \ldots, N-1\}$. Here we decided to present a simplified version in order to emphasize the main idea. The stability proof in [8] uses the techniques from [7], which were already briefly discussed at the end of Sect. 6.1 and in Sect. 6.9, and also applies to non positive definite stage costs under the detectability condition discussed in Sect. 10.3.

While we consider the robustness approaches presented in the Sects. 7.6, 7.7, 7.8 and 7.9 as representative for many commonly used approaches, they are by no means exhaustive; in fact, many more variants and related ideas can be found in the literature. The interested reader may consult, e.g., Rawlings and Mayne [23, Chap. 3] or the survey papers by Magni and Scattolini [16] or by Limón et al. [15] as well as the references therein. Furthermore, an alternative to using tightening constraints was recently presented by Yu, Böhm, Chen and Allgöwer in [26]. Here the error system describing the difference between the nominal and the perturbed system is pre-compensated by an input-to-state stabilizing controller. This, however, comes at the expense that this controller has to be designed before the NMPC scheme can be set up.

All approaches discussed so far incorporate the state constraints as hard constraints into the optimization. An alternative to this approach is by using what is often called *soft* constraints. Here the state constraints are included into the NMPC formulation by including suitable penalty terms in the cost function, which become sufficiently large when the state constraint is violated. This approach is popular, e.g., in robotics in which the penalty terms are closely related to potential fields; see, e.g., Shim, Jin Kim and Sastry [24]. While this approach appears to work well in practice, we are not aware of rigorous NMPC feasibility results in the literature.

We end this discussion by remarking that there is an ample literature on NMPC schemes which explicitly take the effect of disturbances into account in the prediction. This allows for a much more refined treatment of different perturbation structures and one may thus expect a better performance of the closed loop under perturbations under less demanding conditions. However, the price to pay for this better performance is that instead of a "simple" optimal control problem a dynamic game, i.e., a min-max problem has to be solved in each sampling instant. The complexity of solving a dynamic min-max problem is considerably higher than solving an optimal control problem; in particular, the interplay between the control and the perturbation sequences needs to be modeled with care in order to obtain a reasonable solution. For recent surveys on such methods we refer to, e.g., Raimondo, Limón, Lazar, Magni and Camacho [22] (see also the discussion following this article in the same journal) or Rawlings and Mayne [23, Chap. 3].

Problems

1. Consider the feasible set $\mathscr{F}_N$ for a constraint set $\mathbb{X} \subset X$ and an optimization horizon $N \in \mathbb{N}$ according to Definition 7.2. Assume that for a point $x \in \mathbb{X}$ and some $K \in \mathbb{N}$ there exists an admissible control sequence $u \in \mathbb{U}^K(x)$ with $x_u(K, x) \in \mathscr{F}_N$. Prove that $x \in \mathscr{F}_{N+K}$ holds.
2. Consider a symmetric matrix $Q \in \mathbb{R}^{n\times n}$ and a constant $C > 0$ such that the inequality $|x^\top Qy| \le C\|x\|\|y\|$ holds for all $x, y \in \mathbb{R}^n$. Let $\rho > 0$ be given and consider the set $A = \overline{\mathscr{B}_\rho(0)}$.

   (a) Show that
   $$\omega(r) = 2C\rho\, r$$
   is a modulus of continuity of the function $W(x) = x^\top Qx$ on A.

   (b) Compute a modulus of continuity of the function $W(x) = (x^\top Qx)^2$ on A.

3. Verify the following facts that have been used in Example 7.31.

   (a) For $x \in \mathbb{R}^2$ with $x_2 > 0$ and $u \in \mathbb{R}$ with $u < 0$ the step $x^+ = f(x, u)$ defines a clockwise movement.

   (b) For all $c \in (0, 1)$, all circles S_r with $r > r_c = c/\sqrt{1-c^2}$ and all points $x \in S_r \cap \mathbb{X}$ with $x_2 > r$ and $x_1 = c$ the relation $f(x, -1) \notin \mathbb{X}$ holds. Use this fact to conclude that for all initial values $x \in S_r \cap \mathbb{X}$ with $x_2 > r$ it is not possible to move clockwise toward 0.

   (c) For all $c < 1/2$ and $y' \in Y'$ with $\varepsilon > 0$ sufficiently small the inequality $\ell(f(y', -1)) > \ell(y')$ holds.

References

1. Artstein, Z.: Stabilization with relaxed controls. Nonlinear Anal. **7**(11), 1163–1173 (1983)
2. Boccia, A., Grüne, L., Worthmann, K.: Stability and feasibility of state constrained MPC without stabilizing terminal constraints. Syst. Control Lett. **72**, 14–21 (2014)
3. Camilli, F., Grüne, L., Wirth, F.: Control Lyapunov functions and Zubov's method. SIAM J. Control Optim. **47**, 301–326 (2008)
4. Curtain, R.F., Zwart, H.: An Introduction to Infinite-Dimensional Linear Systems Theory. Texts in Applied Mathematics, vol. 21. Springer, New York (1995)
5. De Nicolao, G., Magni, L., Scattolini, R.: On the robustness of receding-horizon control with terminal constraints. IEEE Trans. Autom. Control **41**(3), 451–453 (1996)
6. Grimm, G., Messina, M.J., Tuna, S.E., Teel, A.R.: Examples when nonlinear model predictive control is nonrobust. Automatica **40**(10), 1729–1738 (2004)
7. Grimm, G., Messina, M.J., Tuna, S.E., Teel, A.R.: Model predictive control: for want of a local control Lyapunov function, all is not lost. IEEE Trans. Autom. Control **50**(5), 546–558 (2005)
8. Grimm, G., Messina, M.J., Tuna, S.E., Teel, A.R.: Nominally robust model predictive control with state constraints. IEEE Trans. Autom. Control **52**(10), 1856–1870 (2007)
9. Grüne, L., Nešić, D.: Optimization based stabilization of sampled-data nonlinear systems via their approximate discrete-time models. SIAM J. Control Optim. **42**, 98–122 (2003)
10. Kellett, C.M., Shim, H., Teel, A.R.: Further results on robustness of (possibly discontinuous) sample and hold feedback. IEEE Trans. Autom. Control **49**(7), 1081–1089 (2004)
11. Kerrigan, E.C.: Robust constraint satisfaction: Invariant sets and predictive control. PhD thesis, University of Cambridge (2000)
12. Langson, W., Chryssochoos, I., Raković, S.V., Mayne, D.Q.: Robust model predictive control using tubes. Automatica **40**(1), 125–133 (2004)
13. Lazar, M., Heemels, W.P.M.H.: Predictive control of hybrid systems: input-to-state stability results for sub-optimal solutions. Automatica **45**(1), 180–185 (2009)
14. Limón, D., Alamo, T., Camacho, E.: Input-to-state stable MPC for constrained discrete-time nonlinear systems with bounded additive uncertainties. In: Proceedings of the 41st IEEE Conference on Decision and Control - CDC 2002, Las Vegas, Nevada, pp. 4619–4624 (2002)
15. Limón, D., Alamo, T., Raimondo, D.M., Muñoz de la Peña, D., Bravo, J.M., Ferramosca, A., Camacho, E.F.: Input-to-state stability: a unifying framework for robust model predictive control. In: Magni, L., Raimondo, D.M., Allgöwer, F. (eds.) Nonlinear Model Predictive Control. Lecture Notes in Control and Information Sciences, vol. 384, pp. 1–26. Springer, erlin (2009)
16. Magni, L., Scattolini, R.: Robustness and robust design of MPC for nonlinear discrete-time systems. In: Findeisen, R., Allgöwer, F., Biegler, L.T. (eds.) Assessment and Future Directions of Nonlinear Model Predictive Control. Lecture Notes in Control and Information Sciences, vol. 358, pp. 239–254. Springer, Berlin (2007)
17. Mayne, D.Q.: An apologia for stabilising terminal conditions in model predictive control. Int. J. Control **86**(11), 2090–2095 (2013)
18. Michalska, H., Mayne, D.Q.: Robust receding horizon control of constrained nonlinear systems. IEEE Trans. Autom. Control **38**(11), 1623–1633 (1993)
19. Nešić, D., Teel, A.R., Kokotović, P.V.: Sufficient conditions for stabilization of sampled-data nonlinear systems via discrete-time approximations. Syst. Control Lett. **38**(4–5), 259–270 (1999)
20. Primbs, J.A., Nevistić, V.: Feasibility and stability of constrained finite receding horizon control. Automatica **36**(7), 965–971 (2000)
21. Quincampoix, M.: Differential inclusions and target problems. SIAM J. Control Optim. **30**(2), 324–335 (1992)
22. Raimondo, D.M., Limón, D., Lazar, M., Magni, L., Camacho, E.F.: Min-max model predictive control of nonlinear systems: a unifying overview on stability. Europ. J. Control **15**(1), 5–21 (2009)
23. Rawlings, J.B., Mayne, D.Q.: Model Predictive Control: Theory and Design. Nob Hill Publishing, Madison (2009)

24. Shim, D.H., Jin Kim, H., Sastry, S.: Decentralized nonlinear model predictive control of multiple flying robots. In: Proceedings of the 42nd IEEE Conference on Decision and Control - CDC 2003, Maui, Hawaii, USA, pp. 3621–3626 (2003)
25. Soner, H.M.: Optimal control with state-space constraint. I and II. SIAM J. Control Optim. **24**(3), 552–561, 1110–1122 (1986)
26. Yu, S., Böhm, C., Chen, H., Allgöwer, F.: Robust model predictive control with disturbance invariant sets. In: Proceedings of the American Control Conference - ACC 2010, Baltimore, Maryland, USA, pp. 6262–6267 (2010)

Chapter 8
Economic NMPC

Economic nonlinear model predictive control is the common name for NMPC schemes in which the stage cost does not penalize the distance to a predefined equilibrium, which was one of the key assumptions in Chaps. 5 and 6. Instead, the cost can, in principle, model all kinds of quantities, like energy consumption, yield of a substance, income of a firm, etc., which one would like to minimize or maximize. In such a general setting, it is by no means clear that the moving horizon MPC paradigm yields well-performing closed-loop solutions. It was first observed by Amrit, Angeli, and Rawlings in 2012 (extending earlier work by Diehl, Amrit, and Rawlings) that strict dissipativity is a sufficient systems theoretic property for ensuring proper performance of economic MPC. In this chapter, we will rigorously establish stability as well as averaged and non-averaged performance estimates for strictly dissipative economic MPC problems, both with and without terminal conditions.

8.1 Setting

In terms of the problem formulation, the only difference between economic NMPC and the stabilizing NPMC formulations we discussed in the previous chapters lies in the stage cost ℓ. More precisely, we drop the requirement (3.2) of ℓ being positive definite with respect to an a priori chosen equilibrium (x_*, u_*). In fact, we even entirely drop the idea of stabilizing a prescribed equilibrium (x_*, u_*). This does not mean that we abandon the concept of an equilibrium. On the contrary, equilibria will still play an important role in our analysis. However, instead of a priori identifying a "good" equilibrium (x_*, u_*) and then stabilizing it, we concentrate on optimal equilibria that emerge from the interplay of the stage cost and the dynamics, i.e., which are implicitly defined by the optimal control problem (see Definition 8.2) rather than given exogenously.

L. Grüne and J. Pannek, *Nonlinear Model Predictive Control*,
Communications and Control Engineering, DOI 10.1007/978-3-319-46024-6_8

Example 8.1 An example, which will serve as an illustration for all results in this section, is the 1d discrete-time system with dynamics and stage cost

$$x^+ = 2x + u \quad \text{and} \quad \ell(x,u) = u^2$$

and state and control constraint sets $\mathbb{X} = [-2, 2]$ and $\mathbb{U}(x) = \mathbb{U} = [-3, 3]$, i.e., $\mathbb{Y} = [-2, 2] \times [-3, 3]$. Since the stage cost does not depend on x, the requirement (3.2) is clearly violated.

The uncontrolled system is unstable, hence for initial values $x_0 \neq 0$ the solution will leave the admissible set $\mathbb{X}$ if no control is used. Hence, control action is needed in order to keep the system inside $\mathbb{X}$. Interpreting the stage cost $\ell(x,u) = u^2$ as the energy of the current control action, the control objective can be formulated as "keep the state inside $\mathbb{X}$ with minimal control effort."

Due to the fact that in the general economic setting the stage cost is not sign definite, the infinite horizon functional need not have a finite value. Hence, the infinite horizon closed-loop performance

$$J_\infty^{cl}(x_0, \mu) := \sum_{k=0}^{\infty} \ell(x_\mu(k), \mu(x_\mu(k)))$$

from Definition 4.10 (which in this chapter is considered only in the time-invariant setting) often does not yield a well-defined performance criterion. For this reason, we also consider the finite horizon closed-loop performance

$$J_K^{cl}(x_0, \mu) := \sum_{k=0}^{K-1} \ell(x_\mu(k), \mu(x_\mu(k))) \tag{8.1}$$

and the averaged infinite horizon performance

$$\overline{J}_\infty^{cl}(x_0, \mu) := \limsup_{K\to\infty} \frac{1}{K} J_K^{cl}(x_0, \mu).$$

Throughout this chapter, by $(x^e, u^e) \in \mathbb{Y}$ we denote an equilibrium of the system, i.e., $f(x^e, u^e) = x^e$. Of particular interest are optimal equilibria according to the following definition.

Definition 8.2 An equilibrium $(x^e, u^e) \in \mathbb{Y}$ is called an *optimal equilibrium*, if it yields the lowest value of the cost function among all admissible equilibria, i.e.,

$$\ell(x^e, u^e) \leq \ell(x,u) \qquad \text{for all } (x,u) \in \mathbb{Y} \text{ with } f(x,u) = x.$$

Example 8.3 In Example 8.1, the equilibria are of the form $(x, -x)$ with cost $\ell(x, -x) = x^2$. Thus, the (unique) optimal equilibrium is given by $(x^e, u^e) = (0, 0)$.

The following lemma shows that an optimal equilibrium always exists when f and ℓ are continuous and $\mathbb{Y}$ is compact.

Lemma 8.4 *If the constraint set* $\mathbb{Y} \subset \mathbb{X}\times\mathbb{U}$ *is compact and the maps* $\ell : \mathbb{X}\times\mathbb{U} \to \mathbb{R}$ *and* $f : \mathbb{X} \times \mathbb{U} \to \mathbb{X}$ *are continuous, then there exists an optimal equilibrium, i.e., a pair* $x^e \in \mathbb{X}$, $u^e \in \mathbb{U}$ *with* $f(x^e, u^e) = x^e$ *such that*

$$\ell(x^e, u^e) = \inf\{\ell(x, u) \mid (x, u) \in \mathbb{Y}, f(x, u) = x\}.$$

Proof Since preimages of closed sets under continuous mappings are closed, the set $\{(x, u) \in \mathbb{Y} \mid f(x, u) = x\}$ is closed, hence compact and thus the continuous function ℓ attains a minimum. □

As a consequence of this lemma, in contrast to assuming $\ell(x_*, u_*) < \ell(x, u)$ for all $(x, u) \in \mathbb{Y} \setminus (x_*, u_*)$, which we did in Chaps. 5 and 6, assuming the existence of an optimal equilibrium is not an overly restrictive assumption.

8.2 Averaged Performance with Terminal Conditions

In this and in the following two sections, we choose the optimal control problem ($\mathrm{OCP}_{\mathrm{N,e}}$) in Algorithm 3.10 as

$$\begin{array}{|l|}\hline
\text{minimize} \quad J_N(x_0, u(\cdot)) := \displaystyle\sum_{k=0}^{N-1} \ell(x_u(k, x_0), u(k)) \\
\qquad\qquad\qquad\qquad\qquad + F(x_u(N, x_0)) \\
\text{with respect to } u(\cdot) \in \mathbb{U}^N_{\mathbb{X}_0}(x_0) \text{ subject to} \\
x_u(0, x_0) = x_0, \quad x_u(k+1, x_0) = f(x_u(k, x_0), u(k)) \\ \hline
\end{array} \tag{8.2}$$

where $\mathbb{U}^N_{\mathbb{X}_0}(x_0)$ is defined as in the previous chapters, see Definition 3.2(ii). We note that the only difference to (5.15) lies in the fact that the condition $x_* \in \mathbb{X}_0$ is removed, as we do not a priori fix a particular equilibrium x_*. We also note that the terminal condition is only added to the open-loop functional $J_N(x_0, u)$ used in the NMPC Algorithm 3.10, but not to the closed-loop performance index $J_K^{cl}(x, \mu)$ from (8.1), which is still defined without terminal cost or constraints according to (8.1). As before, the optimal value function is defined by

$$V_N(x) := \inf_{u(\cdot)\in\mathbb{U}^N_{\mathbb{X}_0}(x)} J_N(x, u(\cdot))$$

and we assume the existence of an optimal control sequence for each feasible initial condition x in order to synthesize the NMPC feedback law μ_N according to Algorithm 3.10.

The following assumption links an equilibrium—which will later be chosen as an optimal equilibrium—to the terminal conditions. For its formulation, recall the definition of the feasible sets $\mathbb{X}_N$ from Definition 3.9(i).

Assumption 8.5 (*Terminal conditions*) (a) The set $\mathbb{X}_0$ is bounded and there is an equilibrium $(x^e, u^e) \in \mathbb{Y}$ with $x^e \in \mathbb{X}_0$ and $F(x^e) = 0$, such that for each $x \in \mathbb{X}_0$ there exists $u \in \mathbb{U}$ with $f(x, u) \in \mathbb{X}_0$ and

$$F(f(x,u)) \leq F(x) - \ell(x,u) + \ell(x^e, u^e)$$

(b) There exists $N_0 \in \mathbb{N}$ and $\eta > 0$, such that $\mathbb{X}_{N_0}$ contains the ball $\mathscr{B}_\eta(x^e)$.

Observe that the requirement $F(x^e) = 0$ in part (a) of this assumption can be made without loss of generality because the inequality in (a) is invariant with respect to adding a constant to F.

Obviously, Assumption 8.5(a) is a modification of Assumption 5.9 adapted to the economic NMPC setting. It is easily seen that Lemmas 5.10 and 5.11 remain valid. In particular, under Assumption 8.5, the feasible sets satisfy the inclusion $\mathbb{X}_{N_0} \subseteq \mathbb{X}_{N_1}$ whenever $N_0 \leq N_1$. We also note that Assumption 8.5(a) is satisfied in case of equilibrium terminal conditions, i.e., when setting $\mathbb{X}_0 = \{x^e\}$ and $F \equiv 0$. Hence, we do not treat equilibrium terminal conditions separately. Assumption 8.5(b) is a nondegeneracy condition, which prevents that the feasible sets $\mathbb{X}_N$ have empty interior for any $N \in \mathbb{N}$.

Under these assumptions, we can formulate the first result.

Theorem 8.6 *Consider the NMPC Algorithm 3.10 with optimal control problem* $(\mathrm{OCP}_{\mathrm{N,e}}) = (8.2)$. *Let Assumption 8.5(a) be satisfied, let* $N \geq 2$, *and assume* V_N *is bounded from below on* $\mathbb{X}_N$. *Then, for any* $N \geq 2$ *and any* $x \in \mathbb{X}_N$ *the averaged closed-loop performance satisfies the inequality*

$$\overline{J}^{cl}_\infty(x, \mu_N) \leq \ell(x^e, u^e). \tag{8.3}$$

Proof Similar to the proof of Lemma 5.4, one sees that Assumption 8.5(a) implies the inequality

$$V_N(x) \leq V_{N-1}(x) + \ell(x^e, u^e)$$

for all $x \in \mathbb{X}_{N-1}$. Using the dynamic programming principle, this inequality implies

$$\ell(x, \mu_N(x)) \leq V_N(x) - V_{N-1}(f(x, \mu_N(x))) \leq V_N(x) - V_N(f(x, \mu_N(x))) + \ell(x^e, u^e)$$

and we can conclude

$$\begin{aligned} J_K^{cl}(x_0, \mu_N) &= \sum_{k=0}^{K-1} \ell(x_{\mu_N}(k), \mu_N(x_{\mu_N}(k))) \\ &\le \sum_{k=0}^{K-1} \Big[V_N(x_{\mu_N}(k)) - V_N(x_{\mu_N}(k+1)) + \ell(x^e, u^e) \Big] \\ &= V_N(x_0) - V_N(x_{\mu_N}(K)) + K\ell(x^e, u^e) \\ &\le V_N(x_0) - M + K\ell(x^e, u^e), \end{aligned}$$

where $M \in \mathbb{R}$ is a lower bound on V_N. This yields

$$\overline{J}_\infty^{cl}(x_0, \mu_N) \le \limsup_{K\to\infty} \left(\frac{V_N(x_0)}{K} - \frac{M}{K} + \ell(x^e, u^e) \right) = \ell(x^e, u^e).$$

□

We note that the boundedness assumption on V_N is satisfied if ℓ is continuous, $\mathbb{Y}$ is compact, and F is bounded from below, because in this case both ℓ and F, and thus also V_N, are bounded from below.

Clearly, the estimate from Theorem 8.6 is particularly powerful if $\ell(x^e, u^e)$ is the best, i.e., the smallest possible value that $\overline{J}_\infty^{cl}(x_0, \mu_N)$ can attain. The next definition provides a property that is sufficient for this fact, as the subsequent Proposition 8.9 shows.

Definition 8.7 (*Dissipativity and strict dissipativity*) We say that an optimal control problem with stage cost ℓ is *strictly dissipative* at an equilibrium $(x^e, u^e) \in \mathbb{Y}$ if there exists a *storage function* $\lambda : \mathbb{X} \to \mathbb{R}$ bounded from below and satisfying $\lambda(x^e) = 0$, and a function $\rho \in \mathcal{K}_\infty$ such that for all $(x, u) \in \mathbb{Y}$ the inequality

$$\ell(x, u) - \ell(x^e, u^e) + \lambda(x) - \lambda(f(x, u)) \ge \rho(|x|_{x^e}) \tag{8.4}$$

holds.

We say that an optimal control problem with stage cost ℓ is *dissipative* at (x^e, u^e), if the same conditions hold with $\rho \equiv 0$.

We note that the assumption $\lambda(x^e) = 0$ can be made without loss of generality, because adding a constant to λ does not invalidate (8.4).

The classical physical interpretation of the storage function is that $\lambda(x)$ quantifies the amount of energy stored in the system at state x. The function $s(x, u) := \ell(x, u) - \ell(x^e, u^e)$ is called the *supply rate* and measures the (possibly negative) amount of energy supplied to the system via the input u at state x. With this interpretation, strict dissipativity means that a certain amount of energy, quantified by $\rho(|x|_{x^e})$, is dissipated to the environment in each time step. Of course, in the context of general optimal control problems considered in this chapter, the storage function and the supply rate need not have an energy interpretation.

Example 8.8 (i) Any optimal control problem with stage cost satisfying $\ell(x, u) - \ell(x^e, u^e) \geq \rho(|x - x^e|)$ is strictly dissipative with $\lambda \equiv 0$. Hence, NMPC problems satisfying the conditions of the stability theorems in Chaps. 5 and 6 are always strictly dissipative. In this sense, strict dissipativity extends (and weakens) the assumptions from the previous chapters.

(ii) It is straightforward to check that Example 8.1 is dissipative with $\lambda \equiv 0$ and strictly dissipative with $\lambda(x) = -x^2/2$, both at $(x^e, u^e) = (0, 0)$. Note that the storage function $\lambda = -x^2/2$ is bounded from below since $\mathbb{X}$ is bounded. Indeed, for an unbounded state constraint set $\mathbb{X}$, the system would not be strictly dissipative. In this example, the supply rate $s(x, u) = \ell(x, u) = u^2$ does have an energy interpretation and the storage function $\lambda(x)$ shows that the equilibrium (x^e, u^e) is the state in which the stored energy $\lambda(x)$ becomes maximal.

Proposition 8.9 *For an optimal control problem* ($\mathrm{OCP_N}$) *that is dissipative at* (x^e, u^e), *the point* (x^e, u^e) *is an optimal equilibrium and the inequality*

$$\limsup_{K\to\infty} \frac{1}{K} \sum_{k=0}^{K-1} \ell(x_u(k, x), u(k)) \geq \ell(x^e, u^e) \tag{8.5}$$

holds for all $x \in \mathbb{X}$ *and all admissible control sequences* $u \in \mathbb{U}^\infty(x)$.

Proof Consider an arbitrary equilibrium $(x, u) \in \mathbb{Y}$. Then the identity $x = f(x, u)$ and (8.4) imply

$$\ell(x, u) - \ell(x^e, u^e) = \ell(x, u) - \ell(x^e, u^e) + \lambda(x) - \lambda(f(x, u)) \geq 0,$$

which yields $\ell(x^e, u^e) \leq \ell(x, u)$, and thus (x^e, u^e) is an optimal equilibrium.

Moreover, using again (8.4) and denoting by M a lower bound on λ we have

$$\begin{aligned}\sum_{k=0}^{K-1} \ell(x_u(k, x), u(k)) &\geq \sum_{k=0}^{K-1} \ell(x^e, u^e) - \lambda(x_u(k, x)) + \lambda(x_u(k+1, x)) \\ &= K\ell(x^e, u^e) - \lambda(x) + \lambda(x_u(K, x)) \\ &\geq K\ell(x^e, u^e) - \lambda(x) + M.\end{aligned}$$

for any $u \in \mathbb{U}^\infty(x)$. This yields

$$\limsup_{K\to\infty} \frac{1}{K} \sum_{k=0}^{K-1} \ell(x_u(k, x), u(k)) \geq \limsup_{K\to\infty} \left(\ell(x^e, u^e) - \frac{\lambda(x) - M}{K}\right) = \ell(x^e, u^e).$$

□

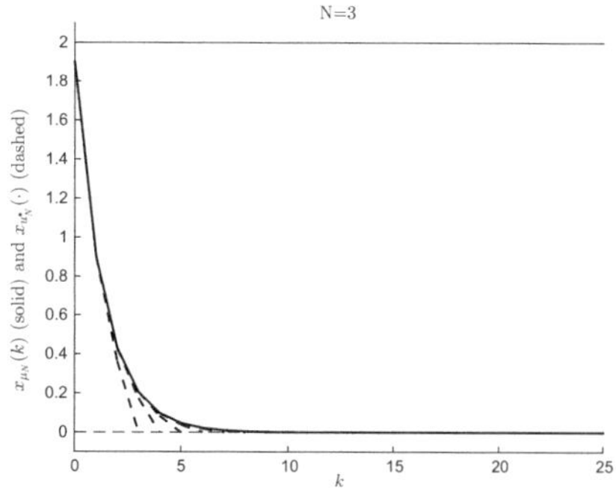

Fig. 8.1 NMPC closed-loop solution (*solid*) and open-loop predictions (*dashed*) for Example 8.1 with terminal constraint $\mathbb{X}_0 = \{0\}$ and horizon $N = 3$. The *solid line* at $x = 2$ indicates the *upper* bound of the admissible set $\mathbb{X}$

The property expressed by inequality (8.5) is known as *optimal operation at steady state*. It has been shown in [17] that under a controllability condition on the system, the converse of Proposition 8.9 is also true, i.e., that optimal operation at a steady state implies dissipativity.

An immediate consequence of Proposition 8.9 is the following corollary.

Corollary 8.10 *Consider the NMPC Algorithm 3.10 with dissipative optimal control problem* ($\mathrm{OCP}_{\mathrm{N,e}}$) = (8.2). *Then for all* $x \in \mathbb{X}_N$

$$\overline{J}_\infty^{cl}(x, \mu_N) = \inf_{u \in \mathbb{U}^\infty(x)} \limsup_{K \to \infty} \frac{1}{K} \sum_{k=0}^{K-1} \ell(x_u(k, x), u(k)).$$

Hence, if dissipativity holds, then Theorem 8.6 ensures infinite horizon averaged optimality of the NMPC closed loop.

Example 8.11 Since Example 8.1 is dissipative (see Example 8.8), the NMPC closed loop must be infinite horizon averaged optimal. Indeed, as Fig. 8.1 shows, the closed-loop solution converges to the optimal equilibrium. Since the control (not shown in the figure) does the same, $\ell(x_{\mu_N}(k), \mu_N(x_{\mu_N}(k))) \to 0$ as $k \to \infty$ follows. This implies $\overline{J}_\infty^{cl}(x, \mu_N) = 0$, which is clearly optimal since $\ell \geq 0$.

8.3 Asymptotic Stability with Terminal Conditions

One might conjecture that optimal operation at the steady state (x^e, u^e) implies that closed-loop solutions satisfying (8.3) must also converge to x^e. However, under the assumptions imposed in Theorem 8.6 and Proposition 8.9, this is not necessarily the case. To see this, it suffices to consider an optimal control problem with $\ell \equiv 0$. Such a problem clearly satisfies all assumptions (with terminal cost $F \equiv 0$ and storage function $\lambda \equiv 0$), yet every trajectory is an optimal trajectory and thus optimal trajectories obviously need not converge to x^e. In order to achieve this—and, in fact, even asymptotic stability of x^e—we need to assume strict dissipativity.

Similar to our reasoning in Chap. 5, we establish asymptotic stability by proving the existence of a Lyapunov function. Here, however, the optimal value function V_N cannot be used for this purpose, because it is not sign definite and thus violates the upper and lower bounds on $V(n, \cdot) = V_N(\cdot)$ in Theorem 4.11. The remedy lies in considering the optimal control problem with stage cost

$$\tilde{\ell}(x, u) := \ell(x, u) - \ell(x^e, u^e) + \lambda(x) - \lambda(f(x, u)) \tag{8.6}$$

and terminal cost

$$\widetilde{F}(x) := F(x) + \lambda(x).$$

These costs are usually called *rotated* or *modified* costs. The name "rotated cost" stems from the fact that for linear f and strictly convex ℓ, the graph of $\tilde{\ell}$ is obtained by rotating the graph of ℓ. The corresponding functional is given by

$$\widetilde{J}_N(x_0, u(\cdot)) = \sum_{k=0}^{N-1} \tilde{\ell}(x_u(k, x_0), u(k)) + \widetilde{F}(x_u(N, x_0))$$

and the optimal value function by

$$\widetilde{V}_N(x_0) := \inf_{u(\cdot) \in \mathbb{U}^N_{\mathbb{X}_0}(x_0)} \widetilde{J}_N(x_0, u(\cdot)).$$

It is an easy exercise to check that the equalities $\tilde{\ell}(x^e, u^e) = 0$ and $\widetilde{F}(x^e) = 0$ and—under Assumption 8.5(a)—that the inequality

$$\widetilde{F}(f(x, u)) \leq \widetilde{F}(x) - \tilde{\ell}(x, u) \tag{8.7}$$

holds for each $x \in \mathbb{X}_0$ and the control u from Assumption 8.5(a). Moreover, for any $x \in \mathbb{X}_N$ and $u \in \mathbb{U}^N_{\mathbb{X}_0}(x)$, one easily verifies the identity

$$\widetilde{J}_N(x, u) = J_N(x, u) + \lambda(x) - N\ell(x^e, u^e). \tag{8.8}$$

Since the last two terms in (8.8) are independent of u, this implies that the optimal trajectories for J_N and $\widetilde{J}_N$ coincide and that the optimal value functions satisfy

$$\widetilde{V}_N(x) = V_N(x) + \lambda(x) - N\ell(x^e, u^e). \tag{8.9}$$

If the optimal control problem is strictly dissipative, $\tilde{\ell}$ satisfies $\tilde{\ell}(x, u) \geq \rho(|x|_{x^e})$ for all $(x, u) \in \mathbb{Y}$ and $\tilde{\ell}(x^e, u^e) = 0$. This immediately implies

$$\widetilde{V}_N(x^e) = 0 \quad \text{and thus} \quad V_N(x^e) = N\ell(x^e, u^e) \tag{8.10}$$

using (8.9) and $\lambda(x^e) = 0$.

The considerations above reveal that with rotated stage cost, the problem satisfies all the properties of the stabilizing NMPC problem of Sect. 5.3. Since the optimal trajectories for J_N and $\widetilde{J}_N$ coincide, so do the NMPC closed-loop trajectories for the two problems. It is therefore not very surprising that $\widetilde{V}_N$ is a Lyapunov function for the NMPC closed-loop system. For the rigorous proof of this property, we need the following continuity assumption on F, λ and V_N in x^e.

Assumption 8.12 (*Continuity of F, λ and V_N at x^e*) There exists γ_F, γ_λ, and $\gamma_V \in \mathcal{K}_\infty$ such that the following properties hold.
(a) For all $x \in \mathbb{X}_0$ it holds that

$$|F(x) - F(x^e)| \leq \gamma_F(|x|_{x^e}).$$

(b) For all $x \in \mathbb{X}$ it holds that

$$|\lambda(x) - \lambda(x^e)| \leq \gamma_\lambda(|x|_{x^e}).$$

(c) For each $N \in \mathbb{N}$ and each $x \in \mathbb{X}_N$ it holds that

$$|V_N(x) - V_N(x^e)| \leq \gamma_V(|x|_{x^e}).$$

Note that γ_V in (c) is independent of N. We will comment at the end of this section on conditions under which (c) can be ensured.

Theorem 8.13 *Consider the NMPC Algorithm 3.10 with strictly dissipative optimal control problem* (OCP$_{\text{N,e}}$) = (8.2). *Let Assumptions 8.5(a) and 8.12 be satisfied. Then the optimal equilibrium x^e is asymptotically stable for the NMPC closed loop on $\mathbb{X}_N$.*

Proof We show that the modified optimal value function $\widetilde{V}_N$ is a Lyapunov function for the closed-loop system in the sense of Theorem 4.11 for $x^{\text{ref}}(n) \equiv x^e$. To this end, we first check inequality (4.14) with $V = \widetilde{V}_N$, $\ell = \tilde{\ell}$ and $\alpha = 1$. As in the proof of Theorem 8.6, from Assumption 8.5(a) we obtain $\ell(x, \mu_N(x)) \leq V_N(x) - V_N(f(x, \mu_N(x))) + \ell(x^e, u^e)$, which we can rewrite as

$$V_N(x) \geq \ell(x, \mu_N(x)) + V_N(f(x, \mu_N(x))) - \ell(x^e, u^e). \tag{8.11}$$

Using (8.9) this implies

$$\begin{aligned}
\widetilde{V}_N(x) &= V_N(x) + \lambda(x) - N\ell(x^e, u^e) \\
&\geq \ell(x, \mu_N(x)) + V_N(f(x, \mu_N(x))) - \ell(x^e, u^e) + \lambda(x) - N\ell(x^e, u^e) \\
&= \ell(x, \mu_N(x)) + \widetilde{V}_N(f(x, \mu_N(x))) - \lambda(f(x, \mu_N(x))) - \ell(x^e, u^e) + \lambda(x) \\
&= \tilde{\ell}(x, \mu_N(x)) + \widetilde{V}_N(f(x, \mu_N(x))),
\end{aligned}$$

i.e., the desired inequality (4.14).

It remains to establish the inequalities

$$\alpha_1(|x|_{x^e}) \le \widetilde{V}_N(x) \le \alpha_2(|x|_{x^e}) \quad \text{and} \quad \tilde{\ell}(x,u) \ge \alpha_3(|x|_{x^e}) \tag{8.12}$$

for $\alpha_1, \alpha_2, \alpha_3 \in \mathscr{K}_\infty$. The third inequality follows immediately from the definition of $\tilde{\ell}$ and strict dissipativity for $\alpha_3 = \rho$ from Definition 8.7. For the inequalities involving α_1 and α_2, we first need to establish a lower bound for $\widetilde{F}$.

To this end, for each $x \in \mathbb{X}_0$ we denote the control u from (8.7) by $\mu_0(x)$. Then (8.7) and strict dissipativity implies

$$\widetilde{F}(f(x,\mu_0(x))) \le \widetilde{F}(x) - \tilde{\ell}(x,\mu_0(x)) \le \widetilde{F}(x) - \rho(|x|_{x^e}).$$

By induction along the closed-loop solution for the feedback law μ_0, we obtain

$$\widetilde{F}(x_{\mu_0}(K,x)) \le \widetilde{F}(x) - \sum_{k=0}^{K-1} \rho(|x_{\mu_0}(k,x)|_{x^e}).$$

This implies that $x_{\mu_0}(K,x) \to x^e$ as $K \to \infty$, because otherwise the sum on the right-hand side of this inequality grows unboundedly, which implies $\widetilde{F}(x_{\mu_0}(K,x)) \to -\infty$ and contradicts Assumption 8.12(a) and (b) since $x_{\mu_0}(K,x)$ is contained in the bounded set $\mathbb{X}_0$. Again by Assumption 8.12(a) and (b), this implies $\widetilde{F}(x_{\mu_0}(K,x)) \to \widetilde{F}(x^e) = 0$ as $K \to \infty$ from which we can finally conclude

$$\widetilde{F}(x) \ge \lim_{K\to\infty} \sum_{k=0}^{K-1} \rho(|x_{\mu_0}(k,x)|_{x^e}) \ge \rho(|x|_{x^e}) \ge 0.$$

From this, the definitions of $\widetilde{J}_N$ and $\widetilde{V}_N$ immediately imply $\widetilde{V}_N(x) \ge \tilde{\ell}(x,\mu_N(x)) \ge \rho(|x|_{x^e})$, and thus the inequality for α_1 in (8.12) with $\alpha_1 = \rho$.

Finally, since $\widetilde{J}_N(x^e,u^e) = 0$, we obtain $\widetilde{V}_N(x^e) = 0$ and the second inequality in (8.12) follows from (8.9) and Assumption 8.12(b) and (c) with $\alpha_2 = \gamma_\lambda + \gamma_V$. □

We end this section by discussing sufficient conditions for the bound on V_N required in Assumption 8.12(c). In the case of equilibrium terminal conditions, i.e., $\mathbb{X}_0 = \{x^e\}$ and $F \equiv 0$, this property can be ensured by the condition that x^e is reachable from every $x \in \mathbb{X}_N$ with suitable bounded costs. In case ℓ and f are continuous, similar to Proposition 5.7 (ii), it is sufficient to assume that the control sequence steering x to x^e is sufficiently close to the constant control with value u^e. For details we refer to [3], particularly to part 2 of Assumption 2 in [3].

In case $\mathbb{X}_0$ contains a neighborhood of x^e, a similar argument as that leading to (5.20) yields the inequality

$$V_N(x) \le F(x) + N\ell(x^e,u^e)$$

while from (8.9) and $\widetilde{V}_N \geq 0$ we obtain

$$V_N(x) \geq -\lambda(x) + N\ell(x^e, u^e).$$

Since from (8.10) we moreover know $V_N(x^e) = N\ell(x^e, u^e)$, this implies Assumption 8.12(c) for $x \in \mathbb{X}_0$ provided Assumption 8.12(a) and (b) hold. For $x \in \mathbb{X}_N \setminus \mathbb{X}_0$, the inequality can be obtained similarly to the proof of Proposition 5.7.

Example 8.14 According to Example 8.8, the optimal control problem from Example 8.1 is strictly dissipative. Moreover, one easily verifies that x^e is reachable in two steps from each $x \in \mathbb{X}$ with cost $4x^2$, which implies the upper bound on V_N for the terminal constraint set $\mathbb{X}_0 = \{0\}$. Hence, we expect the NMPC closed loop to be asymptotically stable, which was already illustrated in Fig. 8.1.

8.4 Non-averaged and Transient Performance with Terminal Conditions

The averaged performance result from Theorem 8.6 provides a useful estimate for large times k. However, it also has two significant weaknesses. First, it does not provide an advantage over a stabilizing NMPC algorithm. Indeed, for any combination of a continuous stage cost and a terminal condition for which the NMPC closed-loop solution converges to x^e and the corresponding control sequence converges to u^e, the value $\ell(x_{\mu_N}(k), \mu_N(x_{\mu_N}(k)))$ converges to $\ell(x^e, u^e)$ from which $\overline{J}^{cl}_\infty(x, \mu_N) = \ell(x^e, u^e)$ follows. Hence, Theorem 8.6 states that the economic NMPC scheme does not perform worse than a stabilizing one. Second, the averaged estimate does not allow any statement about the finite time behavior of the closed-loop trajectory. Indeed, on any finite time interval of arbitrary length, the closed-loop trajectory could behave arbitrarily bad as long as eventually it converges to the equilibrium. Clearly, this is not what we would expect an NMPC closed-loop trajectory to do and it is also not consistent with what we see in numerical simulations, e.g., in Fig. 8.1. Hence in this section, we derive estimates for the non-averaged infinite and finite horizon performance $J^{cl}_\infty(x, \mu_N)$ and $J^{cl}_K(x, \mu_N)$, respectively. For the infinite horizon estimate the additional condition $\ell(x^e, u^e) = 0$ will be imposed, in order to make sure that the infinite sum in $J^{cl}_\infty(x, \mu_N)$ converges. For the finite horizon performance, such a condition is not needed. As we already know that—under the conditions of Theorem 8.13—the equilibrium x^e is asymptotically stable, the finite horizon value $J^{cl}_K(x, \mu_N)$ measures the performance of the solution during the transient phase, i.e., until it reaches a small neighborhood of x^e. This is why we also call this value *transient performance*.

Since $J^{cl}_K(x, \mu_N)$ and $J^{cl}_\infty(x, \mu_N)$ do not involve any terminal constraints or costs, in our analysis we will also need to consider the optimal control problems ($\mathrm{OCP_N}$) and (OCP_∞) without terminal constraints and terminal costs (we only consider the time-invariant infinite horizon problem (OCP_∞) here). In order not to confuse these problems with (8.2), in this section, we denote the functionals and the optimal value functions of the unconstrained problems (OCP_N) and (OCP_∞) by J_N^{uc}, V_N^{uc}, J_∞^{uc} and V_∞^{uc}, respectively. We emphasize that we use the same stage cost ℓ in (8.2), (OCP_N) and (OCP_∞). This implies that if one of the problems is strictly dissipative, then all problems are. If this is the case, we also consider (OCP_N) for the rotated cost $\tilde{\ell}$ and denote the corresponding functional by $\widetilde{J}_N^{uc}$. A straightforward computation reveals that J_N^{uc} and $\widetilde{J}_N^{uc}$ are related by the identity

$$\widetilde{J}_N^{uc}(x, u) = J_N^{uc}(x, u) + \lambda(x) - \lambda(x_u(N, x)) - N\ell(x^e, u^e). \tag{8.13}$$

Observe that compared to (8.8), the additional term $\lambda(x_u(N, x))$ appears here due to the absence of the terminal conditions.

In order to establish our theorems on transient performance, we will need a few preparatory results. The first statement shows that the finite horizon optimal trajectories most of the time stay close to the optimal equilibrium x^e.

Proposition 8.15 *Assume that the optimal control problem* (OCP_N) *is strictly dissipative with bounded storage function λ and $\rho \in \mathcal{K}_\infty$. Then for each $\delta > 0$, there exists $\sigma_\delta \in \mathcal{L}$ such that for all $N, P \in \mathbb{N}$, $x \in \mathbb{X}$ and $u \in \mathbb{U}^N(x)$ with $J_N^{uc}(x, u) \leq N\ell(x^e, u^e)+\delta$, the set $\mathcal{Q}(x, u, P, N) := \{k \in \{0, \ldots, N-1\} \mid |x_u(k, x)|_{x^e} \geq \sigma_\delta(P)\}$ has at most P elements.*

Proof We fix $\delta > 0$ and claim that the assertion holds with $\sigma_\delta(P) := \rho^{-1}((2M + \delta)/P)$ where M is a bound on $|\lambda|$. To prove this claim, assume that there are N, P, x and u such that $J_N^{uc}(x, u) \leq N\ell(x^e, u^e) + \delta$ but $\mathcal{Q}(x, u, P, N)$ contains at least $P + 1$ elements. Then from (8.13) we can estimate

$$\widetilde{J}_N^{uc}(x, u) \leq J_N^{uc}(x, u) + 2M - N\ell(x^e, u^e) \leq 2M + \delta.$$

On the other hand, (8.4), (8.6), and the fact that $\mathcal{Q}(x, u, P, N)$ contains at least $P+1$ elements imply

$$\widetilde{J}_N^{uc}(x, u) \geq \sum_{k=0}^{N-1} \tilde{\ell}(x_u(k, x), u(k)) \geq \sum_{k=0}^{N-1} \rho(|x_u(k, x)|_{x^e}) \geq \sum_{\substack{k\in\{0,\ldots,N-1\}\\ |x_u(k,x)|_{x^e} > \sigma_\delta(P)}} \rho(\sigma_\delta(P))$$
$$\geq (P + 1)\rho(\sigma_\delta(P)) \geq (P + 1)\frac{2M + \delta}{P} > 2M + \delta,$$

which is a contradiction. □

We refer to the property described by Proposition 8.15 as the *turnpike property*. For an illustration, we refer to Fig. 8.2. In fact, there are various variants of the turnpike property known in optimal control, of which the one described by Proposition 8.15 is just a particular version. More information on this property can be found in Sect. 8.8.

We remark that the boundedness assumption on λ can be restrictive in case $\mathbb{X}$ is unbounded. However for bounded subsets of the state constraint set $\mathbb{X}$, it is not a very

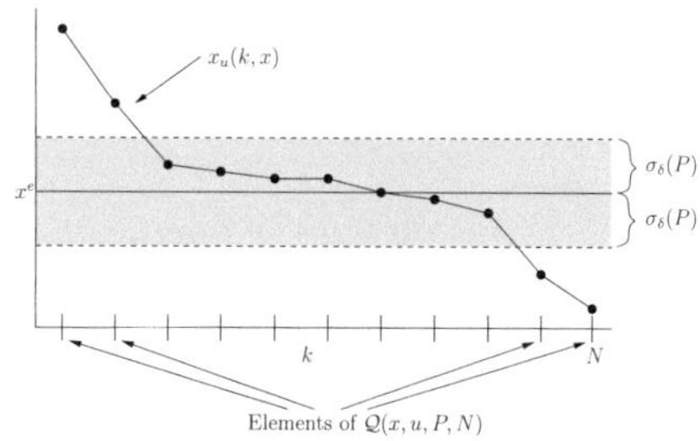

Fig. 8.2 Illustration of the set $\mathcal{Q}(x, u, P, N)$ defined in Proposition 8.15

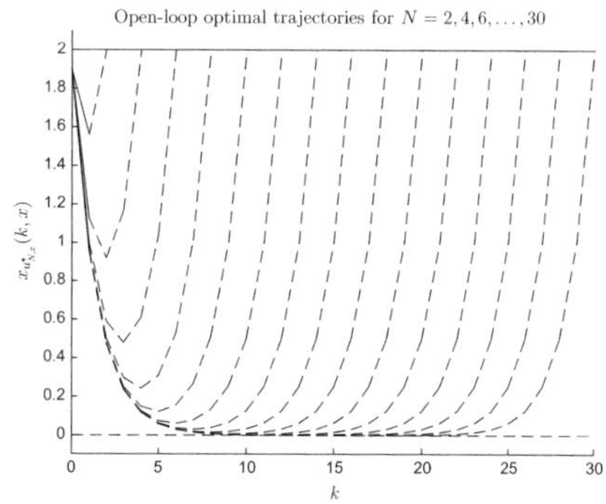

Fig. 8.3 Open-loop optimal trajectories (without terminal conditions) for Example 8.1 with different optimization horizons N. The turnpike property is clearly visible

strong assumption. Hence, it can be assumed to hold if either $\mathbb{X}$ itself is bounded or if near optimal trajectories are guaranteed to stay in a bounded subset of $\mathbb{X}$.

Example 8.16 Since Example 8.1 is strictly dissipative with bounded storage function (cf. Example 8.8), we expect the system to have the turnpike property. The numerical optimal trajectories depicted in Fig. 8.3 support this claim.

Next we derive upper and lower bounds for V_∞^{uc}.

Lemma 8.17 *Assume that the optimal control problem* (OCP$_\text{N}$) *is strictly dissipative with bounded storage function λ, that $\ell(x^e, u^e) = 0$ and that Assumptions 8.5(a) and 8.12 hold. Then there is $C > 0$ such that the inequalities*

$$-C \leq V_\infty^{uc}(x) \leq \gamma_V(|x|_{x^e}) \tag{8.14}$$

hold for all $x \in \bigcup_{N\in\mathbb{N}} \mathbb{X}_N$ with γ_V from Assumption 8.12(c).

Proof For $x \in \mathbb{X}_N$, using the control sequence $u(k) = \mu_N(x_{\mu_N}(k, x))$ induced by the closed loop, from (8.11) with $\ell(x^e, u^e) = 0$ for any $K > 0$, we obtain

$$J_K^{uc}(x, u) = \sum_{k=0}^{K-1} \ell(x_u(k, x), u_k(x)) \leq V_N(x) - V_N(x_u(K, x)).$$

By asymptotic stability of x^e for this solution we obtain $x_u(K,x) \to x^e$, and thus, since $V_N(x^e) = N\ell(x^e,u^e) = 0$ by (8.10), Assumption 8.12(c) yields $V_N(x_u(K,x)) \to 0$ as $K \to \infty$. Using Assumption 8.12(c) and $V_N(x^e) = 0$, this implies

$$V_\infty^{uc}(x) \le \limsup_{K\to\infty} J_K^{uc}(x,u) \le V_N(x) \le \gamma_V(|x|_{x^e}).$$

On the other hand, the fact that $\widetilde{J}_N^{uc}(x,u) \ge 0$, (8.13) and the boundedness of λ imply $J_N^{uc}(x,u) \ge -C$ for some $C \ge 0$ and all x, u and N. This implies $V_\infty^{uc}(x) \ge -C$. □

Using the inequality ensured by this lemma, we can prove an infinite horizon version of the turnpike property from Proposition 8.15.

Proposition 8.18 *Assume that the optimal control problem* ($\mathrm{OCP_N}$) *is strictly dissipative,* $\mathbb{X}$ *is bounded,* $\ell(x^e,u^e) = 0$ *and the inequalities* (8.14) *hold for all* $x \in \bigcup_{N\in\mathbb{N}_0}\mathbb{X}_N$. *Then there exists* $\sigma_\infty \in \mathcal{L}$ *such that for all* $P \in \mathbb{N}$, $x \in \mathbb{X}$ *and* $u \in \mathbb{U}^\infty(x)$ *with* $J_\infty^{uc}(x,u) \le V_\infty^{uc}(x) + 1$, *the set* $\mathcal{Q}(x,u,P,\infty) := \{k \in \mathbb{N}_0 \mid |x_u(k,x)|_{x^e} \ge \sigma_\infty(P)\}$ *has at most* P *elements.*

Proof First note that by Lemma 8.17 and the assumption, we get

$$J_\infty^{uc}(x,u) \le \sup_{x\in\bigcup_{N\in\mathbb{N}}\mathbb{X}_N} V_\infty^{uc}(x) + 1 \le \sup_{x\in\mathbb{X}} \gamma_V(|x|_{x^e}) + 1 =: \delta.$$

Now we can proceed as in the proof of Proposition 8.15: denoting by M a bound on $|\lambda|$, from (8.13) and $\ell(x^e,u^e) = 0$ we obtain

$$\widetilde{J}_\infty^{uc}(x,u) = \limsup_{K\to\infty} \widetilde{J}_K^{uc}(x,u) \le \limsup_{K\to\infty} J_K^{uc}(x,u) + 2M \le \delta + 2M.$$

Setting $\sigma_\infty(K) := \rho^{-1}((2M+\delta)/K)$, the assumption that $\mathcal{Q}(x,u,P,\infty)$ contains more than P elements, again yields a contradiction to this inequality. □

We note that this theorem implies $x_u(k,x) \to x^e$ as $k \to \infty$, because otherwise $\mathcal{Q}(x,u,P,\infty)$ would contain infinitely many elements for sufficiently large $P \in \mathbb{N}$. Using this fact, we can improve the lower bound on V_∞^{uc} from Lemma 8.17.

Lemma 8.19 *Under the assumptions of Proposition 8.18, the inequality* $V_\infty^{uc}(x) \ge -\lambda(x)$ *holds for all* $x \in \bigcup_{N\in\mathbb{N}_0}\mathbb{X}_N$.

Proof Let $u \in \mathbb{U}^\infty(x)$ be such that $J_\infty^{uc}(x,u) \le V_\infty^{uc}(x) + \varepsilon$ for an $\varepsilon \in (0,1)$. As explained above, Proposition 8.18 implies that $x_u(k,x) \to x^e$ as $k \to \infty$. The definition of V_∞^{uc} and (8.13) then imply that

$$\begin{aligned} V_\infty^{uc}(x) + \varepsilon \;&\ge\; \limsup_{K\to\infty} J_K^{uc}(x,u) \\ &=\; \limsup_{K\to\infty}\Big(-\lambda(x) + \underbrace{\widetilde{J}_K^{uc}(x,u)}_{\ge 0} + \underbrace{\lambda(x_u(K,x))}_{\to\lambda(x^e)=0}\Big) \;\ge\; -\lambda(x). \end{aligned}$$

This implies the assertion since $\varepsilon \in (0, 1)$ was arbitrary. □

Our final preparatory result is needed for estimating the finite horizon transient performance. It thus concerns the optimal value of the problem with control functions u that steer a given initial value $x \in \mathbb{X}$ to the closed ball $\overline{\mathscr{B}}_\kappa(x^e)$ with radius $\kappa > 0$ around x^e. In order to simplify the notation, we briefly write

$$\mathbb{U}_\kappa^K(x) := \mathbb{U}^K_{\overline{\mathscr{B}}_\kappa(x^e)}(x) \tag{8.15}$$

using the notation from Definition 3.9 with $\overline{\mathscr{B}}_\kappa(x^e)$ in place of $\mathbb{X}_0$. We remark that Theorem 8.13 yields the existence of a $\beta \in \mathscr{KL}$ such that for all $x \in \mathbb{X}_N$ and all K with $\beta(|x|_{x^e}, K) \leq \kappa$ the control u obtained from the MPC feedback law via $u(k) = \mu_N(x_{\mu_N}(k, x))$ is contained in $\mathbb{U}_\kappa^K(x)$. This, in particular, shows that this set is nonempty for sufficiently large K.

The next lemma shows that the infimum of $J_K^{uc}(x, u)$ over $u \in \mathbb{U}_\kappa^K(x)$ and the corresponding approximately optimal trajectories behave similar to those of the infinite horizon problem. More precisely, part (a) of the following lemma is similar to Lemma 8.17, part (b) to Lemma 8.19, and part (c) to Proposition 8.18. Note that since we only consider finite horizon problems here, we do not need to assume $\ell(x^e, u^e) = 0$.

Lemma 8.20 *Assume that the optimal control problem* (OCP$_\text{N}$) *is strictly dissipative with bounded storage function λ and that Assumptions 8.5(a) and 8.12 hold. Fix $\kappa_0 > 0$ and let β be a $\mathscr{KL}$-function characterizing the asymptotic stability of the closed loop, whose existence is guaranteed by Theorem 8.13. Then for any $\kappa \in (0, \kappa_0]$, any $x \in \bigcup_{N \in \mathbb{N}_0} \mathbb{X}_N$, and $K_0 \in \mathbb{N}$ minimal with $\beta(|x|_{x^e}, K_0) \leq \kappa$, the following holds.*

(a) For all $K \geq K_0$ the inequality

$$\inf_{u \in \mathbb{U}_\kappa^K(x)} J_K^{uc}(x, u) - K\ell(x^e, u^e) \leq \gamma_V(|x|_{x^e}) + \gamma_V(\kappa)$$

holds with $\gamma_V \in \mathscr{K}_\infty$ from Assumption 8.12(c).

(b) For all $K \in \mathbb{N}$ with $\mathbb{U}_\kappa^K(x) \neq \emptyset$ the inequality

$$-\gamma_\lambda(|x|_{x^e}) - \gamma_\lambda(\kappa) \leq \inf_{u \in \mathbb{U}_\kappa^K(x)} J_K^{uc}(x, u) - K\ell(x^e, u^e)$$

holds with γ_λ from Assumption 8.12(b).

(c) If in addition $\mathbb{X}$ is bounded, then there exists $\sigma \in \mathscr{L}$ such that for all $K \geq K_0$, all $P \in \mathbb{N}$ and any $u \in \mathbb{U}_\kappa^K(x)$ with $J_K^{uc}(x, u) \leq \inf_{u \in \mathbb{U}_\kappa^K(x)} J_K^{uc}(x, u) + 1$ there is $k \leq \min\{P, K-1\}$ such that $|x_u(k, x)|_{x^e} < \sigma(\min\{P, K-1\})$.

Proof (a) The proof of this inequality works similarly to the first part of the proof of Lemma 8.17. For $x \in \mathbb{X}_N$, we choose the control u obtained from the MPC feedback law via $u(k) = \mu_N(x_{\mu_N}(k, x))$. By Theorem 8.13 and the choice of K_0, this control lies in $\mathbb{U}_\kappa^K(x)$. As in the proof of Lemma 8.17, from (8.11)—now with $\ell(x^e, u^e) \neq 0$—for this u we get

$$J_K^{uc}(x, u) \leq V_N(x) - V_N(x_u(K, x)) + K\ell(x^e, u^e)$$

and from Assumption 8.12(c) and $|x_u(K, x)|_{x^e} < \kappa$, we obtain the assertion.

(b) Let $\varepsilon > 0$ and take a control $u_\varepsilon \in \mathbb{U}_\kappa^K(x)$ with $\inf_{u\in\mathbb{U}_\kappa^K(x)} J_K^{uc}(x, u) + \varepsilon \geq J_K^{uc}(x, u_\varepsilon)$. Then by (8.13), Assumption 8.12(b) and $\lambda(x^e) = 0$, and recalling that strict dissipativity implies $\widetilde{J}_K^{uc}(x, u_\varepsilon) \geq 0$ we get

$$\begin{aligned}\inf_{u\in\mathbb{U}_\kappa^K(x)} J_K^{uc}(x, u) + \varepsilon \geq J_K^{uc}(x, u_\varepsilon) \\ &= \underbrace{-\lambda(x)}_{\geq -\gamma_\lambda(|x|_{x^e})} + \underbrace{\widetilde{J}_K^{uc}(x, u_\varepsilon)}_{\geq 0} + \underbrace{\lambda(x_{u_\varepsilon}(K, x))}_{\geq -\gamma_\lambda(\kappa)} + K\ell(x^e, u^e) \\ &\geq -\gamma_\lambda(|x|_{x^e}) - \gamma_\lambda(\kappa) + K\ell(x^e, u^e).\end{aligned}$$

This implies (b) since $\varepsilon > 0$ was arbitrary.

(c) The assumptions and (a) imply that Proposition 8.15 can be applied with $\delta = \sup_{x\in\mathbb{X}} \gamma(|x|_{x^e}) + \gamma(\kappa_0) + 1$ for all $x \in \mathbb{X}$ and all $\kappa \in (0, \kappa_0]$. We set $\sigma = \sigma_\delta$ from this proposition. Since the set $\mathcal{Q}(x, u, \min\{P, K-1\}, K)$ has at most $\min\{P, K-1\}$ elements, there exists at least one $k \in \{0, \ldots, \min\{P, K-1\}\}$ with $k \notin \mathcal{Q}(x, u, \min\{P, K-1\}, K)$, which thus satisfies $|x_u(k, x)|_{x^e} \leq \sigma(\min\{P, K-1\})$. □

We now have all the tools to prove the two main theorems of this section. The first theorem gives an upper bound for the non-averaged infinite horizon performance of the NMPC closed-loop trajectory. We recall that when considering the infinite horizon problem, we demand $\ell(x^e, u^e) = 0$. Taking into account the inequality $V_\infty^{uc}(x) \leq J_\infty^{cl}(x, \mu_N)$ that follows immediately from the definition of these functions, the theorem shows that economic MPC delivers an approximately (non-averaged) infinite horizon optimal closed-loop solution for which the approximation error tends to 0 as the horizon N tends to infinity.

Theorem 8.21 *Consider the NMPC Algorithm 3.10 with strictly dissipative optimal control problem* ($\mathrm{OCP_{N,e}}$) = (8.2). *Assume that $\mathbb{X}$ is bounded, that $\ell(x^e, u^e) = 0$ and that Assumptions 8.5 and 8.12 hold. Then there exists $\delta_1 \in \mathcal{L}$ such that the inequalities*

$$J_\infty^{cl}(x, \mu_N) \leq V_N(x) \leq V_\infty^{uc}(x) + \delta_1(N)$$

hold for all $x \in \mathbb{X}_N$.

Proof In order to prove the first inequality, from (8.11) we obtain $\ell(x, \mu_N(x)) \leq V_N(x) - V_N(f(x, \mu_N(x)))$. This implies for any $K \in \mathbb{N}$

$$J_K^{cl}(x, \mu_N) = \sum_{k=0}^{K-1} \ell(x_{\mu_N}(k, x), \mu_N(x_{\mu_N}(k, x))) \leq V_N(x) - V_N(x_{\mu_N}(K, x)). \quad (8.16)$$

Now from Theorem 8.13 we know that $|x_{\mu_N}(k,x)|_{x^e} \le \beta(|x|_{x^e},k) \le \beta(M,k) =: \nu(k)$, where $M := \max_{x,y\in\mathbb{X}} d(x,y)$. Note that $\nu \in \mathcal{L}$. Moreover, by (8.10) we have $V_N(x^e) = N\ell(x^e,u^e) = 0$ and from Assumption 8.12(c) we know the existence of $\gamma_V \in \mathcal{K}$ with $|V_N(x)| = |V_N(x) - V_N(x^e)| \le \gamma_V(|x|_{x^e})$ for all $x \in \mathbb{X}$. Together this yields

$$|V_N(x_{\mu_N}(K,x))| \le \gamma_V(\nu(K)).$$

Since $\gamma_V(\nu(K)) \to 0$ for $K \to \infty$, this inequality together with (8.16) yields the first inequality by letting $K \to \infty$.

For the second inequality, we note that it is sufficient to prove the inequality for all sufficiently large N, because by boundedness of V_N and V_∞^{uc}, for small N the inequality can always be satisfied by choosing $\delta_1(N)$ sufficiently large without violating the requirement $\delta_1 \in \mathcal{L}$. Consider σ_∞ from Proposition 8.18, pick N_0 and η from Assumption 8.5(b), choose N_1 such that $\sigma_\infty(N_1) < \eta$, fix $0 < \varepsilon < 1$ and choose an admissible control u_ε satisfying $J_\infty^{uc}(x,u_\varepsilon) \le V_\infty^{uc}(x) + \varepsilon$. Then for $N \ge 2N_1$, we use Proposition 8.18 with $P = \lfloor N/2 \rfloor$. We thus obtain the existence of $k \in \{0,\dots,P-1\}$ such that $|x_{u_\varepsilon}(k,x)|_{x^e} < \sigma_\infty(P) \le \sigma_\infty(N_1) < \eta$, implying $x_u(k,x) \in \mathbb{X}_{N_1} \subseteq \mathbb{X}_{N_2}$ and thus $u_\varepsilon \in \mathbb{U}^k_{\mathbb{X}_{N_2}}(x)$ for all $N_2 \ge N_1$. Particularly, this holds for $N_2 = N - k$, implying $u_\varepsilon \in \mathbb{U}^k_{\mathbb{X}_{N-k}}(x)$. Now from Assumption 8.12(c) applied to V_{N-k}, we can conclude (again using $V_N(x^e) = 0$)

$$|V_{N-k}(x_{u_\varepsilon}(k,x))| \le \gamma_V(\sigma_\infty(P)).$$

Moreover, Lemma 8.19 and the bound on λ yield

$$\begin{aligned} V_\infty^{uc}(x) + \varepsilon \ge J_\infty^{uc}(x,u_\varepsilon) &\ge J_k^{uc}(x,u_\varepsilon) + V_\infty(x_{u_\varepsilon}(k,x))) \\ &\ge J_k^{uc}(x,u_\varepsilon) - \lambda(x_{u_\varepsilon}(k,x)) \ \ge\ J_k^{uc}(x,u_\varepsilon) - \gamma_\lambda(\sigma_\infty(P)). \end{aligned}$$

Together with the dynamic programming principle (3.15) these inequalities imply

$$\begin{aligned} V_N(x) = &\inf_{u\in\mathbb{U}^k_{\mathbb{X}_{N-k}}(x)} \{J_k^{uc}(x,u) + V_{N-k}(x_u(k,x))\} \ \le\ J_k^{uc}(x,u_\varepsilon) + V_{N-k}(x_{u_\varepsilon}(k,x)) \\ &\le V_\infty^{uc}(x) + \gamma_V(\sigma_\infty(P)) + \gamma_\lambda(\sigma_\infty(P)) + \varepsilon. \end{aligned}$$

Since $\varepsilon > 0$ was arbitrary, this proves the assertion for $\delta_1(N) = \gamma_V(\sigma_\infty(\lfloor N/2 \rfloor)) + \gamma_\lambda(\sigma_\infty(\lfloor N/2 \rfloor))$. □

Since x^e is asymptotically stable for the NMPC closed-loop trajectories, the closed-loop solutions converge toward x^e as $k \to \infty$. More precisely, given a time K, by Theorem 8.13 the solutions are guaranteed to satisfy $x_{\mu_N}(k,x) \in \overline{\mathcal{B}}_\kappa(x^e)$ for all $k \ge K$ and $\kappa = \beta(|x|_{x^e}, K)$ for β from Theorem 8.13. We denote the time span $\{0,\dots,K-1\}$ during which the system is (possibly) outside $\overline{\mathcal{B}}_\kappa(x^e)$ as *transient time* and the related finite horizon functional $J_K^{uc}(x,u)$ as *transient performance*. The next theorem then shows that among all possible trajectories from x to $\overline{\mathcal{B}}_\kappa(x^e)$, the NMPC closed loop has the best transient performance up to error terms vanishing as

$K \to \infty$ and $N \to \infty$. Again, in order to simplify the notation, we use $\mathbb{U}_\kappa^K(x)$ from (8.15). We remark that unlike the previous theorem, here we do not need to assume $\ell(x^e, u^e) = 0$.

Theorem 8.22 *Consider the NMPC Algorithm 3.10 with strictly dissipative optimal control problem* $(\mathrm{OCP_{N,e}}) = (8.2)$*. Assume that $\mathbb{X}$ is bounded and that Assumptions 8.5 and 8.12 hold. Then there exist $\delta_1, \delta_2 \in \mathscr{L}$ such that for all $x \in \mathbb{X}_N$ the inequality*

$$J_K^{cl}(x, \mu_N) \leq \inf_{u \in \mathbb{U}_\kappa^K(x)} J_K^{uc}(x, u) + \delta_1(N) + \delta_2(K)$$

holds with $\kappa = \beta(|x|_{x^e}, K)$ and $\beta \in \mathscr{KL}$ characterizing the asymptotic stability of the closed loop guaranteed by Theorem 8.13.

Proof We can without loss of generality assume $\ell(x^e, u^e) = 0$, because the claimed inequality is invariant under adding constants to ℓ. Moreover, similar to the proof of Theorem 8.21, it is sufficient to prove the inequality for all sufficiently large K and N, because by boundedness of all functions involved for small N and K the inequality can always be achieved by choosing $\delta_1(N)$ and $\delta_2(K)$ sufficiently large. As in the first step of the previous proof, we obtain $|V_N(x_{\mu_N}(K, x))| \leq \gamma_V(\nu(K))$. It is thus sufficient to show the existence of $\delta_1, \tilde{\delta}_2 \in \mathscr{L}$ with

$$V_N(x) \leq \inf_{u \in \mathbb{U}_\kappa^K(x)} J_K^{uc}(x) + \delta_1(N) + \tilde{\delta}_2(K) \tag{8.17}$$

for all $x \in \mathbb{X}_N$ because then the assertion follows from (8.16) with $\delta_2 = \gamma_V \circ \nu + \tilde{\delta}_2$.

In order to prove (8.17), consider σ from Lemma 8.20(c), which we apply with $P = \lfloor N/2 \rfloor$ and pick $u_\varepsilon \in \mathbb{U}_\kappa^K(x)$ with $J_K^{uc}(x, u_\varepsilon) \leq \inf_{u \in \mathbb{U}_\kappa^K(x)} J_K^{uc}(x, u) + \varepsilon$ with an arbitrary but fixed $\varepsilon \in (0, 1)$. This yields the existence of $k \in \{0, \ldots, \lfloor N/2 \rfloor\}$, $k \leq K - 1$ with $|x_{u_\varepsilon}(k, x)|_{x^e} \leq \sigma(\min\{P, K - 1\})$. Since u_ε steers x to $\overline{\mathscr{B}}_\kappa(x^e)$, the shifted sequence $u_\varepsilon(k + \cdot)$ lies in $\mathbb{U}_\kappa^{K-k}(x_{u_\varepsilon}(k, x))$, implying that this set is nonempty. Hence, we can apply Lemma 8.20(b) in order to conclude $J_{K-k}^{uc}(x_{u_\varepsilon}(k, x), u_\varepsilon(k + \cdot)) \geq -\gamma_\lambda(\sigma(\min\{N, K - 1\})) - \gamma_\lambda(\kappa)$. This implies

$$\begin{aligned} \inf_{u \in \mathbb{U}_\kappa^K(x)} J_K^{uc}(x, u) + \varepsilon \geq J_K^{uc}(x, u_\varepsilon) &= J_k^{uc}(x, u_\varepsilon) + J_{K-k}^{uc}(x_{u_\varepsilon}(k, x), u_\varepsilon(k + \cdot)) \\ &\geq J_k^{uc}(x, u_\varepsilon) - \gamma_\lambda(\sigma(\min\{N, K - 1\})) - \gamma_\lambda(\kappa) \end{aligned}$$

Moreover, by choosing N and K sufficiently large we can ensure $\sigma(\min\{P, K-1\}) < \eta$ for η from Assumption 8.5(b), implying $u_\varepsilon \in \mathbb{U}_{\mathbb{X}_Q}^k(x)$ for all $Q \geq N_0$ and N_0 from Assumption 8.5(b). Particularly, choosing $N \geq 2N_0$ implies $N - k \geq N_0$ and thus $u_\varepsilon \in \mathbb{U}_{\mathbb{X}_{N-k}}^k(x)$.

Using this relation, the inequality derived above, the dynamic programming principle (3.15) and Assumption 8.12(c) for V_{N-k} we obtain

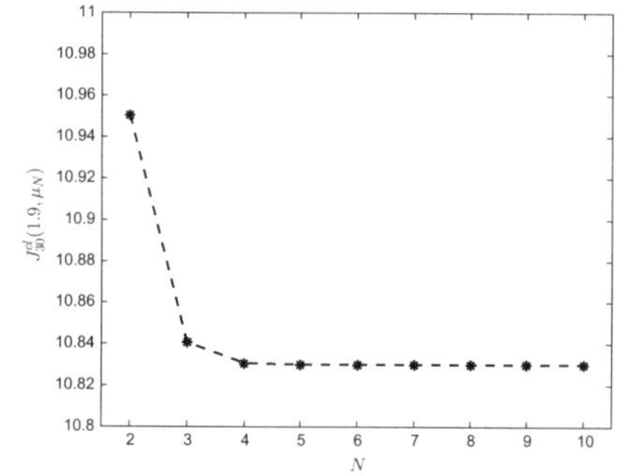

Fig. 8.4 Value of $J_K^{cl}(x, \mu_N)$ for $K = 30$, $x = 1.9$ and varying N with terminal constraint $\mathbb{X} = \{0\}$

$$
\begin{aligned}
V_N(x) &= \inf_{u \in \mathbb{U}^k_{\mathbb{X}_{N-k}}(x)} \{J_k^{uc}(x,u) + V_{N-k}(x_u(k,x))\} \le J_k^{uc}(x,u_\varepsilon) + V_{N-k}(x_{u_\varepsilon}(k,x)) \\
&\le \inf_{u \in \mathbb{U}^K_\kappa(x)} J_K^{uc}(x,u) + \gamma_\lambda(\sigma(\min\{P, K-1\})) + \gamma_\lambda(\kappa) + \varepsilon \\
&\quad + \gamma_V(\sigma(\min\{P, K-1\})).
\end{aligned}
$$

This shows the desired inequality (8.17) for

$$
\delta_1(N) = \gamma_V(\sigma(\lfloor N/2 \rfloor)) + \gamma_\lambda(\sigma(\lfloor N/2 \rfloor)),
$$

and using the choice of κ,

$$
\tilde{\delta}_2(K) = \gamma_V(\sigma(K-1)) + \gamma_\lambda(\sigma(K-1)) \; + \; \gamma_\lambda(\beta(M, K))
$$

with $M = \max_{x,y \in \mathbb{X}} d(x,y)$ and $\beta \in \mathcal{KL}$ characterizing the asymptotic stability of the closed loop. □

Example 8.23 Figure 8.4 illustrates how $J_K^{cl}(x, \mu_N)$ depends on N for Example 8.1. The value $K = 30$ is so large that the effect of the term $\delta_2(K)$ is negligible and not visible in the figure, hence $J_K^{cl}(x, \mu_N)$ converges to $\inf_{u \in \mathbb{U}^K_\kappa(x)} J_K^{uc}(x,u)$ for increasing N.

8.5 Averaged Optimality Without Terminal Conditions

In this and in the subsequent sections, we discuss the case in which we do not impose terminal conditions on the problem, i.e., we consider the NMPC Algorithm with optimal control problem ($\mathrm{OCP_N}$). The corresponding functionals and optimal value functions will, as usual, be denoted by J_N and V_N and their infinite horizon counterparts by J_∞ and V_∞, i.e., we do not use the superscript notation J_N^{uc}, etc., anymore in the sequel. The results are presented in parallel to Sects. 8.2–8.4.

Since we do not impose any terminal conditions, we do not need Assumptions 8.5 and 8.12(a) anymore. However, we still need Part (b) and (a relaxed version of) Part (c) of Assumption 8.12, where the latter now refers to the optimal value function of the unconstrained problem ($\mathrm{OCP_N}$).

Assumption 8.24 (*Continuity of* λ *and* V_N *at* x^e) There exist γ_λ and $\gamma_V \in \mathscr{K}_\infty$ and $\omega \in \mathscr{L}$ such that the following properties hold.
(a) For all $x \in \mathbb{X}$ it holds that

$$|\lambda(x) - \lambda(x^e)| \leq \gamma_\lambda(|x|_{x^e}).$$

(b) For each $N \in \mathbb{N}$ and each $x \in \mathbb{X}$ it holds that

$$|V_N(x) - V_N(x^e)| \leq \gamma_V(|x|_{x^e}) + \omega(N).$$

Note that (b) implies viability of $\mathbb{X}$, which we assume for simplicity in this section. If desired, this condition could be relaxed using techniques similar to those from Sect. 7.3, where the turnpike property replaces the stability assumption (see [8] for details in a continuous time setting). One method of ensuring the continuity from (b) without requiring explicit knowledge of V_N is by assuming strict dissipativity and local controllability around x^e, see [18] or [10, Sect. 6].

We observe that Propositions 8.15 and 8.18 remain valid, as the assumptions, statements and proofs do not involve any terminal constraints or costs. Based on these two propositions, we can prove the following two auxiliary results, which lead to the main result of this section. In what follows, we denote by $u^\star_\infty$ and $u^\star_N$ the optimal control sequences for ($\mathrm{OCP_\infty}$) and ($\mathrm{OCP_N}$), respectively, for initial value $x \in \mathbb{X}$.

Lemma 8.25 *If Assumption 8.24 and the assumptions of Proposition 8.15 hold, then the equation*

$$V_N(x) = J_M(x, u^\star_N) + V_{N-M}(x^e) + R_1(x, M, N) \tag{8.18}$$

holds with $|R_1(x, M, N)| \leq \gamma_V(\sigma_\delta(P)) + \omega(N - M)$ *for all* $x \in X$, *all* $N \in \mathbb{N}$, *all* $P \in \mathbb{N}$ *and all* $M \notin \mathscr{Q}(x, u^\star_N, P, N)$, *with* σ_δ *from Proposition 8.15 with* $\delta = \gamma_V(|x|_{x^e}) + \omega(N)$.

Proof Observe that using the constant control $u \equiv u^e$, we can estimate $V_N(x^e) \leq J_N(x^e, u) = N\ell(x^e, u^e)$. Thus, using Assumption 8.24 we get $J_N(x, u^\star_N) \leq N\ell(x^e, u^e) + \gamma_V(|x|_{x^e}) + \omega(N)$, hence Proposition 8.15 applies to the optimal trajectory with $\delta = \gamma_V(|x|_{x^e}) + \omega(N)$. This in particular ensures $|x_{u^\star_N}(M, x)|_{x^e} \leq \sigma_\delta(P)$ for all $M \notin \mathscr{Q}(x, u^\star_N, P, N)$.

Now the dynamic programming principle (3.16) yields

$$V_N(x) = J_M(x, u^\star_N) + V_{N-M}(x_{u^\star_N}(M, x)).$$

Hence, (8.18) holds with $R_1(x, M, N) = V_{N-M}(x_{u_N^\star}(M, x)) - V_{N-M}(x^e)$. Then for any $P \in \mathbb{N}$ and any $M \notin \mathcal{Q}(x, u_N^\star, P, N)$, this implies $|R_1(x, M, N)| \leq \gamma_V(|x_{u_N^\star}(M, x)|_{x^e}) + \omega(N - M) \leq \gamma_V(\sigma_\delta(P)) + \omega(N - M)$ and thus the assertion. □

Lemma 8.26 *If Assumption 8.24 and the assumptions of Proposition 8.15 hold, then the equation*

$$V_N(x) = V_{N-1}(x) + \ell(x^e, u^e) + R_2(x, N)$$

holds with $|R_2(x, N)| \leq \nu_2(|x|_{x^e}, N) = 2\gamma_V(\sigma_\delta(\lfloor N/2 \rfloor)) + 2\omega(\lfloor N/2 \rfloor - 1)$ *for all* $x \in \mathbb{X}$*, all* $N \in \mathbb{N}$ *and* σ_δ *from Proposition 8.15 with* $\delta = \gamma_V(|x|_{x^e}) + \omega(N - 1)$.

Proof Given $x \in \mathbb{X}$, consider the optimal control $u_{N-1}^\star$ for horizon length $N - 1$ and σ_δ from Proposition 8.15 with $\delta = \gamma_V(|x|_{x^e})$. Then Lemma 8.25 applied with $N - 1$ in place of N and $P = \lfloor N/2 \rfloor$ implies the existence of $M \in \{0, \ldots, \lfloor N/2 \rfloor - 1\}$ with

$$V_{N-1}(x) = J_M(x, u_{N-1}^\star) + V_{N-M-1}(x^e) + R_1(x, M, N - 1)$$

with $|R_1(x, M, N - 1)| \leq \gamma_V(\sigma_\delta(\lfloor N/2 \rfloor)) + \omega(\lfloor N/2 \rfloor - 1)$. The construction in the proof of Lemma 8.25 moreover yields $|x_{u_{N-1}^\star}(M, x)|_{x^e} \leq \sigma_\delta(\lfloor N/2 \rfloor)$. Using $u(k) = u_{N-1}^\star(k)$ for $k = 0, \ldots, M - 1$ and $u(M + k) = u_{N-M}^\star(k)$ with the optimal control $u_{N-M}^\star$ for initial value $x_{u_N^\star}(M, x)$ and horizon $N - M$ for $k = M, \ldots, N - 1$, yields

$$J_N(x, u) = J_M(x, u_{N-1}^\star) + V_{N-M}(x_{u_N^\star}(M, x)) = J_M(x, u_{N-1}^\star) + V_{N-M}(x^e) + \widehat{R}_1(x, M, N)$$

with $|\widehat{R}_1(x, M, N)| \leq \gamma_V(\sigma_\delta(\lfloor N/2 \rfloor)) + \omega(\lfloor N/2 \rfloor)$. Since for initial value x^e, we can always stay at the equilibrium for one step and use the optimal control for initial value x^e for the remaining horizon, we obtain the inequality $V_{N-M}(x^e) \leq \ell(x^e, u^e) + V_{N-M-1}(x^e)$. Together this yields

$$\begin{aligned} V_N(x) \leq J_N(x, u) &= J_M(x, u_{N-1}^\star) + V_{N-M}(x^e) + \widehat{R}_1(x, M, N) \\ &\leq J_M(x, u_{N-1}^\star) + \ell(x^e, u^e) + V_{N-M-1}(x^e) + \widehat{R}_1(x, M, N) \\ &= V_{N-1}(x) + \ell(x^e, u^e) - R_1(x, M, N - 1) + \widehat{R}_1(x, M, N), \end{aligned}$$

and thus the claim with $R_2(x, N) = \widehat{R}_1(x, M, N) - R_1(x, M, N - 1)$. □

Now we can state the theorem on the infinite horizon average performance.

Theorem 8.27 *Consider the NMPC Algorithm 3.1 with strictly dissipative optimal control problem* (OCP$_\mathrm{N}$) *with bounded storage function* λ*. Let Assumption 8.24 hold and assume* V_N *is bounded from below on* $\mathbb{X}$*. Then, for any* $N \geq 2$ *and any* $x \in \mathbb{X}$ *the averaged closed-loop performance satisfies the inequality*

$$\overline{J}_\infty^{cl}(x, \mu_N) \leq \ell(x^e, u^e) + \delta_1(N) \tag{8.19}$$

with $\delta_1(N) \leq 2\gamma_V(\sigma_\delta(\lfloor N/2 \rfloor)) + 2\omega(\lfloor N/2 \rfloor - 1)$ *for* σ_δ *from Proposition 8.15 with* $\delta = \gamma_V(|x|_{x^e}) + \omega(N-1)$ *and* γ_V *and* ω *from Assumption 8.24.*

Proof Abbreviate $x_{\mu_N}(k) = x_{\mu_N}(k, x)$. From the dynamic programming principle (3.16) and Lemma 8.26 applied with $x = x_{\mu_N}(k+1)$, we obtain

$$\begin{aligned}\ell(x_{\mu_N}(k), \mu_N(x_{\mu_N}(k))) &= V_N(x_{\mu_N}(k)) - V_{N-1}(x_{\mu_N}(k+1)) \\ &\leq V_N(x_{\mu_N}(k)) - V_N(x_{\mu_N}(k+1)) + \ell(x^e, u^e) + \nu_2(N).\end{aligned}$$

Thus we obtain

$$\begin{aligned}\overline{J}^{cl}_\infty(x, \mu_N) &= \limsup_{K\to\infty} \frac{1}{K} \sum_{k=0}^{K-1} \ell(x_{\mu_N}(k), \mu_N(x_{\mu_N}(k))) \\ &= \limsup_{K\to\infty} \frac{1}{K} \sum_{k=0}^{K-1} \Big(V_N(x_{\mu_N}(k)) - V_N(x_{\mu_N}(k+1)) + \ell(x^e, u^e) + \nu_2(N) \Big) \\ &= \ell(x^e, u^e) + \nu_2(N) + \limsup_{K\to\infty} \frac{V_N(x_0) - V_N(x_{\mu_N}(K))}{K} \\ &\leq \ell(x^e, u^e) + \nu_2(N) + \limsup_{K\to\infty} \frac{V_N(x_0) + M}{K} \quad = \ell(x^e, u^e) + \nu_2(N)\end{aligned}$$

where $-M$ is a lower bound on V_N on $\mathbb{X}$. This shows the claim with $\delta_1(N) = \nu_2(N)$. □

The difference between this and the corresponding result with terminal conditions is that we get the error term $\delta_1(N)$ on the right-hand side of the estimate, which does, however, tend to 0 as $N \to \infty$.

Example 8.28 Figure 8.5 shows $\overline{J}^{cl}_\infty(x, \mu_N)$ for Example 8.1 depending on N. The plot in the logarithmic scale shows that the value converges to the optimal value $\ell(0,0) = 0$ exponentially fast, hence the error $\delta_1(N)$ also vanishes exponentially fast. This is actually not a coincidence. However, an analysis of the rate of convergence is beyond the scope of this chapter. We refer to [14] for details.

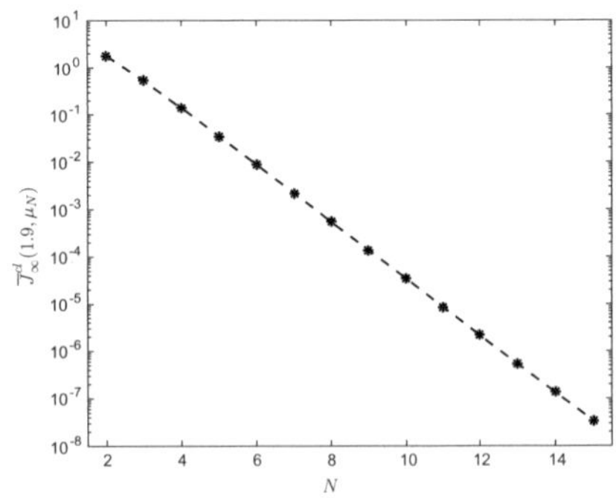

Fig. 8.5 Value of $\overline{J}^{cl}_\infty(x, \mu_N)$ for $x = 1.9$ without terminal conditions depending on N

8.6 Practical Asymptotic Stability Without Terminal Conditions

Now we turn to analyzing the stability properties of the NMPC closed-loop solutions without terminal conditions. As in the case with terminal conditions, our goal is to assume strict dissipativity and to use the optimal value function for the modified stage cost $\tilde{\ell}$ from (8.6) as a Lyapunov function, but now without imposing terminal conditions. The crucial difference is that while the problem with modified stage cost $\tilde{\ell}$ is still a stabilizing problem (hence the theory from Chap. 6 could be applied), without terminal conditions the optimal trajectories of the original and the modified problem no longer coincide.

In order to see why, we refer to the optimal control problem (OCP$_\mathrm{N}$) with stage cost $\tilde{\ell}$ as $(\widetilde{\mathrm{OCP}}_\mathrm{N})$ and, as before, denote the corresponding functional and the optimal value function by $\widetilde{J}_N$ and $\widetilde{V}_N$. Due to the fact that we no longer impose terminal conditions, the relations between V_N and $\widetilde{V}_N$ are not the same as in Sect. 8.3. For J_N and $\widetilde{J}_N$, instead of (8.8) we now have (8.13), which in the notation of this section reads

$$\widetilde{J}_N(x, u) = J_N(x, u) + \lambda(x) - \lambda(x_u(N, x)) - N\ell(x^e, u^e). \tag{8.20}$$

Unfortunately, in contrast to (8.8), this equation does not allow for an easy derivation of a relation between the optimal value functions of the form (8.9), because of the additional u-dependent term $\lambda(x_u(N, x))$ on the right-hand side of (8.20). A first consequence of this fact is that the continuity Assumption 8.24(b) does not immediately carry over to $\widetilde{V}_N$. Hence, we need to introduce this as an independent assumption.

Assumption 8.29 (*Continuity of $\widetilde{V}_N$ at x^e*) There exists $\gamma_{\widetilde{V}} \in \mathscr{K}_\infty$ such that for each $N \in \mathbb{N}$ and each $x \in \mathbb{X}$ it holds that

$$|\widetilde{V}_N(x) - \widetilde{V}_N(x^e)| \leq \gamma_{\widetilde{V}}(|x|_{x^e}).$$

In case strict dissipativity holds, $\tilde{\ell}$ is positive definite w.r.t. the equilibrium x^e, hence we obtain $\widetilde{V}_N(x^e) = 0$ and $\widetilde{V}_N(x) \geq 0$ for all $x \in \mathbb{X}$. Thus, the inequality in Assumption 8.29 is equivalent to $\widetilde{V}_N(x) \leq \gamma_{\widetilde{V}}(|x|_{x^e}) + \omega(N)$, which can be guaranteed using the conditions discussed in Chap. 6.

Unlike continuity, a straightforward check of Definition 8.7 (with storage function $\lambda \equiv 0$) shows that strict dissipativity carries over from (OCP$_\mathrm{N}$) to $(\widetilde{\mathrm{OCP}}_\mathrm{N})$, even with the same ρ. Thus, in particular, all the previous lemmas that apply to (OCP$_\mathrm{N}$) in case of strict dissipativity also apply to $(\widetilde{\mathrm{OCP}}_\mathrm{N})$. As a general rule, we denote all parameters, sets etc., referring to $(\widetilde{\mathrm{OCP}}_\mathrm{N})$ with a tilde, e.g., the set $\mathscr{Q}(x, u, N, P)$ from Proposition 8.15 will be denoted by $\widetilde{\mathscr{Q}}(x, u, N, P)$ when this proposition is applied to $(\widetilde{\mathrm{OCP}}_\mathrm{N})$.

As already mentioned above, from the definition we cannot directly deduce a simple relation like (8.9) between V_N and $\widetilde{V}_N$. The reason why we can still use $\widetilde{V}_N$ as an—at least practical—Lyapunov function lies in the fact that we can still establish an approximate version of (8.9). To this end, we first need the following preparatory lemma.

Lemma 8.30 *If Assumption 8.24 and the assumptions of Proposition 8.15 hold, then the equation*

$$V_N(x^e) = M\ell(x^e, u^e) + V_{N-M}(x^e) - R_3(x, P, N)$$

holds with $0 \le R_3(x, P, N) \le \gamma_V(\sigma_\delta(P)) + \omega(N-M) + \gamma_\lambda(\sigma_\delta(P))$ *for all* $N, P \in \mathbb{N}$ *and for all* $M \notin \mathcal{Q}(x, u_N^\star, N, P)$*, where* $u_N^\star \in \mathbb{U}^N(x^e)$ *is the optimal control of* (OCP$_\text{N}$) *for initial value* x^e *and* σ_δ *is from Proposition 8.15 with* $\delta = \gamma_V(|x|_{x^e}) + \omega(N)$.

Proof The inequality $V_N(x^e) \le M\ell(x^e, u^e) + V_{N-M}(x^e)$ follows from the dynamic programming principle (3.15) using the control $u \equiv u^e$. For the opposite inequality, consider the optimal control $u_N^\star \in \mathbb{U}^N(x^e)$ for initial value x^e. As in the proof of Lemma 8.25, we can apply Proposition 8.15 with $\delta = \gamma_V(|x|_{x^e}) + \omega(N)$ in order to conclude that for each $M \notin \mathcal{Q}(x, u_N^\star, N, P)$ we have

$$\begin{aligned}
V_N(x^e) &= \sum_{k=0}^{M-1} \ell(x_{u_N^\star}(k), u_N^\star(k)) + V_{N-M}(x_{u_N^\star}(M)) \\
&= -\lambda(x^e) + \lambda(x_{u_N^\star}(M)) + M\ell(x^e, u^e) + \sum_{k=0}^{M-1} \underbrace{\tilde{\ell}(x_{u_N^\star}(k), u_N^\star(k))}_{\ge 0} + V_{N-M}(x_{u_N^\star}(M)) \\
&\ge M\ell(x^e, u^e) + V_{N-M}(x^e) + \Big[V_{N-M}(x_{u_N^\star}(M)) - V_{N-M}(x^e)\Big] + \Big[\lambda(x_{u_N^\star}(M)) - \lambda(x^e)\Big] \\
&\ge M\ell(x^e, u^e) + V_{N-M}(x^e) - \gamma_V(\sigma_\delta(P)) - \omega(N-M) - \gamma_\lambda(\sigma_\delta(P)),
\end{aligned}$$

which shows the claim. □

Now we can prove the approximate relation of the form (8.9) between $\widetilde{V}_N$ and V_N.

Lemma 8.31 *If Assumptions 8.24 and 8.29 as well as the assumptions of Proposition 8.15 hold, then the equation*

$$\widetilde{V}_N(x) = V_N(x) + \lambda(x) - V_N(x^e) + R_4(x, N)$$

holds with $|R_4(x, N)| \le \nu_4(|x|_{x^e}, N)$ *with*

$$\begin{aligned}\nu_4(|x|_{x^e}, N) = \max\{&\gamma_V(\tilde{\sigma}_{\tilde{\delta}}(\lfloor N/3\rfloor)) + \gamma_V(\sigma_\delta(\lfloor N/3\rfloor)) + \gamma_{\widetilde{V}}(\tilde{\sigma}_{\tilde{\delta}}(\lfloor N/3\rfloor)) \\ &+ \gamma_\lambda(\sigma_\delta(\lfloor N/3\rfloor)) + \gamma_\lambda(\tilde{\sigma}_{\tilde{\delta}}(\lfloor N/3\rfloor)) + 3\omega(\lfloor N/3\rfloor), \\ &\gamma_{\widetilde{V}}(\sigma_\delta(\lfloor N/3\rfloor)) + \gamma_V(\sigma_\delta(\lfloor N/3\rfloor)) + \gamma_\lambda(\sigma_\delta(\lfloor N/3\rfloor)) \\ &+2\omega(\lfloor N/3\rfloor)\}\end{aligned}$$

with σ_δ *and* $\tilde{\sigma}_{\tilde{\delta}}$ *from Proposition 8.15 applied to* (OCP$_\text{N}$) *and* ($\widetilde{\text{OCP}}_\text{N}$)*, respectively, with* $\delta = \gamma_V(|x|_{x^e}) + \omega(N)$ *and* $\tilde{\delta} = \gamma_{\widetilde{V}}(|x|_{x^e})$.

Proof Fix $x \in \mathbb{X}$ and let $u_N^\star$ and $\tilde{u}_N^\star \in \mathbb{U}^N(x)$ denote the optimal control minimizing $J_N(x,u)$ and $\widetilde{J}_N(x,u)$, respectively. We note that if (OCP$_\text{N}$) is strictly dissipative then ($\widetilde{\text{OCP}}_\text{N}$) is strictly dissipative, too, with bounded storage function $\lambda \equiv 0$ and same $\rho \in \mathcal{K}_\infty$. Moreover, $V_N(x) \le N\ell(x^e,u^e) + \gamma_V(|x|_{x^e}) + \omega(N)$ and $\widetilde{V}_N(x) \le N\tilde{\ell}(x^e,u^e) + \gamma_{\widetilde{V}}(|x|_{x^e})$, since $V_N(x^e) \le N\ell(x^e,u^e)$ and $\widetilde{V}_N(x^e) = 0$. Hence, Proposition 8.15 applies to the optimal trajectories for both problems, yielding $\sigma_\delta \in \mathcal{L}$ and $\mathcal{Q}(x, u_N^\star, P, N)$ for (OCP$_\text{N}$) and $\tilde{\sigma}_{\tilde{\delta}}$ and $\widetilde{\mathcal{Q}}(x, \tilde{u}_N^\star, P, N)$ for ($\widetilde{\text{OCP}}_\text{N}$). For all $M \notin \widetilde{\mathcal{Q}}(x, \tilde{u}_N^\star, P, N)$, we can estimate

$$\begin{aligned}V_N(x) &\le J_M(x, \tilde{u}_N^\star) + V_{N-M}(x_{\tilde{u}_N^\star}(M)) \\ &\le J_M(x, \tilde{u}_N^\star) + V_{N-M}(x^e) + \gamma_V(\tilde{\sigma}_{\tilde{\delta}}(P)) + \omega(N-M) \\ &\le \widetilde{J}_M(x, \tilde{u}_N^\star) - \lambda(x) + \lambda(x^e) + M\ell(x^e,u^e) + V_{N-M}(x^e) + \gamma_V(\tilde{\sigma}_{\tilde{\delta}}(P)) \\ &\quad + \gamma_\lambda(\tilde{\sigma}_{\tilde{\delta}}(P)) + \omega(N-M) \\ &\le \widetilde{V}_N(x) - \widetilde{R}_1(x,P,N) - \lambda(x) \\ &\quad + V_N(x^e) + R_3(x,P,N) + \gamma_V(\tilde{\sigma}_{\tilde{\delta}}(P)) + \gamma_\lambda(\tilde{\sigma}_{\tilde{\delta}}(P)) + \omega(N-M),\end{aligned}$$

where we have applied the dynamic programming principle (3.15) in the first inequality, Proposition 8.15 for ($\widetilde{\text{OCP}}_\text{N}$) and Assumption 8.24(b), respectively, Assumption 8.24(a) and (8.20) in the second and third inequality and Lemma 8.25 (applied to ($\widetilde{\text{OCP}}_\text{N}$), hence with remainder term denoted by $\widetilde{R}_1$) and Lemma 8.30 (applied to (OCP$_\text{N}$)) in the last step. Moreover, $\lambda(x^e) = 0$ and $\widetilde{V}_N(x^e) = 0$ were used.

By exchanging the two optimal control problems and using the same inequalities as above, we get

$$\begin{aligned}\widetilde{V}_N(x) &\le V_N(x) - R_1(x,P,N) + \lambda(x) - V_N(x^e) + \gamma_{\widetilde{V}}(\sigma_\delta(P)) + \gamma_\lambda(\sigma_\delta(P)) \\ &\quad + \omega(N-M)\end{aligned}$$

for all $M \notin \mathcal{Q}(x, u_N^\star, P, N)$. Here we can omit the negative $-R_3$-term. Now, choosing $P = \lfloor N/3\rfloor$, the union $\mathcal{Q}(x, \tilde{u}_N^\star, P, N) \cup \mathcal{Q}(x, u_N^\star, P, N)$ has at most $2N/3$ elements, hence there exists $M \le 2N/3$ for which both inequalities hold. This yields $N - M \ge \lfloor N/3\rfloor$ and thus

$$
\begin{aligned}
|R_1(x,P,N)| &\le \gamma_V(\sigma_\delta(\lfloor N/3\rfloor)) + \omega(\lfloor N/3\rfloor),\\
|\widetilde{R}_1(x,P,N)| &\le \gamma_{\widetilde{V}}(\tilde{\sigma}_{\tilde{\delta}}(\lfloor N/3\rfloor)) + \omega(\lfloor N/3\rfloor) \text{ and}\\
R_3(x,P,N) &\le \gamma_V(\sigma_\delta(\lfloor N/3\rfloor)) + \omega(\lfloor N/3\rfloor) + \gamma_\lambda(\sigma_\delta(\lfloor N/3\rfloor)),
\end{aligned}
$$

which shows the claim. □

The following proposition shows in which sense $\widetilde{V}_N$ is a Lyapunov function for the system. This will be used in the subsequent theorem in order to prove semiglobal practical asymptotic stability of the closed loop.

Proposition 8.32 *Consider the NMPC Algorithm 3.1 with strictly dissipative optimal control problem* (OCP$_\mathrm{N}$) *with bounded storage function λ and $\rho \in \mathscr{K}_\infty$ and let Assumptions 8.24 and 8.29 hold. Then for each $\Theta > 0$ there exists $\delta_1 \in \mathscr{L}$ such that for all $N \ge 2$ the optimal value function $\widetilde{V}_N$ of* ($\widetilde{\mathrm{OCP}}_\mathrm{N}$) *is a Lyapunov function for the closed loop on $S = Y \setminus \mathbb{P}$ for the forward invariant sets $Y = \widetilde{V}_N^{-1}([0,\Theta])$ and $\mathbb{P} = \widetilde{V}_N^{-1}([0,\delta_1(N)])$.*

Proof We have to check that Definition 2.18 is satisfied and that Y and $\mathbb{P}$ are forward invariant. The lower bound in (2.37) follows with $\alpha_1 = \rho$ because strict dissipativity implies $\tilde{\ell}(x,u) \ge \rho(|x|_{x^e})$, and thus

$$
\widetilde{V}_N(x) = \inf_{u\in\mathbb{U}^N(x)} \sum_{k=0}^{N-1} \tilde{\ell}(x_u(k,x),u(k)) \ge \inf_{u\in\mathbb{U}^N(x)} \sum_{k=0}^{N-1} \rho(|x_u(k,x)|_{x^e}) \ge \rho(|x|_{x^e}).
$$

The upper bound in (2.37) follows from Assumption 8.29 and $\widetilde{V}_N(x^e) = 0$ with $\alpha_2 = \gamma_{\widetilde{V}}$.

In order to obtain inequality (2.38), we abbreviate $x^+ = f(x,\mu_N(x))$. Now, for all $x \in Y$ we obtain $\widetilde{V}_N(x) \le \Theta$, which implies $|x|_{x^e} \le \rho^{-1}(\Theta)$. In order to obtain a similar estimate for $|x^+|_{x^e}$, we observe that $\widetilde{V}_N(x) \le \Theta$ implies $V_N(x) \le \Theta - \lambda(x) + M + N\ell(x^e,u^e)$, where $M > 0$ denotes a bound on λ. Thus, Theorem 3.17 and strict dissipativity yield

$$
\begin{aligned}
V_{N-1}(x^+) = V_N(x) - \ell(x,\mu_N(x)) &\le V_N(x) + \lambda(x) - \lambda(x^+) - \ell(x^e,u^e)\\
&\le \Theta - \lambda(x^+) + M + (N-1)\ell(x^e,u^e).
\end{aligned}
$$

This implies

$$
\widetilde{V}_{N-1}(x^+) \le V_{N-1}(x^+) + \lambda(x^+) + M - (N-1)\ell(x^e,u^e) \le \Theta + 2M
$$

and we can conclude that $|x^+|_{x^e} \le \rho^{-1}(\Theta + 2M)$. Hence, we can compute

$$\begin{aligned}
\widetilde{V}_N(x^+) &= V_N(x^+) + \lambda(x^+) - V_N(x^e) + R_4(x^+, N) \\
&= V_{N-1}(x^+) + \ell(x^e, u^e) + \lambda(x^+) - V_N(x^e) + R_2(x^+, N) + R_4(x^+, N) \\
&= V_N(x) - \ell(x, \mu_N(x)) + \ell(x^e, u^e) + \lambda(x^+) - V_N(x^e) \\
&\qquad + R_2(x^+, N) + R_4(x^+, N) \\
&= \widetilde{V}_N(x) \underbrace{-\ell(x, \mu_N(x)) + \ell(x^e, u^e) + \lambda(x^+) - \lambda(x)}_{=-\tilde{\ell}(x,\mu_N(x))} \\
&\qquad + R_2(x^+, N) + R_4(x^+, N) - R_4(x, N).
\end{aligned}$$

where we used Lemma 8.31 for $x = x^+$ for the first equality, Lemma 8.26 for the second, Eq. (3.20) for the third and Lemma 8.31 in the last step. Defining $\nu(N) = \nu_2(\rho^{-1}(\Theta + 2M), N) + 2\nu_4(\rho^{-1}(\Theta + 2M), N)$ with ν_2 and ν_4 from Lemmas 8.26 and 8.31, respectively, we thus obtain

$$\widetilde{V}_N(x^+) \le \widetilde{V}_N(x) - \rho(|x|_{x^e}) + \nu(N) \le \widetilde{V}_N(x) - \chi(\widetilde{V}_N(x)) + \nu(N) \qquad (8.21)$$

for $\chi := \rho \circ \alpha_2^{-1}(r)$. Now we set $\delta_1(N) = \max\{\chi^{-1}(2\nu(N)),\ \chi^{-1}(\nu(N)) + \nu(N)\}$. Then for all $x \in S = Y \setminus \mathbb{P}$ we obtain $\widetilde{V}_N(x) \ge \delta_1(N)$ and thus $\chi(\widetilde{V}_N(x)) \ge 2\nu(N)$, which implies

$$\widetilde{V}_N(x^+) \le \widetilde{V}_N(x) - \chi(\widetilde{V}_N(x))/2 \le \widetilde{V}_N(x) - \chi(\alpha_1(|x|_{x^e}))/2$$

and thus (2.38) with $\alpha_V(r) = \chi(\alpha_1(r))/2$. This inequality also shows forward invariance of Y since $x \in Y$ implies $\widetilde{V}_N(x) \le \Theta$, hence $\widetilde{V}_N(x^+) < \widetilde{V}_N(x) \le \Theta$ and thus $x^+ \in Y$.

Finally to prove forward invariance of $\mathbb{P}$, we recall that $x \in \mathbb{P}$ if and only if $\widetilde{V}_N(x) \le \delta_1(N)$. Now we pick $x \in \mathbb{P}$ and distinguish two cases.

Case 1: $\chi(\widetilde{V}_N(x)) \ge \nu(N)$. In this case from (8.21), we obtain

$$\widetilde{V}_N(x^+) \le \widetilde{V}_N(x) - \chi(\widetilde{V}_N(x)) + \nu(N) \le \widetilde{V}_N(x) \le \delta_1(N).$$

Case 2: $\chi(\widetilde{V}_N(x)) < \nu(N)$. In this case from (8.21) we obtain

$$\begin{aligned}
\widetilde{V}_N(x^+) &\le \widetilde{V}_N(x) - \chi(\widetilde{V}_N(x)) + \nu(N) \ \le\ \widetilde{V}_N(x) + \nu(N) \\
&< \chi^{-1}(\nu(N)) + \nu(N) \ \le\ \delta_1(N).
\end{aligned}$$

Hence, in both cases we get $\widetilde{V}_N(x^+) \le \delta_1(N)$ and thus $x^+ \in \mathbb{P}$, which proves the forward invariance of $\mathbb{P}$. □

We note that for small values of N the inequality $\delta_1(N) \ge \Theta$ may hold, in which case the set S on which $\widetilde{V}_N$ is a Lyapunov function is empty.

The final theorem on practical asymptotic stability is now an easy consequence of Proposition 8.32. To this end, recall the notion of semiglobal practical stability from Definition 6.33(ii).

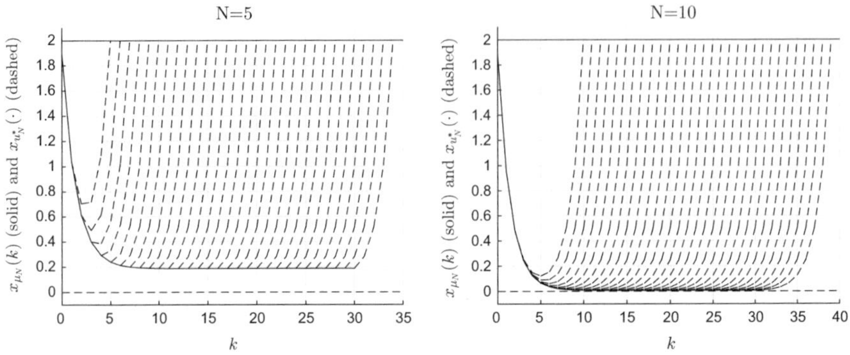

Fig. 8.6 NMPC closed-loop solution (*solid*) and open-loop predictions (*dashed*) for Example 8.1 without terminal conditions and horizon $N = 5$ (left) and $N = 10$ (*right*). The *solid line* at $x = 2$ indicates the *upper* bound of the admissible set $\mathbb{X}$

Theorem 8.33 *Consider the NMPC Algorithm 3.1 with strictly dissipative optimal control problem* (OCP$_\text{N}$) *with bounded storage function* λ *and* $\rho \in \mathcal{K}_\infty$ *and let Assumptions 8.24 and 8.29 hold. Then the equilibrium* x^e *is semiglobally practically asymptotically stable on* $\mathbb{X}$ *with respect to the optimization horizon* N.

Proof Fixing $\Delta > \delta > 0$, the assertion follows immediately from Proposition 8.32, Theorem 2.20 and Lemma 6.34 when choosing $\Theta = \alpha_2(\Delta)$ (implying $\overline{\mathcal{B}}_\Delta(x^e) \subset Y$) and $N_{\delta,\Delta} > 0$ so large that $\rho(\delta(N_{\delta,\Delta})) \leq \delta$ holds for δ from Definition 6.33(ii) and $\delta(N)$ from Proposition 8.32 (implying $\mathbb{P} \subset \overline{\mathcal{B}}_\delta(x^e)$). □

We remark that Theorem 8.33 provides an extension of Theorem 6.37 to the economic case. Indeed, the assumptions of Theorem 6.37 imply those of Theorem 8.33 (particularly, strict dissipativity holds with $\lambda \equiv 0$ and $\rho = \alpha_3$) and the assertions of the two theorems are identical.

Example 8.34 Figure 8.6 shows the trajectories (open-loop dashed, NMPC closed-loop solid) of Example 8.1 without terminal conditions for $N = 5$ and $N = 10$. One clearly sees the practical asymptotic stability of the closed loop and the turnpike phenomenon for the open-loop trajectories.

8.7 Non-averaged and Transient Performance Without Terminal Conditions

Our final results in this chapter concern the adaptation of the results from Sect. 8.4 to the case without terminal conditions. In order to generalize Theorem 8.21, we need the following continuity assumption on the infinite horizon optimal value function and the two subsequent auxiliary results.

Assumption 8.35 (*Continuity of V_∞ at x^e*) There exists $\gamma_{V_\infty} \in \mathscr{K}_\infty$ such that for each $x \in \mathbb{X}$ it holds that

$$|V_\infty(x) - V_\infty(x^e)| \le \gamma_{V_\infty}(|x|_{x^e}).$$

Lemma 8.36 *If Assumption 8.35 and the assumptions of Proposition 8.18 hold, then the equation*

$$V_\infty(x) = J_M(x, u^\star_\infty) + V_\infty(x^e) + R_5(x, M) \tag{8.22}$$

holds with $|R_5(x, M)| \le \gamma_{V_\infty}(\sigma_\infty(P))$ for all $x \in \mathbb{X}$, all $P \in \mathbb{N}$ and all $M \notin \mathscr{Q}(x, u^\star_\infty, P, \infty)$, where $u^\star_\infty \in \mathbb{U}^\infty(x)$ denotes the infinite horizon optimal control for initial value x and σ_∞ is from Proposition 8.18.

Proof The dynamic programming principle (4.6) yields

$$V_\infty(x) = J_M(x, u^\star_\infty) + V_\infty(x_{u^\star_\infty}(M, x)).$$

Hence, (8.22) holds with $R_5(x, M) = V_\infty(x_{u^\star_\infty}(M, x)) - V_\infty(x^e)$. Then for any $P \in \mathbb{N}$ and $M \notin \mathscr{Q}(x, u^\star_\infty, P, \infty)$ we obtain $|R_5(x, M)| \le \gamma_{V_\infty}(\|x_{u^\star_\infty}(M, x) - x^e\|) \le \gamma_{V_\infty}(\sigma_\infty(P))$ and thus the assertion. □

Lemma 8.37 *If Assumptions 8.24 and 8.35 and the assumptions of Propositions 8.15 and 8.18 hold, then the equation*

$$J_M(x, u^\star_\infty) = J_M(x, u^\star_N) + R_6(x, M, N) \tag{8.23}$$

holds with $|R_6(x, M, N)| \le \max\{\gamma_V(\sigma_\delta(P)) + \gamma_V(\sigma_\infty(P)) + 2\omega(N - M), \gamma_{V_\infty}(\sigma_\infty(P)) + \gamma_{V_\infty}(\sigma_\delta(P))\}$ for all $P \in \mathbb{N}$, all $x \in \mathbb{X}$ and all $M \in \{0, \dots, N\} \setminus (\mathscr{Q}(x, u^\star_N, P, N) \cup \mathscr{Q}(x, u^\star_\infty, P, \infty))$, with σ_∞ from Proposition 8.18 and σ_δ from Proposition 8.15 with $\delta = |x|_{x^e}$.

Proof The finite horizon dynamic programming principle (3.15), (3.16) implies that $u = u^\star_N$ minimizes the expression $J_M(x, u) + V_{N-M}(x_u(M, x))$. Together with the error term R_1 from Lemma 8.25 and $\widehat{R}_1(x, M, N) = V_{N-M}(x_{u^\star_\infty}(M, x)) - V_{N-M}(x^e)$ this yields

$$\begin{aligned}
J_M(x, u^\star_N) + V_{N-M}(x^e) &= J_M(x, u^\star_N) + V_{N-M}(x_{u^\star_N}(M, x)) - R_1(x, M, N) \\
&\le J_M(x, u^\star_\infty) + V_{N-M}(x_{u^\star_\infty}(M, x)) - R_1(x, M, N) \\
&= J_M(x, u^\star_\infty) + V_{N-M}(x^e) - R_1(x, M, N) + \widehat{R}_1(x, M, N).
\end{aligned}$$

Similar to the proof of Lemma 8.25 one sees that $|\widehat{R}_1(x, M, N)| \le \gamma_V(\sigma_\infty(P)) + \omega(N - M)$ for all $M \notin \mathscr{Q}(x, u^\star_\infty, P, \infty)$.

Conversely, the infinite horizon dynamic programming principle (4.6) implies that $u^\star_\infty$ minimizes the expression $J_M(x, u^\star_\infty) + V_\infty(x_{u^\star_\infty}(M, x))$. Using the error terms R_5 from Lemma 8.36 and $\widehat{R}_5(x, M, N) = V_\infty(x_{u^\star_N}(M, x)) - V_\infty(x^e)$ we obtain

$$\begin{aligned} J_M(x, u^\star_\infty) + V_\infty(x^e) &= J_M(x, u^\star_\infty) + V_\infty(x_{u^\star_\infty}(M, x)) - R_5(x, M) \\ &\le J_M(x, u^\star_N) + V_\infty(x_{u^\star_N}(M, x)) - R_5(x, M) \\ &= J_M(x, u^\star_N) + V_\infty(x^e) - R_5(x, M) + \widehat{R}_5(x, M, N). \end{aligned}$$

As in the proof of Lemma 8.25 one sees that Proposition 8.15 applies to $x_{u^\star}(\cdot, x)$ with $\delta = \gamma_V(|x|_{x^e})$. Hence, similar to the proof of Lemma 8.36 one obtains $|\widehat{R}_5(x, M, N)| \le \gamma_{V_\infty}(\sigma_\delta(P))$ for all $M \notin \mathcal{Q}(x, u^\star_N, P, N)$. Together with the estimates for R_1 and R_5 from Lemmas 8.25 and 8.36 this yields

$$\begin{aligned} |R_6(x, M, N)| &= |J_M(x, u^\star_\infty) - J_M(x, u^\star_N)| \\ &\le \max\{|R_1(x, M, N)| + |\widehat{R}_1(x, M, N)|, |R_5(x, M)| + |\widehat{R}_5(x, M, N)|\} \\ &\le \max\{\gamma_V(\sigma_\delta(P)) + \gamma_V(\sigma_\infty(P)) + 2\omega(N - M), \gamma_{V_\infty}(\sigma_\infty(P)) + \gamma_{V_\infty}(\sigma_\delta(P))\} \end{aligned}$$

and thus the claim. □

Now we can establish a version of Theorem 8.21 for economic NMPC without terminal conditions. We will discuss after the proof how Theorem 8.38 relates to Theorem 8.21.

Theorem 8.38 *Consider the NMPC Algorithm 3.1 with strictly dissipative optimal control problem* (OCP$_\mathrm{N}$) *with bounded storage function* λ, *assume that* $\ell(x^e, u^e) = 0$ *and* $\mathbb{X}$ *is bounded and let Assumptions 8.24 and 8.35 hold. Then the inequality*

$$J^{cl}_K(x, \mu_N) + V_\infty(x_{\mu_N}(K)) \le V_\infty(x) + K\delta_1(N) \tag{8.24}$$

holds for all $K \in \mathbb{N}$ *and all sufficiently large* $N \in \mathbb{N}$ *with*

$$\begin{aligned} \delta_1(N) := {} & 2\gamma_V(\sigma_\delta(\lfloor (N-1)/8 \rfloor)) + 2\gamma_V(\sigma_\infty(\lfloor (N-1)/8 \rfloor)) \\ & + 2\gamma_{V_\infty}(\sigma_\delta(\lfloor (N-1)/8 \rfloor)) + 4\gamma_{V_\infty}(\sigma_\infty(\lfloor (N-1)/8 \rfloor)) + 4\omega(\lfloor N/2 \rfloor) \end{aligned}$$

with σ_∞ *from Proposition 8.18 and* σ_δ *from Proposition 8.15 with* $\delta = \sup_{x\in\mathbb{X}} |x|_{x^e}$.

Proof We pick $x \in \mathbb{X}$ and abbreviate $x^+ := f(x, \mu_N(x))$. For the corresponding optimal control $u^\star_N$ Corollary 3.16 yields that $u^\star_N(\cdot + 1)$ is an optimal control for initial value x^+ and horizon $N - 1$. Hence, for each $M \in \{1, \ldots, N\}$ we obtain

$$\begin{aligned} \ell(x, \mu_N(x)) &= V_N(x) - V_{N-1}(x^+) = J_N(x, u^\star_N) - J_{N-1}(x^+, u^\star_N(\cdot + 1)) \\ &= J_M(x, u^\star_N) - J_{M-1}(x^+, u^\star_N(\cdot + 1)), \end{aligned}$$

where the last equality follows from the fact that the omitted terms in the sums defining $J_M(x, u_N^\star)$ and $J_{M-1}(x^+, u_N^\star(\cdot + 1))$ coincide. Using Lemma 8.36 for N, x and M and for $N-1, x^+$ and $M-1$, respectively, yields

$$\begin{aligned} V_\infty(x) - V_\infty(x^+) &= J_M(x, u_\infty^\star) + V_\infty(x^e) + R_6(x, M) \\ &\qquad - J_{M-1}(x^+, u_\infty^\star) - V_\infty(x^e) - R_5(x^+, M-1) \\ &= J_M(x, u_\infty^\star) - J_{M-1}(x^+, u_\infty^\star) + R_5(x, M) - R_5(x^+, M-1). \end{aligned}$$

Putting the two equations together and using Lemma 8.37 yields

$$\ell(x, \mu_N(x)) = V_\infty(x) - V_\infty(x^+) + R_7(x, M, N). \tag{8.25}$$

with

$$R_7(x, M, N) = -R_6(x, M, N) + R_6(x^+, M-1, N-1) - R_5(x, M) + R_5(x^+, M-1).$$

From Lemmas 8.36 and 8.37, we obtain the bound

$$\begin{aligned} |R_7(x, M, N)| \le\ & 2\gamma_V(\sigma_\delta(P)) + 2\gamma_V(\sigma_\infty(P)) + 2\gamma_{V_\infty}(\sigma_\delta(P)) + 4\gamma_{V_\infty}(\sigma_\infty(P)) \\ & + 4\omega(N - M) \end{aligned}$$

provided we choose $M \in \{1, \ldots, N\}$ with $M \notin \mathcal{Q}(x, u_N^\star, P, N) \cup \mathcal{Q}(x, u_\infty^\star, P, \infty)$ and $M - 1 \notin \mathcal{Q}(x^+, u_N^\star(\cdot + 1), P, N-1) \cup \mathcal{Q}(x^+, u_\infty^\star(\cdot + 1), P, \infty)$. Since each of the four $\mathcal{Q}$ sets contains at most P elements, their union contains at most $4P$ elements and hence if $N > 8P$ then there is at least one such M with $M \le N/2$.

Thus, choosing $P = \lfloor (N-1)/8 \rfloor$ yields the existence of $M \le N/2$ such that

$$|R_7(x, M, N)| \le \delta_1(N). \tag{8.26}$$

Applying (8.25), (8.26) for $x = x_{\mu_N}(k, x)$, $k = 0, \ldots, K-1$, we can conclude

$$\begin{aligned} J_K^{cl}(x, \mu_N) &= \sum_{k=0}^{K-1} \ell(x_{\mu_N}(k, x), \mu_N(x_{\mu_N}(k, x))) \\ &\le \sum_{k=0}^{K-1} \Big(V_\infty(x_{\mu_N}(k, x)) - V_\infty(x_{\mu_N}(k+1, x)) + \delta_1(N) \Big) \\ &\le V_\infty(x) - V_\infty(x_{\mu_N}(K, x)) + K\delta_1(N). \end{aligned}$$

This proves the claim. □

The interpretation of (8.24) is as follows. If we follow the NMPC closed-loop trajectory up to some time K and then continue by using the infinite horizon optimal trajectory starting at $x_{\mu_N}(K, x)$, then the value of the overall trajectory exceeds the infinite horizon optimal value by at most $K\delta(N)$. Although seemingly different, it is indeed closely related to Theorem 8.21 because of the following fact: the inequality from Theorem 8.21 holds for all $x \in \mathbb{X}$ if and only if

$$J_K^{cl}(x, \mu_N) + V_\infty(x_{\mu_N}(K, x)) \leq V_\infty(x) + \delta_1(N) \tag{8.27}$$

holds for all $x \in \mathbb{X}$ and all $K \geq 1$. This is because $J_K^{cl}(x, \mu_N) + V_\infty(x_{\mu_N}(K, x)) \leq J_\infty^{cl}(x, \mu_N)$ for all $K \geq 1$, hence Theorem 8.21 implies (8.27). Conversely, since the assumptions of Theorem 8.21 imply $V_\infty(x_{\mu_N}(K, x)) \to V_\infty(x^e) = 0$ for $K \to \infty$, the validity of (8.27) for all $K \geq 1$ implies the inequality from Theorem 8.21 by letting $K \to \infty$. Comparing (8.24) with (8.27) one immediately sees the difference between the case with and without terminal conditions: without terminal conditions we get the additional factor K in front of the error term, which implies that for large K the error may increase and that for $K \to \infty$ and fixed N the solution may be far from optimal. A numerical illustration of this effect can be found in Example 8.40(iii), below. However, note that the estimate from Theorem 8.27 shows that the averaged value still behaves well for $K \to \infty$, hence the behavior of the trajectories cannot completely deteriorate.

Finally, we formulate and prove the counterpart of Theorem 8.22 for the case without terminal conditions. To this end, recall the definition of $\mathbb{U}_\kappa^K(x)$ from (8.15).

Theorem 8.39 *Consider the NMPC Algorithm 3.10 with strictly dissipative optimal control problem* (OCP$_{\text{N,e}}$) *with bounded storage function* λ *and* $\rho \in \mathcal{K}_\infty$, *let* $\mathbb{X}$ *be bounded and let Assumptions 8.24 and 8.29 hold. Then there exist* $\delta_1, \delta_2, \delta_3 \in \mathcal{L}$ *such that for all* $x \in \mathbb{X}_N$ *the inequality*

$$J_K^{cl}(x, \mu_N) \leq \inf_{u \in \mathbb{U}_\kappa^K(x)} J_K(x, u) + \delta_1(N) + K\delta_2(N) + \delta_3(K)$$

holds with $\kappa = \max\{\beta(|x|_{x^e}, K), \rho^{-1}(\delta(N))\}$, *with* β *and* δ *from Theorem 8.33.*

Proof First observe that the assumptions of this theorem include those of Theorem 8.33. Hence, from the proof of Theorem 8.33 we obtain the identity

$$\tilde{\ell}(x, \mu_N(x)) = \widetilde{V}_N(x) - \widetilde{V}_N(f(x, \mu_N(x))) + R_2(x, N) + R_4(f(x, \mu_N(x), N) + R_4(x(N))$$

with $|R_2(x, N) + R_4(f(x, \mu_N(x), N) + R_4(x(N))| \leq \nu_2(a, N) + 2\nu_4(a, N) =: \delta_2(N)$, with ν_2 and ν_4 from Lemmas 8.26 and 8.31, respectively, and $a = \sup_{x\in\mathbb{X}} |x|_{x^e}$. Summing this cost along the closed-loop trajectory yields

$$\sum_{k=0}^{K-1} \tilde{\ell}(x_{\mu_N}(k, x), \mu_N(x_{\mu_N}(k, x))) \leq \widetilde{V}_N(x) - \widetilde{V}_N(x_{\mu_N}(K)) + K\delta_2(N). \tag{8.28}$$

Now the dynamic programming principle (3.15) and Assumption 8.29 yield for all $K \in \{1, \ldots, N\}$ and all $u \in \mathbb{U}_\kappa^K(x)$

$$\widetilde{J}_K(x,u) = \underbrace{\widetilde{J}_K(x,u) + \widetilde{V}_{N-K}(x_u(K,x))}_{\geq \widetilde{V}_N(x)} - \underbrace{\widetilde{V}_{N-K}(x_u(K,x))}_{\leq \gamma_{\widetilde{V}}(\kappa)} \geq \widetilde{V}_N(x) - \gamma_{\widetilde{V}}(\kappa). \tag{8.29}$$

Due to the non-negativity of $\tilde{\ell}$, for $K \geq N$ we get $\widetilde{J}_K(x,u) \geq \widetilde{V}_N(x)$ for all $u \in \mathbb{U}^K(x)$. Hence (8.29) holds for all $K \in \mathbb{N}$. Moreover, we have $\widetilde{V}_N(x) \geq 0$. Using (8.28), (8.29) and (8.13) and the definition of δ_2, for all $u \in \mathbb{U}^K_\kappa(x)$ we obtain

$$\begin{aligned}
J_K^{cl}(x,\mu_N(x)) &= \sum_{k=0}^{K-1} \tilde{\ell}(x_{\mu_N}(k,x), \mu_N(x_{\mu_N}(k,x))) - \lambda(x) + \lambda(x_{\mu_N}(K,x)) \\
&\leq \widetilde{V}_N(x) - \widetilde{V}_N(x_{\mu_N}(K,x)) + K\delta_2(N) - \lambda(x) + \lambda(x_{\mu_N}(K,x)) \\
&\leq \widetilde{J}_K(x,u) + \gamma_{\widetilde{V}}(\kappa) - \widetilde{V}_N(x_{\mu_N}(K,x)) + K\delta_2(N) - \lambda(x) + \lambda(x_{\mu_N}(K,x)) \\
&= J_K(x,u) + \gamma_{\widetilde{V}}(\kappa) - \widetilde{V}_N(x_{\mu_N}(K,x)) + K\delta_2(N) - \lambda(x_u(K,x)) + \lambda(x_{\mu_N}(K,x)) \\
&\leq J_K(x,u) + \gamma_{\widetilde{V}}(\kappa) + K\delta_2(N) + 2\gamma_\lambda(\kappa).
\end{aligned}$$

Using the definition of κ we can estimate and define

$$\begin{aligned}
\gamma_{\widetilde{V}}(\kappa) + 2\gamma_\lambda(\kappa) &\leq \underbrace{\sup_{x\in\mathbb{X}} \gamma_{\widetilde{V}}(\beta(|x|_{x^e}, K)) + 2\gamma_\lambda(\beta(|x|_{x^e}, K))}_{=:\delta_3(K)} \\
&\quad + \underbrace{\gamma_{\widetilde{V}}(\rho^{-1}(\delta(N))) + 2\gamma_\lambda(\rho^{-1}(\delta(N)))}_{=:\delta_1(N)},
\end{aligned}$$

which finishes the proof. □

Example 8.40 (i) Figure 8.7 illustrates how $J_K^{cl}(x,\mu_N)$ depends on N in Example 8.1. As in Fig. 8.4, the value $K = 30$ is so large that the effect of the term $\delta_2(K)$ is negligible and not visible in the figure, hence $J_K^{cl}(x,\mu_N)$ converges to $\inf_{u\in\mathbb{U}^K_\kappa(x)} J_K^{uc}(x,u)$ for increasing N.

(ii) We note that the error estimates in Theorems 8.38 and 8.39 depend on the lower bound on the storage function λ, which enters in several of the previous estimates.

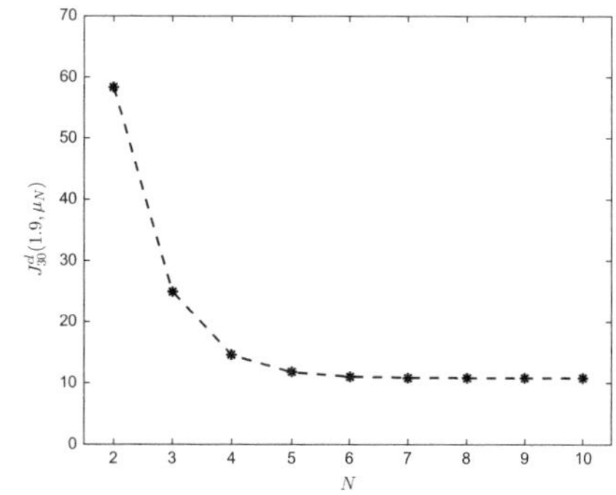

Fig. 8.7 Value of $J_K^{cl}(x,\mu_N)$ for $K = 30$, $x = 1.9$ and varying N without terminal conditions

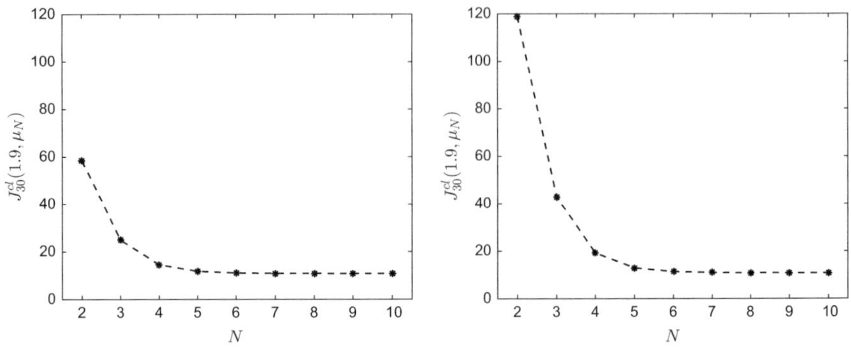

Fig. 8.8 Value of $J_K^{cl}(x, \mu_N)$ for $K = 30$, $x = 1.9$ and varying N without terminal conditions for $\mathbb{X} = [-2, 2]$ on the *left* and $\mathbb{X} = [-3, 3]$ on the *right*

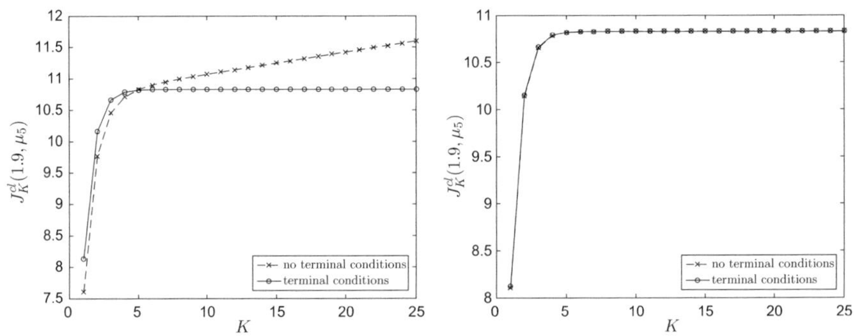

Fig. 8.9 Value of $J_K^{cl}(x, \mu_N)$ for varying K, $x = 1.9$ and $N = 5$ on the *left* and $N = 10$ on the *right*, both with and without terminal conditions $\mathbb{X}_0 = \{0\}$ and $F \equiv 0$

This dependence is actually visible when computing $J_K^{cl}(x, \mu_N)$ via numerical simulations. In Example 8.1, the bound on λ increases with increasing $\mathbb{X}$ (cf. Example 8.8). Figure 8.8 shows that increasing the state constraint set from $\mathbb{X} = [-2, 2]$ to $\mathbb{X} = [-3, 3]$ indeed considerably increases the error, although the optimal trajectories and thus the limiting values for $J_K^{cl}(x, \mu_N)$ for $N \to \infty$ are independent of the choice of $\mathbb{X}$.

(iii) Finally, we observe that the main structural difference between Theorems 8.22 and 8.39 lies in the factor K in the error estimate in Theorem 8.39 without terminal conditions. This predicts a deterioration of the value $J_K^{cl}(x, \mu_N)$ for fixed N and growing K in the case without terminal conditions, which should not appear if terminal conditions are used. This effect can again be seen in numerical simulations for Example 8.1, see Fig. 8.9. Here the increase of $J_K^{cl}(x, \mu_N)$ for increasing K is clearly visible in the left figure, i.e., for $N = 5$. In the right figure, N has been increased to $N = 10$, due to which the $\delta_2(N)$-term in Theorem 8.39 becomes so small that its effect is not visible anymore for the range of K depicted in the figure.

8.8 Notes and Extensions

The first paper to analyze economic NPMC in the sense of this chapter is—to our knowledge—the reference [2] by Angeli, Amrit, and Rawlings from 2009, although special cases for this setting were discussed before in Rawlings, Bonné, Jørgensen, Venkat and Jørgensen [21], and in Rawlings and Amrit [20]. The first property to be established was the averaged infinite horizon optimality under equilibrium terminal constraints. The observation that strict dissipativity is the "right" property in order to ensure asymptotic stability was first made by Diehl, Amrit and Rawlings in [5], where strict duality, i.e., strict dissipativity with a linear storage function, was used. The extension to the nonlinear version of strict dissipativity was then made by Angeli and Rawlings in [4]. The terminal cost of the form of Assumption 8.5 was introduced by Amrit, Rawlings and Angeli in [1] and a consolidated presentation of asymptotic stability and averaged optimality under terminal conditions was given by the same authors in 2012 in [3]. The non-averaged performance result under terminal conditions was finally developed by Grüne and Panin in 2015 [13]. The corresponding results without terminal conditions were developed in 2013 in Grüne [10] (averaged performance) and in 2014 by Grüne and Stieler [14] (asymptotic stability and non-averaged performance).

The results described in this chapter can be extended and modified in various ways. There is a huge body of literature dealing with variants of economic NMPC that combine an economic optimization objective and additional objectives or constraints that ensure stability of the closed loop. For an overview on these approaches, we refer to the survey by Ellis, Duranda and Christofides [7]. In the context of average performance criteria, it appears natural to also consider averaged constraints, which can, for instance, be used to provide a bound on the average energy consumption during the runtime of a process. Results in this direction can be found, e.g., in Müller, Angeli and Allgöwer [16]. In general, it is by no means clear that a system is necessarily optimally operated at an equilibrium. Hence, there have been various attempts to generalize this assumption, for instance to optimal periodic orbits, see, e.g., Angeli, Amrit and Rawlings [2] or Müller and Grüne [18] or Zanon et al. [27]. Finally, some of the results in this chapter have also been transferred to the continuous time setting, see, e.g., Faulwasser et al. [9] and [8].

The fact that (strict) dissipativity turned out to be an important property for ensuring averaged optimality and asymptotic stability (and later also non-averaged performance estimates) has led to a renewed research activity on this classical systems theoretic property, which was introduced by Willems in 1971/1972 [23–25]. Under appropriate controllability conditions, several relations to properties considered in this chapter have been obtained. Particularly, remarkable results are that dissipativity is a necessary and sufficient condition for optimal operation at steady state (8.5), see Müller, Angeli and Allgöwer [17], and that strict dissipativity is a necessary and sufficient condition for the turnpike property described by Proposition 8.15, see Grüne and Müller [12]. Despite the importance of (strict) dissipativity, it is remarkable that parts of the results in this chapter can also be established without this property. This is obtained by directly using the turnpike properties from Propositions 8.15 and 8.18, see Grüne [11].

The turnpike property as described by these propositions is in fact much more than merely an auxiliary property for analyzing NMPC schemes. It has been observed and analyzed in mathematical economy for more than 70 years, starting from von Neumann [19], via Dorfman, Samuelson and Solow [6], and McKenzie [15] to recent surveys, e.g., by Trélat and Zuazua [22] and a monograph by Zaslavski [26]. We refer to these references for further information.

Problems

1. Assume that the stage cost $\ell : X \times U \to \mathbb{R}$ is continuous and μ is a feedback law for which the closed-loop solution satisfies $x_\mu(k) \to x^e$ and $\mu(x_\mu(k)) \to u^e$ as $k \to \infty$. Prove that under these conditions

$$\overline{J}^{cl}_\infty(x, \mu) = \ell(x^e, u^e)$$

holds. Can you also make an statement about $J^{cl}_K(x, \mu)$ under these conditions?
2. Verify that the assumptions of Theorem 6.37 imply those of Theorem 8.33. Is the converse also true?
3. The following optimal control problem is a deterministic version of the Brock–Mirman growth model from mathematical economy.

$$x^+ = u, \quad \ell(x, u) = -\ln(Ax^\alpha - u)$$

with parameters $A \in [1, 10]$, $\alpha \in [0.1, 0.5]$ and $\mathbb{X} = \mathbb{U} = [0.05, 25]$.

   (a) Compute the optimal equilibrium depending on A and α.
   (b) Check that the problem is strictly dissipative with storage function $\lambda(x) = p^e(x - x^e)$ with $p^e = \alpha \sqrt[\alpha-1]{\alpha A}/(1-\alpha)$.

   Hint: Convince yourself that for a continuous and strictly convex function $g : \mathbb{R} \to \mathbb{R}$ with minimum $g(x_{\min}) = 0$ there exists a function $\gamma \in \mathcal{K}_\infty$ with $g(x) \geq \gamma(|x|_{x_{\min}})$ for all $x \in \mathbb{R}$.
4. Use the MATLAB® routine `nmpc.m` in order to implement the economic NMPC problem from Problem 3.

   (a) Implement equilibrium terminal constraints $\mathbb{X}_0 = \{x^e\}$ and verify that the closed-loop solutions indeed converge toward the optimal equilibrium.
   (b) Simulate the closed-loop trajectories without terminal constraints and estimate the limit

$$\lim_{k\to\infty} \ell(x_{\mu_N}(k, x), \mu_N(x_{\mu_N}(k, x))).$$

Compute the distance of this limit to $\ell(x^e, u^e)$ and verify that the distance tends to 0 as N increases. What can you say about the rate of convergence?

(c) Define stabilizing costs ℓ^{stab} that are positive definite w.r.t. x^e and denote the resulting NMPC feedback by μ_N^{stab}. Simulate closed-loop trajectories for varying N and evaluate the summed economic cost $J_K^{cl}(x, \mu_N^{stab})$ (with ℓ from Problem 3) for different N and K along these trajectories. Compare the resulting values with those for the trajectories from (a) and (b).

References

1. Amrit, R., Rawlings, J.B., Angeli, D.: Economic optimization using model predictive control with a terminal cost. Ann. Rev. Control **35**(2), 178–186 (2011)
2. Angeli, D., Amrit, R., Rawlings, J.B.: Receding horizon cost optimization for overly constrained nonlinear plants. In: Proceedings of the 48th IEEE Conference on Decision and Control—CDC 2009, pp. 7972–7977. Shanghai, China (2009)
3. Angeli, D., Amrit, R., Rawlings, J.B.: On average performance and stability of economic model predictive control. IEEE Trans. Autom. Control **57**(7), 1615–1626 (2012)
4. Angeli, D., Rawlings, J.B.: Receding horizon cost optimization and control for nonlinear plants. In: Proceedings of the 8th IFAC Symposium on Nonlinear Control Systems—NOLCOS 2010, pp. 1217–1223. Bologna, Italy (2010)
5. Diehl, M., Amrit, R., Rawlings, J.B.: A Lyapunov function for economic optimizing model predictive control. IEEE Trans. Autom. Control **56**(3), 703–707 (2011)
6. Dorfman, R., Samuelson, P.A., Solow, R.M.: Linear Programming and Economic Analysis. Dover Publications, New York (1987). (Reprint of the 1958 original)
7. Ellis, M., Duranda, H., Christofides, P.D.: A tutorial review of economic model predictive control methods. J. Proc. Control **24**(8), 1156–1178 (2014)
8. Faulwasser, T., Bonvin, D.: On the design of economic NMPC based on approximate turnpike properties. In: Proceedings of the 54th IEEE Conference on Decision and Control—CDC 2015, pp. 4964–4970 (2015)
9. Faulwasser, T., Korda, M., Jones, C.N., Bonvin, D.: Turnpike and dissipativity properties in dynamic real-time optimization and economic MPC. In: Proceedings of the 53rd IEEE Conference on Decision and Control—CDC 2014, pp. 2734–2739 (2014)
10. Grüne, L.: Economic receding horizon control without terminal constraints. Automatica **49**(3), 725–734 (2013)
11. Grüne, L.: Approximation properties of receding horizon optimal control. Jahresbericht der Deutschen Mathematiker Vereinigung **118**(1), 3–37 (2016)
12. Grüne, L., Müller, M.A.: On the relation between strict dissipativity and turnpike properties. Syst. Control Lett. **90**, 45–53 (2016)
13. Grüne, L., Panin, A.: On non-averaged performance of economic MPC with terminal conditions. In: Proceedings of the 54th IEEE Conference on Decision and Control—CDC 2015, pp. 4332–4337. Osaka, Japan (2015)
14. Grüne, L., Stieler, M.: Asymptotic stability and transient optimality of economic MPC without terminal conditions. J. Proc. Control **24**(8), 1187–1196 (2014)
15. McKenzie, L.W.: Optimal economic growth, turnpike theorems and comparative dynamics. In: Handbook of Mathematical Economics, vol. III, Handbooks in Economic, vol. 1, pp. 1281–1355. North-Holland, Amsterdam (1986)
16. Müller, M.A., Angeli, D., Allgöwer, F.: Transient average constraints in economic model predictive control. Automatica **50**(11), 2943–2950 (2014)
17. Müller, M.A., Angeli, D., Allgöwer, F.: On necessity and robustness of dissipativity in economic model predictive control. IEEE Trans. Autom. Control **60**(6), 1671–1676 (2015)

18. Müller, M.A., Grüne, L.: Economic model predictive control without terminal constraints for optimal periodic behavior. Automatica **70**, 128–139 (2016)
19. von Neumann, J.: A model of general economic equilibrium. Rev. Econ. Stud. **13**(1), 1–9 (1945)
20. Rawlings, J.B., Amrit, R.: Optimizing process economic performance using model predictive control. In: Magni, L., Raimondo, D.M., Allgöwer, F. (eds.) Nonlinear Model Predictive Control. Lecture Notes in Control and Information Science, vol. 384, pp. 119–138. Springer, Heidelberg (2009)
21. Rawlings, J.B., Bonné, D., Jørgensen, J.B., Venkat, A.N., Jørgensen, S.B.: Unreachable setpoints in model predictive control. IEEE Trans. Autom. Control **53**(9), 2209–2215 (2008)
22. Trélat, E., Zuazua, E.: The turnpike property in finite-dimensional nonlinear optimal control. J. Differ. Equ. **258**(1), 81–114 (2015)
23. Willems, J.C.: Least squares stationary optimal control and the algebraic Riccati equation. IEEE Trans. Autom. Control **16**(6), 621–634 (1971)
24. Willems, J.C.: Dissipative dynamical systems. I. General theory. Arch. Ration. Mech. Anal. **45**(5), 321–351 (1972)
25. Willems, J.C.: Dissipative dynamical systems. II. Linear systems with quadratic supply rates. Arch. Ration. Mech. Anal. **45**(5), 352–393 (1972)
26. Zaslavski, A.: Turnpike Phenomenon and Infinite Horizon Optimal Control. Springer International Publishing, Heidelberg (2014)
27. Zanon, M., Grüne, L., Diehl, M.: Periodic optimal control, dissipativity and MPC. IEEE Trans. Autom. Control. http://dx.doi.org/10.1109/TAC.2016.2601881. Printed version to appear

Chapter 9
Distributed NMPC

For large-scale systems such as street traffic, cyber- physical production systems or energy grids on an operational level, the MPC approach introduced in Chap. 3 is typically inapplicable in real time. Moreover, communication restrictions or privacy considerations may render the centralized solution of the optimal control problem in each step of the NMPC scheme impossible. To cope with these issues, the optimal control problem is split into subproblems, which are simpler to solve but may be linked by dynamics, cost functions or constraints. As the examples indicate, each subproblem may be seen as an independent unit. If these units are not coordinated, i.e., if there exists no data exchange and if inputs from connected units are considered as disturbances, the problem is referred to as decentralized. Including communication, the problem is called distributed and can again be split into subclasses of cooperative and noncooperative control. Within this chapter, we impose the assumption of flawless communication to analyze both stability and performance of the overall system for the distributed case. Additionally, we briefly sketch how to analyze the robustness of the distributed setting. Last, we discuss basic coordination methods on the tactical control layer to solve the distributed problem and relate these methods to our stability results.

9.1 Background and Problem Formulation

Instead of considering only one system (2.1), in this chapter we consider a set of nonlinear discrete time systems

$$x^p(n+1) = f^p(x^p(n), u^p(n), i^p(n)), \qquad p \in \mathscr{P} := \{1, \ldots, P\}, n \in \mathbb{N}_0 \quad (9.1)$$

where states and controls satisfy $x^p(n) \in X^p$ and $u^p(n) \in U^p$. In analogy to (9.1), we extend all of our notation introduced in Chaps. 2 and 3 by a superscript p to refer to the respective single system. For instance, for all $p \in \mathscr{P}$ the sets X^p and U^p denote the state and control value space of the p-th system, respectively. The second

L. Grüne and J. Pannek, *Nonlinear Model Predictive Control*,
Communications and Control Engineering, DOI 10.1007/978-3-319-46024-6_9

extension within (9.1) is the variable $i^p(n) \in I^p$. The latter will allow us to link the set of systems on all levels and is therefore called neighboring data and neighboring data set respectively. Note that the set depends on the chosen element $p \in \mathscr{P}$ and may also vary over time.

Example 9.1 Reconsider our Example 2.2 with dynamics

$$\begin{pmatrix} x_1(n+1) \\ x_2(n+1) \end{pmatrix} = \begin{pmatrix} x_1(n) + x_2(n) + u(n)/2 \\ x_2(n) + u(n) \end{pmatrix}.$$

We can split the system into two subsystems by choosing $x^1 = x_1$, $x^2 = x_2$ and $u^2 = u$. Leaving u^1 undefined, we obtain

$$\begin{aligned} x^1(n+1) &= x^1(n) + \overbrace{x^2(n) + u^2(n)}^{\text{from subsystem 2}}/2 \\ x^2(n+1) &= x^2(n) + u^2(n). \end{aligned}$$

For our choice of variables, subsystem 2 is independent from subsystem 1, yet to evaluate subsystem 1 we require $i^1(n)$ to contain $x^2(n)$ and $u^2(n)$ from subsystem 2. Note that the connection depends on how the control input from the overall system is assigned to the subsystems. By choosing $u^1 = u$ and leaving u^2 undefined, both subsystems are dependent on the respective other.

The reason to focus on various systems instead of only one is due to the drastic increase in the complexity of control systems observed during the past years, which is driven by a top-down and a bottom-up approach: Regarding top-down, plants and processes are extended by including adjacent components or tasks leading to a larger dynamics. This direction is driven by a central entity to increase the performance of the overall system by integrating boundary components. Typical examples for which this phenomenon occurs are shop floors, production lines [32], or supply chains [11, 12]. In the bottom-up approach, the coupling of systems is driven by information and communication technology and the evolution of cyber-physical systems, in which a central entity may or may not exist. In contrast to embedded systems, cyber-physical system are open in terms of interaction via, e.g., communication or other interfaces. Utilizing the latter, cyber-physical systems can be interfaced with one another to optimize performance in some sense, e.g., energy efficiency, maximal output, shared resources, or capacity utilization. Here, a typical example is given by street traffic [5] or distributed energy generation [34].

Yet, although models are identical, these two lines are subject to a very different way of thinking. The first approach is economically driven and resembles a top-down idea by integrating all aspects into one big problem. Each of these problems can be seen as an island, which is operated optimally, but remains isolated from any other island. Connecting all islands to follow a common goal will render the problem to be too large to solve it centrally, so it needs to be split and coordinated to retain centrally controlled properties like ensuring that a certain performance criterion is met in the best possible way. The second approach, however, is driven by technology and shows a bottom-up structure. Since the systems typically also represent economically independent entities, who cannot divulge information about their dynamics, constraints, and objectives, the splitting of the problem is already fixed and a common goal may initially not exist. The problem in that case is to identify a coordination scheme to improve local properties.

In both cases, the resulting dynamics for the set of systems (9.1) is given by (2.1)

$$x_u(n+1) = f(x_u(n), u(n))$$

with state $x(n) = (x^1(n)^\top, \ldots, x^P(n)^\top)^\top \in X = X^1 \times \cdots \times X^P$ and control $u(n) = (u^1(n)^\top, \ldots, u^P(n)^\top)^\top \in U = U^1 \times \cdots \times U^P$. Within this chapter, we call (2.1) the (overall) system, (9.1) the set of subsystems, and refer to p as a subsystem.

A basic scheme to solve such a problem using NMPC on a local level is given in the following algorithm. Here and in what follows, "local" refers to objects associated to one of the subsystems.

Algorithm 9.2 (*Basic NMPC algorithm for local problems*)
At each sampling time t_n, $n = 0, 1, 2 \ldots$:

(1) For each subsystem $p \in \mathscr{P}$ measure the state $x^p(n)$ of the subsystem and set $x_0^p := x^p(n)$.
(2) For each subsystem $p \in \mathscr{P}$ solve the local optimal control problem incorporating available neighboring data i, denote the obtained admissible local optimal control sequence by $u^{p,\star}(\cdot)$ and communicate with other subsystems.
(3) For each subsystem $p \in \mathscr{P}$ define the NMPC-feedback value $\mu_N^p(x^p(n), i^p) := u^{p,\star}(0)$ and use this control value in the next sampling period.

Within the next Sect. 9.2, we will formally introduce the local optimal control problem for each subsystem. Connecting these subsystems, we present different communication strategies in Sect. 9.3 to complete Algorithm 9.2.

9.2 Classification of Connectedness

Considering the two evolution lines, we can already see that the subsystems in (9.1) may be connected in different ways. Recalling our problems ($\mathrm{OCP_{N,e}}$) and ($\mathrm{OCP^n_{N,e}}$) from Chap. 3, connections may appear in the cost function J_N, the set of admissible control sequences $\mathbb{U}^N_{\mathbb{X}_0}$, and in the dynamics f itself. In order to explicitly state the dependencies between the problems, we must first introduce this decomposition rigorously. Our main tool in this context are the following projections:

Definition 9.3 Given a vector space S, let $\pi : S \to S$ be a linear map which is idempotent, that is $\pi \circ \pi = \pi$. We call π a projection of S onto $\mathrm{Im}(\pi)$ (along $\mathrm{Ker}(\pi)$) where $\mathrm{Im}(\pi)$ and $\mathrm{Ker}(\pi)$ denote the image and kernel of π.

Using a set of projections we define a decomposition of a vector space:

Definition 9.4 Consider a vector space S, a set $\mathscr{P} = \{1, \ldots, P\}$ where $P \in \mathbb{N}$, and a set of projections $(\pi^p)_{p \in \mathscr{P}}$ where $S^p := \operatorname{Im}(\pi^p)$ is a subspace of S for all $p \in \mathscr{P}$ to be given. If the conditions

$$\langle (S^p)_{p \in \mathscr{P}} \rangle = S \text{ and } S^q \cap \langle (S^p)_{p \in \mathscr{P}, p \neq q} \rangle = \{0\} \text{ for all } q \in \mathscr{P}$$

hold, then the set $(S^p)_{p \in \mathscr{P}}$ is called a decomposition of S.

A decomposition allows us to rewrite the dynamics of a control system (2.1) as a set of subsystems operating on the subspaces of the decomposed state and control spaces. Here, the idea is to apply projections to the state and control space to separate the dynamics. To this end, we utilize projections $\pi_X^p : X \to X$ and $\pi_U^p : U \to U$ for all $p \in \mathscr{P}$ such that $\operatorname{Im}(\pi_X^p) = X^p$ and $\operatorname{Im}(\pi_U^p) = U^p$ hold according to Definition 9.4. These projections allow us to split state and control space into three components $[X^p, \widetilde{X}^p, \overline{X}^p]$ and $[U^p, \widetilde{U}^p, \overline{U}^p]$. While $x^p \in X^p$ and $u^p \in U^p$ are our primary variables of interest, we may interpret $\widetilde{x}^p \in \widetilde{X}^p$ and $\widetilde{u}^p \in \widetilde{U}^p$ as the states and controls of neighbors necessary to evaluate the projected dynamic $\pi_X^p \circ f$ correctly. Here, the controls $\widetilde{u}^p \in \widetilde{U}^p$ are computed by different local controllers and need to be communicated to subsystem p to evaluate the dynamics properly. In contrast to that, $\pi_X^p \circ f$ is independent of $\overline{x}^p \in \overline{X}^p$ and $\overline{u}^p \in \overline{U}^p$, which are therefore referred to as independent states and controls. To use the vocabulary of parallel programming, $x^p(n) \in X^p$ and $u^p(n) \in U^p$ are *local or private variables* while $\widetilde{x}^p(n) \in \widetilde{X}^p$, $\overline{x}^p(n) \in \overline{X}^p$, $\widetilde{u}(n) \in \widetilde{U}^p$ and $\overline{u}^p(n) \in \overline{U}^p$ are *interface or public variables.*

Now, we can rewrite the mapping of system (2.1) as

$$f : (X^p \times \widetilde{X}^p \times \overline{X}^p) \times (U^p \times \widetilde{U}^p \times \overline{U}^p) \to (X^p \times \widetilde{X}^p \times \overline{X}^p).$$

Coming back to our introduction of the subsystem dynamics in (9.1), we utilize $\pi_X^p \circ f$ being independent of $\overline{x}^p \in \overline{X}^p$ and $\overline{u}^p \in \overline{U}^p$ to obtain that we only require data from the set $(\widetilde{X}^p \times \widetilde{U}^p)$. Since by definition of a decomposition we have $X = X^1 \times \cdots \times X^P$ and $U = U^1 \times \cdots \times U^P$, there exists an index set $\mathscr{I}^p = \{p_1, \ldots, p_m\} \subset \mathscr{P} \setminus \{p\}$ such that $(\widetilde{X}^p \times \widetilde{U}^p) \subset (X^{p_1} \times \cdots \times X^{p_m}) \times (U^{p_1} \times \cdots \times U^{p_m})$. We call $\mathscr{I}^p$ *neighboring index set* and the required data neighboring data:

Definition 9.5 Consider a neighboring index set $\mathscr{I}^p(n)$ of subsystem $p \in \mathscr{P}$. We call the set

$$i^p(n) = \{(q, n_q, x^q(\cdot), u^q(\cdot)) \mid q \in \mathscr{I}^p(n)\} \in I^p \tag{9.2}$$

neighboring data containing the index of the neighboring subsystem q, the sending time stamp n_q and the planned state and control sequences $x^q(\cdot)$ and $u^q(\cdot)$ computed at time instant n_q to subsystem p. The neighboring data set is given by $I^p = 2^Q$ with $Q = (\mathscr{P} \setminus \{p\}) \times \mathbb{N}_0 \times X^{N+1} \times U^N$.

Within Definition 9.5 the neighboring data is introduced to contain state and control sequences of length N, which is suitable for the evaluation of the dynamics for the MPC open loop. Setting $N = 1$, we can perform an instantaneous evaluation of the dynamics. To illustrate the latter, we reconsider Example 9.1.

Example 9.6 For Example 9.1 we obtain $\mathscr{I}^1(n) = \{2\}$ and $\mathscr{I}^2(n) = \emptyset$. To compute $x_u(n+1)$, we require the information contained in the neighboring data $i^1(n) = \{(2, n, x^2(n), u^2(n))\}$. Note that if we want to compute the trajectory of subsystem 1 over a finite horizon as, e.g., from n to $n + N$ in our problem (OCP_N), then we require the state and control trajectories of subsystem 2, i.e., the neighboring data $i^1(n) = \{(2, n, x^2(\cdot), u^2(\cdot))\}$.

Remark 9.7 For simplicity, we did not include information about the length of the sequences $x^q(\cdot)$, $u^q(\cdot)$ in the neighboring data. By convention, we assume that the horizon length N of each subsystem is identical. As this shall be the case for all subsystems, we implicitly require the horizon to be equal for all subsystems. An extension of (9.2) to

$$i^p(n) = \{(q, n^q, N^q, x^q(\cdot), u^q(\cdot)) \mid q \in \mathscr{I}^p(n)\} \in I^p \tag{9.3}$$

is possible to allow for different optimization horizons in the subsystems.

Remark 9.8 Depending on the implementation of a coordinating algorithm, different neighboring data structures and also different data sources may be used. In Sect. 9.5, the dual decomposition method shown in Algorithm 9.41 sends the Lagrange multiplier λ and external data $z(\cdot)$ in a packet $(0, n, \lambda)$ to coordinate subsystems via a central entity. Also other choices of variables to be exchanged are possible, cf., e.g., [29] where the authors used outputs of the subsystems.

Now, we formally obtain $f^p : X^p \times U^p \times I^p \to X^p$ via

$$f^p(x^p, u^p, (\widetilde{x}^p, \widetilde{u}^p)) := \Upsilon \circ \pi_X^p \circ f({\sigma_{X^p}}^{-1}(x^p, \widetilde{x}^p, 0), {\sigma_{U^p}}^{-1}(u^p, \widetilde{u}^p, 0)) \tag{9.4}$$

for permutations $\sigma_{X^p} : X \to X^p \times \widetilde{X}^p \times \overline{X}^p$ with $\sigma_{X^p}(x) = (x^p, \widetilde{x}^p, \overline{x}^p)$ and $\sigma_{U^p} : U \to U^p \times \widetilde{U}^p \times \overline{U}^p$ with $\sigma_{U^p}(u) = (u^p, \widetilde{u}^p, \overline{u}^p)$ and parameters d, d^p denoting the dimension of the vector spaces X and X^p respectively. As π_X^p is a projection to a subspace of X, we extract the subspace dynamic by applying $\Upsilon : X \to X^p$, which is a bijective map from the image of $\pi(X^p)$ to the span $\langle e_1, \ldots, e_{d^p}\rangle$. Since we can define f^p for all $p \in \mathscr{P}$ in a similar manner, the following holds trivially:

Corollary 9.9 *Suppose a system* (2.1)*, a set* $\mathscr{P} = \{1, \ldots, P\}$ *as well as projections* $(\pi_X^p)_{p\in\mathscr{P}}$, $(\pi_U^p)_{p\in\mathscr{P}}$ *inducing decompositions* $\langle (X^p)_{p\in\mathscr{P}}\rangle$ *and* $\langle (U^p)_{p\in\mathscr{P}}\rangle$ *to be given. Then we can rewrite system* (2.1) *as a set of subsystems* (9.1) *using* (9.4) *for all* $p \in \mathscr{P}$.

Remark 9.10 Note that the choice of the projections is not fixed, so we may want to keep the dimension of the respective subsets $\widetilde{X}^p$, $\widetilde{U}^p$ to be as small as possible.

Additionally, $f^p(\cdot,\cdot,\cdot)$ does not depend on the choice of variables on the subspaces $\overline{X}^p$ and $\overline{U}^p$. Hence, to reduce computing time and communication load, these subspaces should be chosen to be maximal.

To illustrate the implications of a proper choice of the projection, we utilize our basic examples from Chap. 2.

Example 9.11 Reconsider our Example 2.1 with two subsystems and overall dynamics

$$x(n+1) = \begin{pmatrix} x^1(n+1) \\ x^2(n+1) \end{pmatrix} = \underbrace{\begin{pmatrix} 1 & 0 \\ 0 & 1 \end{pmatrix}}_{=:A} \begin{pmatrix} x^1(n) \\ x^2(n) \end{pmatrix} + \underbrace{\begin{pmatrix} u^1(n) \\ u^2(n) \end{pmatrix}}_{=:B} =: f(x(n), u(n)).$$

For this linear example, we can utilize the structure of A to define

$$\pi_X^1 := \begin{pmatrix} 1 & 0 \\ 0 & 0 \end{pmatrix} \quad \text{and} \quad \pi_X^2 := \begin{pmatrix} 0 & 0 \\ 0 & 1 \end{pmatrix}$$

to obtain the state sets $X^1 = \operatorname{Im}(\pi_X^1) = \langle e_1 \rangle \cong \mathbb{R}$ and $X^2 = \operatorname{Im}(\pi_X^2) = \langle e_2 \rangle \cong \mathbb{R}$ and the dynamics of the original subsystems

$$\begin{aligned} x^1(n+1) &= [1, 0] \cdot \pi_X^1 \circ f(x(n), u(n)) = x^1(n) + u^1(n) \\ x^2(n+1) &= [0, 1] \cdot \pi_X^2 \circ f(x(n), u(n)) = x^2(n) + u^2(n). \end{aligned}$$

Example 9.12 Utilizing two instances of Example 2.2, we get

$$\underbrace{\begin{pmatrix} x_1^1(n+1) \\ x_2^1(n+1) \\ x_1^2(n+1) \\ x_2^2(n+1) \end{pmatrix}}_{=\overline{x}(n+1)} = \underbrace{\begin{pmatrix} 1 & 1 & 0 & 0 \\ 0 & 1 & 0 & 0 \\ 0 & 0 & 1 & 1 \\ 0 & 0 & 0 & 1 \end{pmatrix} \begin{pmatrix} x_1^1(n) \\ x_2^1(n) \\ x_1^2(n) \\ x_2^2(n) \end{pmatrix} + \begin{pmatrix} 1/2 & 0 \\ 1 & 0 \\ 0 & 1/2 \\ 0 & 1 \end{pmatrix} \begin{pmatrix} u^1(n) \\ u^2(n) \end{pmatrix}}_{=:f(\overline{x}(n), \overline{u}(n))}.$$

If we consider

$$\pi_X^1 := \begin{pmatrix} 1 & 0 & 0 & 0 \\ 0 & 1 & 0 & 0 \\ 0 & 0 & 1 & 0 \\ 0 & 0 & 0 & 0 \end{pmatrix} \quad \text{and} \quad \pi_X^2 := \begin{pmatrix} 0 & 0 & 0 & 0 \\ 0 & 0 & 0 & 0 \\ 0 & 0 & 0 & 0 \\ 0 & 0 & 0 & 1 \end{pmatrix},$$

then we do not reobtain the original instances but instead the virtual subsystems

$$\overline{x}^1(n+1) = \begin{pmatrix} 1&0&0&0 \\ 0&1&0&0 \\ 0&0&1&0 \end{pmatrix} \cdot \pi_X^1 \circ f(\overline{x}(n), \overline{u}(n)) = \begin{pmatrix} \overline{x}_1^1(n) + \overline{x}_2^1(n) + 1/2\overline{u}^1(n) \\ \overline{x}_2^1(n) + \overline{u}^1(n) \\ \overline{x}_3^1(n) + \overline{x}_1^2(n) + 1/2\overline{u}^2(n) \end{pmatrix}$$

$$\overline{x}^2(n+1) = [0, 0, 0, 1] \cdot \pi_X^2 \circ f(\overline{x}(n), \overline{u}(n)) = \overline{x}_1^2(n) + \overline{u}^2(n).$$

Hence, in order to evaluate subsystem 1, we require the knowledge of the state $\overline{x}_1^2 = x_2^2$ and the control $\overline{u}^2 = u^2$ of subsystem 2, i.e., we must have $(2, n, x_2^2(n), u^2(n)) \in i^1(n)$.

Remark 9.13 Ideally, one wishes to decompose system (2.1) into a set of subsystems (9.1) such that the subsystems are completely independent from each other, i.e.,

$$x^p(n+1) = f^p(x^p(n), u^p(n), \emptyset).$$

Such a decomposition may arise naturally if a set of coupled systems shall be controlled, see, e.g., [9, 17, 35], and may allow for a fully parallel evaluation of the dynamics. Alternatively, the system (2.1) may be decomposed as in [16, 24] such that no control of a neighboring system has to be considered, that is

$$x^p(n+1) = f^p(x^p(n), u^p(n), (\tilde{x}^p(n), \emptyset)).$$

Having dealt with the dynamics, we now focus on the constraints and their local integration. Considering the set of admissible control sequences, our aim is to compute local controls u^p such that none of the constraints imposed by other subsystems are violated, i.e., that they satisfy the condition $u \in \mathbb{U}^N$ locally. To this end, we introduce the notion of partial constraint sets by "projecting" the constraint sets $\mathbb{X}$ and $\mathbb{U}$ to the respective spaces of a subset of systems.

Definition 9.14 Consider the vector space S, the set $\mathcal{P} = \{1, \ldots, P\}$ where $P \in \mathbb{N}$, and a decomposition $(S^p)_{p \in \mathcal{P}}$ of S to be given. For a subset $\mathcal{I}^p = \{p_1, \ldots, p_m\} \subset \mathcal{P} \setminus \{p\}$ we call $\mathcal{Q} = \mathcal{Q}(p) := \{p\} \cup \mathcal{I}^p$ the *partial index set*. Then, the *partial set* is defined as

$$S^{\mathcal{Q}} = S^p \times S^{p_1} \times \cdots \times S^{p_m}$$

and we denote elements of $S^{\mathcal{Q}}$ by $s^{\mathcal{Q}} = (s^p, s^{p_1}, \ldots, s^{p_m})$. Accordingly, given a constraint set $\mathbb{S} \subset S$ the *partial constraint set* is defined by

$$\mathbb{S}^{\mathcal{Q}} := \{s^{\mathcal{Q}} \in S^{\mathcal{Q}} \mid \text{there exists } \tilde{s} \in \mathbb{S} \text{ with } \tilde{s}^i = s \text{ for all } i \in \mathcal{Q}\}.$$

Here, we choose to distinguish between p and $\mathcal{I}^p$ to explicitly account for data from subsystems $q \neq p$ required in the evaluation of local constraints.

Similar to the dynamics, the idea is now to show that considering local evaluations resembles the overall result.

Proposition 9.15 *Consider a vector space S with a set $\mathbb{S} \subset S$, a set $\mathscr{P} = \{1, \ldots, P\}$ where $P \in \mathbb{N}$, and a decomposition $(S^p)_{p \in \mathscr{P}}$ of S in the sense of Definition 9.4 to be given. Then, there exist sets $(\mathscr{I}^p)_{p \in \mathscr{P}}$ such that condition $s \in \mathbb{S}$ is equivalent to*

$$s^{\mathscr{Q}} \in \mathbb{S}^{\mathscr{Q}}, \qquad \text{for all } p \in \mathscr{P} := \{1, \ldots, P\} \text{ with } \mathscr{Q} = \mathscr{Q}(p). \tag{9.5}$$

Proof Follows trivially by choosing $\mathscr{I}^p = \{1, \ldots, P\} \setminus \{p\}$ for all $p \in \mathscr{P}$. □

From Proposition 9.15 it is clear that the constraint $x \in \mathbb{X}$ can be checked locally at subsystem p if the constraint set itself or the respective constraint functions $G_i^S : X \times U \to \mathbb{R}$, $i \in \mathscr{E}^S = \{1, \ldots, p_g\}$ and $H_i^S : X \times U \to \mathbb{R}$, $i \in \mathscr{I}^S = \{p_g + 1, \ldots, p_g + p_h\}$ with $r_g, r_h \in \mathbb{N}_0$ and all required data from other subsystems are available at subsystem p. In contrast to the construction of $(\mathscr{I}^p)_{p \in \mathscr{P}}$ in the proof, we are interested in a set of minimal sets $(\mathscr{I}^p)_{p \in \mathscr{P}}$ to reduce the knowledge requirements of subsystem p. Hence, to use this rather abstract definition to decompose $\mathbb{U}^N_{\mathbb{X}_0}$ effectively, we need to know for each $p \in \mathscr{P}$ which of the subsystems have to be considered. Note that we require this data to evaluate the condition $u \in \mathbb{U}^N_{\mathbb{X}_0}$ in the optimization, i.e., at a fixed closed-loop time instant. To check the constraints for the predictions, we require future open-loop data. To allow for changing network topologies over time like the distribution of cars on a road network, we allow the set of neighboring subsystems to be time variant $\mathscr{I}^p(n)$, cf. Fig. 9.1. Additionally, we allow the case that even if data regarding a subsystem $q \in \mathscr{P} \setminus \{p\}$ is known to subsystem $p \in \mathscr{P}$, that data may be ignored in the calculation of the control. Consequently, the set of neighbors that communicate with system p and the set of neighbors that is used by subsystem p to compute its control may differ. Figure 9.2 visualizes this difference in communication and dependencies via graphs.

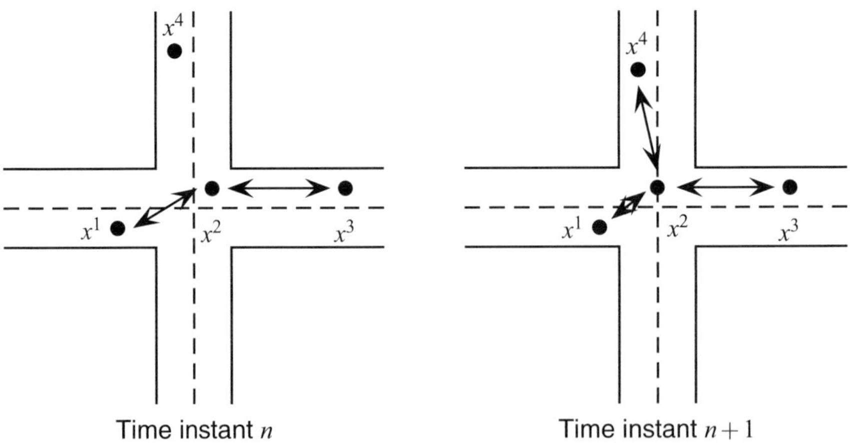

Fig. 9.1 Example of a communication graph between systems x^p, $p \in \mathscr{P}$ at time instants n and $n+1$

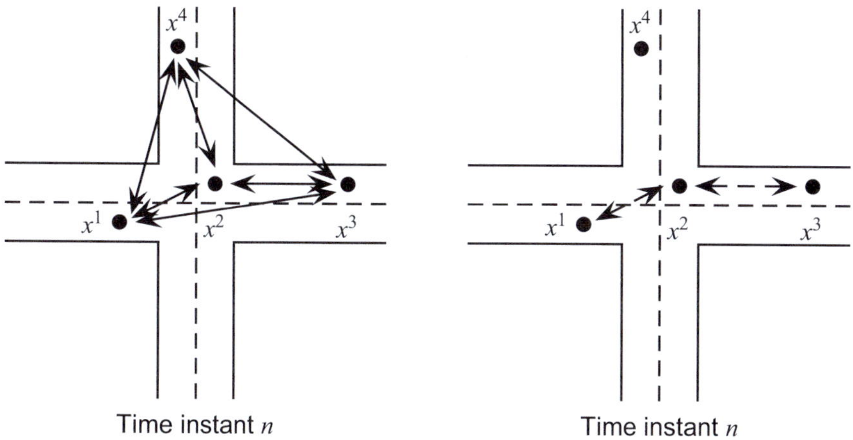

Fig. 9.2 Example of the difference between a communication and a dependency graph between subsystems x^p, $p \in \mathscr{P}$ at time instant n

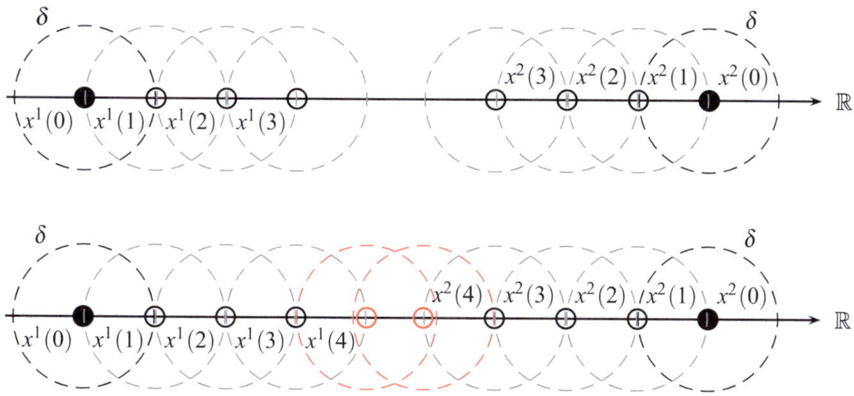

Fig. 9.3 Sketch for Example 9.16 with $N = 4$ (*top*) and $N = 5$ (*bottom*)

Example 9.16 Reconsider our Example 9.11 and assume the controls to be bounded via $u^p(n) \in [-1, 1]$, $p \in \{1, 2\}$. Additionally, we introduce the constraint $|x^1 - x^2| \geq \delta > 0$, which can be seen as a safety distance between the two vehicles. Using initial values $x^1(0) = -4.4 = -x^2(0)$, $N = 4$, $\delta = 1$ and $u^1(n) = 1 = -u^2(n)$ for $n = 0, 1, 2, 3$, the trajectories are sketched in Fig. 9.3(top). In this case, the constraint $|x^1 - x^2| \geq \delta > 0$ is always satisfied, which may be determined by the systems using, e.g., sensor data. Thus, we can choose to drop both the constraint and the respective neighboring data from the local evaluation. However, if we increase the horizon length to $N = 5$, the subsystems are coupled via the constraint $|x^1 - x^2| \geq \delta$.

Remark 9.17 Note that in contrast to links in dynamics, a coordination algorithm may allow for different lengths of neighboring data in case of links in constraints.

In Sect. 9.5, we will consider the method by Richards and How, cf. Algorithm 9.36, which uses fixed order of subsystems and neighboring data of equal lengths N from time steps n for $q < p$ and from time step $n - 1$ for $q > p$. Hence, the constraints induced by neighbors $q > p$ at open-loop time step N are not defined and allow for a certain degree of freedom. In general, this may lead to infeasibility of the solution, yet the scheme of Algorithm 9.36 is arranged such that recursive feasibility can be shown.

Here, we do not specify how the neighboring index set $\mathscr{I}^p(n)$ is obtained, but only specify those data which is needed in order to check the constraints. In Sect. 9.3, we will consider different communication schemes and analyze the respective network loads.

Now, for a given neighboring index set $\mathscr{I}^p(n)$, we can use the induced partial state constraint set and the neighboring data available to subsystem p to define the feasible set and the set of admissible controls from which the control sequence $u^p(\cdot)$ can be chosen in analogy to Definition 3.9. We formulate the definition for terminal constraint set $\mathbb{X}_0 \subseteq \mathbb{X}$, noting that the case without terminal constraints is included by setting $\mathbb{X}_0 = \mathbb{X}$.

Definition 9.18 Suppose a time instant $n \in \mathbb{N}_0$, decompositions $(X^p)_{p\in\mathscr{P}}$, $(U^p)_{p\in\mathscr{P}}$ of X, U and respective projections $\left(\pi_X^p\right)_{p\in\mathscr{P}}$, $\left(\pi_U^p\right)_{p\in\mathscr{P}}$ in the sense of Definition 9.4 to be given, inducing the set of subsystems (9.1) via (9.4). For each subsystem $p \in \mathscr{P}$ with initial value x_0^p and neighboring data i^p from subsystems $q \in \mathscr{I}^p$, we define the *feasible set* via

$$\mathbb{X}_N^p(i^p) := \{x_0^p \in \mathbb{X}^p(i^p) \,|\, \exists \left(u^p(\cdot), (u^q(\cdot))_{q\in\mathscr{I}^p}\right) \in \mathbb{U}^{\mathscr{Q},N}(x_0^p, i^p) \quad \text{with } \left(x_{u^p}^p(N, x_0^p, i^p), (x_{u^q}^q(N, x_0^q, i^q))_{q\in\mathscr{I}^p}\right) \in \mathbb{X}_0^{\mathscr{Q}}(i^p)\} \tag{9.6}$$

where $\mathbb{U}^{\mathscr{Q},N}$ and $\mathbb{X}_0^{\mathscr{Q}}$ with $\mathscr{Q} = \{p\} \cup \mathscr{I}^p$ are the partial control and state constraint sets. Moreover, for each $x_0^p \in \mathbb{X}_N^p$ we call

$$\mathbb{U}_{\mathbb{X}_0^p}^{p,N}(x_0^p, i^p) := \left\{u^p(\cdot) \in \mathbb{U}^{p,N}(x_0^p, i^p) \,|\, \left(x_{u^p}^p(N, x_0^p, i^p), (x_{u^q}^q(N, x_0^q, i^q))_{q\in\mathscr{I}^p}\right) \in \mathbb{X}_0^{\mathscr{Q}}(i^p) \quad \text{with } \left(u^q(\cdot)\right)_{q\in\mathscr{I}^p}\right\} \tag{9.7}$$

set of admissible control sequences for agent p.

As indicated before, the local evaluation of the constraints depends on the neighboring data. Consequently, if to be evaluated exactly, also the feasible set and the set of admissible control sequences require this data. To stress this dependency, which will be of particular importance in terms of communication, the latter sets are parametrized in terms of the neighboring data.

Note that we can extend the notation in Definition 9.18 to cover the time variant case as well. For simplicity of exposition, however, we concentrate on the time invariant case.

Last, we consider the cost function J_N and suppose decompositions $(X^p)_{p\in\mathscr{P}}$, $(U^p)_{p\in\mathscr{P}}$ of X, U to be given. Then, we rewrite the cost function via

$$J_N(x_0, u(\cdot)) = \sum_{p=1}^{P} J_N^p(x_0^p, u^p(\cdot), i^p) \qquad \text{with}$$

$$J_N^p(x_0^p, u^p(\cdot), i^p) = \sum_{k=0}^{N-1} \omega_{N-k}\ell^p(x_{u^p}^p(k, x_0, i^p), u^p(k), i^p) + F^p(x_{u^p}^p(N, x_0, i^p), i^p) \quad (9.8)$$

and refer to J_N^p as the cost function of subsystem $p \in \mathscr{P}$. Note that the separation is always possible since we could simply set J_N^p to $(1/P)J_N$ and ℓ^p to $(1/P)\ell$, respectively.

Example 9.19 Reconsider Example 9.11, now with P subsystems and with cost function

$$J_N(x_0, u(\cdot)) = \sum_{k=0}^{N-1} \underbrace{\left(\sum_{p=1}^{P} \|x_{u^p}^p(k, x_0^p, i^p) - x_*^p\|^2 + \|u^p(k)\|^2\right)}_{=\ell(x_u(k,x_0),u(k))}$$
$$+ \underbrace{\sum_{p=1}^{P} \|x_{u^p}^p(N, x_0^p, i^p) - x_*^p\|^2}_{=F(x_u(N,x_0))}.$$

Then the choice $\ell^p(x^p, u^p, i^p) = \|x^p - x_*^p\|^2 + \|u^p\|^2$, $F^p(x^p, i^p) = \|x^p - x_*^p\|^2$ requires minimal neighboring data, i.e., $I = \emptyset$.

Remark 9.20 As a convention, we assume that summands of the cost function containing x^p or u^p shall be included in the cost function J_N^p. This is typically the case to minimize communication. Note that this means that summands depending on more than one subsystem will appear in more than one local cost function and have to be weighed appropriately in order to ensure (9.8).

Example 9.21 Again consider Example 9.11 with three subsystems each driven by the dynamics from Example 2.1. For the cost function

$$J_N(x_0, u(\cdot)) = \sum_{k=0}^{N-1} \Bigg(\|x^1(k) - x_*^1\|^2 + \|x^1(k) - x^2(k) - \delta\|^2 + \|x^2(k) - x^3(k) - \delta\|^2$$
$$+ \sum_{p=1}^{P} \|u^p(k)\|^2 \Bigg)$$
$$+ \|x^1(N) - x_*^1\|^2 + \|x^1(N) - x^2(N) - \delta\|^2 + \|x^2(N) - x^3(N) - \delta\|^2$$

with $\delta > 0$ the choice

$$\begin{aligned}
\ell^1(x^1, u^1, i^1) &= \|x^1 - x_*^1\|^2 + \|x^1 - x^2 - \delta\|^2 + \|u^1\|^2 \\
\ell^2(x^2, u^2, i^2) &= \|x^1 - x^2 - \delta\|^2 + \|x^2 - x^3 - \delta\|^2 + \|u^2(k)\|^2 \\
\ell^3(x^3, u^3, i^3) &= \|x^2 - x^3 - \delta\|^2 + \|u^3\|^2 \\
F^1(x^1, i^1) &= \|x^1 - x_*^1\|^2 + \|x^1 - x^2 - \delta\|^2 \\
F^2(x^2, i^2) &= \|x^1 - x^2 - \delta\|^2 + \|x^2 - x^3 - \delta\|^2 \\
F^3(x^3, i^3) &= \|x^2 - x^3 - \delta\|^2 + \|x^3 - x_*^3\|^2
\end{aligned}$$

does not satisfy (9.8).

To derive the separation and the required weights, we rewrite the cost function in additive form

$$J_N(x_0, u(\cdot)) = \sum_{k=0}^{N-1} \omega_{N-k} \sum_{j=1}^{m_\ell} \ell_j(x_u(k, x_0), u(k)) + \sum_{j=m_\ell}^{m_F+m_\ell} F_j(x_u(N, x_0)) \quad (9.9)$$

where m_ℓ, m_F denote the number of additive terms of the stage and terminal costs. The idea to obtain (9.8) is to follow Remark 9.20 and assign summands to respective subsystems. Similar to the dynamics, we can then derive the functions ℓ^p and F^p. Last, the reciprocal of the number of assignments of each summand can be used as corresponding weight in ℓ^p and F^p.

Starting with the assignment, for each $p \in \mathscr{P}$ we split the state and control space into three parts using permutations $\sigma_{X^p} : X \to X^p \times \widetilde{X}^p \times \overline{X}^p$, and $\sigma_{U^p} : U \to U^p \times \widetilde{U}^p \times \overline{U}^p$. Now, we identify that only those components j depend on $(x^p, u^p) \in X^p \times U^p$, for which

$$\begin{aligned}
&\ell_j(\sigma_{X^p}^{-1}(x^p, 0, 0), \sigma_{U^p}^{-1}(u^p, 0, 0)) \neq \ell_j(\sigma_{X^p}^{-1}(0, 0, 0), \sigma_{U^p}^{-1}(0, 0, 0)) \\
&F_j(\sigma_{X^p}^{-1}(x^p, 0, 0)) \neq F_j(\sigma_{X^p}^{-1}(x^p, 0, 0))
\end{aligned}$$

holds for some $(x^p, u^p) \in X^p \times U^p$. For each $p \in \mathscr{P}$, we collect the respective components j in the set $\mathscr{M}^p$. For each $j \in \mathscr{M}^p$, we obtain the sets $\overline{X}_j^p$ and $\overline{U}_j^p$ as the maximal subsets such that

$$\begin{aligned}
&\ell_j(\sigma_{X^p}^{-1}(x^p, \widetilde{x}^p, \overline{x}^p), \sigma_{U^p}^{-1}(u^p, \widetilde{u}^p, \overline{u}^p)) = \ell_j(\sigma_{X^p}^{-1}(x^p, \widetilde{x}^p, 0), \sigma_{U^p}^{-1}(u^p, \widetilde{u}^p, 0)) \\
&F_j(\sigma_{X^p}^{-1}(x^p, \widetilde{x}^p, \overline{x}^p)) = F_j(\sigma_{X^p}^{-1}(x^p, \widetilde{x}^p, 0))
\end{aligned}$$

holds for all $(x^p, \widetilde{x}^p, \overline{x}^p) \in X^p \times \widetilde{X}_j^p \times \overline{X}_j^p$, $(u^p, \widetilde{u}^p, \overline{u}^p) \in U^p \times \widetilde{U}_j^p \times \overline{U}_j^p$, which implicitly defines $\widetilde{X}_j^p$ and $\widetilde{U}_j^p$. To derive the functions ℓ^p and F^p, the latter sets allow us to obtain the neighboring index set. To derive $\mathscr{I}^p$, we use the definition of a decomposition for our state and control sets $X = X^1 \times \cdots \times X^P$ and $U = U^1 \times \cdots \times U^P$ and for each $j \in \mathscr{M}^p$ we obtain the index set $\mathscr{I}_j^p = \{p_1, \ldots, p_m\} \subset \mathscr{P} \setminus \{p\}$

as the set with least possible entries satisfying $(\widetilde{X}_j^p \times \widetilde{U}_j^p) \subset (X^{p_1} \times \cdots \times X^{p_m}) \times (U^{p_1} \times \cdots \times U^{p_m})$. Combining the index sets, we obtain the neighboring index set via $\mathscr{I}^p := \bigcup_{j \in \mathscr{M}^p} \mathscr{I}_j^p$.

Recalling Example 9.21, we now set

$$\ell^p(x^p, u^p, i^p) = \sum_{j \in \mathscr{M}^p} \overline{\omega}_j \ell_j(\sigma_{X^p}^{-1}(x^p, \widetilde{x}^p, 0), \sigma_{U^p}^{-1}(u^p, \widetilde{u}^p, 0)) \tag{9.10}$$

$$F^p(x^p, i^p) = \sum_{j \in \mathscr{M}^p} \overline{\omega}_j F_j(\sigma_{X^p}^{-1}(x^p, \widetilde{x}^p, 0)). \tag{9.11}$$

Last, we obtain the weights $\overline{\omega}_j$ such that (9.8) holds from the construction of ℓ^p and F^p. For each $j \in \mathscr{M}^p$ we require data from subsystem p and from $\sharp(\mathscr{I}_j^p)$ neighbors. In turn, each term $j \in \{1, \ldots, m_\ell + m_F\}$ in (9.9) is also contained in the costs (9.10) and (9.11) of $\sharp(\mathscr{I}_j^p) + 1$ subsystems. This connection allows us to derive the following:

Proposition 9.22 *Consider costs* (9.10), (9.11) *and sets* $\mathscr{M}^p$ *and* $\mathscr{I}_j^p$, $j \in \mathscr{M}^p$, *to be given for each* $p \in \mathscr{P}$. *Then* (9.8) *holds for costs* (9.10), (9.11) *with* $\overline{\omega}_j = 1/(\sharp(\mathscr{I}_j^p) + 1)$.

Example 9.23 Reconsider Example 9.21. Setting

$$\begin{aligned}
\ell^1(x^1, u^1, i^1) &= \|x^1 - x_*^1\|^2 + \frac{1}{2}\|x^1 - x^2 - \delta\|^2 + \|u^1\|^2 \\
\ell^2(x^2, u^2, i^2) &= \frac{1}{2}\|x^1 - x^2 - \delta\|^2 + \frac{1}{2}\|x^2 - x^3 - \delta\|^2 + \|u^2(k)\|^2 \\
\ell^3(x^3, u^3, i^3) &= \frac{1}{2}\|x^2 - x^3 - \delta\|^2 + \|u^3\|^2 \\
F^1(x^1, i^1) &= \|x^1 - x_*^1\|^2 + \frac{1}{2}\|x^1 - x^2 - \delta\|^2 \\
F^2(x^2, i^2) &= \frac{1}{2}\|x^1 - x^2 - \delta\|^2 + \frac{1}{2}\|x^2 - x^3 - \delta\|^2 \\
F^3(x^3, i^3) &= \frac{1}{2}\|x^2 - x^3 - \delta\|^2 + \|x^3 - x_*^3\|^2
\end{aligned}$$

yields

$$\begin{aligned}
\sum_{p=1}^{3} \ell^p(x^p, u^p, i^p) &= \ell(x, u) \\
\sum_{p=1}^{3} F^p(x^p, i^p) &= F(x).
\end{aligned}$$

The design procedure can also be applied in the time variant case of problem ($\mathrm{OCP}_{\mathrm{N}}^{\mathrm{n}}$) and ($\mathrm{OCP}_{\mathrm{N,e}}^{\mathrm{n}}$). For simplicity of exposition, we omit this extension here.

Remark 9.24 Note that considering improper weights modifies the shape of the cost function J_N, e.g., if we simply sum the subsystem costs with equal weights, then intensively linked subsystems are overrated. Hence, thinking of search directions in the overall state space, the latter can lead to different results in the coordination.

Remark 9.25 In the so-called additive case we have

$$J_N(x_0, u(\cdot)) = \sum_{k=0}^{N-1} \sum_{p=1}^{P} \ell^p(x_{u^p}^p(k, x_0^p, i^p), u^p(k)) + \sum_{p=1}^{P} F^p(x_{u^p}^p(N, x_0, i^p)),$$

which allows us to split the cost function to

$$J_N^p(x_0^p, u^p(\cdot), i^p) = \sum_{k=0}^{N-1} \ell^p(x_{u^p}^p(k, x_0^p, i^p), u^p(k)) + F^p(x_{u^p}^p(N, x_0)).$$

Hence, each of these parts of the cost function is evaluated by only one subsystem leading to $\overline{\omega}_j = 1$ in Corollary 9.22.

Combining the dynamics, the constraints and the cost functions, we can now formulate the optimal control problem ($\mathrm{OCP_N}$) locally, which give us the following:

$$\begin{aligned}
&\text{minimize} \quad J_N^p(x_0^p, u^p(\cdot), i^p) := \sum_{k=0}^{N-1} \omega_{N-k} \ell^p(x_{u^p}^p(k, x_0^p, i^p), u^p(k), i^p)) \\
&\qquad\qquad\qquad\qquad\qquad\qquad + F^p(x_{u^p}^p(N, x_0, i^p), i^p) \\
&\text{with respect to } u^p(\cdot) \in \mathbb{U}_{\mathbb{X}_0^p}^{p,N}(x_0^p, i^p) \quad \text{subject to} \\
&x_{u^p}^p(0, x_0^p, i^p) = x_0^p, \quad x_{u^p}^p(k+1, x_0^p, i^p) = f^p(x_{u^p}^p(k, x_0^p, i^p), u^p(k), i^p)
\end{aligned} \qquad (\mathrm{OCP}_{\mathrm{N}}^{\mathrm{p}})$$

In the next section, we focus on the integration of ($\mathrm{OCP}_{\mathrm{N}}^{\mathrm{p}}$)and communication into our basic Algorithm 9.2.

9.3 Problem Classes for Different Levels of Connectedness

In the previous section, we derived the local optimal control problem ($\mathrm{OCP}_{\mathrm{N}}^{\mathrm{p}}$). Provided a correct separation of the overall problem was obtained and the respective data from neighboring subsystems is available, these local problems may allow us to obtain the optimal solution of the overall problem ($\mathrm{OCP_{N,e}}$) by solving the local ones.

Clearly, the optimal solution of ($\mathrm{OCP_{N,e}}$) represents a benchmark for other solutions. In the literature, it is also referred to as *centralized* solution. Besides optimality of the open loop, the centralized solution allows us to carry over the concepts of stability and performance of the closed loop to the case of a set of subsystems (9.1). As there is only one system, there is no need for communication, so the only drawback is the complexity of the problem ($\mathrm{OCP_{N,e}}$) for large system sizes.

While the centralized problem is perfectly coordinated in terms of cost function, dynamics, and constraints, the so-called *decentralized* case is the exact opposite. Here, the problem ($\mathrm{OCP_{N,e}}$) is split into subproblems ($\mathrm{OCP^p_N}$), $p \in \mathscr{P}$, which are solved without any exchange of data, i.e., $\mathscr{I}^p(n) = \emptyset = i^p$. As a consequence, the local problem reads

$$
\begin{aligned}
&\text{minimize} \quad J_N^p(x_0, u(\cdot)) := \sum_{k=0}^{N-1} \omega_{N-k} \ell^p(x_{u^p}^p(k, x_0^p), u^p(k))) \\
&\qquad\qquad\qquad\qquad\qquad\qquad + F^p(x_{u^p}^p(N, x_0)) \\
&\text{with respect to } u^p(\cdot) \in \mathbb{U}^{p,N}_{\mathbb{X}_0^p}(x_0^p), \quad \text{subject to} \\
&x_{u^p}^p(0, x_0^p) = x_0^p, \quad x_{u^p}^p(k+1, x_0^p) = f^p(x_{u^p}^p(k, x_0^p), u^p(k))
\end{aligned}
$$

While the communication demand of the decentralized approach is zero, the evaluation of f^p, $\mathbb{U}^{p,N}_{\mathbb{X}_0^p}$ and J_N^p may result in different errors. For one, if the dynamics of subsystems are linked, then the missing state and input information has to be replaced by estimates or default values and thus the evaluation is subject to model errors. In case of links within the set of admissible controls $\mathbb{U}^{p,N}_{\mathbb{X}_0^p}$, the equivalence of constraints depicted in Proposition 9.15 is compromised and may result in infeasibility of the overall solution. Last, a broken connection within the cost function J_N^p may lead to poor performance, and in case of linked dynamics also to loss of stability [8].

Example 9.26 Reconsider Example 9.1. In the decentralized case, the velocity of the vehicle given by x^2 is computed correctly, but replacing the state and input of the second subsystem in the first by the default value 0, the position is constant with $x^1(n) = x_0^1$ for all $n \in \mathbb{N}$. This leads to the modeling error

$$
\begin{aligned}
e(n) &= \|x^{\text{decentralized}}(n) - x^{\text{centralized}}(n)\|^2 = (x_2(n) + u(n)/2)^2 \\
&= \left(x_2(0) + \sum_{k=0}^{n-1} \frac{3}{2} u(k) + u(n)/2 \right)^2.
\end{aligned}
$$

Example 9.27 Given Example 9.16, a decentralized setting disregards the constraint $|x^1 - x^2| \geq \delta$. From Fig. 9.3, we then observe that this broken link causes the overall solution to be infeasible for $N \geq 5$.

Example 9.28 Reconsider Example 9.23. Replacing all unkown terms in the cost function by the default value 0, the cost functions are given by

$$J_N^1(x_0^1, u^1(\cdot)) = \sum_{k=0}^{N-1} \|x_u^1(k, x_0^1)\|^2 + \|u^1(k)\|^2 + \|x_u^1(N, x_0^1)\|^2$$

$$J_N^2(x_0^2, u^2(\cdot)) = \sum_{k=0}^{N-1} \|u^2(k)\|^2$$

$$J_N^3(x_0^3, u^3(\cdot)) = \sum_{k=0}^{N-1} \|u^3(k)\|^2.$$

Hence, the closed-loop solution derived from Algorithm 9.2 with optimal control problem ($\mathrm{OCP}_\mathrm{N}^\mathrm{p}$)will steer subsystem 1 to zero. Since $u^2 \equiv 0 \equiv u^3$ is the optimal solution for subsystems 2 and 3, these subsystems will remain at their initial positions x_0^2 and x_0^3.

Between these two extremes *centralized* and *decentralized*, there are mixtures of connections between the local problems ($\mathrm{OCP}_\mathrm{N}^\mathrm{p}$)with links in dynamics, constraints, and cost function. These local optimal control problems can be sorted into two classes, the so-called *cooperative* and *noncooperative* problems. In the literature, there are different meanings of these terms. Here we refer to a scheme as *noncooperative*, if each subsystem only aims at minimizing its own local cost function without taking care of the objectives of other subsystems. Unlike the decentralized case, the neighboring subsystems may provide information to the other systems, which may be used in the optimization, but as far as the optimization is concerned, all subsystems only "mind their own business." As a consequence, the local controllers optimize to local goals only and anticipate network interactions only locally [4, 11]. The performance of the overall system is optimal at a so-called Nash equilibrium. The latter describes a point, which is by local measures optimal for each of the subsystems, i.e., changing their control strategy would increase their costs. Yet, if more than just one subsystem was to change its control strategy, then an improvement of the costs may be possible both locally and globally. In particular, the Nash equilibrium can be far from the optimal point of operation given by the centralized solution [33]. Considering the isolated cost function can be seen as an egoistic goal, and in the worst case, the respectively computed control may lead to instability of the overall system [26]. Despite these shortcomings, the noncooperative case is of high interest for street traffic [9, 23] to overcome limitations of decentralized control.

In the *cooperative* case, the local controller incorporates the effect of local control choices on the objectives of all subsystems, which are connected to it. To this end, the cost function can, e.g., be a common objective for all subsystems resembling the centralized cost, or the information on other subsystems' goals can be incorporated via information on Lagrange multipliers, as in dual decomposition approaches. In the upcoming Sect. 9.5, we illustrate both ideas by presenting the serial method of Richards and How, and the parallel dual decomposition method in Algorithms 9.2 and 9.41, respectively. Now, the aim of the local controllers is not to reduce the local costs, but to decrease the costs of the overall system. Again, the performance of the overall system is optimal at a Nash equilibrium, which may still be far from the optimum of the centralized solution [33], although under suitable structural assumptions better performance than for the noncooperative case can be expected. Obviously, for evaluating the performance of the common goal, information from the neighboring subsystems is needed, which may not be available in inter-company problems. Moreover, a subsystem may perform better in terms of its local costs if the common goal is ignored, cf., e.g., subsystems 2 and 3 in Examples 9.23 and 9.28. Due to these reasons, cooperative control is typically applied for problems for which there is only one (economic or idealistic) independent entity responsible for the guiding the system. This holds true, e.g., for companies running large plants like in chemical engineering [32] or in supply chains [12].

Remark 9.29 Within the setting of cooperative control, there exists a special case where the dynamics are independent, the constraints are coupled and the subsystems interact using both local and common costs, cf. [31]. This is referred to as *coopetitive* scenario and can be found in, e.g., market economics, where subsystems may compete with some of their neighbors [13].

The cooperative and noncooperative problems are combined in the so-called class of distributed problems. As the properties of centralized and decentralized problems can be analyzed using techniques outlined in the previous Chaps. 5 and 6, we focus on the distributed setting now. A respective algorithm incorporating the different possible connections then reads

Algorithm 9.30 (*Distributed NMPC algorithm for local problems*)
At each sampling time t_n, $n = 0, 1, 2 \ldots$

(1) For each subsystem $p \in \mathscr{P}$ measure the state $x^p(n)$ of the subsystem and set $x_0^p := x^p(n)$.

(2) For each subsystem $p \in \mathscr{P}$ do

(a) Obtain neighboring index set $\mathscr{I}^p(n)$ and collect data i^p.

(b) Solve the local optimal control problem $(\mathrm{OCP}_\mathrm{N}^\mathrm{p})$

$$\begin{aligned}
\text{minimize} \quad & J_N^p(x_0^p, u^p(\cdot), i^p) := \sum_{k=0}^{N-1} \omega_{N-k} \ell^p(x_{u^p}^p(k, x_0^p, i^p), u^p(k), i^p)) \\
& \qquad + F^p(x_{u^p}^p(N, x_0^p, i^p), i^p) \\
\text{with respect to } & u^p(\cdot) \in \mathbb{U}_{\mathbb{X}_0^p}^{p,N}(x_0^p, i^p) \quad \text{subject to} \\
& x_{u^p}^p(0, x_0^p, i^p) = x_0^p, \quad x_{u^p}^p(k+1, x_0^p, i^p) = f^p(x_{u^p}^p(k, x_0^p, i^p), u^p(k), i^p)
\end{aligned}$$

and denote the obtained admissible local optimal control sequence by $u^p(\cdot)$.

(c) Send data $(p, x^p(\cdot), u^p(\cdot))$ to all subsystems $q \in \mathscr{P} \setminus \{p\}$.

until $u^p(\cdot)$ and i^p has converged for all $p \in \mathscr{P}$.

(3) For each subsystem $p \in \mathcal{P}$ define the NMPC-feedback value $\mu_N^p(x^p(n), i^p) := u^p(0)$ and use this control value in the next sampling period.

Remark 9.31 In the linear case, the differences between the cooperative and the noncooperative case can be analyzed and illustrated elegantly using arguments from linear algebra. A very nice compendium can be found in [25, Chaps. 6.2 and 6.3].

To arrive at a converged or approximately converged solution, iterative and non-iterative methods are known in the literature, which we will consider exemplarily in Sect. 9.5. Algorithm 9.30 accounts for both approaches by the convergence requirement in Step (2). We will now use the outlined scheme to analyze stability of the distributed setting.

9.4 Asymptotic Stability and Convergence

In this section, we investigate the stability of the closed-loop trajectories generated by Algorithm 9.30. Since in general we do not assume that the distributed or decentralized optimization in Step (2) of the algorithm yields an optimal control for the overall system, the results will be weaker than those in Chaps. 5 and 6. It will turn out that in general we can only ensure that the closed-loop solution converges to the desired equilibrium, but not that the closed loop is stable. That means, we can only guarantee that $x_{\mu_N}(n, x_0)$ converges to x_*, but not that the distance $|x_{\mu_N}(n, x_0)|_{x_*}$ is bounded from above by a $\mathcal{KL}$-function β. Moreover, since we did not yet specify how exactly the optimization in Step (2) of Algorithm 9.30 works, we formulate the results based on relatively abstract assumptions. Details of the implementation, e.g., whether terminal conditions or not are used in order to guarantee these assumptions, will be discussed in the subsequent section for particular implementations of Step (2).

Here, we use the following notation. With $x_n^p = x^p(n)$ we denote the initial conditions for the optimization in time step n and we write $x_n = (x_n^1, \ldots, x_n^P)$. Note that in the nominal setting the identity $x_n = x_{\mu_N}(n, x_0)$ holds. The symbols i_n^p and u_n^p denote the control and the neighboring data delivered by Step (2) of Algorithm 9.30 upon convergence in the nth time step. To ensure that the optimization delivers a feasible solution, we impose the following assumption.

Assumption 9.32 In Algorithm 9.30, for each $x_n = \left(x_n^1, \ldots, x_n^P\right) \in \mathbb{X}$ measured in Step 1 there exist i^p for all $p \in \mathcal{P}$ such that $\mathbb{U}^{p,N}_{\mathbb{X}_0^p}(x_n^p, i^p) \neq \emptyset$ and Step 2 terminates successfully.

We note that the assumption consists of two statements: on the one hand, it assumes that a feasible solution exists, and on the other hand that the algorithm is able to find it. In the next section, we will see sufficient conditions regarding the problem and the algorithm under which Assumption 9.32 holds.

The open-loop functionals corresponding to the controls delivered by Step (2) of Algorithm 9.30 are abbreviated to

$$J^p_{N,n} := J^p_N(x^p_n, u^p_n(\cdot), i^p_n).$$

Moreover, we write $J_{N,n} := \sum_{p=1}^P J^p_{N,n}$ and recall that $\ell(x,u) = \sum_{p=1}^P \ell^p(x^p, u^p, i^p)$. With this notation, we can formulate the following convergence theorem. In its assumptions, the fact that we cannot expect optimality of the controls computed in Step (2) of Algorithm 9.30 is reflected by assuming a lower bound on $J_{N,n}$ in the first part only. In case we can guarantee a certain quality of the solution, it makes sense to impose an upper bound, which is done in inequality (9.15) in the second part of the theorem.

Theorem 9.33 (Convergence in the distributed setting) *Consider the NMPC Algorithm 9.30 and suppose that Assumption 9.32 holds. Moreover, suppose there exist $\alpha_1, \alpha_3 \in \mathcal{K}_\infty$ and $\alpha > 0$ such that for all initial conditions $x_0 \in \mathbb{X}$, all time instants $n \in \mathbb{N}_0$ and all $(x,u) \in \mathbb{Y}$ the inequalities*

$$\alpha_1(|x|_{x_*}) \leq J_{N,n} \quad \text{and} \quad \alpha_3(|x|_{x_*}) \leq \ell(x,u) \tag{9.12}$$

$$J_{N,n} \geq \alpha\ell(x_n, \mu_N(x_n)) + J_{N,n+1} \tag{9.13}$$

hold.

Then for each $x_0 \in \mathbb{X}$ the nominal NMPC closed-loop trajectory $x_{\mu_N}(n, x_0) = (x^1_{\mu^1_N}(n, x^1_0), \ldots, x^P_{\mu^P_N}(n, x^P_0))$ converges to x_* and the inequality

$$J^{cl}_\infty(x_0, \mu_N) \leq J_{N,0}/\alpha \tag{9.14}$$

holds.

If, in addition, there exists $\alpha_2 \in \mathcal{K}_\infty$ such that for all initial conditions $x_0 \in \mathbb{X}$ and all time instants $n \in \mathbb{N}_0$ the inequality

$$J_{N,n} \leq \alpha_2(|x_n|_{x_*}) \tag{9.15}$$

holds, then there exists $\beta \in \mathcal{KL}$ such that the nominal closed-loop trajectories satisfy

$$|x_{\mu_N}(n, x_0)|_{x_*} \leq \beta(|x_0|_{x_*}, n)$$

for all $x_0 \in \mathbb{X}$ and all $n \in \mathbb{N}_0$.

Proof From the assumptions, it immediately follows that $J_{N,n}$ is nonnegative and satisfies

$$J_{N,n+1} \leq J_{N,n} - \alpha\alpha_3(|x_{\mu_N}(n, x_0)|_{x_*}),$$

where we used that for the nominal closed-loop system the identity $x_n = x_{\mu_N}(n, x_0)$ holds. Hence, $J_{N,n}$ is strictly decreasing in n and bounded from below by 0. Therefore, it converges to some value $J_{N,\infty}$ as $n \to \infty$. Since every convergent sequence is a Cauchy sequence, we have that for every $\varepsilon > 0$ there exists $n_0 \in \mathbb{N}$ such that $\alpha\alpha_3(|x_{\mu_N}(n,x_0)|_{x_*}) \leq |J_{N,n} - J_{N,n+1}| \leq \varepsilon$ holds for all $n \geq n_0$, which yields $\alpha_3(|x_{\mu_N}(n,x_0)|_{x_*}) \to 0$ as $n \to \infty$. Since $\mathcal{K}_\infty$ functions can only converge to 0 if their argument converges to 0, we can conclude $|x_{\mu_N}(n,x_0)|_{x_*} \to 0$ and thus the desired convergence.

Using (9.13) and again $x_n = x_{\mu_N}(n,x_0)$ inductively for $n = 0,\ldots,K-1$, we obtain

$$\alpha \sum_{n=0}^{K-1} \ell(x_{\mu_N}(n,x_0), \mu_N(x_{\mu_N}(n,x_0))) \leq J_{N,0} - J_{N,K} \leq J_{N,0}.$$

Hence (9.14) follows for $K \to \infty$.

Under the additional inequality (9.15), we can construct $\beta \in \mathcal{KL}$ exactly as in the proof of Theorem 2.19. Just as in the proof of this theorem we can then establish the claimed inequality. □

One way of ensuring the inequalities (9.12)–(9.15) is by assuming them to hold for all subsystems. This is made precise in the following corollary:

Corollary 9.34 *Consider the NMPC Algorithm 9.30 and suppose that Assumption 9.32 holds. Moreover, suppose there for each $p \in P$ there exist $\alpha_1^p, \alpha_3^p \in \mathcal{K}_\infty$ and $\alpha > 0$ such that for all initial conditions $x_0 \in \mathbb{X}$, all time instants $n \in \mathbb{N}_0$ and all $x^p \in X^p$, $u^p \in U^p$ and $i^p \in \mathcal{I}^p(n)$ the inequalities*

$$\alpha_1^p(|x|_{x_*}) \leq J_{N,n}^p \quad \text{and} \quad \alpha_3^p(|x|_{x_*}) \leq \ell^p(x^p,u^p,i^p) \tag{9.16}$$

$$J_{N,n}^p \geq \alpha\ell^p(x_n^p, \mu_N^p(x_n^p,i_n^p), i_n^p) + J_{N,n+1}^p \tag{9.17}$$

hold.

Then for each $x_0 \in \mathbb{X}$ the nominal NMPC closed-loop trajectory $x_{\mu_N}(n,x_0) = (x^1_{\mu_N^1}(n,x_0^1),\ldots,x^P_{\mu_N^P}(n,x_0^P))$ converges to x_* and the inequality

$$J_\infty^{cl}(x_0,\mu_N) \leq J_{N,0}/\alpha$$

holds.

If, in addition, for each $p \in P$ there exists $\alpha_2^p \in \mathcal{K}_\infty$ such that for all initial conditions $x_0 \in \mathbb{X}$ and all time instants $n \in \mathbb{N}_0$ the inequality

$$J_{N,n}^p \leq \alpha_2^p(|x_n|_{x_*}) \tag{9.18}$$

holds, then there exists $\beta \in \mathcal{KL}$ such that the nominal closed-loop trajectories satisfy

$$|x_{\mu_N}(n,x_0)|_{x_*} \leq \beta(|x_0|_{x_*}, n)$$

for all $x_0 \in \mathbb{X}$ and all $n \in \mathbb{N}_0$.

Proof Conditions (9.16)–(9.18) imply (9.12)–(9.15) with $\alpha_i = \sum_{p=1}^{P} \alpha_i^p$. Hence, the assertion follows from Theorem 9.33. □

Corollary 9.34 allows us to check the conditions for convergence and asymptotic stability at the subsystem level. While this may simplify the analysis for certain classes of algorithms, it poses considerably stronger conditions. The reason for this is that assumptions on the overall system level allow for increases in $J_{N,n}^p$ for some subsystems $p \in \mathscr{P}$ in some regions of $\mathbb{X}_N^p$ as long as such an increase is compensated by a decrease in $J_{N,n}^q$ for other subsystems $q \in \mathscr{P}$ such that the overall cost $J_{N,n}$ still decreases. In terms of convergence, the impact of the increase in $J_{N,n}^p$ can then be neglected. In some situations, it may be unavoidable that $J_{N,n}^p$ increases for some subsystems $p \in \mathscr{P}$. An example for such a situation is the following:

Example 9.35 Consider the symmetric setting of two cars attempting to cross a one-lane bridge, where the dynamics is given by

$$x^p(n+1) = f^p(x^p(n), u^p(n)) = x^p(n) + u^p(n)$$

with $x^p \in \mathbb{Z}^2$ and $u^p \in \mathbb{U}^p = \{-1, 0, 1\}^2$ for $p \in \{1, 2\}$. We impose the local constraint of the bridge

$$x_1^p = 0 \implies x_2^p = 0$$

and the connecting collision and passing constraints

$$x^1 \neq x^2 \quad \wedge \quad f^1(x^1, u^1) \neq x^2 \quad \wedge \quad f^2(x^2, u^2) \neq x^1,$$

cf. Fig. 9.4 for a sketch. Regarding the cost function, we consider the case without terminal conditions and use $J^p(x^p, u^p(\cdot)) = \sum_{k=0}^{N-1} \|x_{u^p}^p(k, x^p) - x_*^p\|_1^2$ with $x_*^1 = (-2, 0)^\top$ and $x_*^2 = (2, 0)^\top$ with symmetric initial values $x_0^1 = (1, 0)^\top$ and $x_0^2 = (-1, 0)^\top$. In this setting, one car has to hold its position and move aside in the next step—thereby increasing its costs—to let the other car pass. Here, we assume without loss of generality that car $p = 2$ waits and moves aside, cf. Fig. 9.5 for an illustration. Now we analyze the results obtained for the optimal solution. The respective cost function values are given in Table 9.1. We observe that the assumption regarding the decrease of the value function $J_{N,n}^2$ is satisfied for $N \geq 4$. This is because for larger horizons, the additional cost of stepping aside is compensated by being able to cross the bridge afterwards. However, from a computational point of view, it is always desirable to choose the optimization horizon as short as possible and one observes that the subsystems approach their targets also for $N \in \{2, 3\}$, although the cost of System 2 does not decrease in every step. This is because the increase (or nondecrease) in $J_{N,n}^2$ for $n \in \{0, 1\}$ and the respective horizons can be compensated by the reduction in $J_{N,n}^1$, see the last column of Table 9.1, which shows a strict decrease in the combined value function.

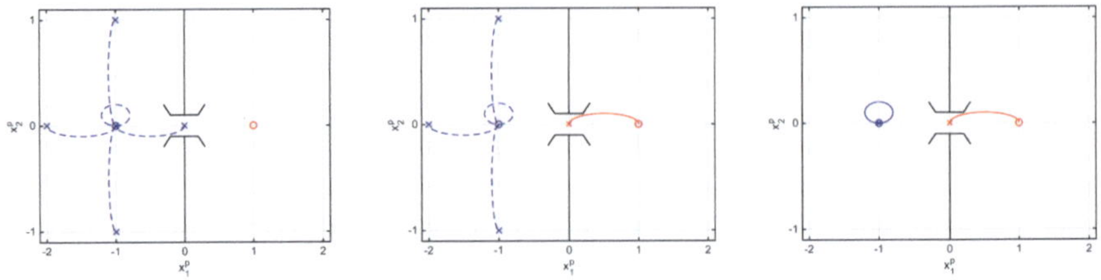

(a) Feasible moves of x^2 if x^1 is off the bridge **(b)** Feasible moves of x^2 if x^1 enters bridge **(c)** Optimal solution

Fig. 9.4 Feasible moves and optimal solution for $p = 1$ (*red*) and $p = 2$ (*blue*)

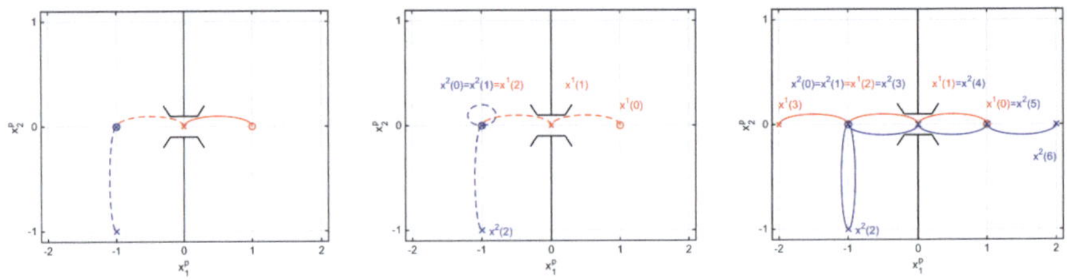

(a) Optimal solution for $n = 1$ and $N = 1$ **(b)** Optimal solution for $n = 0$ and $N = 2$ **(c)** Global optimal solution

Fig. 9.5 Optimal open- and closed-loop trajectories for $p = 1$ (*red*) and $p = 2$ (*blue*)

Table 9.1 Development of cost function

n	$J^2_{N,n}$					$J^1_{N,n} + J^2_{N,n}$
	$N = 2$	$N = 3$	$N = 4$	$N = 5$	$N \geq 6$	$N = 2$
0	18	28	37	41	42	31
1	19	28	32	33	33	24
2	19	23	24	24	24	20
3	13	14	14	14	14	13
4	5	5	5	5	5	5
5	1	1	1	1	1	1
≥ 6	0	0	0	0	0	0

In case of communication errors, we have to take robustness into account to show semiglobal practical asymptotic stability of the system. We have to particularly focus on implications regarding the dynamics and the constraints, which may lead to model errors and infeasibility. We have to assure that viability of the sets $\mathbb{X}^p(i^p)$ as i^p is retained under perturbations, i.e., that the gain is sufficiently strong regarding the dynamics and the constraints. If the latter holds, then each perturbed subsystem is semiglobally practically asymptotically stable in the sense of Definition 7.24. This allows us to apply Corollary 9.34 to show semiglobally practically asymptotic stability of the overall system.

9.5 Communication and Coordination Schemes

After considering convergence and asymptotic stability in an abstract setting, we shift our focus to actual implementations of distributed NMPC schemes. Here, we present two different schemes, which can be seen as prototypes for more advanced methods: The scheme by Richards and How [27, 28] and the dual decomposition approach [2, 16]. Our aim in this section is to verify that Assumption 9.32 and the assumptions of Theorem 9.33 hold for these implementations.

The fundamental difference between the two schemes lies in their coordination and their communication schedule. The scheme by Richards and How is sequential in nature, i.e., the subsystems are in a fixed order $p_1 < \cdots < p_P$. At each sampling instant, the subsystems compute their optimal solutions sequentially in this order and send their results to all other subsystems. Hence, each system is waiting until all systems which are earlier in the order have finished their computations and communicated their results. As a result, each subsystem computes and sends once in every sampling period.

In contrast to that, the dual decomposition approach is parallel in nature. Here, each subsystem computes a new iterate and sends its data to a central server. The server then computes a gradient step for coordination and sends its results to all subsystems. After receiving the new data from the server, the procedure is iterated until a stopping criterion is satisfied. Hence, there are multiple computation and communication instants per sampling period.

The difference in the communication schedule between these methods displayed below in Algorithms 9.36 and 9.41 is depicted in Figs. 9.6 and 9.7, respectively.

While the scheme by Richards and How exhibits communication between the subsystems only, the dual decomposition requires the coordinating server as an additional component. Moreover, there is only one communication instant per sampling period in the scheme by Richards and How, whereas dual decomposition may include multiple communication instants per sampling period.

We first describe the scheme by Richards and How. For simplicity of exposition, we assume that solving the local optimal control problem (OCP_N^p)as well as the

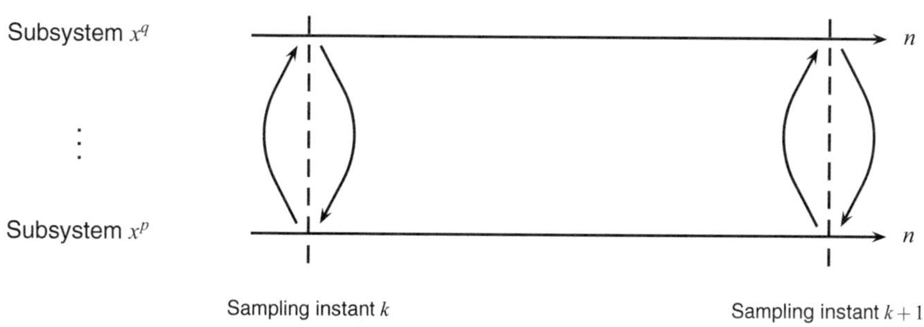

Fig. 9.6 Communication structure for scheme of Richards and How

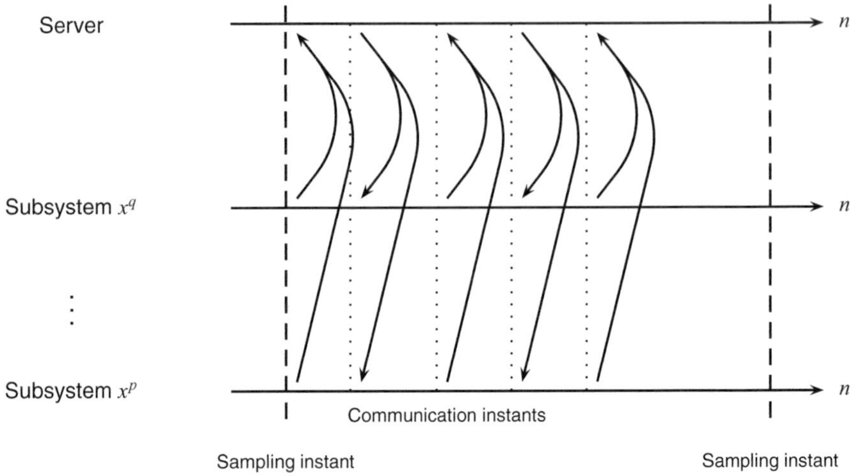

Fig. 9.7 Communication schedule for dual decomposition

communication is instantaneous. In case of significant delays, compensation methods as described in Sect. 10.6 or in [15, 36] may be applied.

Due to the sequential structure, where each subsystem is optimizing and communicating only once per sampling instant, it is not possible to generate neighboring data i^p over the whole optimization horizon N in this scheme. Hence, an evaluation of linked dynamics or costs of subsystem p is not possible if these depend on neighboring data. Indeed, when the pth system in the sequential order determines its control, it may access current neighboring data for all the systems $q < p$, while for $q > p$ only data from the past sampling instant is available. Due to the shift of the horizon, these sequences contain $N - 1$ instead of N elements required to compute the prediction. For this reason—just as in the original paper by Richards and How [27]—we restrict ourselves to systems which are only coupled via constraints. Hence, the neighboring data does not appear as an argument in $x^p_{u^p}$, ℓ and J^p_N.

To cope with missing neighboring data in the evaluation of $\mathbb{U}^{p,N}_{\mathbb{X}^p_0}(x^p_0, i^p)$, we proceed as follows: we define a set of neighboring data $\widetilde{i}^p$ of length N by $\widetilde{I}^p(i^p)$. This set contains all the neighboring data which extends the data from i^p consistently. More precisely, we define that $\widetilde{i}^p = \{(\tilde{q}, \tilde{n}_q, \tilde{x^q}(\cdot), \tilde{u^q}(\cdot)) \,|\, q \in \mathcal{I}^p(n)\} \in \widetilde{I}^p(i^p)$ holds if and only if for all $q \in \mathcal{I}^p(n)$ the identities $\tilde{q} = q$, $\tilde{n}_q = n_q$ as well as $\tilde{u}^q(k) = u^q(k)$ and $\tilde{x}^q(k) = x^q(k)$ hold for all $k = 0, \ldots, N-1$ for which $u^q(k)$ or $x^q(k)$, respectively, is defined. For those i^p containing control sequences which are not defined for all $k = 0, \ldots, N-1$, we extend Definition 9.18 of $\mathbb{U}^{p,N}_{\mathbb{X}^p_0}(x^p_0, i^p)$ via

$$\mathbb{U}^{p,N,ext}_{\mathbb{X}^p_0}(x^p_0, i^p) := \left\{ u^p \in \mathbb{U}^{p,n} \mid \text{there is } \widetilde{i}^p \in \widetilde{I}^p(i^p) \text{ with } u^p \in \mathbb{U}^{p,N}_{\mathbb{X}^p_0}(x^p_0, \widetilde{i}^p) \right\}. \tag{9.19}$$

In words, $u^p \in \mathbb{U}^{p,N,ext}_{\mathbb{X}_0^p}(x_0^p, i^p)$ holds if and only if there exist solutions for the neighboring systems, which are consistent with the available neighboring data, and which together with u^p satisfy all constraints of system p.

Algorithm 9.36 (*Serial Distributed NMPC algorithm, Richards and How*)
Initialization: At each sampling time t_0:

(1) For each subsystem $p \in \mathcal{P}$ measure the state $x^p(0)$ of the subsystem and set $x_0^p := x^p(0)$.
(2) For each subsystem $p \in \mathcal{P}$ find control sequences $u^p \in U^{p,N}$ such that

$$\left(\left(x^1_{u^1}(k, x^1_0), \ldots, x^P_{u^P}(k, x^P_0)\right), \left(u^1(k), \ldots, u^P(k)\right)\right) \in \mathbb{Y} \quad \text{for } k = 0, \ldots, N-1$$
$$\left(x^1_{u^1}(N, x^1_0), \ldots, x^P_{u^P}(N, x^P_0)\right) \in \mathbb{X}_0$$

and send $(p, 0, x^p(\cdot), u^p(\cdot))$ to all other subsystems.
(3) For each subsystem $p \in \mathcal{P}$ define the NMPC-feedback value $\mu_N^p(x_0^p) := u^p(0)$ and use this control value in the next sampling period.

Feedback loop: At each sampling time t_n, $n = 1, 2 \ldots$:

(1) For each subsystem $p \in \mathcal{P}$ measure the state $x^p(n)$ of the subsystem and set $x_0^p := x^p(n)$.
(2) For each subsystem $p \in \mathcal{P}$ do sequentially

(a) Collect neighboring data and set

$$i^p := \left(\left(1, n, x^1(\cdot), u^1(\cdot)\right), \ldots, \left(p-1, n, x^{p-1}(\cdot), u^{p-1}(\cdot)\right),\right.$$
$$\left.\left(p+1, n-1, x^{p+1}(\cdot), u^{p+1}(\cdot)\right), \ldots, \left(P, n-1, x^P(\cdot), u^P(\cdot)\right)\right).$$

(b) Solve the local optimal control problem

$$\begin{aligned}
&\text{minimize} \quad J_N^p(x_0^p, u^p(\cdot)) := \sum_{k=0}^{N-1} \omega_{N-k} \ell^p(x^p_{u^p}(k, x_0^p), u^p(k))) \\
&\qquad\qquad\qquad + F^p(x^p_{u^p}(N, x_0^p, i^p)) \\
&\text{with respect to } u^p(\cdot) \in \mathbb{U}^{p,N,ext}_{\mathbb{X}_0^p}(x_0^p, i^p) \text{ subject to} \\
&x^p_{u^p}(0, x_0^p) = x_0^p, \quad x^p_{u^p}(k+1, x_0^p) = f^p(x^p_{u^p}(k, x_0^p), u^p(k))
\end{aligned}$$

and denote the obtained admissible local optimal control sequence by $u_n^p(\cdot)$.

(c) Send data $(p, n, x^p(\cdot, x_0^p), u^p(\cdot))$ to all subsystems $q \in \mathcal{P} \setminus \{p\}$.

(3) For each subsystem $p \in \mathcal{P}$ define the NMPC-feedback value $\mu_N^p(x_0^p, i^p) := u_n^p(0)$ and use this control value in the next sampling period.

As outlined before, if the systems are coupled not only via the constraints but also via the dynamics or the cost function, then the neighboring data for subsystems $q > p$ is not complete. Filling these gaps with default values or estimates may lead to respective errors. To retain stability, the model error must be small enough to satisfy the conditions outlined at the end of Sect. 9.4. Regarding links in the constraints, Algorithm 9.36 implements a first-come first-serve fashion, i.e., subsystem $p < q$ imposes constraints on subsystem q by choosing its control first. We also note that finding an initial feasible solution in Step (2) of the Initialization can be a difficult task, which may require additional coordination between the subsystems.

An important feature of Algorithm 9.36 is that—given we can find an initial feasible solution—we can show Assumption 9.32 to hold. Here, we start by showing that if the sets of admissible controls $\mathbb{U}^{p,N}_{\mathbb{X}_0^p}(x^p(n), i^p)$ are never empty, then the closed-loop solution will remain feasible.

Proposition 9.37 *Consider Algorithm 9.36 and suppose the initialization is successful and* $\mathbb{U}^{p,N,ext}_{\mathbb{X}_0^p}(x^p(n), i^p) \neq \emptyset$ *holds for all* $n \in \mathbb{N}$. *Then*

$$x(n) = \left(x^1(n), \dots, x^P(n)\right) \in \mathbb{X}$$

holds for all $n \in \mathbb{N}_0$.

Proof The successful initialization implies that the assertion holds for $n = 0$. Now we assume $x(n) \in \mathbb{X}$ to hold for some $n \geq 1$ and $\mathbb{U}^{p,N,ext}_{\mathbb{X}_0^p}(x^p(n), i^p) \neq \emptyset$ for all $p \in \mathscr{P}$. Hence, $u_n^p(\cdot)$ and $\mu^p(x^p(n), i^p)$ are well defined and

$$(x^1_{u^{1,\star}}(1, x^1(n), i^1), \dots, x^P_{u^{P,\star}}(1, x^P(n), i^P)) \in \mathbb{X}$$

holds. By definition of the closed loop, we therefore obtain $x(n+1) \in \mathbb{X}$, which allows to apply the steps iteratively. □

Next, we will utilize the assumptions stated in Theorems 5.5, 5.13, or 6.20 in the context of Algorithm 9.36. The structure of the algorithm will allow us to show that by Step 2 of Algorithm 9.36 we have $\mathbb{U}^{p,N,ext}_{\mathbb{X}_0^p}(x_0^p, i^p) \neq \emptyset$ and the resulting closed-loop solutions converge to x^*. As neighboring data i^p may change from one sampling instant to the next, one of the main problems to be solved in this context is that the set $U^{p,N,ext}_{\mathbb{X}_0^p}(x_0^p, i^p)$ may change as well.

In case terminal conditions are imposed, we will utilize structural properties of $\mathbb{X}_0$ and $\mathbb{U}(x)$ and a local version of the Assumptions 5.1 and 5.9 in order to overcome this problem. These are summarized in the following assumption:

Assumption 9.38 (i) The terminal constraint set is of the form

$$\mathbb{X}_0 = \mathbb{X}_0^1 \times \mathbb{X}_0^2 \times \dots \times \mathbb{X}_0^P \tag{9.20}$$

(ii) The control constraint set is of the form

$$\mathbb{U}(x) = \mathbb{U}^1(x^1) \times \mathbb{U}^2(x^2) \times \cdots \times \mathbb{U}^P(x^P) \tag{9.21}$$

(iii) For each $p \in \mathscr{P}$ and each $x^p \in \mathbb{X}_0^p$ there is $u_{x^p} \in \mathbb{U}^p(x^p)$ with $f^p(x^p, u_{x^p}) \in \mathbb{X}_0^p$ and

$$F^p(f^p(x^p, u_{x^p})) + \ell^p(x^p, u_{x^p}) \leq F^p(x^p). \tag{9.22}$$

We note that this assumption also applies in the case $\mathbb{X}_0 = \{x_*\}$ and $F^p \equiv 0$ for all $p \in \mathscr{P}$, in which case (i) and (iii) are automatically satisfied provided u_* is an admissible control value and $\ell^p(x_*^p, u_*^p) = 0$.

This assumption implies the following: consider a state $x \in \mathbb{X}$ with $x^p \in \mathbb{X}_0^p$ and let $u \in \mathbb{U}(x)$ be a control value with $f(x, u) \in \mathbb{X}_0$. Then, for subsystem p the control value $u^p = u_{x^p} \in \mathbb{U}^p(x^p)$ is admissible, keeps the state x^p inside $\mathbb{X}_0^p$ and satisfies inequality (9.22). This fact will be exploited in the proof of Proposition 9.39(1) and (2).

In the case without terminal conditions, we introduce the following sets of admissible control sequences in order to cope with the time dependency of $\mathbb{U}_{\mathbb{X}_0^p}^{p,N,ext}(x_0^p, i^p)$. Suppose an initial condition x_0^p, neighboring data i^p, a control function $u^p \in \mathbb{U}_{\mathbb{X}_0^p}^{p,K,ext}(x_0^p, i^p)$ to be given. Then, for $K \in \mathbb{N}$ and time indices $l \in \mathbb{N}$, $m \in \mathbb{N}_0$ with $m \leq K$ we define

$$\widetilde{\mathbb{U}}_{\mathbb{X}_0^p}^{p,l,ext}(x_0^p, i^p, u^p, m) :=$$
$$\left\{\tilde{u}^p \in U^{p,l} \mid (u^p(0), \ldots, u^p(m-1), \tilde{u}^p(0), \ldots, \tilde{u}^p(l-1)) \in \mathbb{U}_{\mathbb{X}_0^p}^{p,m+l,ext}(x_0^p, i^p)\right\}.$$

This set contains all control sequences $\tilde{u}^p$ of length l, which are admissible for l steps into the future given that control u^p was used up to time instant m.

Since the respective proofs use similar arguments, we condense the result for Algorithm 9.36 in the following proposition covering the cases with terminal constraints, terminal costs and without terminal conditions.

Proposition 9.39 (Convergence of Serial Distributed NMPC) *Consider Algorithm 9.36 with optimization horizon N and $\omega_k = 1$ for $k = 0, \ldots, N-1$, and suppose the initialization to be successful with resulting control functions u_0^p. Assume that $(x_*, u_*) \in \mathbb{Y}$ is an equilibrium with $\ell(x_*, u_*) = 0$ and that there exist functions $\alpha_3^p \in \mathscr{K}_\infty$ such that $\ell^p(x^p, u^p) \geq \alpha_3^p(|x|_{x_*^p})$ holds for all $x^p \in X^p$, $u^p \in U^p$ and all $p \in \mathscr{P}$. Additionally, suppose either of the following:*

(1) The problem $(\mathrm{OCP}_\mathrm{N}^\mathrm{p})$*uses the terminal conditions $\mathbb{X}_0^p = \{x_*^p\}$ and $F^p \equiv 0$ for all $p \in \mathscr{P}$ and Assumption 9.38(ii) holds.*
(2) The problem $(\mathrm{OCP}_\mathrm{N}^\mathrm{p})$*uses terminal constraints satisfying Assumption 9.38.*

(3) *The problem* ($\mathrm{OCP}_\mathrm{N}^\mathrm{p}$)*does not use terminal constraints, i.e.,* $\mathbb{X}_0 = \mathbb{X}$, *and there exist functions* $B_K \in \mathcal{K}_\infty$ *for* $K = 2, \ldots, N$ *satisfying the following three conditions.*

(i) *Before the minimization in Step (2)(b) of Algorithm 9.36 the inequality*

$$\inf_{\tilde{u}^p \in \widetilde{\mathbb{U}}^{p,l,ext}_{\mathbb{X}_0^p}(x_0^p, i^p, u^p_{n-1}(\cdot+1), m)} J_l(x_{u^p_{n-1}(\cdot+1)}(m, x_0^p), \tilde{u}^p) \leq B_l(|x_{u^p_{n-1}(\cdot+1)}(m, x_0^p)|_{x_*^p})$$

holds for all $l = 2, \ldots, N$ *and* $m = N - l$

(ii) *After the minimization in Step (2)(b) of Algorithm 9.36 the inequality*

$$\inf_{\tilde{u}^p \in \widetilde{\mathbb{U}}^{p,l,ext}_{\mathbb{X}_0^p}(x_0^p, i^p, u_n^p, m)} J_l(x_{u_n^p}(m, x_0^p), \tilde{u}^p) \leq B_l(|x_{u_n^p}(m, x_0^p)|_{x_*^p})$$

holds for all $l = 2, \ldots, N$ *and* $m = N - l$

(iii) *The value* α *from Theorem 6.15 satisfies* $\alpha \in (0, 1]$

Then Assumption 9.32 holds and conditions (9.16) *and* (9.17) *of Corollary 9.34 are satisfied. Hence, the closed-loop solutions are well defined and converge to* x_*.

Proof In Cases (1) and (2) we first show that Assumption 9.32 holds by induction over n. For $n = 0$, this follows immediately from the assumption that the initialization was successful. For $n \to n+1$, note that the choice of the terminal constraints yields $x^p_{u_n^p}(N, x_n^p) \in \mathbb{X}_0^p$. Now by assumption we know that the neighboring data provided by subsystem p has been taken into account in the optimization of all other subsystems. Hence, at sampling instant $n+1$ the control sequence $\tilde{u}^p = (u_n^p(1), \ldots, u_n^p(N-1))$ is still admissible for $x^p_{n+1} = x^p_{u_n^p}(1, x_n^p)$ and yields $x^p_{\tilde{u}^p}(N-1, x^p_{n+1}) = x^p_{u_n^p}(N, x_n^p) \in \mathbb{X}_0^p$. Moreover, all predictions for subsystems $q \neq p$ contained in the neighboring data end up in $\mathbb{X}^q$. From the discussion after Assumption 9.38 it follows that the control $(u_n^p(1), \ldots, u_n^p(N-1), u_{x^p})$ with u_{x^p} from Assumptions 9.38(iii) with $x^p = x^p_{u_n^p}(N, x_n^p)$ is admissible for initial condition x_{n+1}, i.e., it lies in $\mathbb{U}^{p,N,ext}_{\mathbb{X}_0^p}(x_{n+1}, i^p)$. Thus, this set is not empty and the optimization for subsystem p will find an optimal control sequence u^p_{n+1} from this set.

To show that the assumptions of Corollary 9.34 apply, first note that by definition of $J^p_{N,n}$ the lower bound on $J^p_{N,n}$ in (9.16) holds with $\alpha_1^p = \alpha_3^p$. Now inequality (9.17) follows for $\alpha = 1$ with the same arguments as in the proof of Theorems 5.5 and 5.13 using Assumption 9.38(iii) and straightforward extensions of Lemmas 5.4 and 5.12 to the individual subsystems, respectively.

In Case (3), Assumption 9.32 holds because condition (i) implicitly demands that the set $\widetilde{\mathbb{U}}^{p,l,ext}_{\mathbb{X}_0^p}(x_0^p, i^p, u^p_{n-1}(\cdot+1), m)$ is not empty. For $l = N$ and $m = 0$, however, this set coincides with $\widetilde{\mathbb{U}}^{p,N,ext}_{\mathbb{X}_0^p}(x_0^p, i^p)$ which is thus not empty and the algorithm will find u_n^p from this set.

Inequalities in (i) and (ii) ensure that the values $\lambda_k = \ell^p(x_{u_n^p}, u_n^p)$ and $\nu = J_N^p(x_n^p, u_n^p)$ satisfy the constraints (6.11) and (6.12). More precisely, (i) ensures the inequalities (6.12) while (ii) ensure the inequalities (6.11) (for a more detailed explanation of this fact we refer to [17, Theorems 5.2 and 5.3], noting that (i) and (ii) coincide with Conditions (i) and (ii) in [17, Theorem 5.3]). Now, the assertion follows by a straightforward adaptation of the proof of Theorem 6.15. □

Remark 9.40 We can extend these results to the time varying case. To this end, in Case (1) of Proposition 9.39 we replace $\mathbb{X}_0^p = \{x_*^p\}$ by $\mathbb{X}_0^p(n) = \{x^{p,\mathrm{ref}}(n)\}$.

In Case (2) of Proposition 9.39, we can cover the time varying setting by utilizing the tracking reference $x^{p,\mathrm{ref}}$. This allows us to replace $\mathbb{X}_0^p$ by $\mathbb{X}_0^p(n)$ and generalize the conditions in Assumption 5.9 according to Remark 5.17.

Last, in Case (3) of Proposition 9.39, the functionals J_l become time varying. However, this does not affect the reasoning as long as the functions B_K do not depend on the current time.

An obvious advantage of Algorithm 9.36 is that it can be based on local optimal control problem ($\mathrm{OCP_N^p}$)with and without terminal conditions and does not require any particular properties of the involved functions. The method, however, also shows severe drawbacks. For one, it is sequential by nature. In a realistic application, the computing times to solve the problems ($\mathrm{OCP_N^p}$)would add up with the number of subsystems. Hence, the algorithm is well applicable for small P. Although it was shown in [28, Sect. 7] that the computing times using Algorithm 9.36 are lower than the necessary time to solve the respective centralized problem, the scaling problem remains. Second, the conditions needed for obtaining convergence in Proposition 9.39 are very demanding. In the terminal conditioned case, the initialization requires to find a feasible solution for all subsystems $p \in \mathscr{P}$. More precisely, the terminal conditions shift the complete coordination problem into the initialization phase, that is finding a feasible solution for all subsystems which arrives in the terminal set at the end of the prediction horizon. This task will in general require additional coordination between the subsystems which compromises the relatively low coordination effort in Step (2) of the algorithm. In the case without terminal conditions, the solutions do not need to reach the terminal constraint sets, which renders the initialization less demanding. Here, however, the conditions (3)(i) and (ii) in Proposition 9.39 are very demanding. We essentially require each subsystem p to be flexible enough to find an admissible extension of its prediction from the last step, which can be "squeezed" into the already decided moves of the subsystems $q < p$. Moreover, such solutions do not only need to exist, but in addition they also need to satisfy the inequalities demanded in (3)(i) and (ii). While for simple systems this can be rigorously verified (see, e.g., Sect. 6 in [17]), for more complex systems this appears to be a very difficult task. Finally, regardless of whether terminal conditions are imposed, the stability analysis of Algorithm 9.36 has the disadvantage that it relies on Corollary 9.34, i.e., each local value $J_{N,n}$ must decrease. Hence, we cannot verify stability in situations like those illustrated in Example 9.35.

After the discussion of the sequential approach of Richards and How, we now briefly outline the *dual decomposition* method. As already stated before, this method is (almost) completely parallel, i.e., all subproblems are solved at the same time. To introduce the method, we restrict ourselves to problems with $X = \mathbb{R}^d$ and $U = \mathbb{R}^m$, where the set $\mathbb{U}^N_{\mathbb{X}_0}(x_0)$ can be represented by linear equalities, i.e.,

$$\begin{array}{|l|}\hline \text{minimize} \quad J_N(x_0, u(\cdot)) \\ \text{with respect to } u(\cdot) \text{ subject to } Qx_0 + Ru(\cdot) = c \\ \hline \end{array} \tag{9.23}$$

where the matrices Q and R together with the vector c (all of appropriate dimensions) represent the constraints $u(\cdot) \in \mathbb{U}^N_{\mathbb{X}_0}(x_0)$. We note that for representing these constraints in form of an equation, this equation needs to contain all information about the dynamics of the system. Hence, the linear form of the equation in the last line in (9.23) implies that the formulation is limited to linear dynamics and constraints, which we have done in order to simplify the presentation. In general, dual decomposition demands suitable convexity properties of the optimization problem to be solved. Therefore, the approach is not strictly limited to linear dynamics and constraints, but the type of nonlinearities that can be considered is rather limited. As a consequence, nonlinear constraints and dynamics must be approximated by linearizations, which is an obvious drawback of the method.

We also assume the cost function J_N to be additively separable, i.e., that

$$J_N(x_0, u(\cdot)) = \sum_{p=1}^{P} J_N^p(x_0^p, u^p(\cdot))$$

holds, where J_N^p can be evaluated without the knowledge of $x^q(\cdot)$ and $u^q(\cdot)$ for $q \neq p$. This is clearly restrictive, however, the assumption can always be satisfied by introducing auxiliary optimization variables $x^{q,p}(\cdot)$ and $u^{q,p}(\cdot)$ in subsystem p, which represent the state and control of subsystem q, and adding respective constraints which enforce $x^{q,p}(\cdot) = x^q(\cdot)$ and $u^{q,p}(\cdot) = u^q(\cdot)$.

For an illustration of this procedure, consider Example 9.1. Here we have two subsystems with dynamics

$$\begin{aligned} f^1(x^1, x^2, u^2) &= x^1 + x^2 + u^2/2 \\ f^2(x^2, u^2) &= x^2 + u^2 \end{aligned}$$

and thus the knowledge of $x^2(\cdot)$ and $u^2(\cdot)$ is needed in order to compute a prediction $x_u^1(\cdot)$. Introducing the auxiliary state $x^{2,1}$ and control $u^{2,1}$, we can express the new dynamics for subsystem 1 as

$$f^1(x^1, x^{2,1}, u^{2,1}) = x^1 + x^{2,1} + u^{2,1}/2.$$

All elements of the sequences $x^{2,1}(\cdot)$ and $u^{2,1}(\cdot)$ become new optimization variables, which are subject to the additional constraints

$$x^{2,1}(\cdot) = x^2(\cdot), \quad x^{2,1}(\cdot) = x^2(\cdot).$$

This connection is expressed as part of the constraint equation in (9.23).

The basis for applying dual decomposition is the separation of the minimization and constraint satisfaction into two steps. Consequently, the minimization of the J_N^p can be done completely in parallel. The separation is achieved by the so-called dual ascent method, which is based on the Lagrange function

$$L\left(x_0, u(\cdot), \lambda\right) = J_N(x_0, u(\cdot)) + \lambda^\top \left(Qx_0 + Ru(\cdot) - c\right)$$

with dual

$$g(\lambda) = \inf_{u(\cdot)} L\left(x_0, u(\cdot), \lambda\right). \tag{9.24}$$

In the literature, (9.23) is typically referred to as primal problem, whereas minimizing (9.24) is called dual problem. By strong duality [3, Appendix B.4], we have that the optimal cost function values of these two problems are identical. Moreover, we can derive a minimizer $u^\star(\cdot)$ of (9.23) via

$$u^\star(\cdot) = \operatorname*{argmin}_{u(\cdot)} L(x_0, u(\cdot), \lambda^\star),$$

if there exists a unique solution $\lambda^\star = \operatorname{argmin}_\lambda g(\lambda)$.

Instead of solving the two problems (minimization of g and minimization of L) separately, we now compute the solution iteratively using a two stage scheme, the so-called *dual ascent* method. In the first stage, the primal variable is updated to approach a minimizer satisfying the constraints

$$u^{k+1}(\cdot) := \operatorname*{argmin}_{u(\cdot)} L\left(x_u(\cdot, x_0), u(\cdot), \lambda^k\right). \tag{9.25}$$

In the second step, the Lagrange multiplier, i.e., the price of violating the constraints is adjusted using

$$\lambda^{k+1} := \lambda^k + \rho_k \left(Qx_0 + Ru^{k+1}(\cdot) - c\right) \tag{9.26}$$

where $\rho_k > 0$ is the step size. Effectively, (9.26) is a step of a steepest ascent method for solving the dual problem, hence the name *dual ascent*.

Dual decomposition refers to the technique of decomposing the Lagrangian to solve (9.25) in parallel for the subsystems $p \in \mathscr{P}$. To this end, we split the matrices

$$Q = \left[Q^1 \cdots Q^P\right] \quad \text{and} \quad R = \left[R^1 \cdots R^P\right]$$

such that we obtain

$$Qx_0 = \sum_{p=1}^{P} Q^p x_0^p \quad \text{and} \quad Ru(\cdot) = \sum_{p=1}^{P} R^p u^p(\cdot).$$

Then, defining the Lagrangian of the pth subsystem as

$$L^p\left(x_0^p, u^p(\cdot), \lambda\right) = J_N(x_0, u(\cdot)) + \lambda^\top \left(Q^p x_0^p + R^p u^p(\cdot) - \frac{1}{P}c\right) \tag{9.27}$$

one easily sees that

$$L\left(x_0, u(\cdot), \lambda\right) = \sum_{p=1}^{P} L^p\left(x_0^p, u^p(\cdot), \lambda\right).$$

Since the problems of minimizing $L^p\left(x_0^p, u^p(\cdot), \lambda\right)$ with respect to $u^p(\cdot)$ are completely decoupled, they can be carried out in parallel on subsystem level. This leads to the following algorithm:

Algorithm 9.41 (*Parallel Distributed NMPC algorithm, Dual Decomposition*)
At each sampling time t_n, $n = 0, 1, 2\ldots$:
At subsystem:

(1) For each subsystem $p \in \mathcal{P}$ measure the state $x^p(n)$ of the subsystem and set $x_0^p := x^p(n)$, $\lambda^0 := 0$ and $k := 0$.
(2) For each subsystem $p \in \mathcal{P}$ do

(2a) Collect data $(0, n, \lambda^k)$ containing external data and Lagrange multiplier
(2b) Compute a minimizer for the Lagrangian (9.27) of the local optimal control problem

$$u^{p,k+1}(\cdot) := \operatorname*{argmin}_{u^p(\cdot)} L^p\left(x_0^p, u^p(\cdot), \lambda^k\right)$$

and denote the obtained control sequence by $u_n^p(\cdot)$.
(2c) Send data $\left(p, n, x_0^p, u^{p,k+1}(\cdot)\right)$ to central entity.

At central entity:

(3a) Collect subsystem data

$$i^{p,k+1} = \left(p, n, x_0^p, u^{p,k+1}\right) \quad \text{for all } p \in \mathcal{P}.$$

(3b) Update Lagrange multiplier

$$\lambda^{k+1} := \lambda^k + \rho_k \left(Qx_0 + Ru(\cdot)^{k+1} - c \right)$$

(3c) Send Lagrange multiplier as data package $(0, n, \lambda^{k+1})$ to all subsystems $p \in \mathscr{P}$. Set $k := k + 1$ and go to (2) unless a termination criterion is satisfied.

At subsystem:

(4) For each subsystem $p \in \mathscr{P}$ define the NMPC-feedback value $\mu_N^p \left(x_0^p \right) := u_n^p(\cdot)(0)$ and use this control value in the next sampling period.

Under suitable conditions, convergence of the loop in (2), (3) toward a solution of (9.23) can be ensured. Here we do not go into details but only refer to, e.g., [1, Chap. 6] or [2, Sect. 2 and the references therein]. For an analysis of a continuous time version of a dual decomposition algorithm, see also [24].

Provided that the dual decomposition algorithm delivers a solution to the centralized problem, the stability and suboptimality analysis of Algorithm 9.41 is straightforward. Indeed, since the resulting solutions are identical to the centralized ones, each of the stability and suboptimality results from Chaps. 5–8 applied to the centralized problem satisfies the respective assumptions.

From the construction of Algorithm 9.41, we directly see that all subsystems are computing their minimizers in parallel. Therefore, there is no fixed order of subsystems rendering them to be equal, which is different from Algorithm 9.36 where subsystems are ordered. If the dynamics of the subsystems are coupled, then parallel execution of the optimization typically renders neighboring data to be invalid, i.e., for each iterate k of the optimization we have a mismatch between the predicted states $x^q(\cdot)$ and its estimate the $x^{q,p}(\cdot)$ in subsystem p. The coupling via the Lagrange multipliers, however, causes this mismatch to converge to 0 as the iteration proceeds. Moreover, during the initial phase (i.e., the first local optimization at time t_0), the algorithm is not required to find a feasible solution in one step, but for a non-feasible solution feasibility will be restored in the dual decomposition iteration (2)–(3). Hence, the initial step in Algorithm 9.41 is less demanding than in Algorithm 9.36.

Besides the considerable structural assumptions on the problem which are needed in order to guarantee convergence, the drawbacks of Algorithm 9.41 are its communication load and the additional central entity. In contrast to Algorithm 9.36, each subsystem may send data multiple times within each sampling interval, cf. Fig. 9.7. For a large number of participating subsystems, this may lead to congestion in the communication channel and respective delays and packet dropouts. Hence, the scheme may not be real-time computable or result in infeasible solutions. The additional central entity is easy to implement for small problems in plants or local energy networks. Yet, it is particularly difficult to realize for large-scale systems such as in street traffic. Here, a vast number of road side units would have to be deployed to form a backbone communication system, which is a major infrastructure investment.

9.6 Notes and Extensions

Within this chapter, we presented distributed NMPC as a problem driven by two recent developments: on one side, distributed settings such as traffic, production, or energy systems require improved coordination and usage of shared resources due to economic or ecologic pressure. On the other side, centralized solutions become more complex and real-time requirements cause researchers to design distributed implementations to lower the computational burden. In both cases, links in dynamics, costs, and constraints may arise, which need to be considered in order to guarantee stability of the overall system. In what follows, we mention works which extend and complement this chapter analytically and algorithmically.

Within the literature, many articles focus on particular combinations of links in dynamics, costs, and constraints between subsystems. Insights into the analytical part are given, as mentioned, in the compendium of Rawlings and Mayne [25] for the linear case in discrete time, where the authors display interactions and links between game theory, control theory, and distributed optimization. In the work of Farina, Betti, and Scattolini [14], the entire range of links in dynamics, costs, and constraints is considered for the linear continuous time setting. Regarding links in costs and constraints, Camponogara and Scherer [6] extended the linear discrete time case to dynamics with outputs. The introduced output constraints are essential to model systems, for which we cannot measure the entire state such as production or traffic systems. Last, the case of nonseparable costs and coupled constraints was elaborated by Chiang, Nedic and Scaglione [7].

Within this chapter we presented the foundations of distributed MPC in a perfect setting, i.e., with perfect communication and evaluation capabilities, which is also a typical setting within the literature [2, 25, 28]. In particular, we assumed that each subsystem can communicate with any other subsystem without delays, corruptions, or losses of data and as often as required by the coordination scheme. In [9], Dold and Stursberg proposed an extension of the robustness approach by Kerrigan [18] using robust strong feasibility to the distributed case of linear discrete time systems with links in constraints and costs.

From the algorithmic point of view, many new methods have been proposed extending both the serial approach by Richards and How [27, 28] and the dual decomposition approach and its variants [2]. Regarding the latter, main motivation for our description was given by Giselsson and Rantzer [16], who analyzed the usage of dual decomposition for linear MPC problems. Müller, Reble and Allgöwer [21] extended the setting of Richards and How and considered the nonlinear discrete time setting with links in costs and constraints. Instead of solely focusing on stability, they also considered the cooperative tasks of consensus and synchronization. In [29], Savorgnan, Romani, Kozma, and Diehl proposed an extension of multiple shooting in the direction of dual decomposition approach and the methods of multipliers for nonlinear dynamics coupled by inputs and outputs, which allows for structure exploitation and a respective speedup in the computing time. Considering a spatial instead of time separation, Kouzoupis, Quirynen, Houska, and Diehl [19] proposed an alternative to the methods of multipliers to improve distributed solution methods for problems with linked constraints and costs. O'Donoghue, Stathopoulos and Boyd [22] split the operations of the method of multipliers and showed that under certain conditions an implementation using FPGAs is possible, which allows for very high sampling rates.

The last issue which is not covered in this chapter is the optimal choice regarding the partitioning of the overall system. Regarding naturally distributed systems, Scattolini [30] outlined a hierarchical structure to cope with the communication load, but left the subsystem structure intact. Motee and Sayyar-Rodsari [20] describe a method to optimally partition the dynamics, which they balance against the cost of the closed loop.

Problems

1. Reconsider Example 9.11 with dynamics

$$x(n+1) = \begin{pmatrix} x^1(n+1) \\ x^2(n+1) \end{pmatrix} = \begin{pmatrix} 1 & 0 \\ 0 & 1 \end{pmatrix} \begin{pmatrix} x^1(n) \\ x^2(n) \end{pmatrix} + \begin{pmatrix} u^1(n) \\ u^2(n) \end{pmatrix} =: f(x(n), u(n)).$$

   constraints $|x^1 - x^2| \geq \delta > 0$ and costs

$$\begin{aligned} J_N(x_0, u(\cdot)) = \sum_{k=0}^{N-1} \left(\sum_{p=1}^{P} \|x_{u^p}^p(k, x_0^p, i^p) - x_*^p\|^2 + \|u^p(k)\|^2 \right) \\ + \sum_{p=1}^{P} \|x_{u^p}^p(N, x_0^p, i^p) - x_*^p\|^2 \end{aligned}$$

   with $x_0^1 \neq x_0^2$ and $x_*^1 = x_*^2$. Show that Assumption 9.32 holds and explain why Theorem 9.33 still fails to apply.
2. Consider the inverted pendulum from Example 2.10 with dynamics

$$\begin{aligned} \dot{x}_1(t) &= x_2(t) \\ \dot{x}_2(t) &= u(t) \\ \dot{x}_3(t) &= x_4(t) \\ \dot{x}_4(t) &= -\frac{g}{l}\sin(x_3(t)) - u(t)\cos(x_3(t)) - \frac{k_L}{l}x_4(t)|x_4(t)| - k_R\text{sgn}(x_4(t)) \end{aligned}$$

   Extending the single pendulum to a p pendulum (for each pendulum the tip is attached to the end of the previous pendulum) and separating cart and each pendulum into subsystems, show that the Algorithm of Richards and How can only be applied for one order of subsystems.

3. Consider two cars modeled as points in the plane with dynamics

$$x^p(n+1) = x^p(n) + u^p(n)$$

where $x^p(n) \in \mathbb{R}^2$ and $u^p(n) \in [-\delta, \delta] \subset \mathbb{R}^2$. Assume that the constraints are given by $\|x^1(n) - x^2(n)\|_2 \geq 1$ and the cost function is defined via

$$J_N^p(x^p, u^p) = \sum_{k=0}^{N-1} \|x_{u^p}^p(k; x_0^p) - x_*^p\|^2$$

$\|x_*^1 - x_*^2\|_2 \geq 1$. Given $\delta = 1$, what is the minimal horizon N such that for each p we observe a decrease in J_N^p for all feasible initial values? Does the minimal horizon change for smaller δ?

References

1. Bertsekas, D.P.: Nonlinear Programming, 2nd edn, Athena Scientific, Belmont (2003)
2. Boyd, S., Parikh, N., Chu, E., Peleato, B., Eckstein, J.: Distributed optimization and statistical learning via the alternation direction method of multipliers. Found. Trends Mach. Learn. **3**(1), 1–122 (2011)
3. Boyd, S., Vandenberghe, L.: Convex Optimization. Cambridge University Press, Cambridge (2004)
4. Camponogara, E., Jia, D., Krogh, B., Talukdar, S.: Distributed model predictive control. IEEE Control Syst. Mag. **22**, 44–52 (2002)
5. Camponogara, E., de Lima, M.: Distributed optimization for MPC of linear networks with uncertain dynamics. IEEE Trans. Autom. Control **57**(3), 2613–2618 (2012)
6. Camponogara, E., Scherer, H.: Distributed optimization for model predictive control of linear dynamic networks with control-input and output constraints. Autom. Sci. Eng. **8**(1), 233–242 (2011)
7. Chang, T.H., Nedic, A., Scaglione, A.: Distributed constrained optimization by consensus-based primal-dual perturbation method. IEEE Trans. Autom. Control **59**(6), 1524–1538 (2014)
8. Cui, H., Jacobsen, E.: Performance limitations in decentralized control. J. Process Control **7**(12), 485–494 (2002)
9. Dold, J., Stursberg, O.: Distributed predictive control of communicating and platooning vehicles. In: Proceedings of the 48th IEEE Conference on Decision and Control held jointly with the 28th Chinese Control Conference CDC/CCC2009, pp. 561–566 (2009)
10. Dold, J., Stursberg, O.: Distributed Predictive Control of Communicating and Platooning Vehicles. In: Proceedings of the 48th IEEE Conference on Decision and Control held jointly with the 28th Chinese Control Conference CDC/CCC 2009, pp. 561–566 (2009)
11. Dunbar, W.: Distributed receding horizon control of dynamically coupled nonlinear systems. IEEE Trans. Autom. Control **52**(7), 1249–1263 (2007)
12. Dunbar, W., Desa, S.: Assessment and future directions of nonlinear model predictive control, chap. Distributed MPC for Dynamic Supply Chain Management, pp. 607–615. Springer, Berlin (2007)
13. Easley, D., Kleinberg, J.: Networks, Crowds, and Markets: Reasoning About a Highly Connected World. Cambridge University Press, Cambridge (2010)
14. Farina, M., Betti, G., Scattolini, R.: Distributed predictive control of continuous-time systems. Syst. Control Lett. **74**, 32–40 (2014)

15. Findeisen, R., Allgöwer, F.: Computational Delay in Nonlinear Model Predictive Control. In: Proceedings of the International Symposium on Advanced Control of Chemical Processes (2004)
16. Giselsson, P., Rantzer, A.: Distributed Model Predictive Control with Suboptimality and Stability Guarantees. In: Proceedings of the 49th IEEE Conference on Decision and Control CDC2010, pp. 7272–7277 (2010)
17. Grüne, L., Worthmann, K.: A distributed NMPC scheme without stabilizing terminal constraints. In: Johansson, R., Rantzer, A. (eds.) Distributed Decision Making and Control, pp. 261–287. Springer, New york (2012)
18. Kerrigan, E.C.: Robust constraint satisfaction: Invariant sets and predictive control. PhD Thesis, University of Cambridge (2000)
19. Kouzoupis, D., Quirynen, R., Houska, B., Diehl, M.: A block based ALADIN scheme for highly parallelizable direct optimal control. In: Proceedings of the 2016 American Control Conference, Boston, USA, pp. 1124–1129 (2016)
20. Motee, N., Sayyar-Rodsari, B.: Optimal partitioning in distributed model predictive control. In: Proceedings of the 2003 American Control Conference, vol. 6, pp. 5300–5305. IEEE (2003)
21. Müller, M.A., Reble, M., Allgöwer, F.: Cooperative control of dynamically decoupled systems via distributed model predictive control. Int. J. Robust Nonlinear Control **22**(12), 1376–1397 (2012)
22. O'Donoghue, B., Stathopoulos, G., Boyd, S.: A splitting method for optimal control. IEEE Trans. Control Syst. Technol. **21**(6), 2432–2442 (2013)
23. Pannek, J.: Parallelizing a state exchange strategy for noncooperative distributed NMPC. Syst. Control Lett. **62**(1), 29–36 (2013)
24. Rantzer, A.: Dynamic dual decomposition for distributed control. Proc. Am. Control Conf. **2009**, 884–888 (2009)
25. Rawlings, J., Mayne, D.: Model Predictive Control: Theory and Design. Nob Hill Publishing, Madison (2009)
26. Rawlings, J., Stewart, B.: Coordinating multiple optimization-based controllers: New opportunities and challenges. J. Process Control **18**(9), 839–845 (2008)
27. Richards, A., How, J.: A Decentralized Algorithm for Robust Constrained Model Predictive Control. In: Proceedings of the American Control Conference, pp. 4261–4266 (2004)
28. Richards, A., How, J.: Robust distributed model predictive control. Int. J. Control **80**(9), 1517–1531 (2007)
29. Savorgnan, C., Romani, C., Kozma, A., Diehl, M.: Multiple shooting for distributed systems with applications in hydro electricity production. J. Process Control **21**(5), 738–745 (2011)
30. Scattolini, R.: Architectures for distributed and hierarchical model predictive control a review. J. Process Control **19**(5), 723–731 (2009)
31. Singh, V., Atrey, P., Kankanhalli, M.: Coopetitive multi-camera surveillance using model predictive control. Mach. Vis. Appl. **19**(5), 375–393 (2008)
32. Stewart, B., Venkat, A., Rawlings, J., Wright, S., Pannocchia, G.: Cooperative distributed model predictive control. Syst. Control Lett. **59**(8), 460–469 (2010)
33. Venkat, A.: Distributed model predictive control: Theory and applications. PhD Thesis in Chemical Engineering, University of Wisconsin-Madison (2006)
34. Venkat, A., Hiskens, I., Rawlings, J., Wright, S.: Distributed MPC strategies with application to power system automatic generation control. IEEE Trans. Control Syst. Technol. **16**(6), 1192–1206 (2008)
35. Venkat, A., Rawlings, J., Wright, S.: Stability and Optimality of Distributed Model Predictive Control. In: 44th IEEE Conference on Decision and Control and European Control Conference CDC-ECC '05, pp. 6680–6685 (2005)
36. Zavala, V.M., Biegler, L.T.: The advanced-step NMPC controller: optimality, stability and robustness. Automatica **45**(1), 86–93 (2009)

Chapter 10
Variants and Extensions

The results developed so far in this book can be extended in many ways. In this chapter, we present a selection of possible variants and extensions. Some of these introduce new combinations of techniques developed in the previous chapters, others relax some of the previous assumptions in order to obtain more general results or strengthen assumptions in order to derive stronger results. In order to make the presentation concise, we limit ourselves to stabilizing NMPC as presented in Chaps. 5 and 6. Several sections contain algorithmic ideas which can be added on top of the basic NMPC schemes from the previous chapters. Parts of this chapter contain results which are somewhat preliminary and are thus subject to further research. Some sections have a survey-like style and, in contrast to the other chapters of this book, proofs are occasionally only sketched with appropriate references to the literature.

10.1 Schemes with Mixed Terminal Conditions

The previous Chaps. 5 and 6 have featured two extreme cases, namely NMPC schemes with terminal constraints $\mathbb{X}_0$ *and* costs F on the one hand and schemes without both $\mathbb{X}_0$ and F on the other hand. However, it appears natural also to consider terminal conditions of intermediate or mixed type, namely schemes in which (nonequilibrium) terminal constraint sets $\mathbb{X}_0$ but no terminal costs F are used and schemes in which terminal costs F but no terminal constraints sets $\mathbb{X}_0$ are used.

Schemes with terminal constraints $\mathbb{X}_0$ but without terminal costs F appear as a special case of Algorithm 3.10 (or its time varying counterpart 3.11) with ($\mathrm{OCP}_{\mathrm{N,e}}$) = (5.15) and $F \equiv 0$. For this setting, it is not reasonable to expect that Assumption 5.9(ii) holds. Consequently, the argument used in the proof of Theorem 5.13 does not apply; in fact, we are not aware of results in the literature analyzing such schemes with the techniques from Chap. 5.

Fortunately, the stability analysis in Chap. 6 provides a remedy to this problem. Observe that the main structural assumption on the optimal control sequences

L. Grüne and J. Pannek, *Nonlinear Model Predictive Control*,
Communications and Control Engineering, DOI 10.1007/978-3-319-46024-6_10

needed in the fundamental Lemmas 6.10 and 6.11 in Chap. 6 is that each admissible control sequence $u \in \mathbb{U}^N(x)$ can be extended to an admissible control sequence $\hat{u} \in \mathbb{U}^{N+K}(x)$ for each $K \geq 1$. Since Lemma 5.2(i) ensures this property for $\mathbb{U}^N_{\mathbb{X}_0}(x)$ provided $\mathbb{X}_0$ is viable, we can incorporate the terminal constraint set $\mathbb{X}_0$ into the analysis from Chap. 6.

As a consequence, using the optimal value function with terminal constraints $u \in \mathbb{U}^N_{\mathbb{X}_0}(x)$ in Assumption 6.3 or replacing $\mathbb{U}^N(x)$ by $\mathbb{U}^N_{\mathbb{X}_0}(x)$ in Assumption 6.5 and assuming Assumption 5.9(i), i.e., viability of $\mathbb{X}_0$, all results in Chap. 6 carry over to the scheme with terminal constraint set. In particular, the stability results Theorem 6.20, Corollary 6.21, Theorems 6.24 and 6.37 remain valid. However, as in Theorem 5.13, the resulting controller μ_N is only defined on the feasible set $\mathbb{X}_N$ from Definition 3.9.

This combined scheme inherits certain advantages and disadvantages from both schemes. From the terminal constrained scheme, we inherit that the resulting controller μ_N is only defined on the feasible set $\mathbb{X}_N$. On the other hand, as discussed before Lemma 5.3, we do not need to assume viability of $\mathbb{X}$ but only for the terminal constraint set $\mathbb{X}_0$, which thus provides an alternative to the recursive feasibility results without assuming viability of $\mathbb{X}$ in Sects. 7.1, 7.2 and 7.3.

From the unconstrained scheme we inherit the advantage that no terminal cost satisfying Assumption 5.9(ii) needs to be constructed. On the other hand, we need to ensure that the assumptions of one of the mentioned stability results from Chap. 6 hold whose rigorous verification may be involved, cf. also Sect. 6.6. For a more comprehensive discussion on advantages and disadvantages of different NMPC schemes we refer to Sect. 7.4.

A way of imposing terminal constraints without terminal costs which can be found in the literature is via so called *contractive constraints*. Here the terminal constraint set depends on the initial value x_0 of the optimal control problem ($\mathrm{OCP}_{\mathrm{N,e}}$) via

$$\mathbb{X}_0 = \{x \in \mathbb{X} \mid |x|_{x_*} \leq \gamma |x_0|_{x_*}\}$$

for some constant $\gamma \in (0, 1)$, see e.g., the book of Alamir [1] or the works of de Oliveira Kothare and Morari [23] and De Nicolao, Magni and Scattolini [3]. However, for these constraints stability is only guaranteed if either the whole optimal control sequence (as opposed to only the first element) is applied or if the optimization horizon is treated as an optimization variable and the contractivity condition is incorporated into the optimization objective, see [1, Chap. 4]. Since these approaches do not conform with the MPC paradigm used throughout this book, we do not discuss their analysis in detail.

Schemes with terminal cost F but without terminal constraint $\mathbb{X}_0$ have been investigated in several places in the literature, for instance in Grimm, Messina, Tuna and Teel [9] and Jadbabaie and Hauser [19] (for more information on these references see also the discussions at the end of Sect. 6.1 and in Sect. 6.9). In both references, stability results for such schemes are derived in which only positive definiteness of F is assumed. Roughly speaking, these references show that the addition of F does not destroy stability. While the authors emphasize the potential positive effects of adding such costs, they do not rigorously analyze these positive effects. In contrast to this, in the work of Parisini and Zoppoli [28] the specific properties of the terminal cost described in Remark 5.15 were exploited in order to show stability. The proof in [28] uses that under suitable conditions and for sufficiently large optimization horizon N for all initial values from a given region the open-loop optimal trajectories end up in the terminal constraint set without actually imposing this as a condition. The same proof idea has been generalized later by Limón, Alamo, Salas and Camacho [21] for a more general terminal cost.

Here, we outline an approach from Grüne and Rantzer [16] which we combine with the analysis technique from Chap. 6. This approach rigorously shows the positive effect of adding a terminal cost also in the absence of terminal constraints. In contrast to [28] or [21], the stability property is not restricted to sets of initial values for which the open-loop optimal trajectories end up in a terminal constraint set. However, the fact that this happens for a set of initial values around the origin will be used in our proof. We start from a terminal cost function F satisfying Assumption 5.9(ii) with a forward invariant neighborhood $\mathbb{X}_0$ of x_*, however, we will not use $\mathbb{X}_0$ as a terminal constraint set. Instead, we assume that $F \equiv c > 0$ holds on the boundary $\partial\mathbb{X}_0$ with $c \geq \sup_{x\in\mathbb{X}_0} F(x)$. This is, for instance, satisfied if F is constructed from a linearization via linear-quadratic techniques according to Remark 5.15 and $\mathbb{X}_0$ is a sublevel set of F. Then, we may extend F continuously to the whole set $\mathbb{X}$ by setting $F(x) := c$ for all $x \in \mathbb{X} \setminus \mathbb{X}_0$.

With this setting, we obtain the following theorem:

Theorem 10.1 (Stability with mixed terminal conditions) *Let the assumptions of Theorem 6.37 be satisfied for the NMPC Algorithm 3.1 without terminal conditions. Let $F : \mathbb{X} \to \mathbb{R}_0^+$ and assume that Assumption 5.9 holds for some set $\mathbb{X}_0$ containing a ball $\mathcal{B}_\eta(x_*)$ for some $\eta > 0$. Assume, furthermore, that that $F \equiv c$ holds outside $\mathbb{X}_0$ with $c \geq \sup_{x\in\mathbb{X}_0} F(x)$ and that $F(x) \leq \tilde{\alpha}_2(|x|_{x_*})$ holds for all $x \in \mathbb{X}_0$ and some $\tilde{\alpha}_2 \in \mathcal{K}_\infty$. Consider the NMPC Algorithm 3.10 with* (OCP$_{\mathrm{N,e}}$) = (5.15) *for this F but without terminal constraints, i.e., with $\mathbb{X}_0 = \mathbb{X}$ in* (5.15).

Then, the nominal NMPC closed-loop system (3.5) with NMPC feedback law μ_N is semiglobally asymptotically stable on $\mathbb{X}$ with respect to the parameter N in the sense of Definition 6.33(i).

Proof We consider the following three optimal control problems

(a) (5.15) with $\mathbb{X}_0 = \mathbb{X}$, which generates μ_N in this theorem
(b) (5.15) with $\mathbb{X}_0$ from Assumption 5.9 for F, which generates μ_N in Theorem 5.5
(c) (OCP$_{\mathrm{N}}$), which generates μ_N in Theorem 6.20

and denote the respective optimal value functions by $V_N^{(a)}$, $V_N^{(b)}$ and $V_N^{(c)}$. For each $x \in \mathbb{X}$ we obtain the inequalities $V_N^{(c)}(x) \leq V_N^{(a)}(x) \leq V_N^{(c)}(x) + c$ and, for $x \in \mathbb{X}_N$ (where $\mathbb{X}_N$ denotes the feasible set from Definition 3.9 for Problem (b)), we have $V_N^{(a)}(x) \leq V_N^{(b)}(x)$.

In order to show semiglobal asymptotic stability, i.e., Definition 6.33(i), we fix $\Delta > 0$. For an arbitrary $x \in \mathbb{X}$ we consider the optimal control $u^\star$ for Problem (a) (which implies $\mu_N(x) = u^\star(0)$ for μ_N from this theorem) and distinguish two cases:

(i) $x_{u^\star}(N, x) \in \mathbb{X}_0$: This implies $u^\star \in \mathbb{U}^N_{\mathbb{X}_0}(x)$ and hence $x \in \mathbb{X}_N$ and $V_N^{(a)}(x) = V_N^{(b)}(x)$. Using $x_{u^\star}(1, x) = f(x, \mu_N(x)) \in \mathbb{X}_N$ and $V_N^{(a)} \leq V_N^{(b)}$ on $\mathbb{X}_N$, the proof of Theorem 5.5 yields

$$\begin{aligned} V_N^{(a)}(x) &= V_N^{(b)}(x) \geq \ell(x, \mu_N(x)) + V_N^{(b)}(f(x, \mu_N(x))) \\ &\geq \ell(x, \mu_N(x)) + V_N^{(a)}(f(x, \mu_N(x))). \end{aligned} \tag{10.1}$$

This inequality will be used below in order to conclude asymptotic stability. Before we turn to case (ii), we show that case (i) applies to all points $x \in \mathscr{B}_\delta(x_*)$ for some $\delta > 0$.

Since (5.20) shows $V_N^{(b)}(x) \leq F(x)$ on $\mathbb{X}_0$, we obtain $V_N^{(a)}(x) \leq V_N^{(b)}(x) \leq \tilde{\alpha}_2(|x|_{x_*})$ for $x \in \mathscr{B}_\eta(x_*) \subseteq \mathbb{X}_0$. For $\delta = \min\{\eta, \tilde{\alpha}_2^{-1}(c/2)\}$ this implies $V_N^{(a)}(x) \leq c/2$ for all $x \in \mathscr{B}_\delta(x_*)$. On the other hand, $x_{u^\star}(N, x) \notin \mathbb{X}_0$ implies $F(x_{u^\star}(N, x)) = c$ and thus $V_N^{(a)}(x) \geq c$. Hence, case (i) occurs for all $x \in \mathscr{B}_\delta(x_*)$.

(ii) $x_{u^\star}(N, x) \notin \mathbb{X}_0$: This implies $F(x_{u^\star}(N, x)) = c$ and thus $V_N^{(a)}(x) = V_N^{(c)}(x) + c$. This implies that $u^\star$ is an optimal control for $V_N^{(c)}(x)$ and from the proof of Theorem 6.37 we obtain that (5.1), i.e.,

$$V_N^{(c)}(x) \geq \ell(x, \mu_N(x)) + \alpha V_N^{(c)}(f(x, \mu_N(x)))$$

holds for all $x \in Y = S \setminus P$ with S and P chosen as in the proof of Theorem 6.37. The sets S and P are forward invariant and by choosing $N \in \mathbb{N}$ sufficiently large we obtain $\alpha > 0$, $\overline{\mathscr{B}_\Delta(x_*)} \subseteq S$ and $P \subset \mathscr{B}_\delta(x_*)$ for Δ fixed above and δ defined at the end of case (i). Since $V_N^{(a)}(x) = V_N^{(c)}(x) + c$ and $V_N^{(a)}(f(x, \mu_N(x))) \leq V_N^{(c)}(f(x, \mu_N(x))) + c$ we obtain

$$V_N^{(a)}(x) \geq \ell(x, \mu_N(x)) + \alpha V_N^{(a)}(f(x, \mu_N(x))) \tag{10.2}$$

for all $y \in Y$ and some $\alpha > 0$.

Now, the choice of N and P implies that for $x \in S \setminus \mathscr{B}_\delta(x_*)$ inequality (10.2) holds while for $x \in \mathscr{B}_\delta(x_*)$ inequality (10.1) holds. This implies that Theorem 4.14 is applicable with $S(n) = S$ which yields semiglobal practical stability using Lemma 6.34(i). □

Comparing Theorem 10.1 with Theorem 6.37, one sees that the benefit of including the terminal cost F is that here we obtain semiglobal asymptotic stability while without F we can only guarantee semiglobal *practical* asymptotic stability. Loosely speaking, the unconstrained scheme guarantees stability up to the neighborhood $\mathscr{B}_\delta(x_*)$ while F ensures asymptotic stability inside this neighborhood.

10.2 Unconstrained NMPC with Terminal Weights

Our next extension analyzes the effect of inclusion of terminal weights in $(\mathrm{OCP_N})$, i.e., in NMPC schemes without stabilizing terminal conditions. Both in numerical simulations and in practice one can observe that adding terminal weights can improve the stability behavior of the NMPC closed loop. Formally, adding terminal weights can be achieved by replacing the optimization criterion in $(\mathrm{OCP_N})$ by

$$J_N(x_0, u(\cdot)) := \sum_{k=0}^{N-2} \ell(x_u(k, x_0), u(k)) + \omega\ell(x_u(N-1, x_0), u(N-1)) \qquad (10.3)$$

for some $\omega \geq 1$. For $\omega = 1$ we thus obtain the original problem $(\mathrm{OCP_N})$. This extension is a special case of $(\mathrm{OCP_{N,e}})$ in which we specify $\mathbb{X}_0 = \mathbb{X}$, $F \equiv 0$, $\omega_1 = \omega$ and $\omega_2 = \omega_3 = \cdots = \omega_N = 1$. In a similar way, such a terminal weight can be added to the respective time variant problem $(\mathrm{OCP_N^n})$ leading to a special case of $(\mathrm{OCP_{N,e}^n})$. Thus, all results developed in Chap. 3 apply to this problem. Given that the optimal control value $u(N-1)$ in (10.3) will minimize $\ell(x_u(N-1, x_0), u(N-1))$, this approach is identical to choosing $F(x) = \omega\ell^*(x)$ and $N = N-1$ in the terminal cost approach discussed in the previous section, with ℓ^* from (6.2). However, the specific structure of the terminal cost allows for applying different and more powerful analysis techniques which we explain now.

The terminal weight leads to an increased penalization of $\ell(x_u(N-1, x_0), u(N-1))$ in J_N and thus to an increased penalization of the distance of $x_u(N-1, x_0)$ to x_*. Thus, for $\omega > 1$ the optimizer selects a finite time optimal trajectory whose terminal state $x_{u^\star}(N-1, x_0)$ has a smaller distance to x_*. Since our goal is that the NMPC-feedback law μ_N steers the trajectory to x_*, this would intuitively explain better stability behavior.

Formally, however, the analysis is not that easy because in closed loop we never actually apply $u^\star(1), \ldots, u^\star(N-1)$ and the effect of ω on $u^\star(0)$ is not that obvious. Hence, we extend the technique developed in Chap. 6 in order to analyze the effect of ω. To this end, we change the definition (6.7) of B_N to

$$B_N(r) := \sum_{n=0}^{N-2} \beta(r, n) + \omega\beta(r, N-1).$$

With this definition, all results in Sect. 6.3 remain valid for the extended problem. Proposition 6.13 remains valid, too, if we change (6.11) to

$$\sum_{n=k}^{N-2} \lambda_n + \omega\lambda_{N-1} \leq B_{N-k}(\lambda_k), \quad k = 0, \ldots, N-2.$$

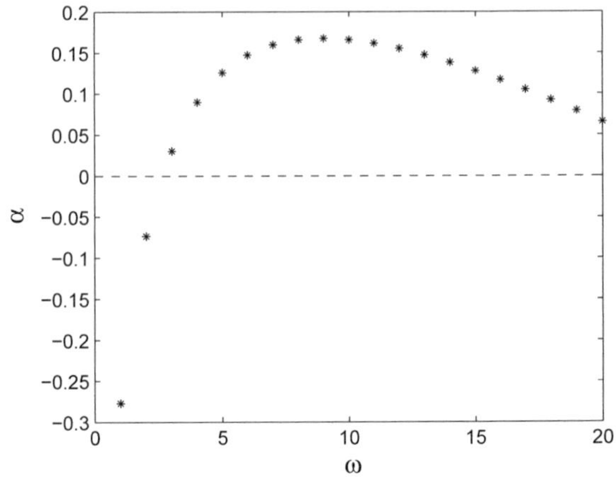

Fig. 10.1 Suboptimality index α depending on terminal weight ω

If, furthermore, in the subsequent statements we replace $\sum_{n=0}^{N-1} \lambda_n$ by $\sum_{n=0}^{N-2} \lambda_n + \omega\lambda_{N-1}$, then it can be shown that Proposition 6.18 remains valid if we replace (6.18) by

$$\underline{\alpha}_N^{\omega} := 1 - \frac{(\gamma_N - 1)(\gamma_2 - \omega)\prod_{i=3}^{N}(\gamma_i - 1)}{\prod_{i=2}^{N}\gamma_i - (\gamma_2 - \omega)\prod_{i=3}^{N}(\gamma_i - 1)}. \tag{10.4}$$

The proof is similar to the proof of Proposition 6.18 and can be found in Grüne, Pannek, Seehafer and Worthmann [15].

With this expression, Theorem 6.20 and its corollaries remain valid, except for the inequalities $V_N(x)/\alpha \le V_\infty(x)/\alpha$ and $CV_N(x) \le CV_\infty(x)$, which do in general no longer hold because of the additional weight which is present in V_N but not in V_∞.

Figure 10.1 shows the values from (10.4) for an exponential β of type (6.4) with $C = 2$ and $\sigma = 0.55$, optimization horizon $N = 5$ and terminal weights $\omega = 1, 2, \ldots, 20$. The figure illustrates that our analysis reflects the positive effect the terminal weight has on the stability: while for $\omega = 1, 2$ we obtain negative values for α and thus stability cannot be ensured, for $\omega \ge 3$ stability is guaranteed. However, one also sees that for $\omega \ge 10$ the value of α is decreasing, again. For more examples for the effect of terminal weights, we refer to [15] and Example 10.14, below.

10.3 Nonpositive Definite Stage Cost

In many regulator problems, one is not interested in driving the whole state to a reference trajectory or point. Rather, often one is only interested in certain output quantities. The following example illustrates such a situation.

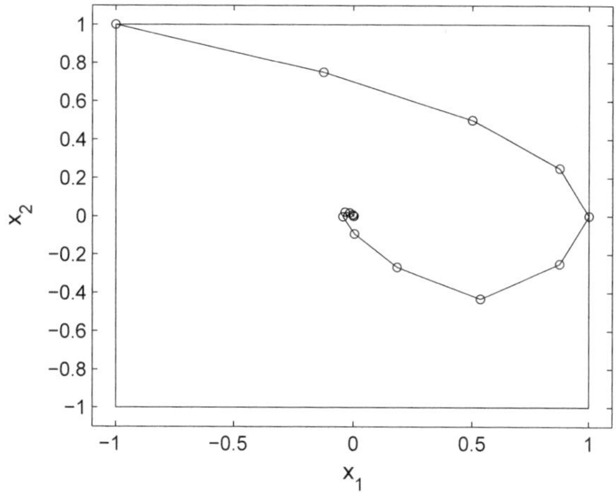

Fig. 10.2 MPC closed-loop trajectory with $N = 5$

Example 10.2 We reconsider Example 2.2, i.e.,

$$\begin{pmatrix} x_1^+ \\ x_2^+ \end{pmatrix} = \begin{pmatrix} x_1 + x_2 + u/2 \\ x_2 + u \end{pmatrix} =: f(x, u)$$

with stage cost

$$\ell(x, u) = x_1^2 + u^2.$$

In contrast to our standing assumption (3.2), no matter how we choose $x_* \in \mathbb{R}^2$, this function does not satisfy $\ell(x, u) > 0$ for all $x \in X$ and $u \in U$ with $x \neq x_*$. Instead, following the interpretation of x_1 and x_2 as position and velocity of a vehicle in a plane, the stage cost only penalizes the distance of the position x_1 from 0 but not the velocity.

However, the only way to put the system at rest with $x_1 = 0$ is to set $x_2 = 0$. Hence, one may expect that the NMPC controller will "automatically" steer x_2 to 0, too. The numerical simulation shown in Fig. 10.2 (performed with optimization horizon $N = 5$ without stabilizing terminal conditions and with state constraints $\mathbb{X} = [-1, 1]^2$ and control constraints $\mathbb{U} = [-1/4, 1/4]$) confirms that this is exactly what happens: the system is perfectly stabilized at $x_* = 0$ even though the stage cost does not "tell" the optimization problem to steer x_2 to 0.

How can this behavior be explained theoretically? The decisive difference of ℓ from this example to ℓ used in the theorems in the previous chapters is that the lower bound $\ell(x, u) \geq \alpha_3(|x|_{x_*})$ imposed in all our results is no longer valid. In other words, the stage cost is no longer positive definite.

One way to cope with this difference would be to interpret the resulting problem as an economic NMPC problem in the sense of Chap. 8. Here, however, we proceed in a more classical way and employ suitable observability and detectability conditions.

For NMPC schemes with stabilizing terminal conditions satisfying Assumption 5.9, the notion of input/output-to-state stability (IOSS) provides a way to deal with this setting. IOSS can be seen as a nonlinear detectability condition which ensures that the state converges to x_* if both the output and the input converge to their steady state values, which can in turn be guaranteed by suitable bounds on ℓ. We sketch this approach for time invariant reference $x^{\mathrm{ref}} \equiv x_*$ with corresponding control value u_* satisfying $f(x_*, u_*) = x_*$.

To this end, we relax the assumptions of Theorem 5.13 as follows: instead of assuming (5.2) we consider an output function $h : X \to Y$ for another metric space Y. In Example 10.2 we have $X = \mathbb{R}^2$, $Y = \mathbb{R}$ and $h(x) = x_1$.

Now, we change (5.2) to

$$\alpha_1(|h(x)|_{y_*}) \le V_N(x) \le \alpha_2(|x|_{x_*}) \quad \text{and} \quad \ell(x,u) \ge \alpha_3(|h(x)|_{y_*}) + \alpha_3(|u|_{u_*}) \tag{10.5}$$

with $y_* = h(x_*)$ and $|h(x)|_{y_*} = d_Y(h(x), y_*)$, where $d_Y(\cdot,\cdot)$ is the metric on Y. Furthermore, we assume that the system with output $y = h(x)$ is IOSS in the following sense: There exist $\beta \in \mathcal{KL}$ and $\gamma_1, \gamma_2 \in \mathcal{K}_\infty$ such that for each $x \in \mathbb{X}$ and each admissible control $u \in \mathbb{U}^\infty(x)$ the inequality

$$|x_u(n,x)|_{x_*} \le \max\left\{\beta(|x|_{x_*}, n),\ \gamma_1\left(\max_{k=0,\dots,n-1} |u(k)|_{u_*}\right),\ \gamma_2\left(\max_{k=0,\dots,n-1} |y(k)|_{y_*}\right)\right\}$$

holds for all $n \in \mathbb{N}_0$ with $y(k) = h(x_u(k,x))$.

With these changed assumptions, the assertion of Theorem 5.13 remains valid. The proof relies on the fact that the function V_N still satisfies

$$V_N(x) \ge \ell(x, \mu_N(x)) + V_N(f(x, \mu_N(x))).$$

This implies that $V_N(x_{\mu_N}(n,x))$ is monotone decreasing in n and since it is bounded from below by 0 it converges to some value as $n \to \infty$, although not necessarily to 0. However, similar to the proof of Theorem 9.33 the convergence of $V_N(x_{\mu_N}(n,x))$ implies convergence of $\ell(x_{\mu_N}(n,x), \mu_N(x_{\mu_N}(n,x))) \to 0$ which by means of the last inequality in (10.5) yields $h(x_{\mu_N}(n,x)) \to 0$ and $\mu_N(x_{\mu_N}(n,x)) \to 0$. Now the IOSS property can be used to conclude asymptotic stability of the closed loop. For more details of this approach, we refer to the book of Rawlings and Mayne [29, Sect. 2.7 and the references therein].

While the approach just sketched relies on stabilizing terminal conditions, the simulation in Example 10.2 shows that asymptotic stability can also be expected without such conditions. For this setting, a stability proof was given in the work of Grimm, Messina, Tuna and Teel [9] and the main result in this reference extends Theorem 6.37. Again, a detectability condition is used, but this time it is formulated via a suitable auxiliary function W: we assume the existence of a function $W : X \to \mathbb{R}_0^+$ which satisfies the inequalities

$$W(x) \le \overline{\alpha}_W(|x|_{x_*})$$
$$W(f(x,u)) - W(x) \le -\alpha_W(|x|_{x_*}) + \gamma_W(\ell(x,u)) \qquad (10.6)$$

for all $x \in \mathbb{X}$, $u \in \mathbb{U}(x)$ and suitable functions $\overline{\alpha}_W, \alpha_W, \gamma_W \in \mathscr{K}_\infty$. In turn, we remove the lower bound $\alpha_3(|x|_{x_*}) \le \ell^*(x)$ for ℓ^* from (6.2) from the assumptions of Theorem 6.37. Observe that whenever this lower bound holds, the detectability condition is trivially satisfied with $W \equiv 0$, $\gamma_W(r) = r$ and $\alpha_W = \alpha_3$. We note that the requirements on the function W are similar but not identical to the conditions on the storage function λ in the strict dissipativity condition (8.4). On the one hand, the requirements $0 \le W(x) \le \alpha_W(|x|_{x_*})$ are stronger than the requirements on λ and do not apply, e.g., to the storage function from Example 8.8(ii). On the other hand, the function γ_W gives additional flexibility in finding W compared to finding λ satisfying (8.4).

Under these modified assumptions, it is shown in [9, Theorem 1] that the semi-global practical stability assertion of Theorem 6.37 remains valid. Furthermore, [9, Corollarys 2 and 3] provide counterparts to Theorems 6.24 and 6.35 which prove semiglobal and "real" asymptotic stability, respectively. In contrast to the IOSS-based result for stabilizing terminal conditions, the proof of [9, Theorem 1] yields a Lyapunov function constructed from the optimal value function V_N and the function W from the detectability condition. In the simplest case, which occurs under suitable bounds on the involved $\mathscr{K}_\infty$-functions, this Lyapunov function is given by $V_N + W$. In general, a weighted sum has to be used.

In Example 10.2, numerical evaluation suggests that the detectability condition is satisfied for $W(x) = \max\{-|x_1x_2| + x_2^2, 0\}/2$ and $\gamma_W(r) = r$. Plots of the difference $W(x) - W(f(x,u)) + \ell(x,u)$ in MAPLE™ indicate that this expression is positive definite and can hence be bounded from below by some function $\alpha_W(|x|_{x_*})$; a rigorous proof of this property is, however, missing up to now.

As discussed in Sect. 6.9, the analysis in [9] uses a condition of the form $V_N(x) \le \alpha_V(r)$ in order to show stability which, compared to our Assumptions 6.3 or 6.5, has the drawback to yield fewer information for the design of "good" stage costs ℓ. Furthermore, suboptimality estimates are not easily available. It would hence be desirable to extend the statement and proof of Theorem 6.20 to the case of nonpositive definite stage costs. One way to achieve is to use the methods developed for economic NMPC from Chap. 8. Indeed, if we are able to find a function $W : X \to \mathbb{R}_0^+$ satisfying (10.6) with $\gamma_W(r) = r$, then W is a storage function characterizing strict dissipativity in the sense of Definition 8.7 and we can apply the theorems from Chap. 8. It would be interesting to see whether a similar reasoning is possible also in case $\gamma_W(r) \ne r$.

In this context, we would like to emphasize once again that even if the stage cost ℓ only depends on an output y, the resulting NMPC-feedback law is still a state feedback law because the full state information is needed in order to compute the prediction $x_u(\cdot, x_0)$ for $x_0 = x(n)$.

10.4 Multistep NMPC-Feedback Laws

Next, we investigate what happens if instead of only the first control value $u^\star(0)$ we implement the first m values $u^\star(0), \ldots, u^\star(m-1)$ before optimizing again. Formally, we can write this NMPC variant as a multistep feedback law

$$\mu_N(x,k) := u^\star(k), \quad k = 0, \ldots, m-1,$$

where $u^\star$ is an optimal control sequence for problem ($\mathrm{OCP}_{\mathrm{N,e}}$) (or one of its variants) with initial value $x_0 = x$. The resulting generalized closed-loop system then reads

$$x(n+1) = f(x(n), \mu_N(x([n]_m), n - [n]_m)), \tag{10.7}$$

where $[n]_m$ denotes the largest product km, $k \in \mathbb{N}_0$, with $km \leq n$. The value $m \in \{1, \ldots, N-1\}$ is called the *control horizon*.

When using stabilizing terminal conditions, the respective stability proofs from Chap. 5 are easily extended to this setting which we illustrate for Theorem 5.13. Indeed, from $V_N(x) \leq V_{N-1}(x)$ one immediately gets the inequality $V_N(x) \leq V_{N-m}(x)$ for each $m \in \{1, \ldots, N-1\}$ and each $x \in \mathbb{X}_{N-m}$. Proceeding as in the proof of Theorem 5.13 using equality (3.20) inductively for $N, N-1, \ldots, N-m+1$ and $V_N(x) \leq V_{N-m}(x)$ one obtains

$$V_N(x) \geq \sum_{k=0}^{m-1} \ell(x_{\mu_N}(k,x), \mu_N(k,x)) + V_N(x_{\mu_N}(m,x)).$$

This shows that V_N is a Lyapunov function for the closed-loop system at the times 0, $m, 2m, \ldots$. Since a similar argument shows that $V_N(x_{\mu_N}(k,x))$ is bounded by $V_N(x)$ for $k = 1, \ldots, m-1$, this proves asymptotic stability of the closed loop.

Without stabilizing terminal conditions, our analysis can be adjusted to the multistep setting, too, by extending Proposition 6.18 as well as the subsequent stability results, accordingly. The respective extension of Formulas (6.18) and (10.4) (including both control horizons $m \geq 1$ and terminal weights $\omega \geq 1$) is given by

$$\underline{\alpha}_{N,m}^{\omega} = 1 - \frac{(\gamma_{m+1} - \omega) \prod_{i=m+2}^{N} (\gamma_i - 1) \prod_{i=N-m+1}^{N} (\gamma_i - 1)}{\left(\prod_{i=m+1}^{N} \gamma_i - (\gamma_{m+1} - \omega) \prod_{i=m+2}^{N} (\gamma_i - 1) \right) \left(\prod_{i=N-m+1}^{N} \gamma_i - \prod_{i=N-m+1}^{N} (\gamma_i - 1) \right)}.$$

Again, the proof proceeds along the lines of the proof of Proposition 6.18 but becomes considerably more involved, cf. the paper by Grüne, Pannek, Seehafer and Worthmann [15].

It is worth noting that these extended stability and performance results remain valid if m is time varying, i.e., if the control horizon is changed dynamically, e.g.,

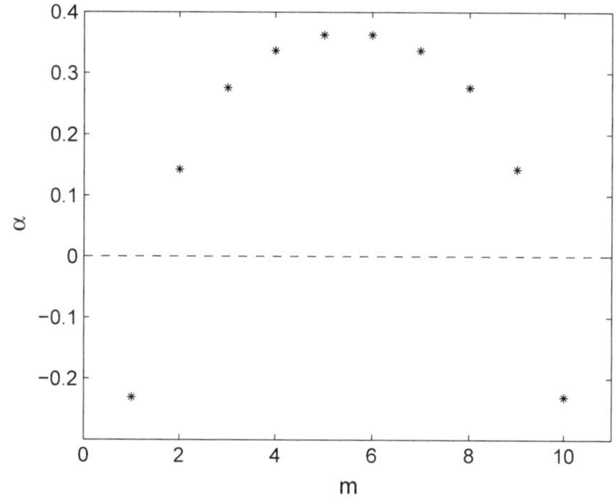

Fig. 10.3 Suboptimality index α depending on control horizon m

by a network induced perturbation. This has interesting applications in NMPC for networked control systems, cf. the work of Grüne, Pannek and Worthmann [17].

Figure 10.3 shows how $\alpha = \underline{\alpha}^{\omega}_{N,m}$ depends on m for an exponential β of type (6.4) with $C = 2$ and $\sigma = 0.75$, optimization horizon $N = 11$, terminal weight $\omega = 1$ and control horizons $m = 1, \ldots, 10$. Here, one observes two facts: first, the α-values are symmetric, i.e., $\underline{\alpha}^{\omega}_{N,m} = \underline{\alpha}^{\omega}_{N,N-m}$ and second, the values increase until $m = (N-1)/2$ and then decrease, again. This is not a particular feature of this example. In fact, it can be rigorously proved for a general class of $\beta \in \mathcal{KL}_0$; see [15] for details.

It is interesting to compare Fig. 10.3 with α-values which have been obtained numerically from an NMPC simulation for the linear inverted pendulum, cf. Example 2.10 and Appendix A.2 or [17] for the precise description of the problem. Figure 10.4 shows the resulting values for a set of different initial values. These values have been computed by Algorithm 10.8 described in Sect. 10.7, below.

While the monotonicity is—at least approximately—visible in this example, the perfect symmetry from Fig. 10.3 is not reflected in Fig. 10.4. A qualitatively similar behavior can be observed for the nonlinear inverted pendulum, see Example 10.14, below. In fact, so far we have not been able to find an example for which the symmetry could be observed in simulations. This may be due to the fact that our stability estimate is tight not for a single system but rather for the whole class of systems satisfying Assumption 6.5, cf. Theorem 6.27. Our numerical findings suggest that the conservativity induced by this "worst case approach" is higher for small m than for large m. This is also supported by Monte Carlo simulations performed by Grüne in [10].

An obvious drawback of multistep NMPC is that the corresponding feedback law is evaluated less often. Hence, the control loop is closed less often which may make the system less robust with respect to perturbations. A remedy to this problem is to re-compute the remaining part of the optimal control sequence in each sampling

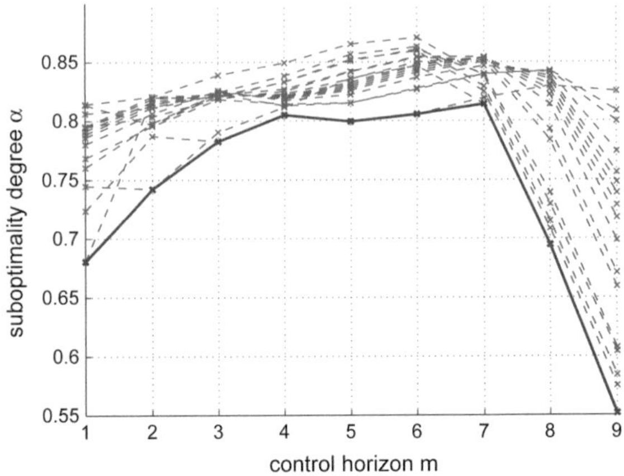

Fig. 10.4 Numerically measured values for α for a linear inverted pendulum and various initial values. The *thick line* represents the minimum

instant, i.e., to update the tail of the optimal control based on the knowledge of the current state. This amounts to solving an optimal control problem with shortened horizon. In contrast to a re-optimization of the problem with full optimization horizon, the computational effort of computing such an update will only be large if the state is perturbed, because in the absence of perturbations the updated control sequence will coincide with the original one. In case the perturbations are sufficiently small, the computational effort can be further reduced by using sensitivity based approximation techniques instead of a full optimization for the updated. For other computational approaches based on sensitivity techniques see Sect. 12.5. Compared to these approaches, the main advantages of the multistep update technique are that it allows for a rigorous analysis of the benefits of the updates using an extension of the techniques from Chap. 6 and that it also works for larger sampling times. For details, we refer to Palma [26] and Grüne and Palma [11] and the discussion at the end of Sect. 12.5.

10.5 Fast Sampling

Let us now turn to the special case of sampled data systems. In this case, according to (2.12) the discrete time solution $x_u(n, x_0)$ represents the continuous time solution $\varphi(t, 0, x_0, v)$ at sampling times $t = nT$. In this setting, it is natural to define the optimization horizon not in terms of the discrete time variable n but in terms of the continuous time t. Fixing an optimization horizon $T_{\text{opt}} > 0$ and picking a sampling period $T > 0$ where we assume for simplicity of exposition that T_{opt} is an integer multiple of T, the discrete time optimization horizon becomes $N = T_{\text{opt}}/T$, cf. also Sect. 3.5.

Having introduced this notation, an interesting question is what happens to stability and performance of the NMPC closed loop if we keep T_{opt} fixed but vary the sampling period T. In particular, it is interesting to see what happens if we sample faster and faster, i.e., if we let $T \to 0$. Clearly, in a practical NMPC implementation we cannot arbitrarily reduce T because we need some time for solving the optimal control problem (OCP$_\text{N}$) or its variants online. Still, in particular, in the case of zero order hold it is often desirable to sample as fast as possible in order to approximate the ideal continuous time control signal as good as possible, cf., e.g., the paper of Nešić and Teel [24], and thus one would like to make sure that this does not have negative effects on the stability and performance of the closed loop.

In the case of equilibrium endpoint constraint from Sect. 5.2 it is immediately clear that the stability result itself does not depend on T, however, the feasible set $\mathbb{X}_N$ may change with T. In the case of zero order hold, i.e., when the continuous time control function v is constant on each sampling interval $[nT, (n+1)T)$, cf. the discussion after Theorem 2.7, it is easily seen that each trajectory for sampling period T is also a trajectory for each sampling period T/k for each $k \in \mathbb{N}$. Hence, the feasible set $\mathbb{X}_{kN}$ for sampling period T/k always contains the feasible set $\mathbb{X}_N$ for sampling period T, i.e., the feasible set cannot shrink for $k \to \infty$ and hence for sampling period T/k we obtain at least the same stability properties as for sampling period T.

In the case of Lyapunov function terminal costs F as discussed in Sect. 5.3 either the terminal costs or the stage costs need to be adjusted to the sampling period T in order to ensure that Assumption 5.9 remains valid. One way to achieve this is to choose a stage cost in integral form (3.4) and the terminal cost F such that the following condition holds: for each $x \in \mathbb{X}_0$ and some $T_0 > 0$ there exists a continuous time control v satisfying $\varphi(t, 0, x, v) \in \mathbb{X}_0$ and

$$F(\varphi(t, 0, x, v)) - F(x) \le -\int_0^t L(\varphi(\tau, 0, x, v), v(\tau))d\tau \tag{10.8}$$

for all $t \in [0, T]$, cf. also the PhD thesis of Findeisen [6, Sect. 4.4.2]. Under this condition one easily checks that Assumption 5.9 holds for ℓ from (3.4) and all $T \le T_0$, provided the control function v in (10.8) is of the form $v|_{[nT,(n+1)T)]}(t) = u(n)(t)$ for an admissible discrete time control sequence $u(\cdot)$ with $u(n) \in U$. If $U = L^\infty([0, T], \mathbb{R}^m)$ then this last condition is not a restriction but if we use some smaller space for U (as in the case of zero order hold, cf. the discussion after Theorem 2.7), then this may be more difficult to achieve, see also [6, Remark 4.7].

Since the schemes from Chap. 6 do not use stabilizing terminal constraints $\mathbb{X}_0$ and terminal costs F, the difficulties just discussed vanish. However, the price to pay for this simplification is that the analysis of the effect of small sampling periods which we present in the remainder of this section is somewhat more complicated. Here, we will first discuss the Controllability Assumption 6.5 under sampling and then extend this assumption.

Fixing T_{opt} and letting $T \to 0$ we obtain that $N = T_{\text{opt}}/T \to \infty$. Looking at Theorem 6.24, this is obviously a good feature, because this theorem states that the larger N becomes, the better the performance will be. However, we cannot directly apply this theorem because we have to take into account that β in the Controllability Assumption 6.5 will also depend on T.

In order to facilitate the analysis, let us assume that in our discrete time NMPC formulation we use a stage cost ℓ that only takes the states $\varphi(nT, 0, x_0, v)$ at the sampling instants and the respective control values into account.[1] For the continuous time system, the controllability assumption can be formulated in discrete time. We denote the set of admissible continuous time control functions (in analogy to the discrete time notation) by $\mathbb{V}^\tau(x)$. More precisely, for the admissible discrete time control values $\mathbb{U}(x) \subseteq U \subseteq L^\infty([0,T],\mathbb{R}^m)$ (recall that these "values" are actually functions on $[0,T]$, cf. the discussion after Theorem 2.7) and any $\tau > 0$ we define

$$\mathbb{V}^\tau(x) := \left\{ v \in L^\infty([0,\tau],\mathbb{R}^m) \;\middle|\; \begin{array}{l} \text{there exists } u \in \mathbb{U}^N(x) \text{ with } N \geq \tau/T + 1 \\ \text{such that } u(n) = v|_{[nT,(n+1)T]}(\cdot + nT) \\ \text{holds for all } n \in \mathbb{N}_0 \text{ with } nT < \tau \end{array} \right\}.$$

Then, the respective assumption reads as follows.

Assumption 10.3 (*Controllability assumption*) We assume that the continuous time system is asymptotically controllable with respect to ℓ with rate $\beta \in \mathcal{KL}_0$, i.e., for each $x \in \mathbb{X}$ and each $\tau > 0$ there exists an admissible control function $v_x \in \mathbb{V}^\tau(x)$ satisfying

$$\ell(\varphi(t,0,x,v_x), v_x(t)) \leq \beta(\ell^*(x), t)$$

for all $t \in [0,\tau]$.

For the discrete time system (2.18) satisfying (2.12) the Controllability Assumption 10.3 translates to the discrete time Assumption 6.5 as

$$\ell(x_{u_x}(n,x), u_x(n)) \leq \beta(\ell^*(x), nT).$$

In the special case of exponential controllability, β in Assumption 10.3 is of the form

$$\beta(r,t) = Ce^{-\lambda t} r \tag{10.9}$$

for $C \geq 1$ and $\lambda > 0$. Thus, for the discrete time system, the Controllability Assumption 6.5 becomes

$$\ell(x_{u_x}(n,x), u_x(n)) \leq Ce^{-\lambda nT}\ell^*(x) = C\left(e^{-\lambda T}\right)^n \ell^*(x)$$

and we obtain a $\mathcal{KL}_0$-function of type (6.4) with C from (10.9) and $\sigma = e^{-\lambda T}$.

Summarizing, if we change the sampling period T then not only the discrete time optimization horizon N but also the decay rate σ in the exponential controllability

[1] Integral costs (3.4) can be treated, too, but this is somewhat more technical, cf. Grüne, von Lossow, Pannek and Worthmann [14, Sect. 4.2].

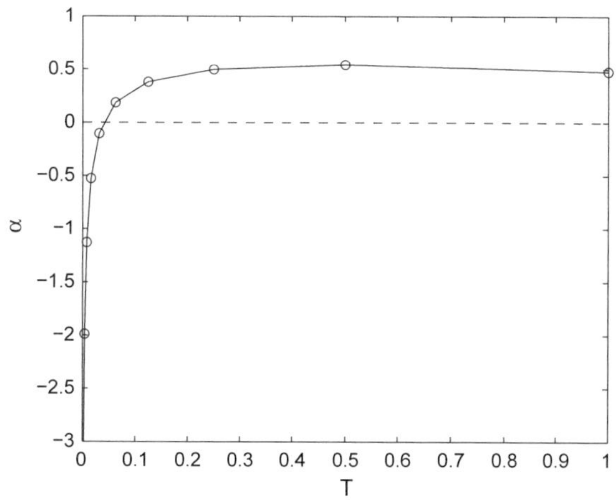

Fig. 10.5 Suboptimality index α from (6.18) for fixed T_{opt} and varying sampling period T

property will change, more precisely we have $\sigma \to 1$ as $T \to 0$. When evaluating (6.18) with the values resulting from (6.7), i.e.,

$$\gamma_k = \sum_{j=0}^{k-1} Ce^{-\lambda jT},$$

it turns out that the convergence $\sigma \to 1$ counteracts the positive effect of the growing optimization horizons $N \to \infty$. In fact, the negative effect of $\sigma \to 1$ is so strong that α diverges to $-\infty$ as $T \to 0$. Figure 10.5 illustrates this fact (which can also be proven rigorously, cf. [14]) for $C = 2$, $\lambda = 1$ and $T_{\mathrm{opt}} = 5$.

This means that whenever we choose the sampling period $T > 0$ too small, then performance may deteriorate and eventually instability may occur. This predicted behavior is not consistent with observations in numerical examples. How can this be explained?

The answer lies in the fact that our stability and performance estimate is only tight for one particular system in the class of systems satisfying Assumption 6.5, cf. Theorem 6.27 and the discussion preceding this theorem, and not for the whole class. In particular, the subclass of sampled data systems satisfying Assumption 6.5 may well behave better than general systems. Thus, we may try to identify the decisive property which makes sampled data systems behave better and try to incorporate this property into our computation of α.

To this end, note that so far we have not imposed any continuity properties of f in (2.1). Sampled data systems, however, are governed by differential equations (2.6) for which we have assumed Lipschitz continuity in Assumption 2.4. Let us assume for simplicity of exposition that the Lipschitz constant in this assumption is independent of r. Then, for a large class of stage costs ℓ the following property for the continuous time system can be concluded from Gronwall's Lemma, cf. [14] for details.

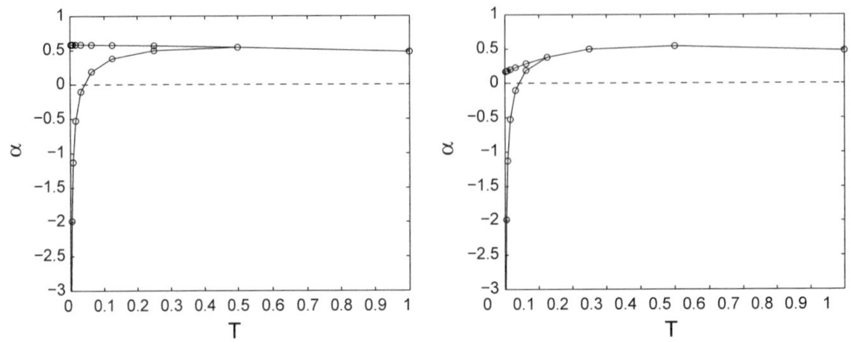

Fig. 10.6 α for fixed T_{opt} and varying sampling period T without Assumption 10.4 (*lower* graphs) and with Assumption 10.4 (*upper* graphs) with $L = 2$ (*left*) and $L = 10$ (*right*)

Assumption 10.4 (*Growth condition*) There exists a constant $L > 0$ such that for each $x \in \mathbb{X}$ and each $\tau > 0$ there exists an admissible control function $v_x \in \mathbb{V}^\tau(x)$ satisfying

$$\ell(\varphi(t, 0, x, v_x), v_x(t)) \leq e^{Lt} \ell^*(x)$$

for all $t \in [0, \tau]$.

The estimates on ℓ induced by this assumption can now be incorporated into the analysis in Chap. 6 by including the additional bound in the computation of B_K in (6.7). As a result, the values γ_k in Formula (6.18) change to

$$\gamma_k = \min\left\{ \sum_{j=0}^{k-1} Ce^{-\lambda jT}, \sum_{j=0}^{k-1} e^{LjT} \right\}.$$

The effect of this change is clearly visible in Fig. 10.6. The α-values from (6.18) no longer diverge to $-\infty$ but rather converge to a finite—and for the chosen parameters also positive—value as $T \to 0$. Again, this convergence behavior can be rigorously proved, for details we refer to [14].

10.6 Compensation of Computation Times

Throughout the previous chapters we assumed that the solution of the optimal control problems ($\text{OCP}_{\text{N,e}}$) and its variants in Step (2) of the Algorithms 3.1, 3.7, 3.10 and 3.11 can be obtained instantaneously, i.e., with negligible computation time. Clearly, this is not possible in general, as the algorithms for solving such problems, cf. Chap. 12 for details, need some time to compute a solution. If this time is large compared to the sampling period T, the computational delay caused by Step (2) is not negligible and

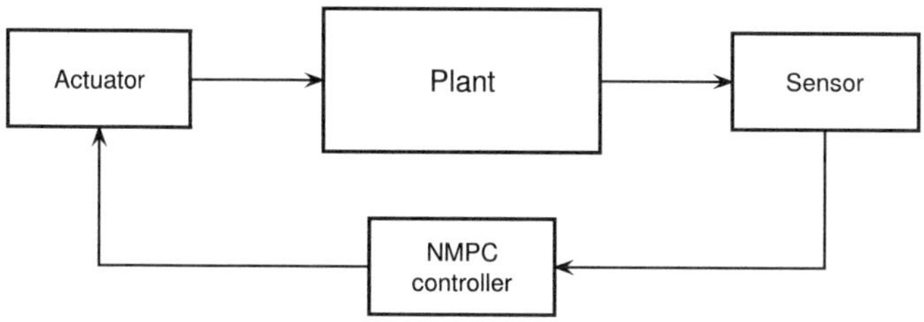

Fig. 10.7 Scheme of the NMPC closed-loop components

needs to be considered. One way for handling these delays would be to interpret them as perturbations and use techniques similar to the robustness analysis in Sects. 7.5–7.9. In this section, we pursue another idea in which a delay compensation mechanism is added to the NMPC scheme.

Taking a look at the structure of the NMPC algorithm from Chap. 3, we see that Steps (1)–(3) correspond to different physical tasks: measuring, computing and applying the control. These tasks are operated by individual components as shown schematically in Fig. 10.7. Note that in the following actuator, sensor and controller are not required to be physically decomposed, however, this case is also not excluded.

While it is a necessity to consider different clocks in a decomposed setting, it may not be the case if the components are physically connected. Here, we assume that every single component possesses its own clock and, for simplicity of exposition, that these clocks are synchronized (see the work of Varutti and Findeisen [31, Sect. III.C] for a possible way to relax this assumption). To indicate that a time instant n is considered with respect to a certain clock, we indicate this by adding indices s for the sensor, c for the NMPC controller and a for the actuator.

The idea behind the compensation approach is to run the NMPC controller component with a predefined time offset. This offset causes the controller to compute a control ahead of time, such that the computed control value is readily available at the time it is supposed to be applied, cf. Fig. 10.8. In this figure, τ_c denotes the actual computational delay and $\tau_c^{\max}$ denotes the predefined offset. In order to be operable, this offset needs to be chosen such that it is larger than the maximal computing time required to solve the optimal control problem in Step (2) of the considered NMPC algorithm. At time n_c, this optimal control problem is solved with a prediction $\tilde{x}(n_a)$ of the initial value $x(n_a)$ based on the available measurement $x(n_c) = x(n_s)$. This prediction is performed using the same model which is used for the NMPC prediction in ($\mathrm{OCP}_{\mathrm{N,e}}$) or its variants, i.e., using (2.1).

In order to perform this prediction, the control values $\mu_N(n, x(n))$, $n \in \{n_s, \ldots, n_a\}$ which are to be applied at the plant during the time interval $[n_s, n_a]$ and which have been computed before by the NMPC controller are needed and are therefore buffered. Thus, we extend the scheme given in Fig. 10.7 by adding the required predictor to the controller. The structure of the resulting scheme is shown in Fig. 10.9.

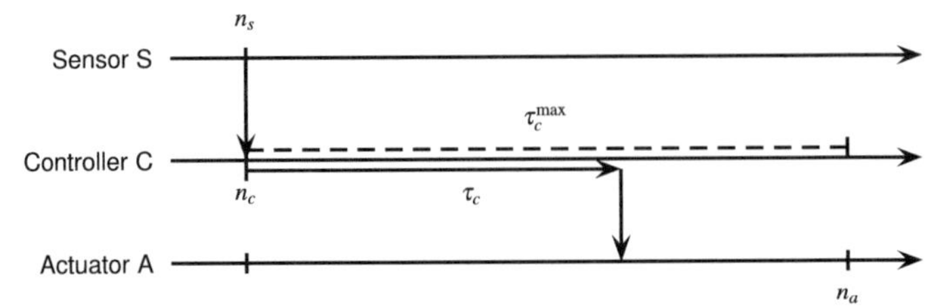

Fig. 10.8 Operation of time decoupled NMPC scheme

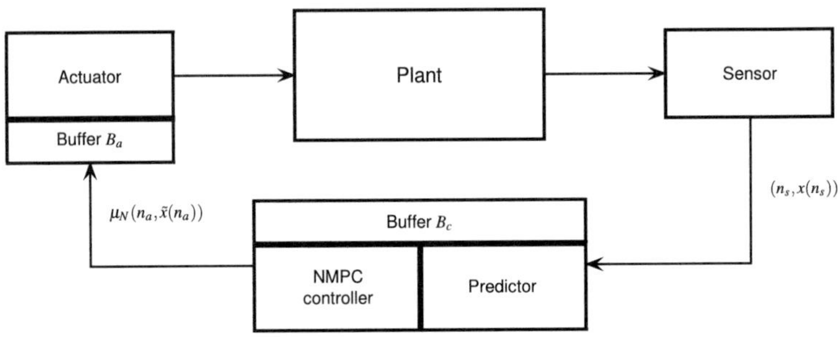

Fig. 10.9 Scheme of the time decoupled NMPC closed-loop components

Observe that in this scheme, we buffer the control values twice: within the predictor, but also at the actuator since the computation of $\mu(n_a, x(n_a))$ will be finished ahead of time if $\tau_c < \tau_c^{\max}$, which is the typical case. Alternatively, one could use only one buffer at the controller and send each control value "just in time". Using two buffers has the advantage that further delays induced, e.g., by network delays between the controller and the actuator can be compensated; see also the discussion at the end of this section.

The corresponding algorithm has the following form. Since all NMPC algorithms stated in Chap. 3 can be modified in a similar manner, we only show the algorithm for the most general form given in Algorithm 3.11:

Algorithm 10.5 (*Time decoupled NMPC algorithm for time varying reference*)
At each sampling time t_n, $n = 0, 1, 2, \ldots$:

(1) Measure the state $x(n_s) := x(n) \in X$ of the system and send pair $(n_s, x(n_s))$ to controller
(2a) Delete pair $(n_c - 1, \mu_N(n_c - 1, x(n_c - 1)))$ from buffer B_c and compute predicted state $\tilde{x}(n_c + \tau_c^{\max})$ from the measured state $x(n_c)$
(2b) Set $\tilde{n} := n_c + \tau_c^{\max}$, $x_0 = \tilde{x}(\tilde{n})$ and solve the optimal control problem

$$\begin{aligned}
&\text{minimize} \quad J_N(\tilde{n}, x_0, u(\cdot)) := \sum_{k=0}^{N-1} \omega_{N-k} \ell(\tilde{n}+k, x_u(k, x_0), u(k)) \\
&\qquad\qquad\qquad\qquad + F(\tilde{n}+N, x_u(N, x_0)) \\
&\text{with respect to} \quad u(\cdot) \in \mathbb{U}^N_{\mathbb{X}_0}(\tilde{n}, x_0), \quad \text{subject to} \\
&x_u(0, x_0) = x_0, \quad x_u(k+1, x_0) = f(x_u(k, x_0), u(k))
\end{aligned} \tag{OCP$^{\mathrm{n}}_{\mathrm{N,e}}$}$$

and denote the obtained optimal control sequence by $u^\star(\cdot) \in \mathbb{U}^N_{\mathbb{X}_0}(\tilde{n}, x_0)$.

(2c) Add pair $(\tilde{n}, \mu_N(\tilde{n}, \tilde{x}(\tilde{n}))) := (\tilde{n}, u^\star(0))$ to Buffer B_c and send it to actuator

(3a) Delete pair $(n_a - \tau_c^{\max} - 1, \mu_N(n_a - \tau_c^{\max} - 1, \tilde{x}(n_a - \tau_c^{\max} - 1)))$ and add received pair $(n_a, \mu_N(n_a, \tilde{x}(n_a)))$ to buffer B_a

(3b) Use $\mu_N(n_a - \tau_c^{\max}, \tilde{x}(n_a - \tau_c^{\max}))$ in the next sampling period

At a first glance, writing this algorithm using three different clocks and sending time stamped information in Steps (1) and (2c) may be considered as overly complicated, given that n_s in Step (1) is always equal to n_c in Step (2a) and n_c in Step (2c) always equals n_a in Step (3a). However, this way of writing the algorithm allows us to easily separate the components—sensor, predictor/controller and actuator—of the NMPC scheme and to assume that the "sending" in Steps (1) and (2c) is performed via a digital network. Then, we can assign Step (1) to the sensor, Steps (2a)–(2c) to the controller and Steps (3a) and (3b) to the actuator. Assuming that all transmissions between the components can be done with negligible delay, we can run these three steps as separate algorithms in parallel. Denoting the real-time by n, the resulting scheduling structure is sketched in Fig. 10.10 for $\tau_c^{\max} = 2$. For comparison, the structure of the NMPC Algorithm 3.11 without prediction is indicated by the dashed lines.

Since the algorithm is already applicable to work in parallel, it can be extended to a more complex networked control context in which transmission delays and packet loss may occur. To this end, such delays have to be considered in the prediction and an appropriate error handling must be added for handling dropouts, see, e.g.,

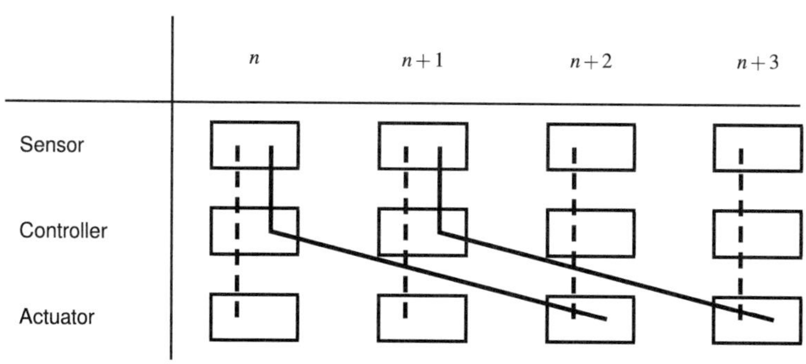

Fig. 10.10 Comparison of scheduling structure between NMPC Algorithms 3.11 (*dashed lines*) and 10.5 (*solid lines*) with $\tau_c^{\max} = 2T$

the paper by Grüne, Pannek and Worthmann [18]. In the presence of transmission delays and dropouts, we cannot expect that all control values are actually available at the actuator when they are supposed to be applied. Using NMPC, this can be compensated easily using the multistep feedback concept and the respective stability results from Sect. 10.4 as presented by Grüne et al. in [17].

Besides [18], which forms the basis for the presentation in this section, model based prediction for compensating computational delay in NMPC schemes has been considered earlier, e.g., in the works of Chen, Ballance and O'Reilly [2] and Findeisen and Allgöwer [7]. Note that the use of the nominal model (2.1) for predicting future states may lead to wrong predictions in case of model uncertainties, disturbances, etc. In this case, the predicted state $\tilde{x}(\tilde{n})$ may differ from the actual state $x(n_a)$ at time $n_a = \tilde{n}$ and hence ($\mathrm{OCP_{N,e}}$) is solved with a wrong initial value. This error is analyzed by Findeisen et al. in [33] and in the paper of Zavala and Biegler [32] a method for correcting this mismatch based on NLP sensitivity techniques is presented, cf. also Sect. 12.5.

10.7 Online Measurement of α

In the analysis of NMPC schemes without stabilizing terminal conditions in Chap. 6, one of the central aims was to establish conditions to rigorously guarantee the existence of $\alpha \in (0, 1]$ such that the inequality

$$V_N(n, x) \geq \alpha \ell(n, x, \mu_N(n, x)) + V_N(n + 1, f(x, \mu_N(n, x))) \qquad (5.1)$$

holds for all $x \in \mathbb{X}$ and $n \in \mathbb{N}_0$. While Theorem 6.15 and Proposition 6.18 provide computational methods for estimating α from the problem data, the assumptions needed for these computations—in particular Assumptions 6.3 or 6.5—may be difficult to check.

In this section, we present methods from Grüne and Pannek [12] and Pannek [27] which allow for the online computation or estimation of α along simulated NMPC closed-loop trajectories. There are several motivations for proceeding this way. First, as already mentioned, it may be difficult to check the assumptions needed for the computation of α using Theorem 6.15 or Proposition 6.18. Although a simulation-based computation of α for a selection of closed-loop trajectories cannot rigorously guarantee stability and performance for all possible closed-loop trajectories, it may still give valuable insight into the performance of the controller. In particular, the information obtained from such simulations may be very useful in order to tune the controller parameters, in particular the optimization horizon N and the stage cost ℓ.

Second, requiring (5.1) to hold for all $x \in \mathbb{X}$ may result in a rather conservative estimate for α. As we will see in Proposition 10.6, below, for assessing the performance of the controller along one closed-loop trajectory it is sufficient that (5.1) holds only for those points $x \in \mathbb{X}$ which are actually visited by this trajectory.

Finally, the knowledge of α may be used for an online adaptation of the optimization horizon N; some ideas in this direction are described in the subsequent Sect. 10.8.

Our first result shows that for assessing stability and performance of the NMPC controller along one specific closed-loop trajectory it is sufficient to find α such that (5.1) holds for the points actually visited by this trajectory.

Proposition 10.6 (A posteriori suboptimality estimate) *Consider the feedback law $\mu_N : \mathbb{N}_0 \times \mathbb{X} \to \mathbb{U}$ computed from Algorithm 3.7 and the closed-loop trajectory $x(\cdot) = x_{\mu_N}(\cdot)$ of (3.9) with initial value $x(0) \in \mathbb{X}$ at initial time 0. If the optimal value function $V_N : \mathbb{N}_0 \times \mathbb{X} \to \mathbb{R}_0^+$ satisfies*

$$V_N(n, x(n)) \geq V_N(n+1, x(n+1)) + \alpha\ell(n, x(n), \mu_N(n, x(n))) \tag{10.10}$$

for some $\alpha \in (0, 1]$ and all $n \in \mathbb{N}_0$, then

$$\alpha V_\infty(n, x(n)) \leq \alpha J_\infty(n, x(n), \mu_N) \leq V_N(n, x(n)) \leq V_\infty(n, x(n)) \tag{10.11}$$

holds for all $n \in \mathbb{N}_0$.

If, in addition, there exist $\alpha_1, \alpha_2, \alpha_3 \in \mathcal{K}_\infty$ such that (5.2) holds for all $(n, x) \in \mathbb{N}_0 \times \mathbb{X}$ with $n \in \mathbb{N}_0$ and $x = x(n)$, then there exists $\beta \in \mathcal{KL}$ which only depends on $\alpha_1, \alpha_2, \alpha_3$ and α such that the inequality

$$|x(n)|_{x^{ref}(n)} \leq \beta(|x(0)|_{x^{ref}(0)}, n)$$

holds for all $n \in \mathbb{N}_0$, i.e., x behaves like a trajectory of an asymptotically stable system.

Proof The proof of (10.11) is similar to the proof of Theorem 4.11.

The existence of β follows with the same construction as in the proof of Theorem 2.19, observing that the definition of β in this proof only depends on α_1, α_2 and $\alpha_V = \alpha\,\alpha_3$ and not on the specific form of $V = V_N$. □

Proposition 10.6 gives us a way to compute α from the data available at runtime and guarantees the performance estimate (10.11) as well as—under the additional assumption that (5.2) holds—asymptotic stability-like behavior for the considered closed-loop trajectory if $\alpha > 0$. Moreover, under this additional assumption (10.10) immediately implies that V_N strictly decreases along the trajectory, i.e., it behaves like a Lyapunov function.

Since the values of α for which (5.1) holds for all $x \in \mathbb{X}$ and for which (10.10) holds along a specific trajectory x_{μ_N} will be different in general, we introduce the following definition:

Definition 10.7 (1) We call $\alpha := \max\{\alpha \mid (5.1) \text{ holds for all } x \in \mathbb{X}\}$ the *global suboptimality degree*.

(2) For fixed $x \in \mathbb{X}$ the maximal value of α satisfying (5.1) for this x is called *local suboptimality degree* in x.

(3) Given a closed-loop trajectory $x_{\mu_N}(\cdot)$ of (3.9) with initial time 0 we call $\alpha := \max\{\alpha \mid (10.10) \text{ holds for all } n \in \mathbb{N}_0 \text{ with } x(\cdot) = x_{\mu_N}(\cdot)\ \}$ the *closed-loop suboptimality degree* along $x_{\mu_N}(\cdot)$.

An algorithm to evaluate α from (10.10) can easily be obtained and integrated into Algorithm 3.7:

Algorithm 10.8 (*NMPC algorithm for time varying reference x^{ref} with a posteriori suboptimality estimate*)

Set $\alpha = 1$. At each sampling time t_n, $n = 0, 1, 2, \ldots$:

(1) Measure the state $x(n) \in X$ of the system
(2) Set $x_0 = x(n)$ and solve the optimal control problem

$$\begin{aligned}&\text{minimize} \quad J_N(n, x_0, u(\cdot)) := \sum_{k=0}^{N-1} \ell(n+k, x_u(k, x_0), u(k))\\ &\text{with respect to } \ u(\cdot) \in \mathbb{U}^N(x_0), \quad \text{subject to}\\ &x_u(0, x_0) = x_0, \quad x_u(k+1, x_0) = f(x_u(k, x_0), u(k))\end{aligned} \tag{$\mathrm{OCP_N^n}$}$$

and denote the obtained optimal control sequence by $u^\star(\cdot) \in \mathbb{U}^N(x_0)$.
(3) Define the NMPC-feedback value $\mu_N(n, x(n)) := u^\star(0) \in U$ and use this control value in the next sampling period
(4) If $n \geq 1$ compute α via

$$\alpha_l = \frac{V_N(n-1, x(n-1)) - V_N(n, x(n))}{\ell(n-1, x(n-1), \mu_N(n-1, x(n-1)))}$$
$$\alpha = \min\{\alpha, \alpha_l\}$$

Proposition 10.6 and Algorithm 10.8 are easily extended to the multistep NMPC case described in Sect. 10.4. In this case, (10.10) is replaced by

$$V_N(n, x(n)) \geq V_N(n+m+1, x(n+1)) + \alpha \sum_{k=0}^{m} \ell(n+k, x_u(k, x(n)), u^\star(k, x(n)))$$

and the definition of α_l in Step (4) is changed, accordingly.

Note that in Step (4) of Algorithm 10.8, the computation of α_l does not provide the value of α in (10.10) for the current time instant n but for $n - 1$. This is why we call α from Algorithm 10.8 an *a posteriori* estimate. The distinction between the current value of α_l and α in Step (4) is required in order to be consistent with Proposition 10.6 since α_l corresponds to the local suboptimality degree in $x(n-1)$

while the suboptimality degree according to Proposition 10.6 is the minimum over all α_l along the closed loop.

While Algorithm 10.8 is perfectly suited in order to evaluate the performance of an NMPC controller via numerical simulations, its a posteriori nature is not suitable if we want to use the estimated α in order to adjust the optimization horizon N. For instance, if we detect that at some time n the value of α in (10.10) is too small—or even negative—then we may want to increase N in order to increase α (see Sect. 10.8 for more details on such procedures). However, in Algorithm 10.8 the value of α in (10.10) only becomes available at time $n+1$, which is too late in order to adjust N.

A simple remedy for this problem is to solve at time n a second optimal control problem ($\mathrm{OCP}_\mathrm{N}^\mathrm{n}$) with initial value $x_u(1, x(n))$ and initial time $n := n+1$. However, since solving the problem ($\mathrm{OCP}_\mathrm{N}^\mathrm{n}$) is the computationally most expensive part of the NMPC algorithm, this solution would be rather inefficient.

In order to obtain an *a priori* estimate with reduced additional computing costs, a few more insights into the local NMPC problem structure are required. The main tool we are going to use is the following lemma.

Lemma 10.9 *Consider the feedback law $\mu_N : \mathbb{N}_0 \times \mathbb{X} \to \mathbb{U}$ computed from Algorithm 3.7 and the closed-loop trajectory $x(\cdot) = x_{\mu_N}(\cdot)$ of (3.9) with initial value $x(0) = x_0 \in \mathbb{X}$ at initial time 0. If*

$$V_N(n+1, x(n+1)) - V_{N-1}(n+1, x(n+1)) \leq (1-\alpha)\ell(n, x(n), \mu_N(n, x(n))) \tag{10.12}$$

holds for some $\alpha \in [0, 1]$ and some $n \in \mathbb{N}_0$, then (10.11) holds for this n.

Proof Using the dynamic programming principle (3.16) with $K = 1$ we obtain

$$\begin{aligned} V_N(n, x(n)) &= \ell(n, x(n), \mu_N(n, x(n))) + V_{N-1}(n+1, x(n+1)) \\ &\geq \ell(n, x(n), \mu_N(n, x(n))) + V_N(n+1, x(n+1)) \\ &\quad - (1-\alpha)\ell(n, x(n), \mu_N(n, x(n))) \\ &= V_N(n+1, x(n+1)) + \alpha\ell(n, x(n), \mu_N(n, x(n))) \end{aligned}$$

Hence, (10.10) holds and Proposition 10.6 guarantees the assertion. □

Now, we would not gain much if we tried to compute α using (10.12) directly, since we would again need the future information $V_N(n+1, x(n+1))$, i.e., the solution of another optimal control problem (in contrast to that $V_{N-1}(n+1, x(n+1))$ is readily available at time n since by the dynamic programming principle it can be computed from $V_N(n, x(n))$ and $\ell(x(n), \mu_N(x(n)))$). There is, however, a way to reduce the size of the additional optimal control problem that needs to be solved. To this end, we introduce the following assumption which will later be checked numerically in our algorithm.

Assumption 10.10 For given $N, N_0 \in \mathbb{N}$, $N \geq N_0 \geq 2$, there exists a constant $\gamma > 0$ such that for the optimal open-loop solution $x_{u^\star}(\cdot, x(n))$ of $(\mathrm{OCP}_\mathrm{N}^\mathrm{n})$ in Algorithm 3.7 the inequalities

$$\frac{V_{N_0}(n+N-N_0, x_{u^\star}(N-N_0, x(n)))}{\gamma+1} \leq \max_{j=N-N_0,\ldots,N-2} \ell(n+j, x_u(n+j, x(n)), \mu_{N-j-1}(n+j, x_{u^\star}(j, x(n))))$$

$$\frac{V_k(n+N-k, x_{u^\star}(N-k, x(n)))}{\gamma+1} \leq \ell(n+N-k, x_{u^\star}(N-k, x(n)), \mu_k(n+N-k, x_{u^\star}(N-k, x(n))))$$

hold for all $k \in \{N_0+1, \ldots, N\}$ and all $n \in \mathbb{N}_0$.

Note that computing γ for which this assumption holds requires only the computation of μ_j for $j = 1, \ldots, N_0 - 1$ in the first inequality, since μ_k in the second inequality can be obtained from $u^\star$ via (3.23). This corresponds to solving $N_0 - 2$ additional optimal control problems which may look like a step backward, but since these optimal control problems are defined on a significantly smaller horizon, the computing costs are actually reduced. In fact, in the special case that ℓ does not depend on u, no additional computations have to be performed, at all. In this assumption, the value N_0 is a design parameter which affects the computational effort for checking Assumption 10.10 as well as the accuracy of the estimate for α obtained from this assumption.

Under Assumption 10.10, we can relate the minimal values of two optimal control problems with different horizon lengths.

Proposition 10.11 *Suppose that Assumption 10.10 holds for $N \geq N_0 \geq 2$. Then*

$$\frac{(\gamma+1)^{N-N_0}}{(\gamma+1)^{N-N_0} + \gamma^{N-N_0+1}} V_N(n, x(n)) \leq V_{N-1}(n, x(n))$$

holds for all $n \in \mathbb{N}_0$.

Proof In the following, we use the abbreviation $x_u(j) := x_u(j, x(n))$, $j = 0, \ldots, N$, since all our calculations use the open-loop trajectory with fixed initial value $x(n)$. Set $\tilde{n} := N - k$. Using the principle of optimality and Assumption 10.10 we obtain

$$V_{k-1}(n+\tilde{n}+1, f(x_u(\tilde{n}), \mu_k(n+\tilde{n}, x_u(\tilde{n})))) \leq \gamma \ell(n+\tilde{n}, x_u(\tilde{n}), \mu_k(n+\tilde{n}, x_u(\tilde{n}))) \quad (10.13)$$

for all $k \in \{N_0+1, \ldots, N\}$ and all $n \in \mathbb{N}_0$.

We abbreviate $\eta_k = \frac{(\gamma+1)^{k-N_0}}{(\gamma+1)^{k-N_0} + \gamma^{k-N_0+1}}$ and prove the main assertion $\eta_k V_k(n + \tilde{n}, x_u(\tilde{n})) \leq V_{k-1}(n+\tilde{n}, x_u(\tilde{n}))$ by induction over $k = N_0, \ldots, N$. By choosing $x_u(0) = x(n)$ with n being arbitrary but fixed we obtain

$$\begin{aligned}
&V_{N_0}(n+N-N_0, x_u(N-N_0)) \\
&\le (\gamma+1) \max_{j=2,\dots,N_0} \ell(n+N-j, x_u(N-j), \mu_{j-1}(n+N-j, x_u(N-j))) \\
&\le (\gamma+1) \sum_{j=2}^{N_0} \ell(n+N-j, x_u(N-j), \mu_{j-1}(n+N-j, x_u(N-j))) \\
&= \frac{1}{\eta_{N_0}} V_{N_0-1}(n+N-N_0, x_u(N-N_0)).
\end{aligned}$$

For the induction step $k \to k+1$ the following holds using (10.13) and the induction assumption

$$\begin{aligned}
&V_k(n+\tilde{n}, x_u(\tilde{n})) \\
&= V_{k-1}(n+\tilde{n}+1, f(x_u(\tilde{n}), \mu_k(n+\tilde{n}, x_u(\tilde{n})))) + \ell(n+\tilde{n}, x_u(\tilde{n}), \mu_k(n+\tilde{n}, x_u(\tilde{n}))) \\
&\ge \eta_k \left(1 + \frac{1-\eta_k}{\gamma+\eta_k}\right) V_k(n+\tilde{n}+1, f(x_u(\tilde{n}), \mu_k(n+\tilde{n}, x_u(\tilde{n})))) \\
&\quad + \left(1 - \gamma \frac{1-\eta_k}{\gamma+\eta_k}\right) \ell(n+\tilde{n}, x_u(\tilde{n}), \mu_k(n+\tilde{n}, x_u(\tilde{n}))) \\
&= \eta_k \frac{\gamma+1}{\gamma+\eta_k} (V_k(n+\tilde{n}+1, f(x_u(\tilde{n}), \mu_k(n+\tilde{n}, x_u(\tilde{n})))) \\
&\quad + \ell(n+\tilde{n}, x_u(\tilde{n}), \mu_k(n+\tilde{n}, x_u(\tilde{n}))))
\end{aligned}$$

using the dynamic programming principle (3.16) with $K = 1$ in the last step. Hence, we obtain $V_k(n+\tilde{n}, x_u(\tilde{n})) \ge \eta_k \frac{\gamma+1}{\gamma+\eta_k} V_{k+1}(n+\tilde{n}, x_u(\tilde{n}))$ with

$$\eta_k \frac{\gamma+1}{\gamma+\eta_k} = \frac{(\gamma+1)^{k-2}}{(\gamma+1)^{k-2}+\gamma^{k-1}} \frac{\gamma+1}{\gamma + \frac{(\gamma+1)^{k-2}}{(\gamma+1)^{k-2}+\gamma^{k-1}}} = \frac{(\gamma+1)^{k-1}}{(\gamma+1)^{k-1}+\gamma^k} = \eta_{k+1}.$$

If we choose $k = N$ then we get $\tilde{n} = 0$. Inserting this into our induction result we can use $x_u(0) = x_u(0, x(n)) = x(n)$ and the assertion holds. □

Finally, we can now use Proposition 10.11 within the NMPC closed loop. This allows us to verify Condition (10.12) and to estimate α directly from Assumption 10.10.

Theorem 10.12 (A priori suboptimality estimate) *Consider $\gamma > 0$ and $N, N_0 \in \mathbb{N}$, $N \ge N_0$ such that $(\gamma+1)^{N-N_0} > \gamma^{N-N_0+2}$ holds. If Assumption 10.10 is fulfilled for these γ, N and N_0, then the estimate* (10.11) *holds for all $n \in \mathbb{N}_0$ where*

$$\alpha := \frac{(\gamma+1)^{N-N_0} - \gamma^{N-N_0+2}}{(\gamma+1)^{N-N_0}}. \tag{10.14}$$

Proof From Proposition 10.11 we know

$$V_N(n, x(n)) - V_{N-1}(n, x(n)) \le \frac{\gamma^{N-N_0+1}}{(\gamma+1)^{N-N_0}} V_{N-1}(n, x(n)).$$

Setting $j = n - 1$, we can reformulate this and obtain

$$V_N(j+1, x(j+1)) - V_{N-1}(j+1, x(j+1)) \leq$$
$$\leq \frac{\gamma^{N-N_0+1}}{(\gamma+1)^{N-N_0}} V_{N-1}(j+1, f(x_u(0, x(j)), \mu_N(j, x_u(0, x(j)))))$$

using the dynamics of the optimal open-loop solution. Now, we can use (10.13) with $k = N$ and get

$$V_N(j+1, x(j+1)) - V_{N-1}(j+1, x(j+1)) \leq \frac{\gamma^{N-N_0+2}}{(\gamma+1)^{N-N_0}} \ell(j, x(j), \mu_N(j, x(j))).$$

Hence, the assumptions of Lemma 10.9 are fulfilled with

$$\alpha = 1 - \frac{\gamma^{N-N_0+2}}{(\gamma+1)^{N-N_0}} = \frac{(\gamma+1)^{N-N_0} - \gamma^{N-N_0+2}}{(\gamma+1)^{N-N_0}}$$

and the assertion follows. □

Similar to the Proposition 10.6, the required values of γ and α are easily computable and allow to extend Algorithm 3.7 in a similar manner as we did in Algorithm 10.8.

Algorithm 10.13 *(NMPC algorithm for time varying reference x^{ref} with a priori suboptimality estimate)*

Set $\alpha = 1$. At each sampling time t_n, $n = 0, 1, 2, \ldots$:

(1) Measure the state $x(n) \in X$ of the system

(2) Set $x_0 = x(n)$ and solve the optimal control problem

$$\begin{aligned} &\text{minimize} \quad J_N(n, x_0, u(\cdot)) := \sum_{k=0}^{N-1} \ell(n+k, x_u(k, x_0), u(k)) \\ &\text{with respect to } u(\cdot) \in \mathbb{U}^N(x_0), \quad \text{subject to} \\ &x_u(0, x_0) = x_0, \quad x_u(k+1, x_0) = f(x_u(k, x_0), u(k)) \end{aligned} \qquad (\mathrm{OCP}_N^n)$$

and denote the obtained optimal control sequence by $u^\star(\cdot) \in \mathbb{U}^N(x_0)$.

(3) Define the NMPC-feedback value $\mu_N(n, x(n)) := u^\star(0) \in U$ and use this control value in the next sampling period

(4) Compute α via

Find the minimal γ which satisfies the inequalities in Assumption 10.10 for the current n and set

$$\alpha = \min\left\{\alpha, \frac{(\gamma+1)^{N-N_0} - \gamma^{N-N_0+2}}{(\gamma+1)^{N-N_0}}\right\}$$

Note that checking the additional condition $(\gamma + 1)^{N-N_0} > \gamma^{N-N_0+2}$ from Theorem 10.12 is unnecessary, since a violation would lead to a negative α in which case asymptotic stability cannot be guaranteed by means of Theorem 10.12, anyway.

Similar to Proposition 10.6, the results from Theorem 10.12 are easily carried over to the multistep NMPC case described in Sect. 10.4 by extending Assumption 10.10.

Example 10.14 To illustrate these results, we consider the inverted pendulum on a cart problem from Example 2.10 with parameters $g = 9.81$, $l = 10$ and $k_R = k_L = 0.01$ and control constraint set $\mathbb{U} = [-15, 15]$. Our aim is to stabilize one of the upright positions $x \in \mathcal{S} := \{((k+1)\pi, 0, 0, 0)^\top \mid k \in 2\mathbb{Z}\}$. For this example, we will provide online measurements of α using Algorithm 10.8. For one fixed initial value and varying terminal weights ω, cf. Sect. 10.2, and control horizons, cf. Sect. 10.4. For a comparison of Algorithms 10.8 and 10.13 we refer to [12, 27].

In order to obtain a suitable cost function, we follow the guidelines from Sect. 6.6 and construct a cost function for which—at least in the first two components—the overshoot of ℓ along a typical stable trajectory becomes small. To this end, we have used the geometry of the vector field of the first two differential equations representing the pendulum, see Fig. 10.11a and shaped the cost function such that it exhibits local maxima at the downward equilibria and "valleys" along the stable manifolds of the upright equilibria to be stabilized. The resulting cost function ℓ is of the integral type (3.4) with

$$\begin{aligned} L(x, u) :=& 10^{-4} u^2 + \Big(3.51 \sin(x_1 - \pi)^2 + 4.82 \sin(x_1 - \pi) x_2 \\ &+ 2.31 x_2^2 + 0.1 \left((1 - \cos(x_1 - \pi)) \cdot (1 + \cos(x_2)^2)\right)^2 \\ &+ 0.01 x_3^2 + 0.1 x_4^2\Big)^2, \end{aligned}$$

cf. Fig. 10.11b. Using the terminal weights from Sect. 10.2, the cost functional becomes

$$J_N(x_0, u) = \sum_{i=0}^{N-2} \ell(x(i), u(i)) + \omega \ell(x(N-1), u(N-1)).$$

This way of adjusting the cost function to the dynamics allows us to considerably reduce the length of the optimization horizon for obtaining stability in the NMPC scheme without stabilizing terminal conditions compared to simpler choices of ℓ. However, for the initial value $x_0 = (2\pi + 1.5, 0, 0, 0)$ and sampling period $T = 0.05$, which have been used in the subsequent computations, we still need a rather large optimization horizon of $N = 70$ to obtain stability of the closed loop.

Since the cost function is 2π–periodic it does not penalize the distance to a specific equilibrium in $\mathcal{S}$; rather, it penalizes the distance to the whole set. For a better comparison of the solutions for different parameters we want to force the algorithm to stabilize one specific upright position in $\mathcal{S}$. To this end, we add box–constraints to $\mathbb{X}$ limiting the x_1-component to the interval $[-\pi + 0.01, 3\pi - 0.01]$. The tolerances

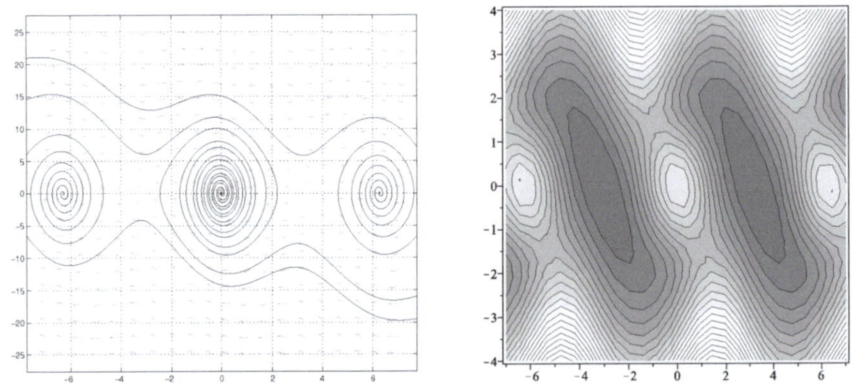

(a) Vector field of the inverted pendulum on a cart example 2.10. Here, the first two components are displayed which represent the pendulum.

(b) Contour plot of the cost function ℓ. Similar to Fig. 10.11a, only the dependency on the first two components is displayed.

Fig. 10.11 Vector field and cost function

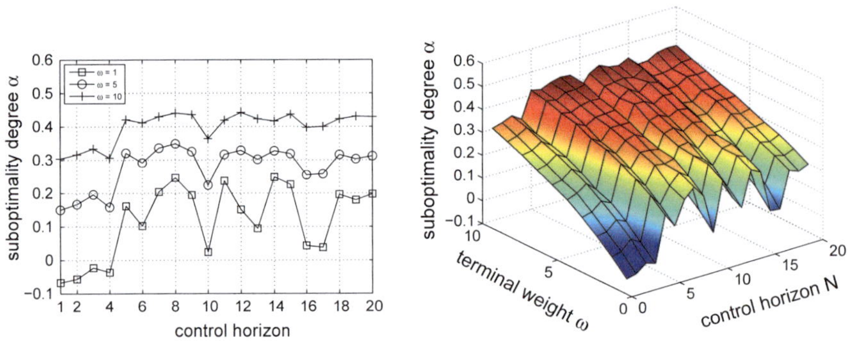

Fig. 10.12 Computed value for $\alpha^{\omega}_{70,m}$ for the nonlinear inverted pendulum Example 2.10 with control horizons $m \in \{1, \ldots, 20\}$ and terminal weights $\omega \in \{1, \ldots, 10\}$

of the optimization routine and the differential equation solver are set to 10^{-6} and 10^{-7}, respectively. The NMPC closed-loop trajectories displayed in Fig. 10.12 are simulated for terminal weights $\omega = 1, \ldots, 10$, cf. Sect. 10.2 and control horizons $m = 1, \ldots, 10$, cf. Sect. 10.4. The resulting α-values from Algorithm 10.8, denoted by $\alpha^{\omega}_{N,m}$, are shown in Fig. 10.12.

Note that for $\omega = 1$ the α values are negative for control horizons $m = 1, \ldots, 4$. Still, larger control horizons exhibit a positive α value such that stability is guaranteed. This is in accordance with the theoretical results from Sect. 10.4, even though these simulation-based results do not share the monotonicity of the theoretical bounds from Fig. 10.3. Additionally, an increase of α can be observed for all control horizons m if ω

is increased. This confirms the stabilizing effect of terminal costs shown theoretically in Sect. 10.2, cf. Fig. 10.1.

Summarizing, these results show that the online measurement of α yields valuable insights into the performance analysis of NMPC schemes without terminal conditions and thus nicely complements the theoretical results from Chap. 6 and Sects. 10.2 and 10.4.

10.8 Adaptive Optimization Horizon

In the previous Sect. 10.7 we have shown how the suboptimality degree α can be computed at runtime of the NMPC scheme without stabilizing terminal constraints. If the horizon length N is not chosen adequately, then it is likely that during runtime a value $\alpha < 0$ is obtained. In this case, stability of the closed loop cannot be guaranteed by Proposition 10.6 or Theorem 10.12. However, the ability to compute α for each point $x(n)$ on the closed-loop trajectory using the techniques from Sect. 10.7 naturally leads to the idea of adapting the optimization horizon N at each time n such that stability and desired performance can be guaranteed. In this section, we will show some algorithms for this purpose, taken from Pannek [27]. Here, we restrict ourselves to the basic idea and refer to [27] for more sophisticated approaches.

The fundamental idea of such an adaptive algorithm is rather simple: introducing a stability and suboptimality threshold $\overline{\alpha} > 0$, at each sampling instant n we prolong the optimization horizon if α for the current horizon is smaller than $\overline{\alpha}$. If $\alpha > \overline{\alpha}$ holds, then we may reduce N in order to save computational time. This leads to the following algorithm.

Algorithm 10.15 (*Adaptive horizon NMPC algorithm for time varying reference*)
Set $N_0 > 0$ and $\overline{\alpha} > 0$. At each sampling time t_n, $n = 0, 1, 2, \ldots$:

(1) Measure the state $x(n) \in X$ of the system
(2) While $\overline{\alpha} > \alpha$

(a) Set $x_0 = x(n)$, $N = N_n$ and solve the optimal control problem

$$\begin{aligned} &\text{minimize} \quad J_N(n, x_0, u(\cdot)) := \sum_{k=0}^{N-1} \ell(n+k, x_u(k, x_0), u(k)) \\ &\text{with respect to} \quad u(\cdot) \in \mathbb{U}^N(x_0), \quad \text{subject to} \\ &x_u(0, x_0) = x_0, \quad x_u(k+1, x_0) = f(x_u(k, x_0), u(k)) \end{aligned} \qquad (\mathrm{OCP}_{\mathrm{N}}^{\mathrm{n}})$$

Denote the obtained optimal control sequence by $u^\star(\cdot) \in \mathbb{U}^N(x_0)$.

(b) Compute α via Proposition 10.6 or Theorem 10.12
(c) If $\alpha > \overline{\alpha}$ call reducing strategy for N_n, else call increasing strategy for N_n; obtain $u^\star(\cdot)$ for the new $N = N_n$ and an initial guess for N_{n+1}

(3) Define the NMPC-feedback value $\mu_N(n, x(n)) := u^\star(0) \in U$ and use this control value in the next sampling period

Here, the initial guess N_{n+1} in Step (2c) will typically be $N_{n+1} = N_n$, however, as we will see below, in the case of reducing N_n the choice $N_{n+1} = N_n - 1$ is more efficient, cf. the discussion after Proposition 10.18.

If this algorithm is successful in ensuring $\alpha \geq \overline{\alpha}$ for each n, then the assumptions of Proposition 10.6 or Theorem 10.12 are satisfied. However, these results require the optimization horizon N to be fixed and hence do not apply to Algorithm 10.15 in which N_n changes with time.

To cope with this issue, we generalize Proposition 10.6 to varying optimization horizons. To this end, for each $x \in \mathbb{X}$ and $N \in \mathbb{N}$ we denote the maximal α from (10.10) by $\alpha(N)$. We then introduce the following assumption which guarantees that for any horizon N satisfying $\alpha(N) \geq \overline{\alpha}$ the controller shows a bounded guaranteed performance if the horizon length is increased.

Assumption 10.16 Given $n \in \mathbb{N}_0$, $x \in \mathbb{X}$, $N < \infty$ and a value $\overline{\alpha} \in (0, 1)$ with $\alpha(N) \geq \overline{\alpha}$, we assume that there exist constants $C_l, C_\alpha > 0$ such that the inequalities

$$C_l \ell(n, x, \mu_N(n, x)) \leq \ell(n, x, \mu_{\widetilde{N}}(n, x)) \frac{V_{\widetilde{N}}(n, x) - V_{\widetilde{N}}(n+1, f(x, \mu_N(n, x))}{V_{\widetilde{N}}(n, x) - V_{\widetilde{N}}(n+1, f(x, \mu_{\widetilde{N}}(n, x))} \tag{10.15}$$

$$C_\alpha \alpha(N) \leq \alpha(\widetilde{N}) \tag{10.16}$$

hold for all $\widetilde{N} \geq N$.

The reason for Assumption 10.16 is that it is possible that the performance of the controller μ_N may not improve monotonically as N increases; see Di Palma and Magni [4]. Consequently, we cannot expect $\alpha(\widetilde{N}) \geq \alpha(N)$ for $\widetilde{N} > N$. Still, we need to ensure that $\alpha(\widetilde{N})$ does not become too small compared to $\alpha(N)$, in particular, $\alpha(\widetilde{N})$ should not drop below zero if the horizon length is increased; this is ensured by (10.15). Furthermore, we need an estimate for the dependence of $\ell(n, x, \mu_N(n, x))$ on N which is given by (10.16). Unfortunately, for both inequalities so far we were not able to provide sufficient conditions in terms of the problem data, like, e.g., boundedness or a controllability conditions similar to Assumptions 6.3 or 6.5. Still, numerical evaluation for several examples showed that these inequalities are satisfied and that C_l and C_α attain reasonable values.

Using Assumption 10.16, we obtain a stability and performance estimate of the closed loop in the context of changing horizon lengths similar to Proposition 10.6. Since the closed-loop control resulting from Algorithm 10.15 now depends on a sequence of horizons $(N_n)_{n \in \mathbb{N}_0}$ we obtain a sequence of control laws $(\mu_{N_n})_{n \in \mathbb{N}_0}$. The closed-loop trajectory generated by this algorithm is then given by

$$x(n+1) = f(x(n), \mu_{N_n}(n, x(n))). \tag{10.17}$$

Theorem 10.17 *Consider the sequence of feedback laws* (μ_{N_n}) *computed from Algorithm 10.15 and the corresponding closed-loop trajectory* $x(\cdot)$ *from* (10.17). *Assume that for optimal value functions* $V_{N_n} : \mathbb{N}_0 \times \mathbb{X} \to \mathbb{R}_0^+$ *of* $(\mathrm{OCP}_\mathrm{N}^\mathrm{n})$ *with* $N = N_n$ *the inequality*

$$V_{N_n}(n, x(n)) \geq V_{N_n}(n+1, x(n+1)) + \overline{\alpha}\ell(n, x(n), \mu_{N_n}(n, x(n))) \qquad (10.18)$$

holds for all $n \in \mathbb{N}_0$ *and that Assumption 10.16 is satisfied for all triplets* $(n, x, N) = (n, x(n), N_n)$, $n \in \mathbb{N}_0$, *with constants* $C_l^{(n)}$, $C_\alpha^{(n)}$. *Then*

$$\alpha_C V_\infty(n, x(n)) \leq \alpha_C J_\infty^{cl}(n, x(n), \mu_{(N_n)}) \leq V_{N^\star}(n, x(n)) \leq V_\infty(n, x(n)) \quad (10.19)$$

holds for all $n \in \mathbb{N}_0$ *where* $\alpha_C := \min_{i \in \mathbb{N}_{\geq n}} C_\alpha^{(i)} C_l^{(i)} \overline{\alpha}$.

Proof Given $(i, x(i), N_i)$ for some $i \in \mathbb{N}_0$, Assumption 10.16 for $(n, x, N) = (i, x(i), N_i)$ guarantees $\alpha(N_i) \leq \alpha(\widetilde{N})/C_\alpha^{(i)}$ for $\widetilde{N} \geq N_i$. Choosing $\widetilde{N} = N^\star$, we obtain $\overline{\alpha} \leq \alpha(N_i) \leq \alpha(N^\star)/C_\alpha^{(i)}$ using the relaxed Lyapunov inequality (10.18). Multiplying by the stage cost $\ell(i, x(i), \mu_{N_i}(i, x(i)))$, we can conclude

$$\begin{aligned}
\overline{\alpha}\ell(i, x(i), \mu_{N_i}(i, x(i))) &\leq \frac{\alpha(N^\star)}{C_\alpha^{(i)}} \ell(i, x(i), \mu_{N_i}(i, x(i))) \\
&= \frac{V_{N^\star}(i, x(i)) - V_{N^\star}(i+1, f(x(i), \mu_{N^\star}(i, x(i)))}{C_\alpha^{(i)} \ell(i, x(i), \mu_{N^\star}(i, x(i)))} \ell(i, x(i), \mu_{N_i}(i, x(i))) \\
&\leq \frac{V_{N^\star}(i, x(i)) - V_{N^\star}(i+1, f(x(i), \mu_{N_i}(i, x(i)))}{C_\alpha^{(i)} C_l^{(i)}}
\end{aligned}$$

using (10.18) and (10.15). In particular, the latter condition allows us to use an identical telescope sum argument as in the proof of Proposition 10.6 since it relates the closed-loop varying optimization horizon to a fixed one. Hence, summing the stage costs along the closed-loop trajectory reveals

$$\alpha_C \sum_{i=n}^{K} \ell(i, x(i), \mu_{N_i}(i, x(i))) \leq V_{N^\star}(n, x(n)) - V_{N^\star}(K+1, x(K+1))$$

where we defined $\alpha_C := \min_{i \in [n, \ldots, K]} C_\alpha^{(i)} C_l^{(i)} \overline{\alpha}$. Since $V_{N^\star}(K+1, x(K+1)) \geq 0$ holds, we can neglect it in the last inequality. Taking K to infinity yields

$$\alpha_C V_\infty^{\mu_{(N_i)}}(n, x(n)) = \alpha_C \lim_{K \to \infty} \sum_{i=n}^{K} \ell(i, x(i), \mu_{N_i}(i, x(i))) \leq V_{N^\star}(n, x(n)).$$

Since the first and the last inequality of (10.19) hold by definition of V_N and V_∞, the assertion follows. □

If the conditions of this theorem hold, then stability-like behavior of the closed loop can be obtained analogously to Proposition 10.6.

Having shown the analytical background, we now present adaptation strategies which can be used for increasing or reducing the optimization horizon N in Step (2c) of Algorithm 10.15. For simplicity of exposition, we restrict ourselves to two simple strategies and consider a posteriori estimates based variants only. Despite their simplicity, these methods have shown to be reliable and fast in numerical simulations. A more detailed analysis, further methods, and comparisons can be found in [27]. The following proposition yields the basis for a strategy for reducing N_n

Proposition 10.18 *Consider the optimal control problem* $(\mathrm{OCP}_\mathrm{N}^\mathrm{n})$ *with initial value* $x_0 = x(n)$, $N_n \in \mathbb{N}$, *and denote the optimal control sequence by* $u^\star$. *For fixed* $\overline{\alpha} \in (0, 1)$, *suppose there exists an integer* $\overline{i} \in \mathbb{N}_0$, $0 \leq \overline{i} < N$ *such that*

$$\begin{aligned} &V_{N_n-i}(n+i+1, x_{u^\star}(i+1, x(n))) + \overline{\alpha}\ell(n+i, x_{u^\star}(i, x(n)), \mu_{N_n-i}(n+i, x_{u^\star}(i, x(n)))) \\ &\quad \leq V_{N_n-i}(n+i, x_{u^\star}(i, x(n))) \end{aligned} \tag{10.20}$$

holds for all $0 \leq i \leq \overline{i}$. *Then, setting* $N_{n+i} = N_n - i$ *and* $\mu_{N_{n+i}}(n+i, x(n+i)) = u^\star(i)$ *for* $0 \leq i \leq \overline{i} - 1$, *inequality* (10.18) *holds for* $n = n, \ldots, n + \overline{i} - 1$.

Proof The proof follows immediately from the fact that for $\mu_{N_{n+i}}(n+i, x(n+i)) = u^\star(i)$ the closed-loop trajectory (10.17) satisfies $x(n+i) = x_{u^\star}(i, x(n))$. Hence, (10.18) follows from (10.20). □

Observe that Proposition 10.18 is quite similar to the results from Sect. 10.4, since $\mu_{N_{n+i}}(n+i, x(n+i))$ as defined in this theorem coincides with the multistep feedback law from Sect. 10.4. Thus, Proposition 10.18 guarantees that if $\overline{i} > 1$, then the multistep NMPC feedback from Sect. 10.4 can be applied with $m = \overline{i}$ steps such that the suboptimality threshold $\overline{\alpha}$ can be guaranteed. With the choice $N_{n+i} = N_n - i$, due to the principle of optimality we obtain that the optimal control problems within the next $\overline{i} - 1$ NMPC iterations are already solved since $\mu_{N_{n+i}}(n+i, x(n+i))$ can be obtained from the optimal control sequence $u^\star(\cdot) \in \mathbb{U}^N(x(n))$ computed at time n. This implies that the most efficient way for the reducing strategy in Step (2c) of Algorithm 10.15 is not to reduce N_n itself but rather to reduce the horizons N_{n+i} by i for the subsequent sampling instants $n+1, \ldots, n+\overline{i}$, i.e., we choose the initial guess in Step (2c) as $N_{n+1} = N_n - 1$. Still, if the a posteriori estimate is used, the evaluation of (10.20) requires the solution of an additional optimal control problem in each step in order to compute $V_{N_n-i}(n+i+1, x_{u^\star}(i+1, x(n)))$.

In contrast to this efficient and simple shortening strategy, it is quite difficult to obtain efficient methods for prolonging the optimization horizon N in Step (2c) of Algorithm 10.15. In order to understand why this is the case, we first introduce the basic idea behind any such prolongation strategy: at each sampling instant we iteratively increase the horizon N_n until (10.18) is satisfied and use this horizon for the next NMPC step. In order to ensure that iteratively increasing N_n will eventually lead to a horizon for which (10.18) holds, we make the following assumption.

Assumption 10.19 Given $\overline{\alpha} \in (0, 1)$, for all $x_0 \in \mathbb{X}$ and all $n \in \mathbb{N}_0$ there exists a finite horizon length $\overline{N} = \overline{N}(n, x_0) \in \mathbb{N}$ such that (10.18) holds with $\alpha(N_n) \geq \overline{\alpha}$ for $x(n) = x_0$ and $N_n \geq \overline{N}$.

Assumption 10.19 can be seen as a performance assumption which requires the existence of a horizon length N_n such that the predefined threshold $\overline{\alpha}$ can be satisfied. If no such horizon exists, no prolongation strategy can be designed which can guarantee closed-loop suboptimality degree $\alpha > \overline{\alpha}$. Assumption 10.19 is, for instance, satisfied if the conditions of Theorem 6.24 hold.

The following proposition shows that under this assumption any iterative strategy which increases the horizon will terminate after finitely many steps with a horizon length N for which the desired local suboptimality degree holds.

Proposition 10.20 *Consider the optimal control problem* ($\mathrm{OCP}^{\mathrm{n}}_{\mathrm{N}}$) *with initial value* $x_0 = x(n)$ *and* $N_n \in \mathbb{N}$. *For fixed* $\overline{\alpha} \in (0, 1)$ *suppose that Assumption 10.19 holds. Then, any algorithm which iteratively increases the optimization horizon* N_n *and terminates if* (10.18) *holds will terminate in finite time with an optimization horizon* N_n *for which* (10.18) *holds. In particular, Theorem 10.17 is applicable provided Assumption 10.16 holds.*

Proof The proof follows immediately from Assumption 10.19. □

Unfortunately, if (10.18) does not hold it is in general difficult to assess by how much N_n should be increased such that (10.18) holds for the increased N_n. The most simple strategy of increasing N_n by one in each iteration shows satisfactory results in practice, however, in the worst case it requires us to check (10.18) $\overline{N} - N_n + 1$ times at each sampling instant. In contrast to the shortening strategy, the principle of optimality cannot be used here to establish a relation between the optimal control problems for different N_n and, moreover, these problems may exhibit different solution structures which makes it a hard task to provide a suitable initial guess for the optimization algorithm, see also Sect. 12.5.

In order to come up with more efficient strategies, different methods have been developed [27] which utilize the structure of the suboptimality estimate itself to determine by how much N_n should be increased. Compared to these methods, however, the performance of the simple strategy of increasing N_n by one is still acceptable. In the following example we illustrate the performance of this strategy for Example 2.11.

Example 10.21 For the ARP system (2.19)–(2.26) we have already analytically derived a continuous time tracking feedback in (2.28). However, this feedback law performs poorly under sampling, in particular, for the sampling period $T = 0.2$ which we consider here we obtain an unstable closed-loop sampled data system.

In order to obtain a sampled data feedback law which shows better performance we use the digital redesign technique proposed by Nešić and Grüne in [25]: given a signal $v(t)$ to track, we numerically simulate the continuous time controlled system in order to generate the output x^{ref} which in turn will be used as the reference trajectory for an NMPC tracking problem. The advantage of proceeding this way compared to the direct formulation of an NMPC tracking problem lies in the fact that—according to our numerical experience—the resulting NMPC problem is much easier to solve and in particular requires considerably smaller optimization horizons in order to obtain a stable NMPC closed loop.

Specifically, we consider the piecewise constant reference function

$$v(t) = \begin{cases} 10, & t \in [0, 5) \cup [9, 10) \\ 0, & t \in [5, 9) \cup [10, 15) \end{cases}$$

for the x_5–component of the trajectory of the system. In order to obtain short transient times for the continuous time feedback, we set the design parameters c_i in (2.28) to $(c_0, c_1, c_2, c_3) = (10000, 3500, 500, 35)$. Then, we incorporate the resulting trajectory displayed in Fig. 10.13 as reference $x^{\mathrm{ref}}(\cdot)$ in the NMPC algorithm. Since our goal is to track the reference with the x_5–component of the trajectory, we use the simple quadratic cost function

$$J(x_0, u) = \sum_{j=0}^{N} \int_{t_j}^{t_{j+1}} |x_{5,u}(t, x_0) - x_{5,\mathrm{ref}}(t)| dt$$

within the adaptive horizon NMPC Algorithm 10.15. Moreover, we select the sampling period $T = 0.2$ and fix the initial value $x(0) = (0, 0, 0, 0, 10, 0, 0, 0)$ for both the continuously and the sampled–data controlled system.

Using the a posteriori and a priori estimation techniques within the adaptive NMPC Algorithm 10.15, we obtain the evolutions of horizons N_n along the closed loop for the suboptimality bound $\overline{\alpha} = 0.1$ as displayed in Fig. 10.14. Comparing the horizons chosen by the a priori and the a posteriori estimates, one sees that the a posteriori algorithms yields smaller optimization horizons which makes the resulting scheme computationally more efficient, however, at the expense that the evaluation of the a posteriori criterion itself is computationally more demanding; see also Fig. 10.15.

It is also interesting to compare these horizons to the standard NMPC Algorithm 3.7 with fixed N which needs a horizon of $N = 6$ in order to guarantee $\alpha \geq \overline{\alpha}$ along the closed loop. Here, one observes that the required horizon N_n for the adaptive NMPC approach is typically smaller than $N = 6$ for both the a posteriori and

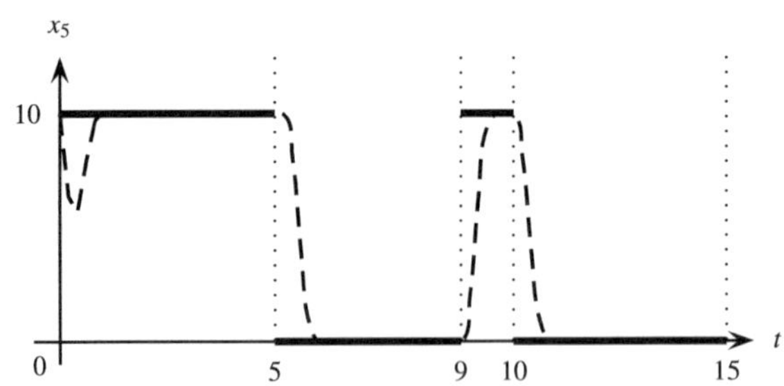

Fig. 10.13 The figure displays the reference function for the continuous time feedback (*solid*) and state trajectory using the continuous time feedback (*dashed*). The latter will be used as reference function within the NMPC algorithm

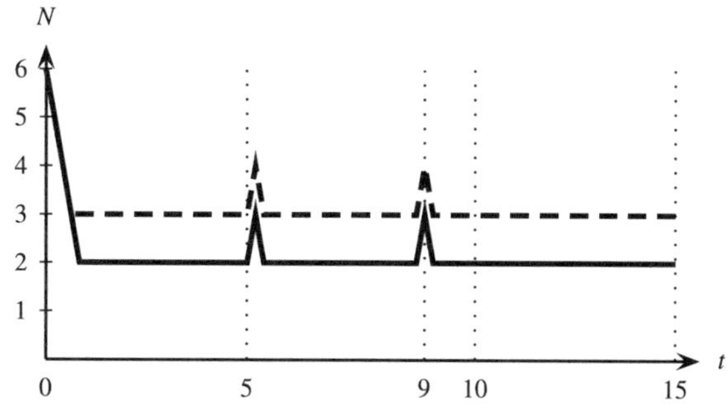

Fig. 10.14 Optimization horizons computed by the adaptive NMPC Algorithm 10.15 for the ARP problem using the a posteriori estimate (*solid*) and the a priori estimate (*dashed*)

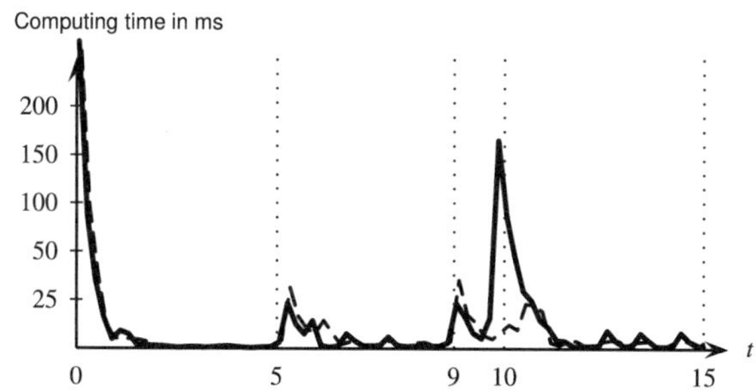

Fig. 10.15 Computing times of the adaptive NMPC Algorithm 10.15 for the ARP problem using the a posteriori estimate (*solid*) and the a priori estimate (*dashed*)

the a priori estimate based variant. One also observes that the horizon is increased at the jump points of the reference function $v(\cdot)$, which is the behavior one would expect in a "critical" situation and nicely reflects the ability of the adaptive horizon algorithm to adapt to the new situation.

Although the algorithm chooses to modify the horizon length throughout the run of the closed loop, one can barely see a difference between the resulting x_5 trajectories and the (dashed) reference trajectory given in Fig. 10.13. For this reason, we do not display the closed-loop solutions. Instead, we additionally plotted the computing times of the two adaptive NMPC variants in Fig. 10.15. Again, one can immediately see the spikes in the graph right at the points in which $v(\cdot)$ jumps. This figure also illustrates the disadvantage of the algorithm of having to solve multiple additional optimal control problems whenever N is increased, which clearly shows up in the higher computation times at these points, in particular for the computationally more expensive a posteriori estimation.

While the adaptive optimization horizon algorithm produces good results in this example, we would like to mention that there are other examples—e.g., the swing-up of the inverted pendulum—for which the algorithm performs less convincing. We conjecture that a better understanding of Assumption 10.16 may provide the insight needed in order to tell the situations in which the adaptive algorithms provides good results from those in which it does not.

10.9 Nonoptimal NMPC

In the case of limited computational resources and/or fast sampling, the time available for solving the optimization problems (OCP_N) or its variants may not be sufficient to obtain an arbitrary accurate solution. Typically, the algorithms for solving these problems, i.e., for obtaining $u^\star$ and thus $\mu_N(x(n)) = u^\star(0)$, work iteratively[2] and with limited computation time may we may be forced to terminate this algorithm prior to convergence to the optimal control sequence $u^\star$.

It is therefore interesting to derive conditions which ensure stability and performance estimates for the NMPC closed loop in this situation. To this end, we modify Algorithm 3.1 as follows.

Algorithm 10.22 We replace Steps (2) and (3) of Algorithm 3.1 (or its variants) by the following:

(2') For initial value $x_0 = x(n)$, given an initial guess $u_n^0(\cdot) \in U^N$ we iteratively compute $u_n^j(\cdot) \in U^N$ by an iterative optimization algorithm such that

$$J_N(x_0, u_n^{j+1}(\cdot)) \leq J_N(x_0, u_n^j(\cdot)).$$

We terminate this iteration after $j^* \in \mathbb{N}$ iterations, set $u_n(\cdot) := u_n^{j^*}(\cdot)$ and $\widetilde{V}_N(n) := J_N(x_0, u_n(\cdot))$.

(3') Define the NMPC-feedback value $\mu_N(x(n)) := u_n(0) \in U$ and use this control value in the next sampling period

One way to ensure proper operation of such an algorithm is by assuming that the sampling period is so small such that the optimal control from sampling instant $n-1$ is still "almost optimal" at time n. In this case, one iteration starting from $u_n^0 = u_{n-1}$, i.e., $j^* = 1$, may be enough in order to be sufficiently close to an optimal control, i.e., to ensure $J_N(x(n), u_n^1) \approx V_N(x(n))$. This procedure is, e.g., investigated by Diehl, Findeisen, Allgöwer, Bock and Schlöder in [5].

An alternative but conceptually similar idea is presented in work of Graichen and Kugi [8]. In this reference, a sufficiently large number of iterations j^* is fixed and conditions are given under which the control sequences $u_n^{j^*}$ become more and more optimal as n increases, i.e., they satisfy $J_N(x(n), u_n^{j^*}) \approx V_N(u^\star)$ for sufficiently large n. Using suitable bounds during the transient phase in which this approximate optimality does not yet hold then allows the authors to conclude stability estimates.

While these results use that $u_n^{j^*}$ is close to $u^\star$ in an appropriate sense, here we investigate the case in which $u_n^{j^*}$ may be far away from the optimal solution. As we will see, asymptotic stability in the sense of Definition 2.14 is in general difficult to establish in this case. However, it will still be possible to prove the following weaker property, which we already encountered in Theorem 9.33.

[2]For more information on these algorithms see Chap. 12 and for numerical aspects of the theory in this section in particular Sect. 12.6.

Definition 10.23 Given a set $S \subseteq \mathbb{X}$, we say that the NMPC closed loop (2.5) is *attractive on* S if for each $x \in S$ the convergence

$$\lim_{k\to\infty} x_{\mu_N}(k, x) = x_*$$

holds.

Contrary to asymptotic stability, a merely attractive solution x_{μ_N} which starts close to the equilibrium x_* may deviate far from it before it eventually converges to x_*. In order to exclude this undesirable behavior, one may wish to require the following stability property in addition to attraction.

Definition 10.24 Given a set $S \subseteq \mathbb{X}$, we say that the NMPC closed loop (2.5) is *stable on* S if there exists $\alpha_S \in \mathcal{K}$ such that the inequality

$$|x_{\mu_N}(k, x)|_{x_*} \le \alpha_S(|x|_{x_*})$$

holds for all $x \in S$ and all $k = 0, 1, 2, \ldots$.

It is well-known that under suitable regularity conditions attractivity and stability imply asymptotic stability; see, e.g., the book of Khalil [20, Chap. 4]. Since this is not the topic of this book, we will not go into technical details here and rather work with the separate properties attractivity and stability in the remainder of this section.

The following variant of Proposition 10.6 will be used in order to ensure attractivity and stability.

Proposition 10.25 *Consider the solution $x(n) = x_{\mu_N}(n, x_0)$ of the NMPC closed loop* (2.5)*, a set $S \subseteq \mathbb{X}$ and a value $\alpha \in (0, 1]$. Assume that ℓ satisfies*

$$\ell^*(x) \ge \alpha_3(|x|_{x_*}) \tag{10.21}$$

for some $\alpha_3 \in \mathcal{K}_\infty$ and all $x \in S$ and that for each $x_0 \in S$ there exists a function $\widetilde{V}_N : \mathbb{N}_0 \to \mathbb{R}_0^+$ which for all $n \in \mathbb{N}_0$ satisfies

$$\widetilde{V}_N(n) \ge \widetilde{V}_N(n+1) + \alpha\ell(x(n), \mu_N(x(n))). \tag{10.22}$$

Then the closed loop (2.5) *is attractive on S and the inequality*

$$J_\infty^{cl}(x_0, \mu_N) \le \widetilde{V}_N(0) \tag{10.23}$$

holds for $J_\infty^{cl}(x_0, \mu_N)$ from Definition 4.10.

If, in addition, there exists $\tilde{\alpha}_2 \in \mathcal{K}_\infty$ *independent of* x_0 *such that the functions* $\widetilde{V}_N$ *satisfy*

$$\widetilde{V}_N(0) \le \tilde{\alpha}_2(|x(0)|_{x_*}), \tag{10.24}$$

then the closed loop (2.5) *is stable on* S.

Proof Iterating inequality (10.22) for $n = 0 \dots, k$ and using $\widetilde{V}_N(n) \ge 0$ yields

$$\sum_{n=0}^{k} \ell(x(n), \mu_N(x(n))) \le \widetilde{V}_N(0) - \widetilde{V}_N(k+1) \le \widetilde{V}_N(0).$$

Letting $k \to \infty$ we obtain

$$J_\infty^{cl}(x, \mu_N) = \lim_{k\to\infty} \sum_{n=0}^{k} \ell(x(n), \mu_N(x(n))) \le \widetilde{V}_N(0),$$

i.e., (10.23). Now, nonnegativity of ℓ implies $\lim_{n\to\infty} \ell(x(n), \mu_N(x(n))) = 0$ and thus (10.21) implies $x(n) \to 0$, i.e., attractivity.

In order to prove stability under the additional assumption (10.24), observe that (10.22) together with the nonnegativity of $\widetilde{V}_N$ and (10.21) implies

$$\widetilde{V}_N(n) \ge \alpha\ell(x(n), \mu_N(x(n))) \ge \alpha\, \alpha_3(|x(n)|_{x_*}) =: \tilde{\alpha}_1(|x(n)|_{x_*}).$$

Furthermore, (10.22) implies that $\widetilde{V}_N(n)$ is decreasing in n. Using these properties, stability immediately follows from

$$|x(n)|_{x_*} \le \tilde{\alpha}_1^{-1}(\widetilde{V}_N(n)) \le \tilde{\alpha}_1^{-1}(\widetilde{V}_N(0)) \le \tilde{\alpha}_1^{-1}(\tilde{\alpha}_2(|x_0|_{x_*})) =: \alpha_S(|x_0|_{x_*}).$$

□

The precise conditions on u_n^j and u_n in Algorithm 10.22 which ensure attractivity, stability and suboptimality estimates now depend on whether stabilizing terminal conditions are used or not. We first consider the case of stabilizing terminal conditions which was investigated, e.g., by Michalska and Mayne [22], Scokaert, Mayne and Rawlings [30] and Rawlings and Mayne [29, Sect. 2.8] which all use conceptually similar ideas. Here, we follow the latter reference.

The approach in [29, Sect. 2.8] can be written as a variant of Theorem 5.13. In particular, we assume that Assumption 5.9 is satisfied. In order to obtain a more convenient notation, on the terminal constraint set $\mathbb{X}_0$ we define a map $\kappa : \mathbb{X}_0 \to \mathbb{U}$ which assigns to each $x \in \mathbb{X}_0$ the control value $u_x \in \mathbb{U}(x)$ from Assumption 5.9(ii). With this notation, the corresponding theorem reads as follows.

Theorem 10.26 (Stability for nonoptimal NMPC) *Assume that the conditions of Theorem 5.13 are satisfied. Consider Algorithm 3.10 with Steps (2) and (3) replaced* *by Steps (2') and (3') of Algorithm 10.22 under the following assumptions for a set* $S \subseteq \mathbb{X}_N$.

(i) *For* $n = 0$*, we are able to find an admissible initial guess* $u_0^0(\cdot) \in \mathbb{U}^N_{\mathbb{X}_0}(x_0)$ *for each initial value* $x_0 = x(0) \in S$.

(ii) *For* $n = 1, 2, \ldots$*, the initial guess* $u_n^0(\cdot)$ *is chosen as* $u_n^0(k) = u_{n-1}(k+1)$, $k = 0, \ldots, N-2$ *and* $u_n^0(N-1) = \kappa(x_{u_n^0}(N-1, x_0))$.

(iii) *For all* $n = 0, 1, 2, \ldots$ *the control sequences* $u_n(\cdot) = u_n^{j^*}(\cdot)$ *satisfy* $u_n^{j^*}(\cdot) \in \mathbb{U}^N_{\mathbb{X}_0}(x_0)$*, i.e., they are admissible.*

Then the NMPC closed loop (2.5) is attractive on S and the inequality

$$J_\infty^{cl}(x_0, \mu_N) \leq \widetilde{V}_N(0)$$

holds. If, in addition, there exists $\tilde{\alpha}_3 \in \mathscr{K}_\infty$ *such that the inequality* $J_N(x_0, u_0^0(\cdot)) \leq \tilde{\alpha}_3(|x_0|_{x_*})$ *holds for* $u_0^0(\cdot)$ *from (i), then 2.5 is also stable on S.*

Proof First note that (i) ensures that u_0^0 is admissible at time $n = 0$ and that (iii) ensures that u_n^0 in (ii) is admissible for $n = 1, 2, \ldots$, cf. also Lemma 5.10(i).

We abbreviate $x(n) = x_{\mu_n}(n)$. Then, (ii) and the same computation as in the proof of Lemma 5.12 yield the inequality $J_N(x(n+1), u_{n+1}^0(\cdot)) \leq J_{N-1}(x(n+1), u_n(\cdot+1))$ for each $n \geq 0$. On the other hand, the definition of J_N in Algorithm 3.10 implies

$$\widetilde{V}_N(n) = J_N(x(n), u_n(\cdot)) = \ell(x(n), u_n(0)) + J_{N-1}(f(x(n), u_n(x)), u_n(\cdot + 1)).$$

The identities $f(x(n), u_n(x)) = x(n+1)$, $u_n(0) = \mu_N(x(n))$ and the inequality $\widetilde{V}_N(n+1) \leq J_N(x(n+1), u_{n+1}^0(\cdot))$ then lead to

$$\widetilde{V}_N(n) \geq \ell(x(n), u_n(0)) + J_N(x_0, u_n^0(\cdot)) \geq \ell(x(n), \mu_N(x(n))) + \widetilde{V}_N(n+1),$$

i.e., (10.22). Now all properties follow directly from Proposition 10.25. □

Remark 10.27 (i) If the assumptions of Proposition 5.14(ii) hold, then for $x_0 \in \mathbb{X}_0$, the additional stability condition $J_N(x_0, u_0^0(\cdot)) \leq \tilde{\alpha}_3(|x|_{x_*})$ can be guaranteed if we define $u_0^0(\cdot)$ by $u_0^0(k) := \kappa(x_{u_0^0}(k, x_0))$, $k = 0, \ldots, N-1$. From Assumption 5.9(ii) it follows that this choice implies $J_N(x_0, u_0^0(\cdot)) \leq F(x_0) \leq \tilde{\alpha}_2(|x|_{x_*})$ and thus the desired inequality follows with $\tilde{\alpha}_3 = \tilde{\alpha}_2$. Hence, this choice guarantees stability locally around x_*.

One may also apply this definition to u_n^0 in (ii) for those n in which $x(n) \in \mathbb{X}_0$ holds. This way, stability is ensured at least for the tail of the resulting closed-loop trajectory. If we use this choice of u_n^0 and do not perform the iterative optimization in Step (2') of Algorithm 10.22, i.e., if we choose $j^* = 0$, then we obtain an algorithm similar to the so called dual mode strategy from [22].

(ii) Iterative optimizations algorithms are usually designed such that the intermediate results satisfy the desired constraints as soon as the algorithm has succeeded in finding an admissible solution, see Sect. 12.6 for details. Since condition (ii) in

Theorem 10.26 ensures that we already initialize the iterative optimization with an admissible solution, most common optimization algorithms will yield solutions $u_n^{j^*}(\cdot)$ satisfying condition (iii) of Theorem 10.26 regardless of how j^* is chosen.

(iii) Theorem 10.26 yields attractivity for arbitrary $j^* \in \mathbb{N}_0$. In particular, it applies to $j^* = 0$, i.e., to the case in which we do not optimize at all. This means that attractivity follows readily from the stabilizing terminal conditions and the particular construction of the initial guesses. An important consequence of this property is that we can fix j^* a priori, e.g., determined by the available computation time, which makes this approach suitable for real-time NMPC schemes.

Without stabilizing terminal conditions, stability is inherited from optimality and we can no longer expect attractivity or stability for arbitrary j^*. Instead, we need to make sure that $u_n^{j^*}$ is at least "good enough" to ensure (10.22). This is the idea of the following algorithm for determining j^* taken from Grüne and Pannek [13].

Algorithm 10.28 Given $\alpha \in (0, 1)$, in Step (2') of Algorithm 10.22 we iterate over $u_n^j(\cdot) \in \mathbb{U}^N(x(n))$ for $j = 1, 2, \ldots$ until the termination criterion

$$J_N(x(n), u_n^{j^*}(\cdot) \leq \widetilde{V}_N(n-1) - \alpha\ell(x(n-1), u_{n-1}(0)) \tag{10.25}$$

is satisfied.

The following theorem shows attractivity, suboptimality and stability for this algorithm.

Theorem 10.29 (Stability for nonoptimal NMPC) *Consider a set $S \subseteq \mathbb{X}_N$, $\alpha \in (0, 1]$ and Algorithm 3.1 with Steps (2) and (3) replaced by Steps (2') and (3') of Algorithm 10.22. Assume that Algorithm 10.28 is used in Step (2') of Algorithm 10.22 and that* (10.25) *is feasible for each $n \in \mathbb{N}$, i.e., that for each $n \in \mathbb{N}$ there exists $u_n^{j^*} \in \mathbb{U}^N(x(n))$ such that* (10.25) *holds. Assume furthermore that* (10.21) *holds for the stage cost ℓ.*

Then the NMPC closed loop (2.5) *is attractive on S and the inequality*

$$J_\infty^{cl}(x, \mu_N) \leq \widetilde{V}_N(0)$$

holds. If, in addition, there exists $\tilde{\alpha}_3 \in \mathcal{K}_\infty$ such that the inequality $J_N(x_0, u_0^0(\cdot)) \leq \tilde{\alpha}_3(|x|_{x_*})$ holds for the initial guess $u_0^0(\cdot)$ in Step (2') of Algorithm 10.22 for each $x(0) \in S$, then (2.5) *is also stable on S.*

Proof Under the stated assumptions, all properties follow directly from Proposition 10.25. □

Remark 10.30 In contrast to what was observed in Remark 10.27(iii) for the terminal conditioned scheme, here we cannot in general fix j^* a priori. Indeed, the number of iterations of the optimization algorithm which are needed until (10.25) is satisfied depends on various factors—particularly on the choice of u_{n-1} and u_n^0—and is in

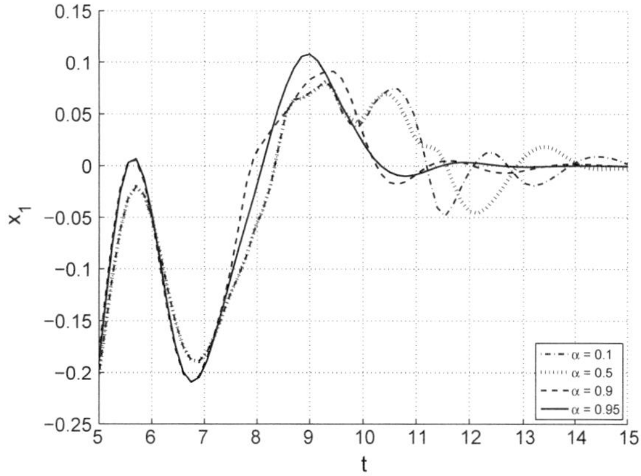

Fig. 10.16 Angle of the pendulum x_1 for varying α

general unknown before the optimization is started. We assume that for sufficiently small sampling periods techniques similar to the works of Diehl, Findeisen, Allgöwer, Bock and Schlöder [5] or Graichen and Kugi [8] can be used in order to bound the number of needed iterations when setting $u_n^0 = u_{n-1}$ but this has not yet been investigated rigorously.

In the general case, the feasibility assumption for (10.25) in Theorem 10.29 may not even be satisfied. Before we investigate this issue, we illustrate the performance of this algorithm by a numerical example.

Example 10.31 We consider the nonlinear pendulum from Example 2.10, where the task is now to stabilize the downward equilibrium $x_* = (0, 0, 0, 0)^T$. Figures 10.16 and 10.17 below show parts of the closed-loop trajectories of x_1 and x_3 using Algorithms 10.22 and 10.28 in Step (2') for varying α. The stage cost is of type (3.4) with

$$L(x, u) = 100 \sin^2(0.5 x_1) + x_2^2 + 10.0 x_3^2 + x_4^2 + u^2,$$

and sampling period $T = 0.15$ and the NMPC algorithm was run with optimization horizon $N = 17$ and input constraints $\mathbb{U} = [-1, 1]$ using a recursive discretization and a line–search SQP method to solve the resulting optimization problem, see Chap. 12 for details on such methods.

One can see clearly from Figs. 10.16 and 10.17 that the closed-loop system is stable for all values of α. Moreover, one can nicely observe the improvement of the closed-loop behavior visible in the decreasing time until the system comes to rest for increasing values of α.

This is also reflected in the total closed-loop costs: While for $\alpha = 0.1$ the costs sum up to $V_\infty^{\tilde{\mu}_N}(x_0) \approx 2512.74$, we obtain a total cost of $V_\infty^{\tilde{\mu}_N}(x_0) \approx 2485.83$ for $\alpha = 0.95$. Note that the majority of the costs, i.e., approximately 2435, is accumulated on the interval $[0, 5]$ on which the trajectories for different α are almost identical and which is therefore not displayed in Figs. 10.16 and 10.17. However, the choice

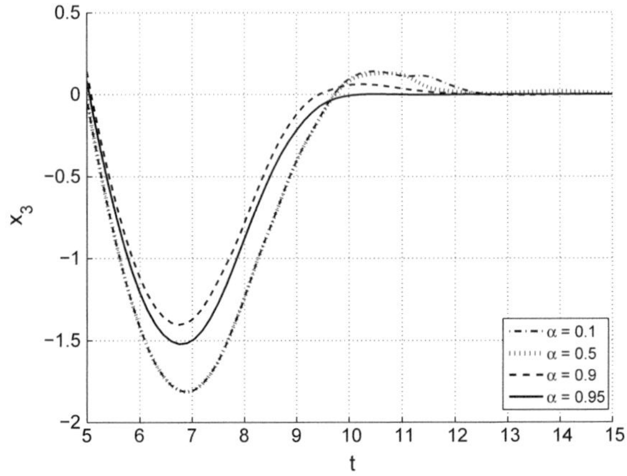

Fig. 10.17 Position of the cart x_3 for varying α

of α has a visible impact on the closed-loop performance in the remaining part of the interval.

Regarding the computational cost, the total number of SQP steps which are executed during the run of the NMPC procedure reduces from 455 for $\alpha = 0.95$ and 407 for $\alpha = 0.9$, to 267 and 246 for $\alpha = 0.5$ and $\alpha = 0.1$, respectively. Hence, we obtain an average of approximately 2.5 – 4.5 optimization iterations per MPC step over the entire interval [0, 15], while using standard termination criteria 9.5 optimization iterations per NMPC step are required.

A closer look at the numerical simulation in this example reveals that for each α there were some sampling instants n at which it was not possible to satisfy the suboptimality based termination criterion (10.25). In this case, we simply iterated the SQP optimization routine until convergence.

While this fact is not visible in Figs. 10.16 and 10.17 and obviously does not affect stability and performance in our example, this observation raises the question whether (10.25) is feasible, i.e., whether at time n we can ensure the existence of $u_n^{j^*}$ such that (10.25) is satisfied regardless of how u_{n-1} was chosen, before. In order to analyze this question, let us suppose that Assumption 6.3 holds. Then, observing that for optimal controls (10.25) coincides with (5.1), Theorem 6.15 yields that (10.25) is feasible if u_{n-1} is an optimal control sequence and α in (10.25) is smaller than α from (6.14). However, even with this choice of α in (10.25), condition (10.25) may not be feasible for nonoptimal control sequences u_{n-1}.

In order to understand why this is the case we investigate how Proposition 6.13—which provides the crucial ingredient for deriving (6.14)—changes if the optimal control sequence $u^\star$ in this proposition is replaced by a nonoptimal control sequence u_{n-1}. To this end, we fix $n \in \mathbb{N}$ and set $x = x_{\mu_N}(n)$ and $u = u_{n-1}$. Now, first observe that the inequalities in (6.12) remain valid regardless of the optimality of $u^\star$. All inequalities in (6.11), however, require optimality of the control sequence $u^\star$ generating the λ_n. In order to maintain at least some of these inequalities we can pick an optimal control sequence $u^\star$ for initial value $x_u(1, x)$ and horizon length $N-1$ and define a control sequence $\tilde{u}$ via $\tilde{u}(0) = u(0)$, $\tilde{u}(n) = u^\star(n-1)$, $n = 1, \ldots, N-1$. Then, abbreviating

$$\begin{aligned}\tilde{\lambda}_n &= \ell(x_{\tilde{u}}(n,x), \tilde{u}(n)), \quad n = 0, \ldots, N-1 \quad \text{and} \\ \tilde{\nu} &= V_N(x_u(1,x)) = V_N(x_{\tilde{u}}(1,x)),\end{aligned} \tag{10.26}$$

we arrive at the following version of Proposition 6.13.

Proposition 10.32 *Let Assumption 6.3 hold. Then the inequalities*

$$\sum_{n=k}^{N-1} \tilde{\lambda}_n \le B_{N-k}(\tilde{\lambda}_k) \quad \text{and} \quad \tilde{\nu} \le \sum_{n=0}^{j-1} \tilde{\lambda}_{n+1} + B_{N-j}(\tilde{\lambda}_{j+1}) \tag{10.27}$$

hold for $k = 1, \ldots, N-2$ and $j = 0, \ldots, N-2$.

Proof Analogous to the proof of Proposition 6.13. □

The subtle but crucial difference of (10.27) to (6.11), (6.12) is that the left inequality in (10.27) is not valid for $k = 0$. As a consequence, $\tilde{\lambda}_0$ does not appear in any of the inequalities, thus for any $\tilde{\lambda}_1, \ldots, \tilde{\lambda}_n$ and $\tilde{\nu}$ satisfying (10.27) and any $\delta > 0$ the values $\delta\tilde{\lambda}_1, \ldots, \delta\tilde{\lambda}_n$ and $\delta\tilde{\nu}$ satisfy (10.27), too. Hence, unless (10.27) implies $\tilde{\nu} \le \sum_{n=0}^{N-1} \tilde{\lambda}_n$—which is a very particular case—replacing (6.11), (6.12) in (6.14) by (10.27) will lead to the optimal value $\alpha = -\infty$. Consequently, feasibility of (10.25) cannot be concluded for any positive α.

The following example shows that this undesirable result is not simply due to an insufficient estimate for α but that infeasibility of (10.25) can indeed happen.

Example 10.33 Consider the 1d system

$$x^+ = x/2 + u \tag{10.28}$$

with $\ell(x,u) = |x|$, input constraint $u \ge 0$ and optimization horizon $N = 3$. A simple computation using $u_x \equiv 0$ shows that for this system Assumption 6.5 is satisfied with $\beta(r,k) = C\sigma^k r$ with $C = 1$ and $\sigma = 1/2$. Hence, Corollary 6.21 applies to the B_N from (6.7) and we can use (6.18) in order to compute that for $N = 3$ inequality (5.1) holds for $\alpha = 7/8$. If u_{n-1} in the termination criterion (10.25) is chosen as the optimal control $u^\star$, then (10.25) implies that (5.1) is feasible for this α.

For $x(n-1) = 0$, it is obvious that the control $u^\star \equiv 0$ is optimal. Using the nonoptimal control given by $u_{n-1}(0) = \varepsilon > 0$ and $u_{n-1}(1) = u_{n-1}(2) = 0$ and yields the trajectory $x_{u_{n-1}}(0) = x(n-1) = 0$, $x_{u_{n-1}}(k) = \varepsilon 2^{-k+1}$, $k = 1, 2$, which implies $x(n) = \varepsilon$ and

$$J_3(x(n-1), u_{n-1}) = \sum_{k=0}^{1} \varepsilon 2^{-k} = 3\varepsilon/2.$$

On the other hand, for the initial value $x(n) = \varepsilon$ it is easily seen that for each control u_n the inequality

$$J_3(x(n), u_n) \geq \sum_{k=0}^{2} \varepsilon 2^{-k} = 7\varepsilon/4 > 3\varepsilon/2 = J_N(x(n-1), u_{n-1})$$

holds. Hence, for this choice of u_{n-1} the inequality (10.25) is not feasible for any $\alpha > 0$.

Clearly, in order to rigorously ensure attraction and guaranteed performance one should derive conditions which exclude these situations and we briefly discuss two possible approaches for this purpose.

One way to guarantee feasibility of (10.25) is to add the missing inequality in (10.27) (i.e., the left inequality for $k = 0$) as an additional constraint in the optimization. This guarantees feasibility of (10.25) for any α smaller than the value from (6.18). One drawback of this approach is that—similar to the terminal conditioned case—an additional constraint in the optimization is needed which needs to be ensured for all $j \geq 1$ or at least for j^*. This makes the optimization more demanding, since in contrast to Remark 10.27(ii) here we do not have a canonical candidate for an admissible solution which can be used for initializing the iterative optimization. Another drawback is that the value $B_N(\tilde{\lambda}_0)$ from Assumption 6.3 needs to be determined either by an a priori analysis or by a try-and-error procedure.

Another way to guarantee feasibility is to choose ℓ in such a way that there exists $\gamma > 0$ for which

$$\gamma \ell(x, u) \geq \ell^*(f(x, u)) \tag{10.29}$$

holds for all $x \in X$ and all $u \in U$. Then from (10.29) and Assumption 6.3 for $x = f(x(n-1), \tilde{u}_{n-1}(0))$ we get

$$\sum_{k=0}^{N-1} \tilde{\lambda}_k \leq \tilde{\lambda}_0 + B_{N-1}(\ell^*(f(x(n-1), \tilde{u}_{n-1}(0)))) \leq \tilde{\lambda}_0 + B_{N-1}(\gamma \tilde{\lambda}_0).$$

Replacing $\beta(r, 0)$ by $\max\{\beta(r, t), \tilde{\beta}(r, t)\}$ with $\tilde{\beta}(r, 0) = \beta(\gamma r, 0) + r$ and $\tilde{\beta}(r, k) = \beta(\gamma r, k)$ for $k \geq 1$, this right hand side is $\leq B_N(\tilde{\lambda}_0)$ which again yields the left inequality in (10.27) for $k = 0$ and thus feasibility of (10.25). Note that (10.29) holds for our example (10.28) if we change $\ell(x, u) = |x|$ to $\ell(x, u) = |x| + |u|/\gamma$. For this ℓ and the points and control sequences considered in the example, we obtain

$$J_3(x(n-1), u_{n-1}) = 3\varepsilon/2 + \varepsilon = 5\varepsilon/2$$

from which one computes that (10.25) is now feasible.

The advantage of this method is that no additional constraints have to be imposed in the optimization. Its disadvantages are that constructing ℓ satisfying (10.29) may be complicated for more involved dynamics and that the bounds B_K from Assumption 6.3 will in general increase for the re-designed ℓ. As outlined in Sect. 6.6, this may lower the NMPC closed-loop performance and cause the need for larger optimization horizons N in order to obtain stability.

An in depth study of these approaches and in particular their algorithmic implementation and numerical evaluation will be the topic of further research.

References

1. Alamir, M.: Stabilization of Nonlinear Systems using Receding-horizon Control Schemes. Lecture Notes in Control and Information Sciences, vol. 339. Springer, London (2006)
2. Chen, W., Ballance, D.J., O'Reilly, J.: Model predictive control of nonlinear systems: computational delay and stability. IEE Proc. Control Theory Appl. **147**(4), 387–394 (2000)
3. De Nicolao, G., Magni, L., Scattolini, R.: Stability and robustness of nonlinear receding horizon control. In: Nonlinear Predictive Control, pp. 3–23. Birkhäuser (2000)
4. Di Palma, F., Magni, L.: On optimality of nonlinear model predictive control. Syst. Control Lett. **56**(1), 58–61 (2007)
5. Diehl, M., Findeisen, R., Allgöwer, F., Bock, H.G., Schlöder, J.P.: Nominal stability of the real-time iteration scheme for nonlinear model predictive control. IEE Proc. Control Theory Appl. **152**, 296–308 (2005)
6. Findeisen, R.: Nonlinear Model Predictive Control: A Sampled-Data Feedback Perspective., PhD thesis, University of Stuttgart, VDI-Verlag, Düsseldorf (2004)
7. Findeisen, R., Allgöwer, F.: Computational delay in nonlinear model predictive control. In: Proceedings of the International Symposium on Advanced Control of Chemical Processes, Hong Kong, China, Paper No. 561 (2003)
8. Graichen, K., Kugi, A.: Stability and incremental improvement of suboptimal MPC without terminal constraints. IEEE Trans. Automat. Control **55**, 2576–2580 (2010)
9. Grimm, G., Messina, M.J., Tuna, S.E., Teel, A.R.: Model predictive control: for want of a local control Lyapunov function, all is not lost. IEEE Trans. Autom. Control **50**(5), 546–558 (2005)
10. Grüne, L.: Worst case vs. average performance estimates for unconstrained NMPC schemes. PAMM 10, 607–608 (2010)
11. Grüne, L., Palma, V.G.: Robustness of performance and stability for multistep and updated multistep MPC schemes. Discrete Cont. Dyn. Syst. A **35**(9), 4385–4414 (2015)
12. Grüne, L., Pannek, J.: Practical NMPC suboptimality estimates along trajectories. Syst. Control Lett. **58**(3), 161–168 (2009)
13. Grüne, L., Pannek, J.: Analysis of unconstrained NMPC schemes with incomplete optimization. In: Proceedings of the 8th IFAC Symposium on Nonlinear Control Systems - NOLCOS 2010, Bologna, Italy, pp. 238–243 (2010)
14. Grüne, L., von Lossow, M., Pannek, J., Worthmann, K.: MPC: implications of a growth condition on exponentially controllable systems. In: Proceedings of the 8th IFAC Symposium on Nonlinear Control Systems - NOLCOS 2010, Bologna, Italy, pp. 385–390 (2010)
15. Grüne, L., Pannek, J., Seehafer, M., Worthmann, K.: Analysis of unconstrained nonlinear MPC schemes with varying control horizon. SIAM J. Control Optim. **48**, 4938–4962 (2010)
16. Grüne, L., Rantzer, A.: On the infinite horizon performance of receding horizon controllers. IEEE Trans. Automat. Control **53**, 2100–2111 (2008)
17. Grüne, L., Pannek, J., Worthmann, K.: A networked unconstrained nonlinear MPC scheme. In: Proceedings of the European Control Conference - ECC 2009, Budapest, Hungary, pp. 91–96 (2009)
18. Grüne, L., Pannek, J., Worthmann, K.: A prediction based control scheme for networked systems with delays and packet dropouts. In: Proceedings of the 48th IEEE Conference on Decision and Control - CDC 2009, Shanghai, China, pp. 537–542 (2009)

19. Jadbabaie, A., Hauser, J.: On the stability of receding horizon control with a general terminal cost. IEEE Trans. Automat. Control **50**(5), 674–678 (2005)
20. Khalil, H.K.: Nonlinear Systems, 3rd edn. Prentice Hall, Upper Saddle River (2002)
21. Limón, D., Alamo, T., Salas, F., Camacho, E.F.: On the stability of constrained MPC without terminal constraint. IEEE Trans. Automat. Control **51**(5), 832–836 (2006)
22. Michalska, H., Mayne, D.Q.: Robust receding horizon control of constrained nonlinear systems. IEEE Trans. Automat. Control **38**(11), 1623–1633 (1993)
23. de Oliveira Kothare, : S.L., Morari, M.: Contractive model predictive control for constrained nonlinear systems. IEEE Trans. Automat. Control **45**(6), 1053–1071 (2000)
24. Nešić, D., Teel, A.R.: A framework for stabilization of nonlinear sampled-data systems based on their approximate discrete-time models. IEEE Trans. Automat. Control **49**(7), 1103–1122 (2004)
25. Nešić, D., Grüne, L.: A receding horizon control approach to sampled-data implementation of continuous-time controllers. Syst. Control Lett. **55**, 660–672 (2006)
26. Palma, V.G.: Robust Updated MPC Schemes. PhD thesis, Universität Bayreuth (2015). https://epub.uni-bayreuth.de/2056/
27. Pannek, J.: Receding Horizon Control: A Suboptimality-based Approach. PhD thesis, University of Bayreuth, Germany (2009)
28. Parisini, T., Zoppoli, R.: A receding-horizon regulator for nonlinear systems and a neural approximation. Automatica **31**(10), 1443–1451 (1995)
29. Rawlings, J.B., Mayne, D.Q.: Model Predictive Control: Theory and Design. Nob Hill Publishing, Madison (2009)
30. Scokaert, P.O.M., Mayne, D.Q., Rawlings, J.B.: Suboptimal model predictive control (feasibility implies stability). IEEE Trans. Autom. Control **44**(3), 648–654 (1999)
31. Varutti, P., Findeisen, R.: Compensating network delays and information loss by predictive control methods. In: Proceedings of the European Control Conference - ECC 2009. Budapest, Hungary, pp. 1722–1727 (2009)
32. Zavala, V.M., Biegler, L.T.: The advanced-step NMPC controller: optimality, stability and robustness. Automatica **45**(1), 86–93 (2009)
33. Findeisen, R., Grüne, L., Pannek, J., Varutti, P.: Robustness of prediction based delay compensation for nonlinear systems. In: IFAC Proceedings Volumes **44**(1) (18th IFAC World Congress), pp. 203–208 (2011)

Chapter 11
Numerical Discretization

This chapter is particularly devoted to sampled data systems, which need to be discretized in order to be able to solve the optimal control problem within the NMPC algorithm numerically. We present suitable methods, discuss the convergence theory for one step methods and give an introduction into step size control algorithms. Furthermore,we explain how these methods can be integrated into NMPC algorithms, investigate how the numerical errors affect the stability of the NMPC controller derived from the numerical model and show which kind of robustness is needed in order to ensure a practical kind of stability.

11.1 Basic Solution Methods

In order to define the setting, we start by summarizing the main concepts from Sect. 2.2. As already mentioned there, in most applications the discrete time system (2.1) is obtained from sampling a continuous time system

$$\dot{x}(t) = f_c(x(t), v(t)) \tag{2.6}$$

with $x(t) \in \mathbb{R}^d$ and $v(t) \in \mathbb{R}^m$. More precisely, given a subset $U \subseteq L^\infty([0, T], \mathbb{R}^m)$, i.e., each $u \in U$ is a continuous time control function defined on the sampling interval $[0, T]$, we define the discrete time dynamics f in (2.1) by

$$x^+ = f(x, u) := \varphi(T, 0, x, u) \tag{2.8}$$

where $\varphi(T, 0, x, u)$ is the solution of (2.6) with $v = u$ satisfying the initial condition $\varphi(0, 0, x, u) = x$. Here, we tacitly assume that for all admissible initial values $x \in \mathbb{X} \subseteq X = \mathbb{R}^d$ and all admissible controls $u \in \mathbb{U}(x) \subseteq U$ the solution $\varphi(t, 0, x, u)$

L. Grüne and J. Pannek, *Nonlinear Model Predictive Control*,
Communications and Control Engineering, DOI 10.1007/978-3-319-46024-6_11

exists for $t \in [0, T]$. This way we obtain a discrete time system (2.1) whose solutions for each control sequence $u(\cdot) \in \mathbb{U}^N(x)$ satisfy

$$\varphi(t_n, t_0, x_0, v) = x_u(n, x_0), \quad n = 0, 1, 2, \ldots, N, \tag{2.7}$$

for all sampling times $t_n = nT$, $n = 0, \ldots, N$, and the continuous time control function v given by

$$v(t) = u(n)(t - t_n) \quad \text{for almost all } t \in [t_n, t_{n+1}] \text{ and all } n = 0, \ldots, N-1, \tag{2.13}$$

cf. Theorem 2.7.

Since a closed formula for f defined in (2.8) will only be available in exceptional cases, it is in general necessary to use numerical schemes in order to compute a numerical approximation of f. This way, instead of an analytical formula we obtain an algorithm which can be used in order to compute the predictions needed in the optimal control problem and its variants. For the exposition in this chapter we restrict ourselves to sampling with zero order hold in which each element $u(n)$ of the control sequence is a constant function from $[0, T]$ to $\mathbb{R}^m$. This amounts to defining

$$U := \{u : [0, T] \to \mathbb{R}^m \mid \text{there exists } u_0 \in \mathbb{R}^m \text{ with } u(t) = u_0 \text{ for all } t \in [0, T]\}.$$

Observe that each element in U is uniquely defined by the value $u_0 \in \mathbb{R}^m$. Accordingly, we identify U with $\mathbb{R}^m$ and regard each $u \in U$ as a value in $\mathbb{R}^m$. Henceforth, we will again use the symbol u (instead of u_0) for this value. The resulting continuous time control function v in (2.7) is then piecewise constant on the sampling intervals, cf. also Fig. 2.3 and the discussion after Theorem 2.7. Recall from Remark 2.8 that the overlap of the sampling intervals at the sampling times t_n does not pose a problem in the definition of v in (2.13).

In the following, we give an introduction into numerical methods for ordinary differential equations and their analysis. In particular, we give details on so-called one step methods and show convergence results and requirements. Moreover, we sketch the basic idea of the very useful step size control algorithms. These algorithms allow us to externally define an error tolerance level for the solution and produce an adaptive time grid which is computationally much more efficient than using a sufficiently fine uniform grid, a requirement that is frequently found in the sampled data literature. For references to textbooks which cover the material presented here more comprehensively and in more detail we refer to Sect. 11.6.

For computing f in (2.8) it is sufficient to solve (2.6) on the interval $[0, T]$ on which $u \in U$ in (2.8) is constant. Hence, the right-hand side in (2.8) does not depend on t and—more importantly—does not exhibit discontinuities on the interval $[0, T]$. For this reason, standard numerical techniques can be applied. Still, the solution depends on the constant control value u which will be reflected in the subsequent notation.

Before we can develop solution methods for ordinary differential equations, we need to define some general concepts. As we have pointed out before, the fundamental idea of almost all numerical solution methods is to replace the analytic solution $\varphi(t, 0, x, u)$ for $t \in [0, T]$ by an approximation. Throughout the rest of this chapter, we denote this approximation by $\tilde{\varphi}(t, 0, x, u)$. The following definition states for which t such an app roximation is defined and what convergence of such an approximation means.

Definition 11.1 (i) A set $\mathscr{G} = \{\tau_0, \tau_1, \ldots, \tau_M\}$ of time instants with $0 = \tau_0 < \tau_1 < \ldots < \tau_M = T$ is called a *time grid* on the interval $[0, T]$. The values $h_i := \tau_{i+1} - \tau_i$ and $\overline{h} := \max_{i=0,\ldots,M-1} h_i$ are called *step sizes* and *maximal step size*, respectively.

(ii) A function $\tilde{\varphi} : \mathscr{G} \times \mathscr{G} \times \mathbb{R}^d \times U \to \mathbb{R}^d$ is called *grid function.*

(iii) Assume that the solution $\varphi(t; \tau_0, x_0, u)$ of (2.6) exists for $t \in [0, T]$. Then a family of grid functions $\tilde{\varphi}_j$, $j \in \mathbb{N}$, on time grids $\mathscr{G}_j$ on the interval $[0, T]$ with maximal step sizes $\overline{h}_j$ is called *(discrete) approximation* of $\varphi(t; \tau_0, x_0, u)$ (2.6), if it is *convergent*, i.e.,

$$\max_{\tau_i \in \mathscr{G}_j} \|\tilde{\varphi}_j(\tau_i; \tau_0, x_0, u) - \varphi(\tau_i; \tau_0, x_0, u)\| \to 0 \quad \text{as } \overline{h}_j \to 0.$$

The convergence of the approximation is said to be of *order* $p > 0$ if for all compact sets $K \subset \mathbb{R}^d$, $Q \subset U$ there exists a constant $M > 0$ such that

$$\max_{\tau_i \in \mathscr{G}_j} \|\tilde{\varphi}_j(\tau_i; \tau_0, x_0, u) - \varphi(\tau_i; \tau_0, x_0, u)\| \leq M\overline{h}_j^p \tag{11.1}$$

holds for all $x_0 \in K$, all $u \in Q$ and all sufficiently fine grids $\mathscr{G}_j$ on $[0, T]$.

Less technically speaking, an approximation $\tilde{\varphi}(\tau_i, 0, x, u)$ is a grid function defined on $\mathscr{G}$ which approximates the values of the true solution at the grid points and becomes the more accurate the finer the grid becomes. Moreover, the larger the order of convergence p is, the faster the approximation will converge toward the exact solution for $\overline{h} \to 0$.

The most simple class of numerical methods to compute a discrete approximation satisfying Definition 11.1 are so-called one step methods. Although simple to design, these methods are nonetheless well suited even for rather complicated problems. One step methods compute the grid function $\tilde{\varphi}$ iteratively via

$$\tilde{\varphi}(\tau_0; \tau_0, x_0, u) := x_0, \quad \tilde{\varphi}(\tau_{i+1}; \tau_0, x_0, u) := \Phi(\tilde{\varphi}(\tau_i; \tau_0, x_0, u), u, h_i) \tag{11.2}$$

for $i = 0, \ldots, M - 1$ starting from the given initial value x_0. Here Φ is a mapping

$$\Phi : \mathbb{R}^d \times \mathbb{U} \times \mathbb{R} \to \mathbb{R}^d$$

which should be easy to implement and cheap to evaluate on a computer and, of course, provide a convergent approximation in the sense of Definition 11.1(iii).

In order to design such a map Φ, we use that the solution of the differential equation (2.6) for two consecutive grid points τ_i and τ_{i+1} satisfies the integral equation

$$\varphi(\tau_{i+1}; \tau_0, x_0, u) = \varphi(\tau_i; \tau_0, x_0, u) + \int_{\tau_i}^{\tau_{i+1}} f_c(\varphi(t; \tau_0, x_0, u), u)dt.$$

Approximating the integral expression by the rectangle rule we obtain

$$\int_{\tau_i}^{\tau_{i+1}} f_c(\varphi(t; \tau_0, x_0, u), u)dt \approx (\tau_{i+1} - \tau_i) f_c(\varphi(\tau_i; \tau_0, x_0, u), u) = h_i f_c(\varphi(\tau_i; \tau_0, x_0, u), u)$$

Inserting this approximation into the above integral equation then yields

$$\varphi(\tau_{i+1}; \tau_0, x_0, u) \approx \varphi(\tau_i; \tau_0, x_0, u) + h_i f_c(\varphi(\tau_i; \tau_0, x_0, u), u).$$

Now we define an approximate solution $\tilde{\varphi}$ by requiring that it exactly solves this approximate equation, i.e.,

$$\tilde{\varphi}(\tau_{i+1}; \tau_0, x_0, u) = \tilde{\varphi}(\tau_i; \tau_0, x_0, u) + h_i f_c(\tilde{\varphi}(t; 0, x_0, u), u). \tag{11.3}$$

This is exactly the iteration in (11.2) with

$$\Phi(x, u, h) := x + hf_c(x, u).$$

This one step method is called the Euler scheme. Now, if we assume $\tilde{\varphi}(\tau_i; \tau_0, x_0, u) \approx \varphi(\tau_i; \tau_0, x_0, u)$, then we see that

$$\begin{aligned}\tilde{\varphi}(\tau_{i+1}; \tau_0, x_0, u) &\approx \varphi(\tau_i; 0, x_0, u) + h_i f_c(\varphi(\tau_i; \tau_0, x_0, u), u)\\ &\approx \varphi(\tau_i; \tau_0, x_0, u) + \int_{\tau_i}^{\tau_{i+1}} f_c(\varphi(t; \tau_0, x_0, u), u)dt = \varphi(\tau_{i+1}; \tau_0, x_0, u)\end{aligned}$$

which suggests that this method yields an approximation in the sense of Definition 11.1(iii). Formally, we will prove this property for general one step methods in Theorem 11.5 below. Before we turn to the convergence analysis, we present an important class of solution methods which follow from a generalization of the Euler approximation idea to solve the integral equation.

The idea to generalize the Euler method is to use a higher order approximation for the integral. For example, one can approximate the integral by the trapezoidal rule instead of the rectangle rule, which leads to the approximation

$$\varphi(\tau_{i+1}; \tau_0, x_0, u) \approx \varphi(\tau_i; \tau_0, x_0, u) + \frac{h_i}{2}\Big(f_c(\varphi(\tau_i; \tau_0, x_0, u), u) + f_c(\varphi(\tau_{i+1}; \tau_0, x_0, u), u)\Big).$$

When trying to use this approximation in order to define $\tilde{\varphi}$ analogous to (11.3), above, we run into the problem that the unknown value $\tilde{\varphi}(\tau_{i+1}; \tau_0, x_0, u)$ appears on the right hand side. We can avoid this if we use the Euler scheme in order to approximate

$$f_c(\varphi(\tau_{i+1}; \tau_0, x_0, u), u) \approx f_c(\varphi(\tau_i; \tau_0, x_0, u) + h_i f_c(\varphi(\tau_i; \tau_0, x_0, u))).$$

Proceeding this way we end up with the so-called Heun method

$$\Phi(x, v, h) := x + \frac{h}{2}\big(f_c(x, u) + f_c(x + hf_c(x, u), u)\big).$$

Observe that in this formula the value $f_c(x, u)$ appears twice and that the scheme uses nested evaluations of the vector field f_c. The formalism of Runge–Kutta methods now gives a systematic way to formalize this nested structure. We first illustrate this formalism using the Heun method, for which it reads

$$\begin{aligned} k_1 &:= f_c(x, u) \\ k_2 &:= f_c(x + hk_1, u) \\ \Phi(x, u, h) &:= x + h\left(\frac{1}{2}k_1 + \frac{1}{2}k_2\right) \end{aligned}$$

The advantage of this formalism is that one can easily add new function evaluations or modify the weighted combination. This leads to the following general form.

Definition 11.2 An s-stage (explicit) Runge–Kutta method is given by

$$\begin{aligned} k_i &:= f\left(x + h\sum_{j=1}^{i-1} a_{ij}k_j\right) \quad \text{for } i = 1, \ldots, s \\ \Phi(x, u, h) &:= x + h\sum_{i=1}^{s} b_i k_i. \end{aligned}$$

The value $k_i = k_i(x, u, h)$ is called the ith stage of the method.

The methods thus defined depend on the parameters a_{ij} and b_i. If the vector field explicitly depends on t—which is not the case in our setting—then additional

Table 11.1 Butcher tableaus for the Euler, Heun, and classical Runge–Kutta method (left to right)

$$\begin{array}{c|c} 0 & \\ \hline & 1 \end{array} \qquad \begin{array}{c|cc} 0 & & \\ 1 & 1 & \\ \hline & \frac{1}{2} & \frac{1}{2} \end{array} \qquad \begin{array}{c|cccc} 0 & & & & \\ \frac{1}{2} & \frac{1}{2} & & & \\ \frac{1}{2} & 0 & \frac{1}{2} & & \\ 1 & 0 & 0 & 1 & \\ \hline & \frac{1}{6} & \frac{2}{6} & \frac{2}{6} & \frac{1}{6} \end{array}$$

parameters c_i are used in the definition. More compactly, these parameters are written as so-called *Butcher tableaus* of the form

$$\begin{array}{c|ccccc} c_1 & & & & & \\ c_2 & a_{21} & & & & \\ c_3 & a_{31} & a_{32} & & & \\ \vdots & \vdots & \vdots & \ddots & & \\ c_s & a_{s1} & a_{s2} & \cdots & a_{s\,s-1} & \\ \hline & b_1 & b_2 & \cdots & b_{s-1} & b_s \end{array}$$

Table 11.1 shows Butcher tableaus corresponding to the Euler scheme (left), the Heun scheme (middle) and the so-called classical Runge–Kutta scheme with $s = 4$ stages proposed by Carl Runge and Martin Kutta in 1895 (right).

Remark 11.3 Models based on partial differential equations, like the one discussed in Example 6.32, require discretization techniques different from the one discussed here. In particular, apart from the discretization in time also a discretization in space has to be performed. Popular techniques for this purpose are finite difference or finite element methods and the interested reader is referred to the large amount of textbooks on this topic, like, e.g., the books by LeVeque [9] or Braess [1] respectively.

11.2 Convergence Theory

Having defined one step methods we now show that the resulting approximations actually converge toward the solution. To this end, we define the error at time $\tau_i \in \mathcal{G}$ as

$$e(\tau_i) := \|\tilde{\varphi}(\tau_i; \tau_0, x, u) - \varphi(\tau_i; \tau_0, x, u)\|.$$

The main idea to show convergence is to use the triangle inequality in order to separate the error sources in the iteration (11.2) into the error caused by the previously accumulated error (a) and the local error (b). Abbreviating $\varphi(\tau_i) = \varphi(\tau_i; \tau_0, x_0, u)$ and $\tilde{\varphi}(\tau_i) = \tilde{\varphi}(\tau_i; \tau_0, x_0, u)$, this leads to the estimate

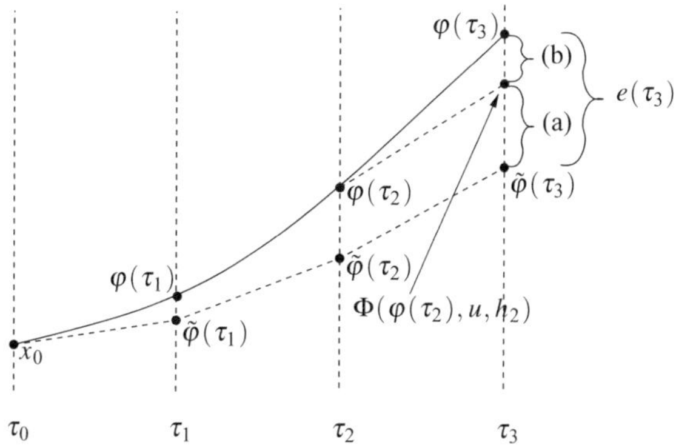

Fig. 11.1 Illustration of the separation of errors

$$\begin{aligned} e(\tau_{i+1}) &= \|\tilde{\varphi}(\tau_{i+1}) - \varphi(\tau_{i+1})\| \;=\; \|\Phi(\tilde{\varphi}(\tau_i), u, h_i) - \varphi(\tau_{i+1})\| \\ &\leq \underbrace{\|\Phi(\tilde{\varphi}(\tau_i), u, h_i) - \Phi(\varphi(\tau_i), u, h_i)\|}_{\text{accumulated error (a)}} + \underbrace{\|\Phi(\varphi(\tau_i), u, h_i) - \varphi(\tau_{i+1})\|}_{\text{local error (b)}} . \end{aligned} \tag{11.4}$$

The idea is sketched in Fig. 11.1 for $i = 2$.

In order to prove convergence we will use the following conditions which guarantee that both errors (a) and (b) remain small.

Definition 11.4 (i) A one step method satisfies the *Lipschitz condition* if for all compact subsets $K \subset \mathbb{R}^d$ and $Q \subset U$ there exists a constant $\Lambda > 0$ such that for all sufficiently small $h > 0$ the inequality

$$\|\Phi(x_1, u, h) - \Phi(x_2, u, h)\| \leq (1 + \Lambda h)\|x_1 - x_2\| \tag{11.5}$$

holds for all $x_1, x_2 \in K$ and all $u \in Q$.

(ii) A one step method is called *consistent* with *order of consistency* $p > 0$ if for all compact subsets $K \subset \mathbb{R}^d$ and $Q \subset U$ there exists a constant $C > 0$ such that for all sufficiently small $h > 0$ the inequality

$$\|\Phi(x, u, h) - \varphi(h; 0, x, u)\| \leq C h^{p+1} \tag{11.6}$$

holds for all $x \in K$ and all $u \in Q$.

Inequality (11.5) guarantees that the propagation of previous errors within a one step method, i.e., term (a), stays bounded. The consistency condition (11.6), on the other hand, ensures that the local error (b) remains small.

One can easily show that the previously introduced Euler approximation as well as all explicit Runge–Kutta methods satisfy the Lipschitz condition (11.5) if the vector field f_c satisfies the Lipschitz condition from Assumption 2.4. The consistency condition (11.6), on the other hand, cannot be checked that easily in general. In order to verify that a method Φ exhibits an order of consistency $p \geq 1$, one utilizes the Taylor approximation of the method with respect to the step size h in $h = 0$, i.e.,

$$\Phi(x,u,h) = x + \sum_{i=1}^{p} \frac{h^i}{i!} \frac{\partial^i}{\partial h^i}\bigg|_{h=0} \Phi(x,u,h) + O(h^{p+1}) \tag{11.7}$$

and compares it to the Taylor approximation of the exact solution $\varphi(h;0,x,u)$ with respect to h in $h = 0$. It turns out that this Taylor approximation can be computed without actually using the—in general unknown—solution φ. To this end, we use the higher order Lie derivative $L^i_{f_c}$, $i \in \mathbb{N}_0$, with respect to the vector field f_c which for arbitrary smooth vector fields $g : \mathbb{R}^d \times U \to \mathbb{R}^d$ is defined inductively by

$$L^0_{f_c} g(x,u) = g(x,u), \qquad L^i_{f_c} g(x,u) = \left(\frac{\partial}{\partial x} L^{i-1}_{f_c} g(x,u)\right) f_c(x,u).$$

Using the Lie derivative, the Taylor approximation of φ reads

$$\varphi(h;0,x,u) = x + \sum_{i=1}^{p} \frac{h^i}{i!} L^{i-1}_{f_c} f_c(x,u) + O(h^{p+1}). \tag{11.8}$$

Then, if the p summands in (11.7) and (11.8) coincide, the scheme is consistent with order p. In particular, if Φ can be written as

$$\Phi(x,u,h) = x + h\psi(x,u,h)$$

with a continuous function ψ satisfying $\psi(x,u,0) = f_c(x,u)$, then it follows that the order of consistency is at least $p = 1$.

Using this technique one can show that the order of consistency of the classical Runge–Kutta method is $p = 4$. More generally, the comparison of the summands can be used in order to derive conditions on the coefficients of arbitrary Runge–Kutta schemes for any consistency order $p \geq 1$. Unfortunately, the number of these condition grows exponentially with p, hence for $p \geq 10$ it is almost impossible to use them for constructing appropriate Runge–Kutta methods.

Note that in order to guarantee the order of consistency p the vector field f_c needs to be p times continuously differentiable with respect to x in order to ensure that approximation (11.8) holds. If the vector field depends on t, then it also needs to be p times continuously differentiable with respect to t on the interval $[\tau_i, \tau_{i+1}]$ if we want to apply (11.6) on this interval. This is no problem as long as $[\tau_i, \tau_{i+1}] \subseteq [0, T]$ which is always the case in this section. However, if we consider sampled data systems, i.e., (2.6) with v from (2.13) on an interval $[\tau_i, \tau_{i+1}]$ with $t_n \in (\tau_i, \tau_{i+1})$ for some sampling time t_n, this becomes a major issue since the control v and thus the map $t \mapsto f_c(x, v(t))$ is in general discontinuous and thus in particular nonsmooth at the sampling times. We will discuss this issue in Sect. 11.4, below.

After discussing the assumptions and how these assumptions can be checked, we are now ready to state the main result of this section.

Theorem 11.5 *If a one step method Φ satisfies the Lipschitz condition* (11.5) *and the consistency condition* (11.6) *with order p, then the approximation $\tilde{\varphi}$ from 11.2 is convergent in the sense of Definition 11.1(iii) with order of convergence p.*

Proof We will show (11.1) for each grid $\mathcal{G}$ on $[0, T]$ with $\overline{h} > 0$ sufficiently small. For simplicity of notation, we drop the index j in (11.1). To this end, fix two compact sets $K \subset \mathbb{R}^d$ and $Q \subset U$. Then the set

$$K_1 := \{\varphi(t; 0, x_0, u) \mid t \in [0, T], x_0 \in K, u \in Q\}$$

is again compact, since φ is continuous in all variables and images of compact sets under continuous maps are again compact. We choose some $\delta > 0$ and consider the compact set

$$K_2 := \overline{\mathcal{B}}_\delta(K_1) = \bigcup_{x \in K_1} \overline{\mathcal{B}}_\delta(x)$$

which contains exactly those points $x \in \mathbb{R}^d$ which have a distance less or equal δ to a point on a solution $x(t; 0, x_0, u)$ with $x_0 \in K$ and $u \in Q$. Let $\Lambda > 0$ and $C > 0$ be the constants in the Lipschitz condition (11.5) and the consistency condition (11.6), respectively, for $K = K_2$ and the set Q fixed above.

We first prove (11.1) under the following condition, which we will verify afterwards.

$$\begin{aligned}&\text{For all grids } \mathcal{G} \text{ with sufficiently small } \overline{h} > 0, \text{ all initial}\\ &\text{values } x_0 \in K \text{ and all } u \in Q \text{ the grid function } \tilde{\varphi} \text{ from}\\ &(11.2) \text{ satisfies } \varphi(\tau_i, \tau_0, x_0, u) \in K_2 \text{ for all } \tau_i \in \mathcal{G}.\end{aligned} \tag{11.9}$$

For proving (11.1) we choose $x_0 \in K$ and $u \in Q$ and abbreviate $\varphi(t) = \varphi(t; \tau_0, x_0, u)$ and $\tilde{\varphi}(\tau_i) = \tilde{\varphi}(\tau_i; \tau_0, x_0, u)$. With

$$e(\tau_i) := \|\tilde{\varphi}(\tau_i) - \varphi(\tau_i)\|$$

we denote the error at time $\tau_i \in \mathcal{G}$. Then from (11.4) we obtain

$$\begin{aligned} e(\tau_{i+1}) &\le \|\Phi(\tilde{\varphi}(\tau_i), u, h_i) - \Phi(\varphi(\tau_i), u, h_i) \; + \; \|\Phi(\varphi(\tau_i), u, h_i) - \varphi(\tau_{i+1})\|[0] \\ &\le (1 + \Lambda h_{i-1})\|\tilde{\varphi}(\tau_{i-1}) - \varphi(\tau_{i-1})\| \; + \; C h_{i-1}^{p+1}[0] \\ &= (1 + \Lambda h_{i-1}) e(\tau_{i-1}) + C h_{i-1}^{p+1}, \end{aligned}$$

using (11.5) and (11.6) for $K = K_2$ in the second inequality. These inequalities apply since the construction of K_1 and K_2 implies $\varphi(\tau_i) \in K_1 \subset K_2$ and (11.9) ensures $\tilde{\varphi}(\tau_i) \in K_2$.

By induction over i we now show that this inequality implies the estimate

$$e(\tau_i) \leq C\overline{h}^p \frac{1}{\Lambda}(\exp(\Lambda(\tau_i - \tau_0)) - 1).$$

For $i = 0$ this inequality follows immediately. For $i - 1 \to i$ we use

$$\exp(\Lambda h_i) = 1 + \Lambda h_i + \frac{\Lambda^2 h_i^2}{2} + \ldots \geq 1 + \Lambda h_i$$

which together with the induction assumption yields

$$\begin{aligned}
e(\tau_i) &\leq (1 + \Lambda h_{i-1}) e(\tau_{i-1}) + C h_{i-1}^{p+1} [0] \\
&\leq (1 + \Lambda h_{i-1}) C\overline{h}^p \frac{1}{\Lambda}(\exp(\Lambda(\tau_{i-1} - \tau_0)) - 1) + h_{i-1} \underbrace{C h_{i-1}^p}_{\leq C\overline{h}^p} [0] \\
&= C\overline{h}^p \frac{1}{\Lambda}\Big(h_{i-1}\Lambda + (1 + \Lambda h_{i-1})(\exp(\Lambda(\tau_{i-1} - \tau_0)) - 1)\Big)[0] \\
&= C\overline{h}^p \frac{1}{\Lambda}\Big(h_{i-1}\Lambda + (1 + \Lambda h_{i-1}) \exp(\Lambda(\tau_{i-1} - \tau_0)) - 1 - \Lambda h_{i-1}\Big)[0] \\
&= C\overline{h}^p \frac{1}{\Lambda}\Big((1 + \Lambda h_{i-1}) \exp(\Lambda(\tau_{i-1} - \tau_0)) - 1\Big)[0] \\
&\leq C\overline{h}^p \frac{1}{\Lambda}\Big(\exp(\Lambda h_{i-1}) \exp(\Lambda(\tau_{i-1} - \tau_0)) - 1\Big)[0] \\
&= C\overline{h}^p \frac{1}{\Lambda}(\exp(\Lambda(\tau_i - \tau_0)) - 1).
\end{aligned}$$

Since $\tau_0 = 0$ this implies (11.1) with $M = C(\exp(\Lambda T) - 1)/\Lambda$.

It remains to show that condition (11.9) is satisfied. We show that this assumption holds for all grids $\mathcal{G}$ whose maximal step size satisfies

$$C\overline{h}^p \leq \frac{\delta\Lambda}{\exp(\Lambda(T - \tau_0)) - 1}.$$

To this end, we consider a numerical solution $\tilde{\varphi}(\tau_i) = \tilde{\varphi}(\tau_i, \tau_0, x_0, u)$ for some $x_0 \in K$ and $u \in Q$ and show $\tilde{\varphi}(\tau_i) \in K_2$ by induction. Since $\tilde{\varphi}(\tau_0) = x_0 \in K \subset K_2$ the assertion holds for $i = 0$.

For the induction step $i-1 \to i$ assume that the induction assumption $\tilde{\varphi}(\tau_k) \in K_2$ holds for $k = 0, 1, \ldots, i-1$. We have to show $\tilde{\varphi}(\tau_i) \in K_2$. Observe, that for the inequality

$$e(\tau_i) \leq C\overline{h}^p \frac{1}{\Lambda}(\exp(\Lambda(T - \tau_0)) - 1)$$

to hold it is sufficient that $\tilde{\varphi}(\tau_k) \in K_2$ holds for $k = 0, 1, \ldots, i-1$. By choice of $\overline{h}$ we thus obtain $e(\tau_i) \leq \delta$, i.e.,

$$\|\tilde{\varphi}(\tau_i) - \varphi(\tau_i)\| \leq \delta.$$

Since by construction of K_1 we have $\varphi(\tau_i) \in K_1$, it follows that $\tilde{\varphi}(\tau_i) \in \overline{B}_\delta(\varphi(\tau_i)) \subset K_2$, i.e., the desired property. □

11.3 Adaptive Step Size Control

The convergence theorem from the previous section shows that the presented one step methods are applicable to solve the underlying continuous time dynamics of the form (2.6) of a problem. Yet, so far we can only guarantee those methods to exhibit small errors if each time step h_i in the grid $\mathscr{G}$ is sufficiently small since the error bound in (11.1) depends on $\overline{h} = \max h_i$. In the literature, it is occasionally proposed to use the grid induced by the sampling times as computational grid, i.e., to choose $\tau_n = t_n = nT$. This, however, results in $\overline{h} = T$ and thus requires the sampling period T to be small in order to obtain an accurate approximation. Apart from the fact that it may not be desirable to use very small sampling periods, there are subtle pitfalls regarding stability of the closed-loop system when the accuracy of the approximate model and the sampling rate are linked, see the discussion in Sect. 11.6.

A way to avoid linking $\overline{h}$ and T is to use a constant step size $h_i \equiv \overline{h}$ with $\overline{h} = T/K$ for some $K \in \mathbb{N}$. Adjusting $\overline{h}$ appropriately, we can make the error term in (11.1) arbitrarily small without changing T. This, however, leads to equidistant grids which are known to be computationally inefficient since they do not reflect the properties of the solution. A much more efficient way is to choose the time steps h_i adapted to the solution, i.e., we allow for large h_i if the error is small and use small h_i when large errors are observed. However, we surely do not want to manually adapt the step sizes to every situation the NMPC controller may face since this would render such an algorithm to be inapplicable.

In order to obtain an efficient way to construct an adaptive grid $\mathscr{G}$, we consider step size control algorithms. Such methods are well established in the numerics of ordinary differential equations. In this section, we explain the central idea behind step size control algorithms. The key idea is to use two different one step methods Φ_1, Φ_2 with different orders of consistency $p_1 < p_2$ in order to compute a step length $h_i = \tau_{i+1} - \tau_i$ at time τ_i for the next time step which guarantees a predefined local error bound tol_{ODE}. Here, by $p_1 < p_2$ we mean that for $\Phi = \Phi_1$ the inequality (11.6) cannot hold for $p = p_2$, i.e., no matter how C is chosen (11.6) will be violated for all sufficiently small h. As in the previous sections, we consider the solution of (2.6) on one sampling interval $[0, T]$ on which the control u is constant.

In (11.4) we used the auxiliary term $\Phi(\varphi(\tau_i; \tau_0, x, u), u, h_i)$ in order to quantify the local error. Since the value of $\varphi(\tau_i; 0, x, u)$ is not available at runtime of a one step method, we cannot use it to guarantee the local error (a) to satisfy

$$\|\Phi(\varphi(\tau_i; \tau_0, x, u), u, h_i) - \varphi(\tau_i; \tau_0, x, v)\| \leq \text{tol}_{\text{ODE}}.$$

To circumvent this problem, in the triangle inequality for estimating $e(\tau_{i+1})$ we insert the term $\varphi(\tau_{i+1}; \tau_i, \tilde{\varphi}(\tau_i; \tau_0, x, u), u)$ instead of $\Phi(\varphi(\tau_i; \tau_0, x, u), u, h_i)$. Using that by the cocycle property we have $\varphi(\tau_{i+1}; \tau_0, x, u) = \varphi(\tau_{i+1}; \tau_i, \varphi(\tau_i; \tau_0, x, u), u)$, this leads to the inequality

$$\|\tilde{\varphi}(\tau_{i+1}; \tau_0, x, u) - \varphi(\tau_{i+1}; \tau_0, x, u)\| \leq$$
$$\leq \|\Phi(\tilde{\varphi}(\tau_i; \tau_0, x, u), u, h_i) - \varphi(\tau_{i+1}; \tau_i, \tilde{\varphi}(\tau_i; \tau_0, x, u), u)\|$$
$$+ \|\varphi(\tau_{i+1}; \tau_i, \tilde{\varphi}(\tau_i; \tau_0, x, u), u) - \varphi(\tau_{i+1}; \tau_i, \varphi(\tau_i; \tau_0, x, u), u)\|.$$

In this sum, the second term essentially depends on the error of the approximation at time instant τ_i, which is independent of the choice of $h_i = \tau_{i+1} - \tau_i$. Hence, for choosing h_i we only consider the first summand. More precisely, we attempt to choose h_i such that the tolerable error bound

$$\|\Phi(\tilde{\varphi}(\tau_i; \tau_0, x, u), u, h_i) - \varphi(\tau_{i+1}; t_i, \tilde{\varphi}(\tau_i; \tau_0, x, u), u)\| \leq \text{tol}_{\text{ODE}}$$

is satisfied.

When trying to implement this method, one faces the problem that the value $\varphi(\tau_{i+1}; t_i, \tilde{\varphi}(\tau_i; \tau_0, x, u), u)$ is not known. This is where the idea of using two methods Φ_1 and Φ_2 with different orders of consistency $p_2 > p_1$ is used. Setting $\Phi = \Phi_1$ and approximating $\varphi(\tau_{i+1}; t_i, \tilde{\varphi}(\tau_i; \tau_0, x, u), u)$ by the more accurate method Φ_2 one can show the following theorem.

Theorem 11.6 *Consider two one step methods Φ_1, Φ_2 with orders of consistency p_1, p_2 satisfying $p_2 \geq p_1 + 1$. Then there exist constants $k_1, k_2 > 0$ such that for all sufficiently small $h_i > 0$ the computable error*

$$\bar{\varepsilon} := \|\Phi_1(\tilde{\varphi}(\tau_i; 0, x, u), u, h_i) - \Phi_2(\tilde{\varphi}(\tau_i; 0, x, u), u, h_i)\| \tag{11.10}$$

and the local error of the one step method Φ_1

$$\varepsilon := \|\Phi_1(\tilde{\varphi}(\tau_i; 0, x, u), u, h_i) - \varphi(\tau_{i+1}; \tau_i, \tilde{\varphi}(\tau_i; 0, x, u), u)\|$$

satisfy the inequality

$$k_1\varepsilon \leq \bar{\varepsilon} \leq k_2\varepsilon.$$

Proof First we define the errors

$$\eta_{i,j} := \Phi_j(\tilde{\varphi}(\tau_i; 0, x, u), u, h_i) - \varphi(\tau_{i+1}; \tau_i, \tilde{\varphi}(\tau_i; 0, x, u), u)$$

for both one step methods Φ_j, $j = 1, 2$. By Definition 11.4(ii) we obtain the local error bounds $\varepsilon_{i,j} := \|\eta_{i,j}\| \leq C_j h_i^{p_j+1}$. Using $p_2 \geq p_1 + 1$ and the fact that this implies $\varepsilon_{i,1} \geq C h_i^{p_2+1}$ for all $C > 0$ and all sufficiently small $h_i > 0$, we can conclude $\theta := \varepsilon_{i,2}/\varepsilon_{i,1} < 1$ if h_i is chosen sufficiently small since $\theta \to 0$ as $h_i \to 0$. We fix $\theta_0 < 1$, consider $h_i > 0$ such that $\theta < \theta_0 < 1$ holds and define

$$\bar{\eta} := \Phi_1(\tilde{\varphi}(\tau_i; 0, x, u), u, h_i) - \Phi_2(\tilde{\varphi}(\tau_i; 0, x, u), u, h_i) = \eta_{i,1} - \eta_{i,2}.$$

Then we have

$$(1-\theta)\varepsilon_{i,1} = (1-\theta)\|\eta_{i,1}\| = \left(1 - \frac{\|\eta_{i,1}-\overline{\eta}\|}{\|\eta_{i,1}\|}\right)\|\eta_{i,1}\| =$$
$$= \|\eta_{i,1}\| - \|\eta_{i,1}-\overline{\eta}\| \le \|\overline{\eta}\| = \overline{\varepsilon}$$

which yields the lower bound $k_1 = 1-\theta_0$ and

$$\overline{\varepsilon} = \|\overline{\eta}\| \le \|\eta_{i,1}\| + \|\eta_{i,1}-\overline{\eta}\| = \left(1 + \frac{\|\eta_{i,1}-\overline{\eta}\|}{\|\eta_{i,1}\|}\right)\|\eta_{i,1}\| =$$
$$= (1+\theta)\|\eta_{i,1}\| = (1+\theta)\varepsilon_{i,1}$$

giving the upper bound $k_2 = 1+\theta_0$. □

Using Theorem 11.6 we can now compute a suitable step size h_i if we additionally assume that the local error is of the form $\varepsilon_{i,1} \approx c_i h_i^{p_1+1}$ for small h_i. Note that for Runge–Kutta methods this assumption is satisfied if the vector field f is p_1+2 times continuously differentiable. In this case, c_i is given by the coefficient of the $h_i^{p_1+1}$ term in the Taylor approximation of the method.

For small step sizes it follows from the proof of Theorem 11.6 that $k_1 \approx k_2 \approx 1$, i.e., $\overline{\varepsilon} \approx \varepsilon_{i,1} \approx c_i h_i^{p_1+1}$ which gives us the estimate $\overline{c}_i \approx \overline{\varepsilon}/h_i^{p_1+1}$ for the coefficient c_i. Hence, the error tolerance $\mathrm{tol}_{\mathrm{ODE}}$ is satisfied (approximately) for the step size

$$\mathrm{tol}_{\mathrm{ODE}} = \overline{c}_i h_{i,\mathrm{new}}^{p_1+1} = \frac{\overline{\varepsilon}}{h_i^{p_1+1}} h_{i,\mathrm{new}}^{p_1+1} \iff h_{i,\mathrm{new}} = \sqrt[p_1+1]{fac\frac{\mathrm{tol}_{\mathrm{ODE}}}{\overline{\varepsilon}}}\, h_i \tag{11.11}$$

Since all these equalities are only satisfied approximately, a security factor $fac \in (0,1)$ has been introduced to compensate for these approximation errors. For this factor, $fac = 0.9$ is a typical choice in many algorithms.

A schematic implementation of a one step scheme with adaptive step size is given in Algorithm 11.7, below. This algorithm combines the iteration (11.2) with the computation of the step size h_i described above. Here, we solve (2.6) on one sampling interval $[0, T]$ using the length T of the sampling interval as an initial choice for the first step size h_0. For large T, one may alternatively choose $h_0 < T$. In each step the error $\overline{\varepsilon}$ is computed. If $\overline{\varepsilon}$ exceeds the tolerance $\mathrm{tol}_{\mathrm{ODE}}$, then the step is rejected and repeated using the new step size from (11.11). If $\overline{\varepsilon}$ maintains the desired tolerance, then the step is accepted and the new step size from (11.11) is used as an initial choice for the next time step.

Algorithm 11.7 Suppose an initial value x, a control value u, a tolerance $\mathrm{tol}_{\mathrm{ODE}}$, and sampling period T are given.

(1) Set $\tilde{\varphi}(0; 0, x, v) = x$, $i = 0$, $\tau_0 = 0$, $h_0 = T$
(2) If $\tau_i = T$ stop; If $\tau_i + h_i > T$ set $h_i = T - \tau_i$

(3) Set $\tau_{i+1} = \tau_i + h_i$ and compute $\Phi_1(\tilde{\varphi}(\tau_{i+1}; t_j, x, v), v, h_i)$, $\Phi_2(\tilde{\varphi}(\tau_{i+1}; t_j, x, v), v, h_i)$
(4) Compute $\overline{\varepsilon}$ and $h_{i,\text{new}}$ according to (11.10), (11.11)
(5) If $\overline{\varepsilon} > \text{tol}_{\text{ODE}}$ set $h_i = h_{i,\text{new}}$ and goto (3)
(6) If $\overline{\varepsilon} \leq \text{tol}_{\text{ODE}}$ set $\tilde{\varphi}(\tau_{i+1}; \tau_0, x, u) = \Phi_2(\tilde{\varphi}(\tau_{i+1}; \tau_0, x, u), u, h_i)$, $h_{i+1} = h_{i,\text{new}}$, $i = i + 1$ and goto (2)

In practical implementations, this basic algorithm is often refined in various ways. For instance, the new step size may be derived on the basis of a weighted sum of the absolute and the relative error instead of using only the absolute error as above. Upper and lower bounds for the time step h_i as well as for the ratio between h_i and h_{i+1} are also frequently used in practice.

Although the evaluation of two methods Φ_1 and Φ_2 and their possibly repeated evaluation in every step seems to be computationally more demanding, step size control algorithms are usually much more efficient than the use of equidistant time grids. This is due to two different aspects: on the one hand, there typically exist regions which allow for larger time steps and thus allow for a faster progress of the adaptive iteration procedure. On the other hand, the additional effort of simultaneously evaluating two methods can be reduced significantly by embedding these methods into each other. This means that the less accurate Runge–Kutta method Φ_1 uses the same stages k_i, cf. Definition 11.2, as the more accurate methods Φ_2 and thus the stages k_i only need to be evaluated once for both methods. One standard embedded method is the Dormand–Prince method of order (4)5, also called DoPri5, in which Φ_1 has order $p_1 = 4$ and Φ_2 is of order $p_2 = 5$. The Butcher tableau is displayed in Table 11.2. The second last line specifies the coefficients b_i for Φ_1 and the last line the b_i for Φ_2.

With the same induction as in the proof of Theorem 11.5, one sees that if the local errors maintain the tolerances $\text{tol}_{\text{ODE}} = \varepsilon h_i$ for some $\varepsilon > 0$, then the overall error at time T can be estimated as $e(T) \leq \varepsilon(\exp(\Lambda T) - 1)/\Lambda$ and thus scales linearly with ε. It should, however, be mentioned that adaptive step size selection schemes usually do not *rigorously* maintain the specified error tolerance. The reason for this is that Theorem 11.6 and the derivation of (11.11) require h_i to be sufficiently small. Suitable upper bounds which quantify this "sufficiently small" are, however, difficult to obtain without an extensive a priori analysis of the individual system and can therefore not be enforced in practice. Hence, the step size selection algorithm may select large step sizes for which the error estimation is no longer valid and thus the desired accuracy is no longer guaranteed. Thus, in general only equidistant grids with sufficiently small maximal step size $\overline{h}$ provide rigorous error bounds. Still, numerical experience shows that in the vast majority of examples error estimation-based adaptive step size algorithms like Algorithm 11.7 perform very reliably.

Table 11.2 Butcher tableau of the DoPri(4)5 method

0							
$\frac{1}{5}$	$\frac{1}{5}$						
$\frac{3}{10}$	$\frac{3}{40}$	$\frac{9}{40}$					
$\frac{4}{5}$	$\frac{44}{45}$	$-\frac{56}{15}$	$\frac{32}{9}$				
$\frac{8}{9}$	$\frac{19372}{6561}$	$-\frac{25360}{2187}$	$\frac{64448}{6561}$	$-\frac{212}{729}$			
1	$\frac{9017}{3168}$	$-\frac{355}{33}$	$\frac{46732}{5247}$	$\frac{49}{176}$	$-\frac{5103}{18656}$		
1	$\frac{35}{384}$	0	$\frac{500}{1113}$	$\frac{125}{192}$	$-\frac{2187}{6784}$	$\frac{11}{85}$	
	$\frac{35}{384}$	0	$\frac{500}{1113}$	$\frac{125}{192}$	$-\frac{2187}{6784}$	$\frac{11}{84}$	0
	$\frac{5179}{57600}$	0	$\frac{7571}{16695}$	$\frac{393}{640}$	$-\frac{92097}{339200}$	$\frac{187}{2100}$	$\frac{1}{40}$

11.4 Using the Methods Within the NMPC Algorithms

Looking at the NMPC Algorithm 3.11 and its variants, we see that in every iteration an optimal control problem has to be solved. To this end, the optimization algorithm needs to be able to compute the solution x_u and to evaluate the functional J_N. In fact, there are various ways for incorporating x_u into the optimization algorithm, for details see Sect. 12.1. However, no matter which method from this section we use, we need to be able to evaluate $\varphi(T, 0, x, u)$ in (2.8) numerically.

To this end, we replace the unknown map $\varphi(T, 0, x, u)$ in (2.8) by its approximation $\tilde{\varphi}(T, 0, x, u)$ from Algorithm 11.7. This way we end up with the definition

$$x^+ = f(x, u) := \tilde{\varphi}(T, 0, x, u). \qquad (11.12)$$

Iterating this map according to (2.2), which amounts to calling Algorithm 11.7 N times with initial values $x_u(n, x)$ and control values $u(n)$, $n = 0, \ldots, N-1$, we can then obtain an approximate predicted solution trajectory. Proceeding this way, one should keep in mind that the numerical scheme provides only an approximation of the exact solution. The effects of the approximation errors will be discussed in Sect. 11.5, below.

When the stage cost ℓ is defined via the integral formula (3.4) with running cost L, then we can efficiently include the numerical evaluation of the integral

$$\ell(x,u) = \int_0^T L(\varphi(t,0,x,u),u)dt$$

into the computation of $\tilde{\varphi}$. Here, we have removed the argument t from u because—following the convention in this chapter—u is constant on the sampling interval $[0,T]$. In order to compute the integral, consider the augmented ordinary differential equation

$$\dot{\overline{x}}(t) = \overline{f}(\overline{x}(t),u) \tag{11.13}$$

with

$$\overline{x}(t) = \begin{pmatrix} x(t) \\ y(t) \end{pmatrix} \in \mathbb{R}^d \times \mathbb{R} \quad \text{and} \quad \overline{f}(\overline{x},u) = \begin{pmatrix} f_c(x,u) \\ L(x,u) \end{pmatrix}.$$

Solving (11.13) with initial condition $\overline{x} = (x,0)$ we obtain the solution

$$\overline{\varphi}(T,x,u) = \begin{pmatrix} \varphi(T,x,u) \\ \ell(x,u) \end{pmatrix}.$$

Thus, solving (11.13) numerically yields a numerical solution whose first n components equal $\tilde{\varphi}(T,x,u)$ and whose $(n+1)$st component approximates $\ell(x,u)$. Proceeding this way we avoid the use of a separate numerical integration formula, in particular, we do not have to store the intermediate values $\tilde{\varphi}(\tau_i,x,u)$ for a subsequent numerical integration of L. Furthermore, the adaptive step size algorithm ensures that ℓ is approximated with the same accuracy as the solution φ.

As we will see in detail in Sect. 12.1, one way to incorporate the dynamics of the system into the numerical optimization algorithm is to externally compute the whole trajectory $x_u(\cdot,x_0)$, an approach called *recursive elimination*. In order to compute this trajectory, instead of defining f via (11.12) and then iterating f according to (2.2) one could apply a numerical one step method directly on the interval $[0,NT]$. This way we obtain a numerical approximation of x_u (and of J_N if we include the computation of ℓ) on $[0,NT]$ invoking Algorithm 11.7 only once. However, this has to be done with care. As already mentioned, in order to guarantee consistency with order p of the numerical schemes, it is important that the map $(t,x) \mapsto f_c(x,v(t))$ in (2.6) is p times continuously differentiable. Formally, this can be shown by extending the Formulas (11.7) and (11.8) to time varying vector fields f_c.

However, we can also give an informal explanation of this fact: when considering the solution of (2.6) with zero-order hold, then the control function v is discontinuous at the sampling times t_n. Consequently, the solution $\varphi(t,0,x,v)$ is not differentiable for $t = t_n$, as sketched in Fig. 11.2. Since we cannot approximate nonsmooth functions by a Taylor approximation, Formula (11.8) will not hold if we replace $\varphi(h,0,x,u)$ by $\varphi(\tau_i + h_i, \tau_i, x, v)$ with $t_n \in (\tau_i, \tau_i + h_i) = (\tau_i, \tau_{i+1})$ for some sampling time t_n. Thus, we have to make sure that this situation does not happen.

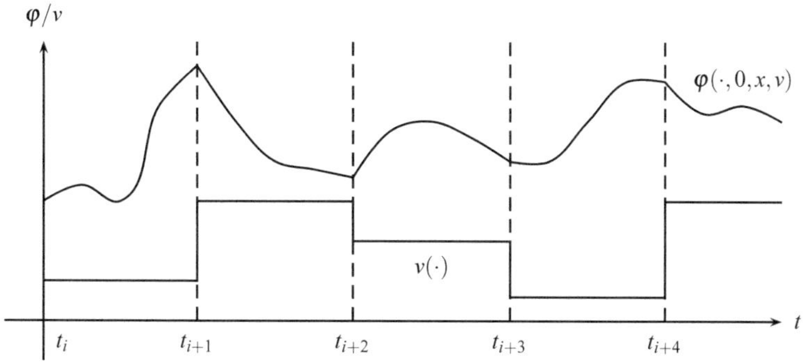

Fig. 11.2 Approximation of the sampled data solution

Defining the set of sampling times

$$\mathcal{T} := \{t_n \in \mathbb{R} \mid t_n = nT, n = 0, \dots, N\} \tag{11.14}$$

and using the time grid

$$\mathcal{G} := \{\tau_i \in [0, NT] \mid \tau_i \text{ is a discretization time in the one step method}\}, \tag{11.15}$$

in order to exclude the existence of i and n with $t_n \in (\tau_i, \tau_i + h_i) = (\tau_i, \tau_{i+1})$ we need to make sure that the inclusion $\mathcal{T} \subset \mathcal{G}$ holds. This assumption is not very restrictive, however, in order to ensure it we need to appropriately adjust Algorithm 11.7.

11.5 Numerical Approximation Errors and Stability

Defining the discrete dynamics f via the numerical approximation $\tilde{\varphi}$, cf. (11.12), introduces errors in the predictions x_u in the optimal control problems and its variants. In this section, we shift our focus from analyzing the effects of these errors on the open loop toward their effect on the closed loop. To this end, we utilize the techniques from Sects. 7.5–7.9 and – similar to these sections – restrict ourselves to constant reference $x^{\text{ref}} \equiv x_*$ in order to simplify the exposition. For the extension to time varying x^{ref} we refer to the remarks following the main results in Sects. 7.5–7.9.

As a general assumption we suppose that for each $\varepsilon > 0$ we can compute a numerical approximation $\tilde{\varphi}^\varepsilon$ which satisfies

$$\|\tilde{\varphi}^\varepsilon(T, 0, x, u) - \varphi(T, 0, x, u)\| \leq \varepsilon \tag{11.16}$$

for some $\varepsilon > 0$, all $x \in \mathbb{X}$ and $u \in \mathbb{U}(x)$. As discussed at the end of Sect. 11.3, such an estimate is rigorously ensured for $\tilde{\varphi}^\varepsilon$ generated by one step methods on equidistant grids with sufficiently small $\overline{h} > 0$ but can typically also be expected for $\tilde{\varphi}$ from the adaptive step size Algorithm 11.7 by adjusting the tolerance $\mathrm{tol}_{\mathrm{ODE}}$ appropriately. Observe that since (11.1) only holds for x and u from compact sets, in the case of equidistant grids we may have to adjust $\overline{h} > 0$ to x and u in order to ensure (11.16) if $\mathbb{X}$ or $\mathbb{U}(x)$ are noncompact. In case of Algorithm 11.7 the step size will be automatically adjusted by the step size selection mechanism.

Given that $\tilde{\varphi}^\varepsilon$ is an approximation of the true solution φ it seems natural to consider $\tilde{\varphi}^\varepsilon$ as a perturbed version of φ. However, since by definition the model used in the NMPC algorithm—i.e., the numerical approximation $\tilde{\varphi}^\varepsilon$—is the nominal model, we need the converse interpretation in order to apply the results from Sects. 7.5–7.9. In what follows we show that the closed loop obtained from the exact sampled data system (2.8) can be considered as a perturbed system in the sense of Sect. 7.5. To this end, we consider the following setting.

The NMPC algorithm is run with the numerically approximated discrete dynamics

$$f(x, u) = f^\varepsilon(x, u) := \tilde{\varphi}^\varepsilon(T, 0, x, u) \tag{11.17}$$

as a nominal model. The resulting NMPC-feedback law is denoted by $\mu_N^\varepsilon x_{\mu_N^\varepsilon}^\varepsilon$ and $\tilde{x}_{\mu_N^\varepsilon}^\varepsilon$ we denote the corresponding *nominal* and *perturbed NMPC closed-loop trajectory* from (3.5) and (7.7) with $f = f^\varepsilon = \tilde{\varphi}^\varepsilon$ and $\mu_N = \mu_N^\varepsilon$, respectively, i.e.,

$$x_{\mu_N^\varepsilon}^\varepsilon(n+1) = f^\varepsilon(x_{\mu_N^\varepsilon}^\varepsilon(n), \mu_N^\varepsilon(x_{\mu_N^\varepsilon}^\varepsilon(n)))$$

and

$$\tilde{x}_{\mu_N^\varepsilon}^\varepsilon(n+1) = f^\varepsilon(\tilde{x}_{\mu_N^\varepsilon}^\varepsilon(n), \mu_N^\varepsilon(\tilde{x}_{\mu_N^\varepsilon}^\varepsilon(n) + e(n))) + d(n).$$

The closed-loop system obtained from applying the numerically computed NMPC-feedback μ_N^ε to the exact model $f = \varphi$ from (2.8) according to (3.5), i.e.,

$$x_{\mu_N^\varepsilon}^{ex}(n+1) = f(x_{\mu_N^\varepsilon}^{ex}(n), \mu_N^\varepsilon(x_{\mu_N^\varepsilon}^{ex}(n))),$$

will be called the *exact closed-loop system*. The resulting trajectories will be denoted by $x_{\mu_N^\varepsilon}^{ex}$.

Note that the same NMPC-feedback law μ_N^ε—computed from $f = f^\varepsilon = \tilde{\varphi}^\varepsilon$—is used in (3.5) for generating $x_{\mu_N^\varepsilon}^\varepsilon$ and $x_{\mu_N^\varepsilon}^{ex}$. The difference between the two trajectories only lies in the map f in (3.5) which is given by $f = f^\varepsilon = \tilde{\varphi}^\varepsilon$ for $x_{\mu_N^\varepsilon}^\varepsilon$ and by $f = \varphi$ for $x_{\mu_N^\varepsilon}^{ex}$. Using this notation we obtain the following result.

Lemma 11.8 (Perturbed solution) *Consider the discrete time dynamics $f = f^\varepsilon$ from* (11.17) *obtained from a numerical approximation $\tilde{\varphi}^\varepsilon$ satisfying* (11.16)*, an NMPC-feedback law μ_N^ε with $\mu_N^\varepsilon(x) \in \mathbb{U}(x)$ and the solution $x_{\mu_N^\varepsilon}^\varepsilon$ of the corresponding closed-loop system* (3.5)*. Consider, furthermore, the solution $x_{\mu_N^\varepsilon}^{ex}$ of the exact closed-loop system.*

Then for each $x_0 \in \mathbb{X}$ *there exists a perturbation sequence* $d(\cdot) \in (\mathbb{R}^d)^\infty$ *with* $\|d(n)\| \leq \varepsilon$ *such that the solution* $\tilde{x}^\varepsilon_{\mu^\varepsilon_N}(n, x_0)$ *of the perturbed system* (7.7) *with* $f = f^\varepsilon$ *and* $e \equiv 0$ *satisfies*

$$x^{ex}_{\mu^\varepsilon_N}(n, x_0) = \tilde{x}^\varepsilon_{\mu^\varepsilon_N}(n, x_0)$$

for all $n \in \mathbb{N}_0$.

Proof Define

$$d(n) := \varphi(T, 0, x^{ex}_{\mu^\varepsilon_N}(n, x_0), \mu^\varepsilon_N(x^{ex}_{\mu^\varepsilon_N}(n, x_0))) - \tilde{\varphi}^\varepsilon(T, 0, x^{ex}_{\mu^\varepsilon_N}(n, x_0), \mu^\varepsilon_N(x^{ex}_{\mu^\varepsilon_N}(n, x_0)))$$

for all $n \in \mathbb{N}_0$. Then (11.16) with $x = x^{ex}_{\mu^\varepsilon_N}(n, x_0)$ and $u = \mu^\varepsilon_N(x^{ex}_{\mu^\varepsilon_N}(n, x_0))$ implies $\|d(n)\| \leq \varepsilon$ for all $n \in \mathbb{N}_0$. We show the desired identity by induction over n. For $n = 0$ we obtain $x^{ex}_{\mu^\varepsilon_N}(0, x_0) = x_0 = \tilde{x}^\varepsilon_{\mu^\varepsilon_N}(0, x_0)$. For $n \to n + 1$ assume that $x^{ex}_{\mu^\varepsilon_N}(n, x_0) = \tilde{x}^\varepsilon_{\mu^\varepsilon_N}(n, x_0)$ holds. Then we get

$$\begin{aligned}
x^{ex}_{\mu^\varepsilon_N}(n+1, x_0) &= \varphi(T, 0, x^{ex}_{\mu^\varepsilon_N}(n, x_0), \mu^\varepsilon_N(x^{ex}_{\mu^\varepsilon_N}(n, x_0))) \\
&= \tilde{\varphi}^\varepsilon(T, 0, x^{ex}_{\mu^\varepsilon_N}(n, x_0), \mu^\varepsilon_N(x^{ex}_{\mu^\varepsilon_N}(n, x_0))) + d(n) \\
&= f^\varepsilon(x^{ex}_{\mu^\varepsilon_N}(n, x_0), \mu^\varepsilon_N(x^{ex}_{\mu^\varepsilon_N}(n, x_0))) + d(n) \\
&= f^\varepsilon(\tilde{x}^\varepsilon_{\mu^\varepsilon_N}(n, x_0), \mu^\varepsilon_N(\tilde{x}^\varepsilon_{\mu^\varepsilon_N}(n, x_0))) + d(n) \quad = \quad \tilde{x}^\varepsilon_{\mu^\varepsilon_N}(n+1, x_0).
\end{aligned}$$

This shows the assertion. □

Lemma 11.8 shows that the closed-loop solution for the discrete time model obtained from the exact sampled data system (2.8) can be interpreted as a perturbed solution of the discrete time model obtained from the numerical approximation (11.17). The size of the perturbation $d(\cdot)$ directly corresponds to the numerical error (11.16).

This lemma enables us to use all results from Sects. 7.5–7.9 in order to conclude stability properties for $x^{ex}_{\mu^\varepsilon_N}$. The appropriate stability property is given by the following definition, cf. Definition 7.24.

Definition 11.9 Consider the exact closed-loop system (2.5) with $f = \varphi$ from (2.8) with μ^ε_N computed from $f = f^\varepsilon = \tilde{\varphi}^\varepsilon$ from (11.17) satisfying (11.16) for some $\varepsilon > 0$. Given a set $A \subseteq \mathbb{X}$ such that the optimal control problem defining μ^ε_N is feasible for all $x_0 \in A$, we say that x_* is *semiglobally practically asymptotically stable on* A *with respect to the numerical error* ε if there exists $\beta \in$ such that the following property holds: for each $\delta > 0$ and $\Delta > \delta$ there exists $\overline{\varepsilon} > 0$, such that for each initial value $x_0 \in A$ with $|x_0|_{x_*} \leq \Delta$ and each $\varepsilon \in (0, \overline{\varepsilon}]$ the solution $x^{ex}_{\mu^\varepsilon_N}(\cdot, x_0)$ satisfies $x^{ex}_{\mu^\varepsilon_N}(k, x_0) \in A$ and

$$|x^{ex}_{\mu^\varepsilon_N}(k, x_0)|_{x_*} \leq \max\{\beta(|x_0|_{x_*}, k),\ \delta\}$$

for all $k \in \mathbb{N}_0$.

The following theorem now gives conditions under which this stability property holds.

Theorem 11.10 (Stability for perturbed solution) *Consider the NMPC-feedback laws μ_N^ε obtained from one of the NMPC algorithms from Theorems 7.26 and 7.36 or 7.41 with $f = f^\varepsilon = \tilde{\varphi}^\varepsilon$ from* (11.17). *Assume that* (11.16) *holds and that there is $\varepsilon_0 > 0$ such that one of the following assumptions is satisfied for all $\varepsilon \in (0, \varepsilon_0]$.*

(i) In case of Theorem 7.26, assume that α, α_1, α_2, α_3 in Theorem 4.11 as well as ω_V and ω_f can be chosen independently of $\varepsilon > 0$.
(ii) In case of Theorem 7.36, assume that α, α_1, $\tilde{\alpha}$ in Theorem 6.20, β from Assumption 7.35 and η in Definition 7.33 can be chosen independently of $\varepsilon > 0$.
(iii) In case of Theorem 7.41, assume that α, α_1, α_2, α_3 in Theorem 4.11, δ, γ, ε' in Assumption 7.38 and the bound on f as well as the moduli of continuity of f and ℓ can be chosen independently of $\varepsilon > 0$.

Then the exact closed-loop system (2.5) *with $f = \varphi$ from* (2.8) *is semiglobally practically asymptotically stable with respect to ε from* (11.16) *in the sense of Definition 11.9 on the set A specified in the respective theorem.*

Proof The respective theorems ensure semiglobal practical asymptotic stability for all perturbed trajectories $\tilde{x}^\varepsilon_{\mu_N^\varepsilon}$ with respect to $\overline{d}$ and $\overline{e}$ in the sense of Definition 7.24. An inspection of the proofs of the respective theorems then reveals that the uniformity assumptions (i)–(iii) guarantee that for given δ and Δ the bounds $\overline{d}$ and $\overline{e}$ and the function $\beta \in$ in Definition 7.24 are independent of $\varepsilon > 0$.

Fixing δ and Δ we thus find $\overline{d} > 0$ such that each perturbed solution $\tilde{x}^\varepsilon_{\mu_N^\varepsilon}$ with perturbations $\|d(n)\| \leq \overline{d}$ and $e \equiv 0$ satisfies the conditions of Definition 7.24 for all $\varepsilon \in (0, \varepsilon_0]$. Setting $\overline{\varepsilon} = \min\{\overline{d}, \varepsilon_0\}$ and using that by Lemma 11.8 the exact closed-loop trajectory $x^{ex}_{\mu_N^\varepsilon}$ equals one of the trajectories $\tilde{x}^\varepsilon_{\mu_N^\varepsilon}$ with $\overline{d} = \varepsilon$ and $\overline{e} = 0$, we obtain that $x^{ex}_{\mu_N^\varepsilon}$ satisfies the conditions of Definition 11.9 for the given δ and Δ and all $\varepsilon \in (0, \overline{\varepsilon}]$. This yields the assertion. □

Note that Theorem 11.10 only guarantees the stability of the discrete time closed-loop system (2.5) with f from (2.8) but not for the sampled data closed loop (2.30). In order to conclude stability properties of (2.30) the techniques from Sect. 2.4 can be used. While Theorem 2.27 and its assumptions are formulated for the case of "real" asymptotic stability, its statement, and proof can be straightforwardly extended to the semiglobal practical setting of Definition 11.9. Recall from Remark 4.13 that the assumptions of Theorem 2.27 are satisfied for suitable integral costs (3.4). Although we have not rigorously analyzed the effect of the error induced by the numerical approximation of such integral costs, we conjecture that the estimates in Remark 4.13 remain valid in a suitable approximate sense if these errors are sufficiently small.

Since numerical approximations are used in virtually all NMPC algorithms for sampled data systems, Theorem 11.10 implies that all such algorithms need appropriate robustness—either inherently as in case (i) or by an appropriate design of the state constraints as in cases (ii) and (iii) of Theorem 11.10—in a uniform way with respect to ε in order to ensure semiglobal practical stability in the presence of numerical errors. In practice, however, this is hardly ever rigorously ensured. The reason for this is that for good numerical methods numerical errors are usually very small compared to other error sources like model errors, external perturbations, etc. Although even very small errors may in the worst case be destabilizing, as illustrated by Example 7.31, it is not very likely that this indeed happens and—also according to our experience—such phenomena are hardly ever observed in simulations or practical examples. Hence, unless robustness is needed in order to cope with error sources which are significantly larger than the numerical errors discussed in this chapter, for most practical purposes it seems justified to neglect the robustness issue, provided, of course, the numerical errors are indeed sufficiently small. Still, one has to keep in mind that proceeding this way does not rigorously ensure stability of the exact closed-loop system.

11.6 Notes and Extensions

The material contained in Sects. 11.1–11.3 can be found in many textbooks on numerical analysis for ordinary differential equations, like, e.g., the books by Deuflhard an Bornemann [2], Hairer, Nørsett and Wanner [8] or Stoer and Bulirsch [11]. Clearly, the presentation in this chapter cannot replace any of these textbooks and aims at giving an introduction into the subject rather than a comprehensive treatment.

Among the many topics we have not covered in this chapter we would in particular like to mention stiff problems and differential algebraic equations (DAEs), often called descriptor systems in systems theory. While stiff problems "look" like normal ordinary differential equations, they are very difficult to solve with the explicit methods presented in Sect. 11.1. For stiff equations, which often appear when modeling technical systems, an adaptive step size algorithm like Algorithm 11.7 will typically select very small time steps even though the solution is almost constant. Explaining the precise mathematical reasons for this behavior goes beyond these notes, but we would at least like to mention that so-called implicit methods perform much better for stiff equations. DAEs are ordinary differential equations with additional algebraic constraints, often given implicitly. DAEs appear as models, e.g., in mechanics and electrical engineering and NMPC is perfectly suited for handling DAES, however, the solution methods presented in this chapter do not apply to such equations and specialized numerical schemes are needed, which are again often of the implicit type. While also covered in some standard textbooks, there is a large amount of literature particularly devoted to stiff and DAE problems, as, e.g., Hairer and Wanner [7], and we refer the reader to such books for more details.

As Examples 2.12 and 6.32 show, NMPC is also suitable for infinite-dimensional systems generated by controlled PDEs. NMPC for PDEs requires the solution of an optimal control problem for PDEs in each step. The monograph by Troeltzsch [12] provides a good introduction into such problems. A simple way to approach this problem numerically is to proceed similar as described for the ordinary differential equations in this chapter with an additional spatial discretization by, e.g., a finite difference method (which is what we used in Example 6.32), see, e.g., LeVeque [9] or a finite element method, see, e.g., Braess [1]. However, it is by no means clear whether this is the most efficient way of approaching the problem numerically; in fact, the development of suitable numerical schemes is currently a very active research area. Furthermore, we are not aware of a rigorous analysis of the effects of spatial discretization errors in NMPC controller design.

The need to use numerical approximations and the consequences for the stability analysis discussed in Sect. 11.5 are largely ignored in the NMPC literature. An exception to this rule are the papers by Gyurkovics and Elaiw [5, 6], which are in the same spirit as cases (i) and (iii) of Theorem 11.10 in the sense that they exploit uniform continuity properties, in particular of the optimal value function V_N. However, these results require Lyapunov function terminal costs and do not consider state constraints as in cases (ii) and (iii) of Theorem 11.10.

More generally, the problem considered in Sect. 11.5 can be seen as a special case of a nonlinear controller design based on approximate models. A comprehensive treatment of this topic in a rather general setting can be found in Nešić and Teel [10]. An application to infinite horizon optimal control based feedback design was given in Grüne and Nešić [4]. The idea to treat numerical errors as perturbations is classical in numerical analysis. In a control theoretic framework this idea was used extensively in the monograph Grüne [3]. All these approaches are similar to our approach in the sense that the stability property of the approximate system is required to be robust in some suitable sense, that the robustness can be quantified and that this quantitative measure of the robustness is independent of the numerical accuracy. In all cases the obtained stability property is semiglobal practical stability, just as in Theorem 11.10. State constraints are, however, again not considered in these references.

Nešić and Teel [10] also nicely illustrate the pitfalls of feedback design based on approximate models by means of simple examples and discuss the case in which the numerical accuracy is linked to the sampling period T. Roughly speaking, in this case uniform continuity of the Lyapunov function under consideration is not sufficient in order to ensure stability of the exact closed-loop system. Rather, a stronger property like Lipschitz continuity with Lipschitz constant independent of the numerical accuracy ε is needed in this case.

There are numerous issues related to numerical errors we have not addressed in this chapter. For instance, numerical errors may lead to the situation that the inequalities in Assumption 5.9(ii) or Assumptions 6.3 or 6.5 are only satisfied up to an error term ε, which has to be taken into account in the results relying on these assumptions. While we conjecture that in both cases the respective proofs can be modified in order to obtain at least semiglobal practical asymptotic stability of $x^{\varepsilon}_{\mu^{\varepsilon}_n}$, we are not aware of respective results in the literature. Hence, this area certainly offers a number of open questions for future research.

Problems

1. Prove that the solution $\varphi(t, 0, x_0, u)$ of (2.6) with $t \in [0, T]$ and constant control function u satisfies the integral equation

$$\varphi(\tau_{i+1}; \tau_0, x_0, u) = \varphi(\tau_i; \tau_0, x_0, u) + \int_{\tau_i}^{\tau_{i+1}} f_c(\varphi(t; \tau_0, x_0, u), u)dt.$$

for all $\tau_i, \tau_{i+1} \in [0, T]$ with $\tau_{i+1} > \tau_i$.

2. Prove that the Euler and the Heun scheme satisfy the Lipschitz condition (11.5) if the vector field f_c satisfies the Lipschitz condition from Assumption 2.4.
3. Given the control system $\dot{x}(t) = x(t) + u(t)$ with stage cost $\ell(x, u) = x^2 + u^2$.
   (a) Consider the NMPC Algorithm 3.1 with $N = 2$ and f generated by the Euler method with $\mathscr{G} = \mathscr{T}$ for (11.14) and (11.15). Prove that the control $\mu_N(x)$ converges tends to zero as $T \to 0$ for each $x \in \mathbb{R}$.
   (b) Consider the same situation as in (a) but with the grid

$$\mathscr{G} := \{\tau_i = iT/k \mid i = 0, \ldots, Nk\}$$

   with $k \in \mathbb{N}$. Does the control value $\mu_N(x)$ converge if $T > 0$ is fixed and k tends to infinity?

4. Consider the differential equation

$$\begin{aligned} \dot{x}_1(t) &= -x_2(t) \\ \dot{x}_2(t) &= x_1(t) \end{aligned}$$

   whose solution shall be used to generate a time varying reference for an NMPC algorithm.

   (a) Using a transformation to polar coordinates, compute the analytical solution of the system.
   (b) Show that the numerical solution of the system using Euler's method will deviate from the analytical solution from (a) for every step size $h > 0$ and every initial value $x_0 \neq (0, 0)^\top$.
   (c) Applying the transformation to polar coordinates, show that the occurring error from (b) can be avoided if the resulting differential equation is solved using Euler's method.

5. Consider the continuous time control system

$$\begin{aligned} \dot{x}_1(t) &= -x_2(t) + v(t) \\ \dot{x}_2(t) &= x_1(t) \end{aligned}$$

where u shall be computed via NMPC to track the (exact) time varying reference solution from Problem 4.

(a) Show that this system is (uniformly) asymptotically controllable in the sense of Definition 4.2 for control functions which are piecewise constant on each interval $[iT, (i+1)T)$ for arbitrary sampling time $T > 0$.
(b) Consider the approximate discrete time system (11.12) with $\tilde{\varphi}$ obtained from applying the Euler method with step size $h = T/k$ for arbitrary $k \in \mathbb{N}$ to the (non transformed) differential equation. Show that this approximate system is not asymptotically controllable regardless how $T > 0$ and $k \in \mathbb{N}$ are chosen.

Hint for (b): A necessary condition for asymptotic controllability is that the reference is a solution of the system.

References

1. Braess, D.: Finite Elements, 3rd edn. Cambridge University Press, Cambridge (2007) (Theory, fast solvers, and applications in elasticity theory. Translated from the German by Larry L. Schumaker)
2. Deuflhard, P., Bornemann, F.: Scientific Computing with Ordinary Differential Equations. Texts in Applied Mathematics, vol. 42. Springer, New York (2002) (Translated from the1994 German original by Werner C. Rheinboldt)
3. Grüne, L.: Asymptotic Behavior of Dynamical and Control Systems under Perturbation and Discretization. Lecture Notes in Mathematics, vol. 1783. Springer, Berlin (2002)
4. Grüne, L., Nešić, D.: Optimization based stabilization of sampled-data nonlinear systems via their approximate discrete-time models. SIAM J. Control Optim. **42**, 98–122 (2003)
5. Gyurkovics, E., Elaiw, A.M.: Stabilization of sampled-data nonlinear systems by receding horizon control via discrete-time approximations. Automatica **40**(12), 2017–2028 (2004)
6. Gyurkovics, E., Elaiw, A.M.: Conditions for MPC based stabilization of sampled-data nonlinear systems via discrete-time approximations. In: Findeisen, R., Allgöwer, F., Biegler, L.T. (eds.) Assessment and Future Directions of Nonlinear Model Predictive Control. Lecture Notes in Control and Information Sciences, vol. 358, pp. 35–48. Springer, Berlin (2007)
7. Hairer, E., Wanner, G.: Solving Ordinary Differential Equations. II, Springer Series in Computational Mathematics, vol. 14, 2nd edn. Springer, Berlin (1996)
8. Hairer, E., Nørsett, S.P., Wanner, G.: Solving Ordinary Differential Equations. I, 2nd edn. Springer Series in Computational Mathematics, vol. 8, 2nd edn. Springer, Berlin (1993)
9. LeVeque, R.J.: Finite Difference Methods for Ordinary and Partial Differential Equations. SIAM, Philadelphia (2007)
10. Nešić, D., Teel, A.R.: A framework for stabilization of nonlinear sampled-data systems based on their approximate discrete-time models. IEEE Trans. Automat. Control **49**(7), 1103–1122 (2004)
11. Stoer, J., Bulirsch, R.: Introduction to Numerical Analysis. Texts in Applied Mathematics, vol. 12, 3rd edn. Springer, New York (2002) (Translated from the German by R. Bartels, W. Gautschi and C. Witzgall)
12. Tröltzsch, F.: Optimal Control of Partial Differential Equations. Graduate Studies in Mathematics, vol. 112. American Mathematical Society, Providence (2010) (Theory, methods and applications. Translated from the 2005 German original by Jürgen Sprekels)

Chapter 12
Numerical Optimal Control of Nonlinear Systems

In this chapter, we focus on numerically solving the constrained finite horizon nonlinear optimal control problems occurring in each iterate of the NMPC procedure. To this end, we first state standard discretization techniques to obtain a nonlinear optimization problem in standard form. Utilizing this form, we outline basic versions of the two most common solution methods for such problems, that is, Sequential Quadratic Programming (SQP) and Interior Point Methods (IPM). Furthermore, we investigate interactions between the differential equation solver, the discretization technique, and the optimization method and present several NMPC specific details concerning the warm start of the optimization routine. Finally, we discuss NMPC variants relying on inexact solutions of the finite horizon optimal control problem.

12.1 Discretization of the NMPC Problem

The most general NMPC problem formulation is given in Algorithm 3.11 and will be the basis for this chapter. In Step (2) of Algorithm 3.11 we need to solve the optimal control problem

$$
\begin{aligned}
&\text{minimize} \quad J_N(n, x_0, u(\cdot)) := \sum_{k=0}^{N-1} \omega_{N-k} \ell(n+k, x_u(k, x_0), u(k)) \\
&\qquad\qquad\qquad\qquad + F_J(n+N, x_u(N, x_0)) \\
&\text{with respect to} \quad u(\cdot) \in \mathbb{U}^N_{\mathbb{X}_0}(n, x_0), \quad \text{subject to} \\
&x_u(0, x_0) = x_0, \quad x_u(k+1, x_0) = f(x_u(k, x_0), u(k))
\end{aligned}
\tag{$\mathrm{OCP}^{\mathrm{n}}_{\mathrm{N,e}}$}
$$

L. Grüne and J. Pannek, *Nonlinear Model Predictive Control*,
Communications and Control Engineering, DOI 10.1007/978-3-319-46024-6_12

We will particularly emphasize the case in which the discrete time system (2.1) is induced by a sampled data continuous time control systems

$$\dot{x}(t) = f_c(x(t), v(t)), \tag{2.6}$$

however, all results also apply to discrete time models not related to a sampled data system. Throughout this chapter we assume $X = \mathbb{R}^d$ and $U = \mathbb{R}^m$. Furthermore, we rename the terminal cost F in ($\mathrm{OCP}^{\mathrm{n}}_{\mathrm{N,e}}$) to F_J because we will use the symbol F with a different meaning, below.

So far, in Chap. 11 We have shown how the solution $x_u(k, x_0)$ of the discrete time system (2.1) in the last line of ($\mathrm{OCP}^{\mathrm{n}}_{\mathrm{N,e}}$) can be obtained and evaluated using numerical methods for differential equations, but not how the minimization problem ($\mathrm{OCP}^{\mathrm{n}}_{\mathrm{N,e}}$) can be solved.

The purpose of this chapter is to fill this gap. In particular, we first show how problem ($\mathrm{OCP}^{\mathrm{n}}_{\mathrm{N,e}}$) can be reformulated to match the *standard problem in nonlinear optimization*

$$\begin{array}{ll} \text{minimize} & F(z) \\ \text{with respect to } z \in \mathbb{R}^{n_z} & \\ \text{subject to } G(z) = 0 \text{ and } H(z) \geq 0 & \end{array} \tag{NLP}$$

with maps $F : \mathbb{R}^{n_z} \to \mathbb{R}$, $G : \mathbb{R}^{n_z} \to \mathbb{R}^{r_g}$ and $H : \mathbb{R}^{n_z} \to \mathbb{R}^{r_h}$.

Even though ($\mathrm{OCP}^{\mathrm{n}}_{\mathrm{N,e}}$) is already a discrete time problem, the process of converting ($\mathrm{OCP}^{\mathrm{n}}_{\mathrm{N,e}}$) into (NLP) is called *discretization*. Here, we will stick with this commonly used term even though in a strict sense we only convert one discrete problem into another.

As we will see, the (NLP) problem related to ($\mathrm{OCP}^{\mathrm{n}}_{\mathrm{N,e}}$) can be formulated in different ways. The first variant, called *full discretization*, incorporates the dynamics (2.1) as additional constraints into (NLP). This approach is very straightforward but causes large computing times for solving the problem (NLP) due to its dimensionality, unless special techniques for handling these constraints can be used on the optimization algorithm level, cf. the paragraph on condensing in Sect. 12.4, below. The second approach is designed to deal with this dimensionality problem. It eliminates the additional constraints by recursively computing $x_u(k, x_0)$ from the dynamics (2.1) outside of the optimization problem (NLP) and is hence called *recursive elimination*. Proceeding this way, the dimension of the optimization variable z and the number of constraints p is reduced significantly. However, in this method it is difficult to incorporate preexisting knowledge on optimal solutions, as derived, e.g., from the reference, to the optimizer. Furthermore, computing $x_u(k, x_0)$ on large time intervals may lead to a rather sensitive dependence of the solution on the control $u(\cdot)$, which may cause numerical problems in the algorithms for solving (NLP). To overcome these problems, we introduce a third problem formulation, called *multiple shooting*, which is closely related to concept of shooting methods for differential equations and can be seen as a compromise between the two other methods.

After we have given the details of these discretization techniques, methods for solving the optimization problem (NLP) and their applicability in the NMPC context will be discussed in the subsequent sections. In order to illustrate certain effects, we will repeatedly consider the following example throughout this chapter.

Example 12.1 Consider the inverted pendulum on a cart problem from Example 2.10 with initial value $x_0 = (2, 2, 0, 0)$, sampling period $T = 0.1$ and cost functional

$$J_N(x_0, u) := \sum_{i=0}^{N-1} \ell(x(i), u(i))$$

where the stage costs ℓ are of the integral type (3.4) with

$$\begin{aligned} L(x, u) := \Big(& 3.51 \sin(x_1 - \pi)^2 + 4.82 \sin(x_1 - \pi) x_2 \\ & + 2.31 x_2^2 + 2 \left((1 - \cos(x_1 - \pi)) \cdot (1 + \cos(x_2)^2) \right)^2 \\ & + x_3^2 + x_4^2 + u^2 \Big)^2 . \end{aligned}$$

We impose constraints on the state and the control vectors by defining

$$\begin{aligned} \mathbb{X} &:= \mathbb{R} \times \mathbb{R} \times [-5, 5] \times [-10, 10] \\ \mathbb{U} &:= [-5, 5] \end{aligned}$$

but we do not impose stabilizing terminal conditions. All subsequent computations have been performed with the default tolerance 10^{-6} in the numerical solvers involved, cf. Sect. 12.4 for details on these tolerances.

As we have seen in the previous chapters, the horizon length can be a critical component for the stability of an NMPC controlled system. In particular, the NMPC closed loop may be unstable if the horizon length is too short as shown in Fig. 12.1. In

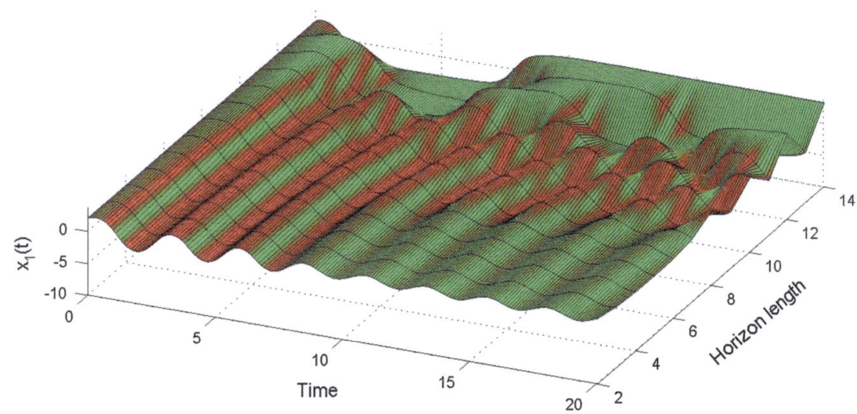

Fig. 12.1 Closed-loop trajectories $x_1(\cdot)$ for small optimization horizons N

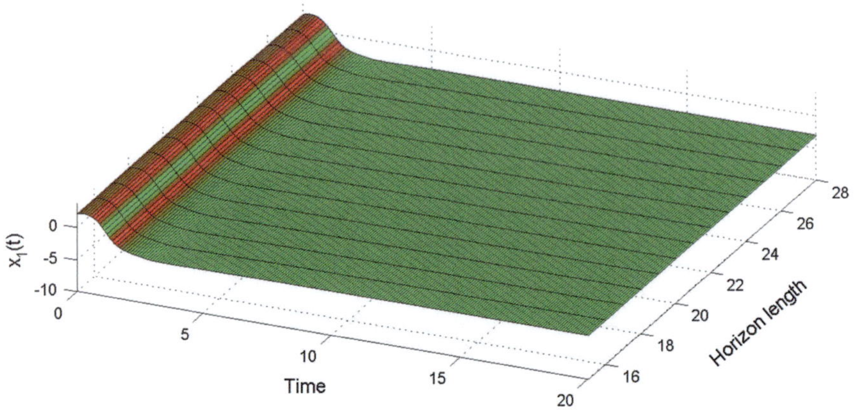

Fig. 12.2 Closed-loop trajectories $x_1(\cdot)$ for medium size optimization horizons N

this and in the subsequent figures we visualize closed-loop trajectories for different optimization horizons as a surface in the (t, N)-plane. Looking at the figure one sees that for very small N the trajectory simply swings around the downward equilibrium. For $N \geq 12$, the NMPC controller is able to swing up the pendulum to one of the upright equilibria $(\pi, 0, 0, 0)$ or $(-\pi, 0, 0, 0)$ but is not able to stabilize the system there.

As expected from Theorem 6.24, for larger optimization horizons the closed-loop solution tends toward the upright equilibrium $(-\pi, 0, 0, 0)$ as shown in Fig. 12.2.

If we further increase the optimization horizon, then it can be observed that the algorithm chooses to stabilize the upright equilibrium $(\pi, 0, 0, 0)$ instead of $(-\pi, 0, 0, 0)$ as illustrated in Fig. 12.3. Moreover, for some horizon lengths N, stability is lost. Particularly, for N between 50 and 55 the behavior is similar to that for $N \in \{12, 13, 14\}$: the controller is able to swing up the pendulum to an upright position but unable to stabilize it there. As a consequence, it appears that the NMPC algorithm cannot decide which of the upward equilibria shall be stabilized and the trajectories repeatedly move from one to another, i.e., the x_1-component of the closed-loop trajectory remains close to one of these equilibria only for a short time.

This behavior contradicts what we would expect from the theoretical result from Theorem 6.24 and can hence only be explained by numerical problems in solving ($\mathrm{OCP}^{\mathrm{n}}_{\mathrm{N,e}}$) due to the large optimization horizon. Since by now we do not have the means to explain the reason for this behavior, we postpone this issue to Sect. 12.4. In particular, the background of this effect will be discussed after Example 12.28 where we also outline methods to avoid this effect.

As a result, we obtain that numerically the set of optimization horizon N for which stability can be obtained is not only bounded from below—as expected from the theoretical result in Theorem 6.24—but also from above. Unfortunately, for a general example it is a priori not clear if the set of numerically stabilizing horizons

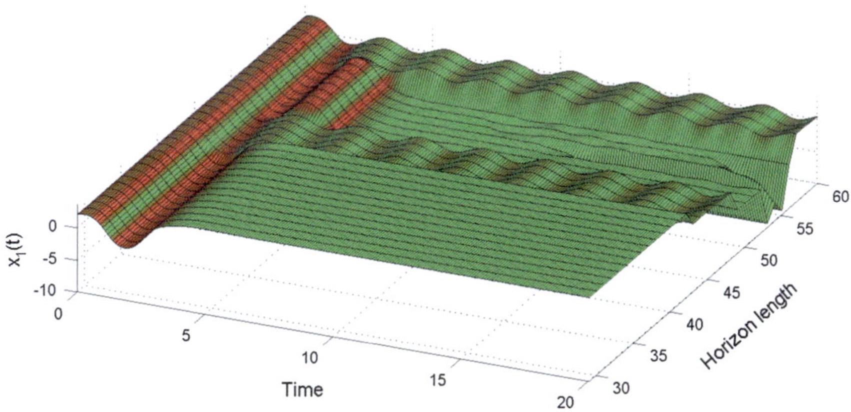

Fig. 12.3 Closed-loop trajectories $x_1(\cdot)$ for large optimization horizons N

is nonempty, at all. Furthermore, as we will see in this chapter, also the minimal optimization horizon which numerically stabilizes the closed-loop system may vary with the used discretization technique. This is due to the fact that for N close to the theoretical minimal stabilizing horizon and for the NMPC variant without stabilizing terminal conditions considered here, the value α in (5.1) is close to 0 and hence small numerical inaccuracies may render α negative and thus destabilize the system. Since different discretizations lead to different numerical errors, it is not entirely surprising that the minimal stabilizing horizon in the numerical simulation depends on the chosen discretization technique, as Example 12.2 will show.

Full Discretization

In order to obtain an optimization problem in standard form (NLP) the full discretization technique is the simplest and most common one. Recall that the discrete time trajectory $x_u(k, x_0)$ in ($\mathrm{OCP}^{\mathrm{n}}_{\mathrm{N,e}}$) is defined by the dynamics (2.1) via

$$x_u(0, x_0) = x_0, \quad x_u(k+1, x_0) = f(x_u(k, x_0), u(k)), \tag{2.2}$$

where in the case of a continuous time system the map f is obtained from a numerical approximation via (11.12).

Clearly, each control value $u(k)$, $k \in \{0, \ldots, N-1\}$ is an optimization variable in ($\mathrm{OCP}^{\mathrm{n}}_{\mathrm{N,e}}$) and will hence also be an optimization variable in (NLP). The idea of the full discretization is now to treat each point on the trajectory $x_u(k, x_0)$ as an additional independent d-dimensional optimization variable. This implies that we need additional conditions which ensure that the resulting optimal choice of the variables $x_u(k, x_0)$ obtained from solving (NLP) is a trajectory of (2.1). To this end, we include the equations in (2.2) as equality constraints in (NLP). This amounts to rewriting (2.2) as

$$x_u(k+1, x_0) - f(x_u(k, x_0), u(k)) = 0 \quad \text{for } k \in \{0, \dots, N-1\} \tag{12.1}$$
$$x_u(0, x_0) - x_0 = 0 \tag{12.2}$$

Next, we have to reformulate the constraints $u(\cdot) \in \mathbb{U}^N_{\mathbb{X}_0}(x_0)$. According to Definition 3.2 these conditions can be written explicitly as

$$\begin{aligned} & x_u(k, x_0) \in \mathbb{X} && k \in \{0, \dots, N\} \\ \text{and} \quad & u(k) \in \mathbb{U}(x_u(k, x_0)) && k \in \{0, \dots, N-1\} \end{aligned}$$

and in the case of stabilizing terminal conditions we get the additional condition

$$x_u(N, x_0) \in \mathbb{X}_0.$$

For simplicity of exposition, we only consider the case of time-invariant state constraints. The setting is, however, easily extended to the case of time-varying constraints $\mathbb{X}^k$ as introduced in Sect. 7.8.

Here and in the following, we assume $\mathbb{X}$, $\mathbb{U}(x)$ and—if applicable—$\mathbb{X}_0$ to be characterized by Definition 3.6, i.e., by a set of functions $G_i^S : \mathbb{R}^d \times \mathbb{R}^m \to \mathbb{R}$, $i \in \mathcal{E}^S = \{1, \dots, p_g\}$, and $H_i^S : \mathbb{R}^d \times \mathbb{R}^m \to \mathbb{R}$, $i \in \mathcal{I}^S = \{1, \dots, p_h\}$ via equality and inequality constraints of the form

$$G_i^S(x_u(k, x_0), u(k)) = 0, \quad i \in \mathcal{E}^S,\ k \in K_i \subseteq \{0, \dots, N\} \tag{12.3}$$
$$H_i^S(x_u(k, x_0), u(k)) \geq 0, \quad i \in \mathcal{I}^S,\ k \in K_i \subseteq \{0, \dots, N\}. \tag{12.4}$$

The index sets K_i, $i \in \mathcal{E}^S \cup \mathcal{I}^S$ in these constraints formalize that some of the conditions may not be required for all times $k \in \{0, \dots, N\}$. For instance, the terminal constraint condition $x_u(N, x_0) \in \mathbb{X}_0$ is only required for $k = N$, hence the sets K_i corresponding to the respective conditions would be $K_i = \{N\}$. In order to simplify the notation we have included $u(N)$ into these conditions even though the functional J_N in ($\mathrm{OCP}^{\mathrm{n}}_{\mathrm{N,e}}$) does not depend on this variable. However, since the functions H_i^S and G_i^S in (12.3) and (12.4) do not need to depend on u, this can be done without loss of generality.

Summarizing, we obtain the constraint function G in (NLP) from Eqs. (12.1), (12.2), (12.3) and H in (NLP) from Eq. (12.4). The remaining components of the optimization problem is the optimization variable, which is defined as

$$z := (x_u(0, x_0)^\top, \dots, x_u(N, x_0)^\top, u(0)^\top, \dots, u(N-1)^\top)^\top \tag{12.5}$$

and the cost function F, which we obtain straightforwardly as

$$F(z) := \sum_{k=0}^{N-1} \omega_{N-k} \ell(n+k, x_u(k, x_0), u(k)) + F_J(n+N, x_u(N, x_0)). \tag{12.6}$$

Hence, the fully discretized problem $(\mathrm{OCP}^{\mathrm{n}}_{\mathrm{N,e}})$ is of the form

$$\begin{aligned}
&\text{minimize} \quad F(z) := \sum_{k=0}^{N-1} \omega_{N-k}\ell(n+k, x_u(k,x_0), u(k)) \\
&\qquad\qquad\qquad\qquad + F_J(n+N, x_u(N,x_0)) \\
&\text{with respect to} \\
&z := (x_u(0,x_0)^\top, \ldots, x_u(N,x_0)^\top, u(0)^\top, \ldots, u(N-1)^\top)^\top \in \mathbb{R}^{n_z} \\
&\text{subject to } G(z) = \begin{bmatrix} \left[G_i^S(x_u(k,x_0), u(k))\right]_{i\in\mathcal{E}^S, k\in K_i} \\ [x_u(k+1,x_0) - f(x_u(k,x_0), u(k))]_{k\in\{0,\ldots,N-1\}} \\ x_u(0,x_0) - x_0 \end{bmatrix} = 0 \\
&\text{and } H(z) = \left[H_i^S(x_u(k,x_0), u(k))\right]_{i\in\mathcal{I}^S, k\in K_i} \geq 0
\end{aligned}$$

Similar to Definition 3.6 we write the equality and inequality constraints as $G = (G_1, \ldots, G_{r_g})$ and $H = (H_{r_g+1}, \ldots, H_{r_g+r_h})$ with $r_g := (N+2)\cdot d + \sum_{i\in\mathcal{E}^S} \sharp K_i$ and $r_h := \sum_{i\in\mathcal{I}^S} \sharp K_i$ where $\sharp K_i$ denotes the number of elements of the set K_i. The corresponding index sets are denoted by $\mathcal{E} = \{1, \ldots, r_g\}$ and $\mathcal{I} = \{r_g+1, \ldots, r_g + r_h\}$.

The advantage of the full discretization is the simplicity of the transformation from $(\mathrm{OCP}^{\mathrm{n}}_{\mathrm{N,e}})$ to (NLP). Unfortunately, it results in a high-dimensional optimization variable $z \in \mathbb{R}^{(N+1)\cdot d + N\cdot m}$ and a large number of both equality and inequality constraints r_g and r_h. Since computing times and accuracy of solvers for problems of type (NLP) depend massively on the size of these problems, this is highly undesirable. One way to solve this problem is to handle the additional constraints by special techniques in the optimization algorithm, cf. the paragraph on condensing in Sect. 12.4, below. Another way is to reduce the number of optimization variables directly in the discretization procedure, which is what we describe next.

Recursive Elimination

In the previous section, we have seen that the full discretization of $(\mathrm{OCP}^{\mathrm{n}}_{\mathrm{N,e}})$ leads to a high-dimensional optimization problem (NLP). The discretization technique which we present now avoids this problem and minimizes the number of components within the optimization variable z as well as in the equality constraints G.

The methodology of the recursive elimination is inspired by the (hierarchical) divide and conquer principle. According to this principle, the problem is broken down into subproblems which can then be treated by specialized solution methods. The fundamental idea of the recursive elimination is to decouple the dynamics of the control system from the optimization problem (NLP).

At the control system level in the hierarchy displayed in Fig. 12.4, a specialized solution method—for instance a numerical solver for an underlying ordinary differential equation—can be used to evaluate the dynamics of the system for given control sequences $u(\cdot)$. These control sequences correspond to values that are required by the solver for problem (NLP). Hence, the interaction between these two components consists in sending control sequences $u(\cdot)$ and initial values x_0 from the (NLP) solver to the solver of the dynamics which in turn sends computed state sequences $x_u(\cdot, x_0)$ back to the (NLP) solver, cf. Fig. 12.5.

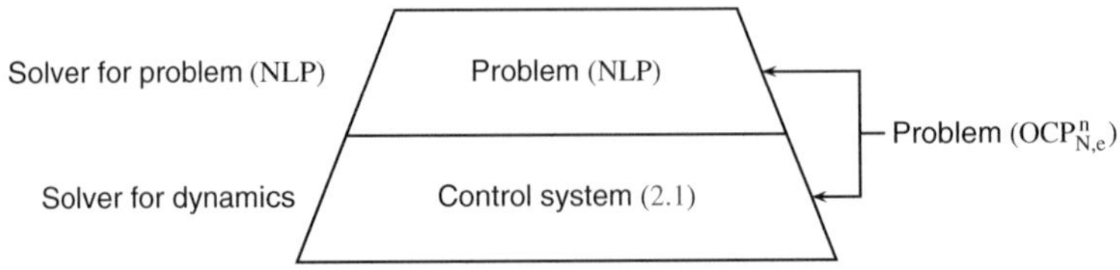

Fig. 12.4 Hierarchy

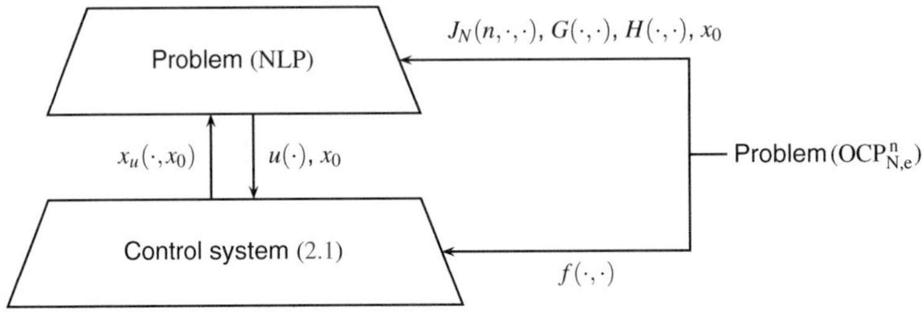

Fig. 12.5 Communication of data between the elements of the computing hierarchy

Formally, the optimization variable z reduces to

$$z := (u(0)^\top, \ldots, u(N-1)^\top)^\top \tag{12.7}$$

The constraint functions $G_i^S : \mathbb{R}^d \times \mathbb{R}^m \to \mathbb{R}, i \in \mathscr{E}^S$, and $H_i^S : \mathbb{R}^d \times \mathbb{R}^m \to \mathbb{R}, i \in \mathscr{I}^S$ according to (12.3), (12.4) can be evaluated after computing the solution $x_u(\cdot, x_0)$ by the solver for the dynamics. This way we do not have to consider Eqs. (12.1), (12.2) in the constraint function G. Hence, the equality constraints in (NLP) are given by $G = [G_i^S]$ with G_i^S from (12.3). The inequality constraints H which are given by (12.4) and the cost function F from (12.6) remain unchanged compared to the full discretization. In total, the problem with recursive elimination (OCP$^{\mathrm{n}}_{\mathrm{N,e}}$) takes the form

$$\begin{aligned}
&\text{minimize} \quad F(z) := \sum_{k=0}^{N-1} \omega_{N-k}\ell(n+k, x_u(k,x_0), u(k)) \\
&\qquad\qquad\qquad\qquad + F_J(n+N, x_u(N,x_0)) \\
&\text{with respect to } z := (u(0)^\top, \ldots, u(N-1)^\top)^\top \in \mathbb{R}^{n_z} \\
&\text{subject to } G(z) = \left[G_i^S(x_u(k,x_0), u(k))\right]_{i\in\mathscr{E}^S, k\in K_i} = 0 \\
&\text{and } H(z) = \left[H_i^S(x_u(k,x_0), u(k))\right]_{i\in\mathscr{I}^S, k\in K_i} \geq 0
\end{aligned}$$

Taking a look at the dimensions of the optimization variable and the equality constraints, we see that using the recursive elimination the optimization variable consists of $N \cdot m$ scalar components and the number of equality constraints is reduced to the number of conditions in (12.3), that is, $r_g := \sum_{i\in\mathscr{E}^S} \sharp K_i$. Hence, regarding the number of optimization variables and constraints, this discretization is optimal.

Still, this method has some drawbacks compared to the full discretization. As we will see in the following sections, the algorithms for solving (NLP) proceed iteratively, i.e., starting from an initial guess z_0 they compute a sequence z_k converging to the minimizer $z^\star$. The convergence behavior of this iteration can be significantly improved by providing a good initial guess z_0 to the algorithm. If, for instance, the initial value x_0 is close to the reference solution x^{ref}, then x^{ref} itself is likely to be such a good initial guess. However, since in the recursive elimination the trajectory $x_u(k, x_0)$ is not a part of the optimization variable z, there is no easy way to use this information.

Another drawback is that the solution $x_u(k, x_0)$ may depend very sensitively on the control sequence $u(\cdot)$, in particular when N is large. For instance, a small change in $u(0)$ may lead to large changes in $x_u(k, x_0)$ for large k and consequently in $F(z)$, $G(z)$, and $H(z)$, which may cause severe problems in the iterative optimization algorithm. In the full discretization, the control value $u(0)$ only affects $x_u(0, x_0)$ and thus those entries in F, G, and H corresponding to $k = 0$, i.e., the functions $F(z)$, $G(z)$, and $H(z)$ depend much less sensitively on the optimization variables.

For these reasons, we now present a third method, which can be seen as a compromise between the full discretization and the recursive elimination.

Multiple Shooting

The idea of the so-called multiple shooting as introduced in Bock [5] is derived from the solution of boundary value problems of differential equations, see, e.g., Stoer and Bulirsch [40]. Within these boundary value problems one tries to find initial values for trajectories which satisfy given terminal conditions. The term shooting has its origins in the similarity of this problem to a cannoneers problem of finding the correct setting for a cannon to hit a target.

The idea of this discretization is to include some components of some state vectors $x_u(k, x_0)$ as independent optimization variables in the problem. These variables are treated just as in the full discretization except that we do not do this for all $k \in \{0, \ldots, N-1\}$ but only for some time instants and that we do not necessarily include all components of the state vector $x_u(k, x_0)$ as additional optimization variables but only some. These new variables are called the *shooting nodes* and the corresponding times will be called the *shooting times*.

Proceeding this way, we may then provide useful information, e.g., from the reference trajectory $x^{\text{ref}}(\cdot)$ as described at the end of the discussion of the recursive elimination for obtaining a good initial guess for the iterative optimization. Much like the cannoneer we aim at hitting the reference trajectory, the only difference is that we do not only want to hit the target at the end of the (finite) horizon but to stay close to it for the entire time interval which we consider within the optimization.

This motivates to set the state components at the shooting nodes to a value which may violate the dynamics of the system (2.1) but is closer to the reference trajectory. This situation is illustrated in Fig. 12.6 and gives a good intuition why the multiple shooting initial guess is preferable from an optimization point of view.

Unfortunately, we have to give up the integrity of the dynamics of the system to achieve this improvement in the initial guess, i.e., the trajectory pieces starting

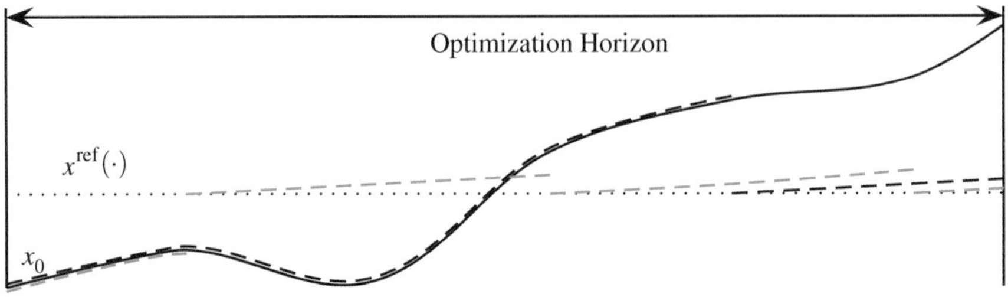

Fig. 12.6 Resulting trajectories for initial guess u using no multiple shooting nodes (*solid*), one shooting node (*dashed*) and three shooting nodes (*gray dashed*)

in the respective shooting nodes can in general not be "glued" together to form a continuous trajectory. In order to solve this problem, we have to include additional equality constraints similar to (12.1), (12.2).

For the formal description of this method, we denote the vector of multiple shooting nodes by $s := (s_1, \ldots, s_{r_s}) \in \mathbb{R}^{r_s}$ where s_i is the new optimization variable corresponding to the ith multiple shooting node. The shooting times $\varsigma : \{1, \ldots, r_s\} \to \{0, \ldots, N\}$ and indices $\iota : \{1, \ldots, r_s\} \to \{1, \ldots, d\}$ then define the time and the component of the state vector corresponding to s_i via

$$x_u(\varsigma(j), x_0)_{\iota(j)} = s_j.$$

This means that the components $x_u(k, x_0)_i$, $i = 1, \ldots, d$, of the state vector are determined by the iteration

$$x_u(k+1, x_0)_i = s_j$$

if there exists $j \in \{1, \ldots, r_s\}$ with $\varsigma(j) = k+1$ and $\iota(j) = i$ and

$$x_u(k+1, x_0)_i = f(x_u(k, x_0), u(k))_i$$

with initial condition $x_u(0, x_0)_i = s_j$ if there exists $j \in \{1, \ldots, r_s\}$ with $\varsigma(j) = 0$ and $\iota(j) = i$ and $x_u(0, x_0)_i = (x_0)_i$, otherwise. The shooting nodes s_i now become part of the optimization variable z which hence reads

$$z := (u(0)^\top, \ldots, u(N-1)^\top, s^\top)^\top. \tag{12.8}$$

As in the full discretization we have to ensure that the optimal solution of (NLP) leads to values s_i for which the state vectors $x_u(k, x_0)$ thus defined form a trajectory of (2.1). To this end, we define the continuity condition for all shooting nodes s_j, $j \in \{1, \ldots, r_s\}$ with $\varsigma(j) \geq 1$ analogously to (12.1) as

$$s_j - f\,(x_u(\varsigma(j) - 1, x_0), u(\varsigma(j) - 1)))_{\iota(j)} = 0, \tag{12.9}$$

and for all $j \in \{1, \ldots, r_s\}$ with $\varsigma(j) = 0$ analogously to (12.2) as

$$s_j - (x_0)_{\iota(j)} = 0. \tag{12.10}$$

These so-called shooting constraints are included as equality constraints in (NLP). Since the conditions (12.9) and (12.10) are already in the form of equality constraints which we require for our problem (NLP), we can achieve this by defining G to consist of the equalities (12.3), (12.9), and (12.10). As for the recursive elimination, the set of inequality constraints as well as the cost function are still identical to those given by the full discretization. As a result, the multiple shooting discretization of problem ($\mathrm{OCP}^{\mathrm{n}}_{\mathrm{N,e}}$) is of the form

$$\begin{aligned}
&\text{minimize} \quad F(z) := \sum_{k=0}^{N-1} \omega_{N-k}\ell(n+k, x_u(k,x_0), u(k)) \\
&\qquad\qquad\qquad\qquad + F_J(n+N, x_u(N,x_0)) \\
&\text{with respect to } z := (u(0)^\top, \ldots, u(N-1)^\top, s^\top)^\top \in \mathbb{R}^{n_z} \text{ subject to} \\
&G(z) = \begin{bmatrix} \left[G_i^S(x_u(k,x_0), u(k))\right]_{i\in\mathcal{E}^S, k\in K_i} \\ \left[s_j - f\left(x_u(\varsigma(j)-1, x_0), u(\varsigma(j)-1)\right)\right)_{\iota(j)}\right]_{j\in\{1,\ldots,r_s\},\varsigma(j)\geq 1} \\ \left[s_j - (x_0)_{\iota(j)}\right]_{j\in\{1,\ldots,r_s\},\varsigma(j)=0} \end{bmatrix} = \\
&0 \\
&\text{and } H(z) = \left[H_i^S(x_u(k,x_0), u(k))\right]_{i\in\mathcal{I}^S, k\in K_i} \geq 0
\end{aligned}$$

Comparing the size of the optimization problem discretized by multiple shooting to the one with recursive elimination, we see that the dimension of the optimization variable and the number of equality constraints is increased by r_s. An appropriate choice of this number as well as for the values of the shooting nodes and times is crucial in order to obtain an improvement of the NMPC closed loop, as the following example shows.

Example 12.2 Consider the inverted pendulum on a cart problem from Example 12.1. For this example, numerical experience shows that the most critical trajectory component is the angle of the pendulum x_1. In the following, we discuss and illustrate the efficient use and effect of multiple shooting nodes for this variable.

(i) If we define every sampling instant in every dimension of the problem to be a shooting node, then multiple shooting coincides with the full discretization. Proceeding this way slows down the optimization process significantly due to the enlarged dimension of the optimization variable. Unless the computational burden can be reduced by exploiting the special structure of the shooting nodes in the optimization algorithm, cf. the paragraph on condensing on Sect. 12.4, below, this implies that the number of shooting nodes should be chosen as small as possible. On the other hand, using shooting nodes we may be able to significantly improve the initial guess of the control. Therefore, in terms of the computing time, a balance between improving the initial guess and the number of multiple shooting nodes must be found.

(ii) If the state trajectories evolve slowly, i.e., in regions where the dynamics is slow, the use of shooting nodes may obstruct the optimization since there may not exist a trajectory which links two consecutive shooting nodes. In this case, the optimization routine wastes iteration steps trying to adapt the value of the shooting nodes and may be unable to find a solution. Ideally, the shooting nodes are chosen close to the optimal transient trajectory, which is, however, usually not known at runtime. Using shooting nodes on the reference, instead, may be a good substitute but only if the initial value is sufficiently close to the reference or if the shooting times are chosen sufficiently large in order to enable the trajectory to reach a neighborhood of the reference.

(iii) Using inappropriate shooting nodes may not only render the optimization routine slow but may lead to unstable solutions even if the problem without shooting nodes showed stable behavior. On the other hand, choosing good shooting nodes may have a stabilizing effect.

In the following, we illustrate the effects mentioned in (iii) for the horizons $N = 14$, 15 and 20. We compute the NMPC closed-loop trajectories for the inverted pendulum on a cart problem on the interval $[0, 20]$ where in each optimization we use one shooting node for the first state dimension x_1 at different times $\varsigma(1) \in \{0, \ldots, N-1\}$, and the corresponding initial value to $x_1 = s_1 = -\pi$. In a second test, we use two shooting nodes for the first state dimension with different $(\varsigma(1), \varsigma(2)) \in \{0, \ldots, N-1\}^2$ again with $s_1 = s_2 = -\pi$. Here, the closed-loop costs in the following figures are computed by numerically evaluating

$$\sum_{k=0}^{200} \ell(x_{\mu_N}(k, x_0), \mu_N(x_{\mu_N}(k, x_0)))$$

in order to approximate J_∞^{cl} from Definition 4.10. Darker colors in the Figs. 12.7b–12.9b indicate higher closed-loop costs.

As we can see from Fig. 12.7a, the state trajectory is stabilized for $N = 14$ if we add a shooting node for the first differential equation. Hence, using a single shooting node, we are now able to stabilize the problem for a reduced optimization horizon N. Yet, the stabilized equilibrium is not identical for all values $\varsigma(1)$, i.e., for $\varsigma(1) \in \{t_0, \ldots, t_2\}$ the equilibrium $(\pi, 0, 0, 0)$ is chosen which in our case corresponds to larger closed-loop costs in comparison to the solutions approaching $(-\pi, 0, 0, 0)$. Similarly, all solutions converge to an upright equilibrium if we use two shooting nodes. Here, Fig. 12.7b shows the corresponding closed-loop costs which allow us to see that for small values of $\varsigma(i)$, $i = 1, 2$, the x_1 trajectory converges toward $(\pi, 0, 0, 0)$.

As we see, for $N = 14$ the shooting nodes may change the closed-loop behavior. A similar effect happens for the horizon $N = 15$. For the case without shooting nodes, the equilibrium $(-\pi, 0, 0, 0)$ is stabilized, cf. Fig. 12.2. If shooting nodes are considered, we obtain the results displayed in Fig. 12.8.

Here, we observe that choosing the shooting time $\varsigma(1) \in \{0, 1, 3\}$ results in stabilizing $(\pi, 0, 0, 0)$ while for all other cases the x_1 trajectory converges toward $(-\pi, 0, 0, 0)$. Still, for $\varsigma(1) \in \{5, \ldots, 9\}$ the solutions differ significantly from the

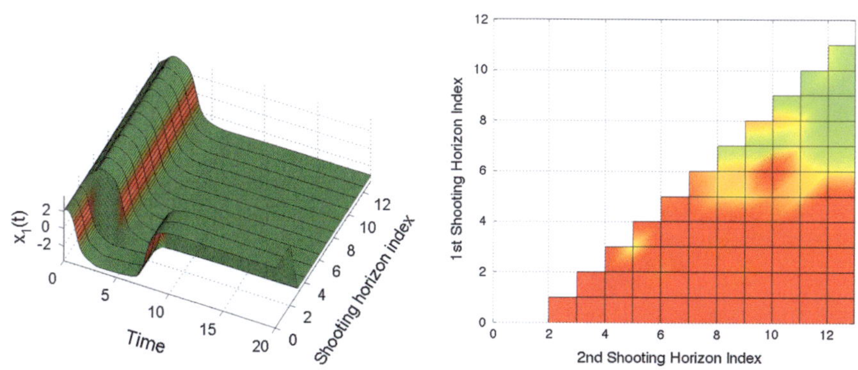

(a) $x_1(\cdot)$ trajectories using one shooting node
(b) Closed-loop costs using two shooting nodes

Fig. 12.7 Results for a varying shooting nodes for horizon length $N = 14$

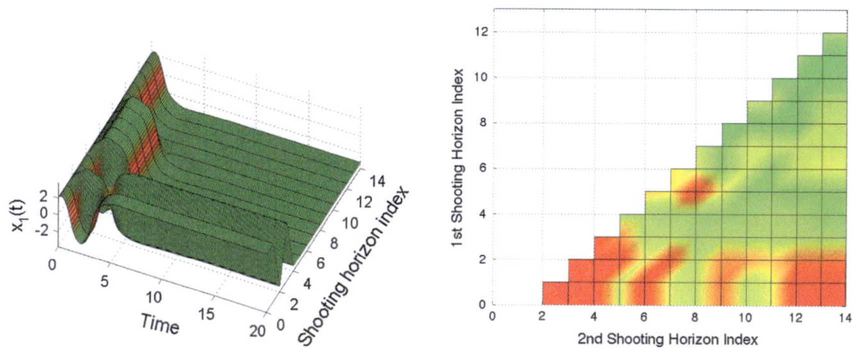

(a) $x_1(\cdot)$trajectories using one shooting node
(b) Closed-loop costs using two shooting nodes

Fig. 12.8 Results for a varying shooting nodes for horizon length $N = 15$

solution without shooting nodes which also affects the closed-loop costs. As indicated by Fig. 12.8b, a similar effect can be experienced if more than one shooting node is used. Hence, by our numerical experience, the effect of stabilizing a chosen equilibrium by adding a shooting node can only be confirmed if $\varsigma(1)$ is set to a time instant close to the end of the optimization horizon.

This is further confirmed by Fig. 12.9 which illustrates the results for $N = 20$. Here, one also sees that adding shooting nodes which may also lead to instability of the closed loop.

Moreover, we like to stress the fact that adding shooting nodes to a problem may cause the optimization routine to stabilize a different equilibrium than intended. Although all equilibria with $x_1 = (2k + 1)\pi, k \in \mathbb{Z}$, correspond to the same physical upright position and yield the same value in the stage cost, one may expect that setting

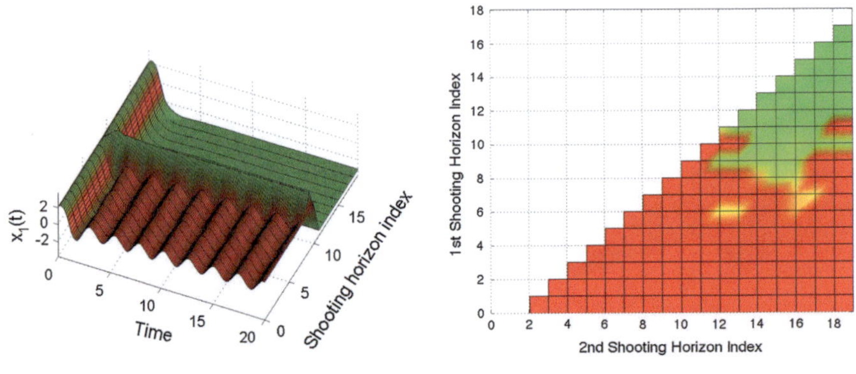

(a) $x_1(\cdot)$ trajectories using one shooting node

(b) Closed-loop costs using two shooting nodes

Fig. 12.9 Results for a varying shooting nodes for horizon length $N = 20$

shooting nodes to the value $x_1 = -\pi$ forces the closed-loop trajectory to approach this particular equilibrium. As we have seen in Figs. 12.7a, 12.8a and 12.9a, this is not always the case since some trajectories approach $(\pi, 0, 0, 0)$, even though this leads to higher closed-loop costs. In particular, Fig. 12.8 shows nicely that even though the equilibrium $(-\pi, 0, 0, 0)$ is stabilized without using shooting nodes, fixing a shooting node $x_1 = -\pi$ obstructs the computation of the optimal solution. Hence, the values of the shooting nodes have to be selected with care since different choices may result in closed-loop trajectories with different costs and computing times and may have both stabilizing or destabilizing effects.

12.2 Unconstrained Optimization

Now that we have discretized the optimal control problem ($\text{OCP}^{\text{n}}_{\text{N,e}}$) and transformed it into a nonlinear optimization problem (NLP) in standard form, our aim is to compute a minimizer z, which then gives us an optimal control u for our original problem. In this section, we discuss the foundations of all optimization techniques; details for methods like the popular Sequential Quadratic Programming (SQP) or Interior-Point Methods (IPM) are then given in the subsequent section. On the one hand, this allows us to characterize the main principles of such algorithms. On the other hand, it reveals an abstract method, exposes the computationally expensive parts of such algorithms and allows for rearranging the ordering of these steps to reduce the computational cost.

Although the problem (NLP) is actually a constrained optimization problem, in this section we present solution methods for dealing with unconstrained optimization problems since the basic principles, as we will see in Sect. 12.3, are similar to those in constrained optimization. Since there do not exist any restrictions on the optimization variable z, the unconstrained optimization problem is a special case of the standard (NLP) problem and is given by

$$\begin{array}{|ll|}\hline \text{minimize} & F(z) \\ \text{with respect to } z \in \mathbb{R}^{n_z} & \\ \hline\end{array}$$

Due to the sheer size of such a problem, in practice we need to solve it on a computer. The computer, however, cannot deal with this problem in an abstract way, it can only evaluate the given functions like the cost function F and possibly its derivative for finitely many z. Hence, the goal in constructing a solution method for any (NLP) problem is to find a strategy for choosing these evaluation points in order to reliably identify a solution of (NLP). While the identification of and the search strategy for a solution are primary goals for a solution method, one still has to keep in mind some secondary goals like computing time, required memory storage, and the number of required function evaluations whose detailed analysis is beyond the scope of this chapter but can be found in the standard literature for nonlinear optimization.

Focussing on the primary goals, we first need to characterize a solution of a problem (NLP) and how it can be checked whether a given point $z \in \mathbb{R}^{n_z}$ is a solution. In principle, we are interested in so-called global minimizers.

Definition 12.3 A point $z^\star \in \mathbb{R}^{n_z}$ is a *global minimizer* of the function $F : \mathbb{R}^{n_z} \to \mathbb{R}$ if $F(z^\star) \leq F(z)$ holds for all $z \in \mathbb{R}^{n_z}$.

Unfortunately, for general nonlinear—and in particular nonconvex—problems such global minimizers are hard to find in practice since we only have local knowledge of the function F and its derivative $\frac{d}{dz}F$. Due to this local knowledge and our intention to evaluate only a small number of vectors z, we cannot cover the entire definition space of F. As a consequence, if we construct an algorithm under these restrictions we can never be sure if we reached a global minimizer. Nevertheless we are often able to identify a so-called local minimizer.

Definition 12.4 A point $z^\star \in \mathbb{R}^{n_z}$ is a *local minimizer* of the function $F : \mathbb{R}^{n_z} \to \mathbb{R}$ if there exists a neighborhood $\mathcal{N}$ of $z^\star$ such that $F(z^\star) \leq F(z)$ holds for all $z \in \mathcal{N}$.

In some cases not all hopes for identifying a global minimizer are lost since we may have additional information on the function F. For example, if F is convex, then every local minimizer is also a global minimizer. But even if we know that F is convex, we still need to find such a minimizer. In order not to have to check all values z in a certain area, we will assume the function F to be at least twice continuously differentiable, which allows us to use a more practicable way of locating a minimizer using Taylor's theorem. Here and in the following we denote derivatives using the following notation which is common in nonlinear optimization. For a continuously differentiable function $g = (g_1, \ldots, g_p) : \mathbb{R}^{n_z} \to \mathbb{R}^p$ we use the gradient notation for the *Jacobian matrix*

$$\nabla_z g(z) = \begin{pmatrix} \frac{\partial g_1}{\partial z_1} & \cdots & \frac{\partial g_p}{\partial z_1} \\ \vdots & & \vdots \\ \frac{\partial g_1}{\partial z_n} & \cdots & \frac{\partial g_p}{\partial z_n} \end{pmatrix}$$

which we abbreviate to ∇g if there is no ambiguity. For a twice continuously differentiable function $g : \mathbb{R}^{n_z} \to \mathbb{R}$ we write the so-called *Hessian* as

$$\nabla^2_{zz} g(z) = \begin{pmatrix} \frac{\partial^2 g}{\partial z_1 z_1} & \cdots & \frac{\partial^2 g}{\partial z_1 z_{n_z}} \\ \vdots & & \vdots \\ \frac{\partial^2 g}{\partial z_{n_z} z_1} & \cdots & \frac{\partial^2 g}{\partial z_{n_z} z_{n_z}} \end{pmatrix}$$

which we abbreviate to $\nabla^2 g$ if there is no danger of confusion.

Theorem 12.5 *(Taylor's theorem for order 1 and 2) Consider a function $F : \mathbb{R}^{n_z} \to \mathbb{R}$ which is continuously differentiable and a direction vector $d \in \mathbb{R}^{n_z}$. Then we have*

$$F(z + d) = F(z) + \nabla F(z + td)^\top d \tag{12.11}$$

for some $t \in (0, 1)$. If F is twice continuously differentiable, then we also have

$$F(z + d) = F(z) + \nabla F(z)^\top d + \frac{1}{2} d^\top \nabla^2 F(z + td) d \tag{12.12}$$

for some $t \in (0, 1)$.

Proof Using the fundamental theorem of calculus, we have

$$F(z + d) = F(z) + \int_0^1 \frac{d}{dt} F(z + td) dt.$$

By the mean value theorem, there exist a $t \in (0, 1)$ with

$$\int_0^1 \frac{d}{dt} F(z + td) dt = \frac{d}{dt} F(z + td) = \nabla F(z + td)^\top d,$$

where we used the chain rule for the second equality. This shows (12.11). By partial integration, we further obtain

$$\int_0^1 \frac{d}{dt} F(z + td) dt = \left. \frac{d}{dt} \right|_{t=0} F(z + dt) + \int_0^1 (1 - t) \frac{d^2}{dt^2} F(z + td) dt$$

and again using the mean value theorem we get

$$\int_0^1 (1 - t) \frac{d^2}{dt^2} F(z + td) dt = \frac{d^2}{dt^2} F(z + t'd) \int_0^1 (1 - t) dt = \frac{1}{2} \frac{d^2}{dt^2} F(z + t'd)$$

for some $t' \in (0, t)$. Since by the chain rule we have

$$\left.\frac{d}{dt}\right|_{t=0} F(z+dt) = \nabla F(z)^\top d \quad \text{and} \quad \frac{1}{2}\frac{d^2}{dt^2}F(z+t'd) = \frac{1}{2}d^\top \nabla^2 F(z+t'd)d$$

this shows (12.12). □

The advantage of Taylor's theorem is that it allows us to introduce knowledge on the gradient $\nabla F(z^\star)$ and the Hessian $\nabla^2 F(z^\star)$ into the search for a local minimizer $z^\star$. In particular, first-order necessary conditions are derived very easily.

Theorem 12.6 *Consider a vector $z^\star \in \mathbb{R}^{n_z}$ and a function $F : \mathbb{R}^{n_z} \to \mathbb{R}$ where F is continuously differentiable in an open neighborhood of $z^\star$ and $z^\star \in \mathbb{R}^{n_z}$ is a local minimizer of F. Then we have $\nabla F(z^\star) = 0$.*

Proof Suppose $\nabla F(z^\star) \neq 0$ and set $d := -\nabla F(z^\star)$. Then, we get $d^\top \nabla F(z^\star) = -\|\nabla F(z^\star)\|^2 < 0$. Since ∇F is continuous in a neighborhood of $z^\star$, there exists a scalar $T > 0$ such that $d^\top \nabla F(z^\star + td) < 0$ holds for all $t \in [0, T]$. By (12.11), for any $\bar{t} \in (0, T]$ we have $F(z^\star + \bar{t}d) = F(z^\star) + \bar{t}d^\top \nabla F(z^\star + td)$ for some $t \in (0, \bar{t})$. This implies $F(z^\star + \bar{t}d) < F(z^\star)$ for all $\bar{t} \in (0, T]$ which contradicts the local minimizer property of $z^\star$. □

In a similar manner, information on the Hessian can be used to derive second-order necessary conditions from Eq. (12.12).

Theorem 12.7 *Consider a vector $z^\star \in \mathbb{R}^{n_z}$ and a function $F : \mathbb{R}^{n_z} \to \mathbb{R}$ where F is twice continuously differentiable in an open neighborhood of $z^\star$ and $z^\star \in \mathbb{R}^{n_z}$ is a local minimizer of F. Then we have $\nabla F(z^\star) = 0$ and the Hessian $\nabla^2 F(z^\star)$ is positive semidefinite.*

Proof From Theorem 12.6 we know that $\nabla F(z^\star) = 0$. Now, suppose $\nabla^2 F(z^\star)$ is not positive semidefinite and choose a vector d such that $d^\top \nabla^2 F(z^\star)d < 0$ holds. Using continuity of $\nabla^2 F(z^\star)$ in a neighborhood of $z^\star$, we know that there exists a scalar $T > 0$ such that $d^\top \nabla^2 F(z^\star + td)d < 0$ holds for all $t \in [0, T]$. Hence, using (12.12), for any $\bar{t} \in (0, T]$ and some $t \in (0, \bar{t})$ we obtain

$$F(z^\star + \bar{t}d) = F(z^\star) + \bar{t}\nabla F(z^\star)^\top d + \frac{1}{2}\bar{t}d^\top \nabla^2 F(z^\star + td)d\bar{t} < F(z^\star).$$

Similar to the proof of Theorem 12.6, F is strictly decreasing along the direction d which contradicts the local minimizer property of $z^\star$. □

The results from Theorems 12.6 and 12.7 reveal guidelines to what we are looking for, i.e., which properties a local minimizer must fulfill. However, these results cannot be used to identify a local minimizer once we have found a candidate satisfying the previous conditions. In order to perform such a check, the following theorem can be used.

Theorem 12.8 *Consider a vector* $z^\star \in \mathbb{R}^{n_z}$ *and a function* $F : \mathbb{R}^{n_z} \to \mathbb{R}$ *where* F *is twice continuously differentiable in an open neighborhood of* $z^\star$. *If* $\nabla F(z^\star) = 0$ *and* $\nabla^2 F(z^\star)$ *is positive definite, then* $z^\star$ *is a local minimizer of* F.

Proof Due to F being twice continuously differentiable there exists a radius $r > 0$ such that $\nabla^2 F(z)$ is positive definite for all $z \in \{z \mid \|z - z^\star\| < r\}$. Now take any vector $d \in \mathbb{R}^{n_z}$ with $\|d\| < r$, then we have $z^\star + d \in \{z \mid \|z - z^\star\| < r\}$ and

$$\begin{aligned} F(z^\star + d) &= F(z^\star) + d^\top \nabla F(z^\star) + \frac{1}{2} d^\top \nabla^2 F(z^\star + td)d \\ &= F(z^\star) + \frac{1}{2} d^\top \nabla^2 F(z^\star + td)d \end{aligned}$$

for some $t \in (0, 1)$. Since $(z^\star + td) \in \{z \mid \|z - z^\star\| < r\}$, we have $d^\top \nabla^2 F(z^\star + td)d > 0$ and therefore $F(z^\star + d) > F(z^\star)$ holds showing the assertion. □

Before we consider constraints we give a short description of the standard strategies for nonlinear optimization, the line-search and the trust-region strategy. Both methods have in common that they approximately compute local minimizers by iteratively computing values z_k converging to $z^\star$. Hence, an initial guess z_0 needs to be supplied by the user for starting the iteration. A good initial guess, i.e., a vector close to a minimizer, can usually only be obtained by utilizing knowledge on the process. If such knowledge is not at hand, the starting point can be chosen arbitrarily, however, the convergence speed of the sequence z_k toward a minimizer is drastically reduced in general.

Within the line-search strategy, the approximation method computes a direction d_k in its kth step and searches along the vector d_k starting from the current iterate z_k for a new iterate $z_{k+1} = z_k + \alpha_k d_k$ with lower cost function value $F(z_{k+1})$. Here, the direction d_k is typically obtained from minimizing a model function m_k which catches the local behavior of the cost function F at the current iterate z_k and is easy to minimize numerically. Often, quadratic functions of the form

$$m_k(z_k + d_k) = F(z_k) + d_k^\top \nabla F(z_k) + \frac{1}{2} d_k^\top B_k d_k$$

are used for this purpose, where B_k is either the Hessian $\nabla^2 F(z_k)$ or an approximation of it. The corresponding step length $\alpha_k > 0$ can then, e.g., be determined by solving the one-dimensional minimization problem

$$\min_{\alpha_k > 0} F(z_k + \alpha_k d_k). \tag{12.13}$$

Once an (approximated) solution to the step length problem has been found, a new search direction and a new step length are computed and the scheme is applied iteratively.

In contrast to the line-search approach, the trust-region method takes the only local approximation properties of the model function m_k into account when minimizing this function in order to determine d_k. Since m_k can only be guaranteed to be a good approximation close to z_k—i.e., it can only be "trusted" in a neighborhood of z_k—the search region for a minimizer of m_k is restricted to a so-called trust-region which is usually given by a ball $\mathscr{B}_\Delta(z_k)$. Hence, the problem consists in computing a suitable next iteration candidate by solving

$$\min_{d_k} m_k(z_k + d_k) \qquad \text{where } z_k + d_k \in \mathscr{B}_\Delta(z_k). \tag{12.14}$$

If the candidate $z_k + d_k$ does not show a sufficient decrease in the cost function F, then the trust-region is considered to be too large. Hence, the radius Δ is reduced and the new minimization problem (12.14) is solved again. Like in the line-search approach, the model function m_k in (12.14) is often generated by quadratic functions.

Taking an abstract look on both line-search and trust-region method, the difference between those two approaches lies in the ordering of the basic steps, i.e., finding a search direction and a suitable step length. While the line-search method fixes the search direction first and then computes a step length α_k, the trust-region method first defines the maximal step length Δ and then searches for a minimizer using the model m_k.

12.3 Constrained Optimization

So far we have dealt with unconstrained optimization problems and shown the fundamental results and basic algorithmic ideas which can be used to solve such problems. In the NMPC algorithm, however, we face the constrained nonlinear optimization problem

$$\boxed{\begin{array}{ll} \text{minimize} & F(z) \\ \text{with respect to } z \in \mathbb{R}^{n_z} & \\ \text{subject to } G_i(z) = 0 \text{ for all } i \in \mathscr{E} \text{ and } H_i(z) \geq 0 \text{ for all } i \in \mathscr{I} & \end{array}} \tag{NLP}$$

in every step of the NMPC iteration where the functions F, G, and H are defined by one of the discretizations of the problem ($\mathrm{OCP}^{\mathrm{n}}_{\mathrm{N,e}}$) described in Sect. 12.1. Note that all three discretizations lead to a problem of type (NLP), hence all subsequent results hold for either of these discretizations.

The index sets in (NLP) are given by $\mathscr{E} = \{1, \ldots, r_g\}$ and $\mathscr{I} = \{r_g + 1, \ldots, r_g + r_h\}$, respectively, and the functions G_i and H_i are called equality and inequality constraints, respectively. These constraints induce the following feasible set, which will be important for our upcoming analysis.

Definition 12.9 For a problem (NLP) the set

$$\Omega = \{z \mid G_i(z) = 0, \; i \in \mathscr{E}; \; H_i(z) \geq 0, \; i \in \mathscr{I}\} \tag{12.15}$$

is called the *feasible set* and the elements $z \in \Omega$ are called *feasible points*.

Since a minimizer for the problem (NLP) has to be an element of Ω by definition, we have to modify the definition of a local minimizer in the context of constrained optimization problems which we want to approximate later:

Definition 12.10 A point $z^\star \in \mathbb{R}^{n_z}$ is a *local minimizer* of the problem (NLP) if there exists a neighborhood $\mathcal{N}$ of $z^\star$ such that $F(z^\star) \leq F(z)$ holds for all $z \in \mathcal{N} \cap \Omega$.

In a similar way as for unconstrained optimization problems, we now want to derive necessary and sufficient conditions which will allow us to construct numerical methods to compute a local minimizer $z^\star$ of a problem (NLP). As we have seen in the previous Sect. 12.2, the mathematical background of the necessary and sufficient conditions given in Theorems 12.6, 12.7 and 12.8 is Taylor's Theorem 12.5 stating results for a linear or quadratic approximation. In constrained optimization, the functions G and H will now also be replaced by suitable approximations. Here, we use linear approximations

$$G(z+d) \approx G(z) + \nabla G(z)^\top d \quad \text{and} \quad H(z+d) \approx H(z) + \nabla H(z)^\top d$$

for this purpose. This, however, only makes sense if the geometry of the feasible set Ω is—at least locally—reflected properly when G and H are replaced by approximations. To this end so-called constraint qualifications are imposed, which we are going to deal with next.

Before we state the popular linear independent constraint qualification (LICQ), we need some definitions. We first introduce the tangent cone $T_\Omega(z)$ to the feasible set Ω.

Definition 12.11 A vector $v \in \mathbb{R}^{n_z}$ is called *tangent vector* to Ω at a point $z \in \Omega$ if there exists a sequence of feasible points $(z_k)_{k\in\mathbb{N}}$ with $z_k \to z$, $z_k \in \Omega$ and a sequence of positive scalars $(t_k)_{k\in\mathbb{N}}$ with $t_k \to 0$ such that

$$\lim_{k\to\infty} \frac{z_k - z}{t_k} = v \tag{12.16}$$

holds. The set of all tangent vectors to Ω at z is called the *tangent cone* and is denoted by $T_\Omega(z)$.

Note that T_Ω only depends on the geometry of Ω. The set $T_\Omega(z)$ can be seen as a local approximation of all feasible directions at a given feasible point $z \in \Omega$. The feasible directions are all vectors $d \in \mathbb{R}^{n_z}$ for which $z + \alpha d \in \Omega$ holds for all sufficiently small $\alpha > 0$ and the definition of T_Ω implies that each feasible direction is contained in $T_\Omega(z)$. Conversely, for each element $v \in T_\Omega(z)$ and each $\varepsilon > 0$ there exists a feasible direction d with $\|d - v\| < \varepsilon$.

Obviously, all equality constraints G_i restrict these feasible directions but not necessarily all inequality constraints: if $H_i(z) > 0$ holds, then since H_i is continuous we get $H_i(z + \alpha d) > 0$ for all $d \in \mathbb{R}^{n_z}$ provided $\alpha > 0$ is sufficiently small. If, however, $H_i(z) = 0$ holds, then an arbitrarily small change of z in the "wrong" direction may lead to $H_i(z + \alpha d) < 0$. Hence, all inequality constraints H_i with $H_i(z) = 0$ and all equality constraints G_i may restrict feasible moving directions. These constraints are called *active* and their indices are characterized by the following definition.

Definition 12.12 The *active set* $\mathscr{A}(z)$ at any feasible point z consists of the equality constraint indices from $\mathscr{E}$ together with the indices of the inequality constraints $i \in \mathscr{I}$ where $H_i(z) = 0$ holds, that is, $\mathscr{A}(z) := \mathscr{E} \cup \{i \in \mathscr{I} \mid H_i(z) = 0\}$.

Using the active set we can now define a set of "linearized" feasible directions obtained from the linearizations of H.

Definition 12.13 For a feasible point $z \in \Omega$ and the active set $\mathscr{A}(z)$ we call the set

$$\mathscr{F}(z) = \left\{ v \in \mathbb{R}^{n_z} \;\middle|\; \begin{array}{l} v^\top \nabla G_i(z) = 0 \text{ for all } i \in \mathscr{E} \text{ and} \\ v^\top \nabla H_i(z) \geq 0 \text{ for all } i \in \mathscr{A}(z) \cap \mathscr{I} \end{array} \right\} \tag{12.17}$$

the *set* (or cone) *of linearized feasible directions*.

Note that in general we have $T_\Omega(z) \subseteq \mathscr{F}(z)$, see Fletcher [20, Lemma 9.2.1]. For the proof of necessary optimality conditions based on the linearizations of the G_i and H_i as well as for using the linearized G_i and H_i in our algorithms it is now important to see that these sets coincide. The idea of constraint qualifications is to guarantee that these sets indeed coincide, i.e., that the geometry of T_Ω is captured by the linearizations of G_i and H_i. Although there is quite a range of different constraint qualifications, see, e.g., the book of Mangasarian [30], the linear independence constraint qualification is probably the most popular one.

Definition 12.14 Consider a feasible point z and the active set $\mathscr{A}(z)$. Suppose that F, G and H are continuously differentiable. If the elements of the gradient set $\{\nabla G_i(z) \mid i \in \mathscr{E}\} \cup \{\nabla H_i(z) \mid i \in \mathscr{A}(z) \cap \mathscr{I}\}$ are linearly independent then we say that the *linear independence constraint qualification* (LICQ) holds.

Under this condition we obtain $T_\Omega(z) = \mathscr{F}(z)$, see Fletcher [20, Lemma 9.2.1].

Now we want to proceed as in the unconstrained case, i.e., give characterizations of minimizers of the cost function F amongst the feasible points $z \in \Omega$. Note that in the constrained case we cannot simply use Taylor's Theorem 12.5 to conclude that if $z^\star \in \Omega$ is a local minimizer for the problem (NLP) then we have $\nabla F(z^\star) = 0$.

Example 12.15 In order to see that Taylor's Theorem 12.5 cannot be used in the constrained case, consider the example of minimizing $F(z) = z$ over $\Omega = [-1, 1]$. Obviously $z = -1$ is a local minimizer, yet we have $\nabla F(-1) = 1$.

The problem with applying Theorem 12.6 in the constrained case is that no boundaries of the constraints sets are included in the analysis in this theorem. In constrained optimization, however, we often face the situation of a minimizer lying on the boundary of the feasible set Ω, cf. Example 12.15. To deal with this matter, the following auxiliary function $L : \mathbb{R}^{n_z} \times \mathbb{R}^{r_g + r_h} \to \mathbb{R}$, the so called *Lagrangian* is introduced. For its definition, we combine the constraints G_i and H_i into one function $C : \mathbb{R}^{n_z} \to \mathbb{R}^{r_g+r_h}$ given by

$$C : z \mapsto \begin{bmatrix} (G_i(z))_{i\in\mathscr{E}} \\ (H_i(z))_{i\in\mathscr{I}} \end{bmatrix}$$

and define as a modification of the cost function F by

$$L(z, \lambda) := F(z) - \lambda^\top C(z). \tag{12.18}$$

The idea behind this definition is that the additional term $-\lambda^\top C(z)$ penalizes violations of the state constraints. The vector $\lambda \in \mathbb{R}^{r_g+r_h}$ is called Lagrange multiplier.

Similar to Theorem 12.6, we can now state a first-order necessary optimality condition—usually called *KKT (Karush–Kuhn–Tucker) condition*—in the constrained case using the Lagrangian (12.18) which will serve as a guideline to find local minimizers, see Fletcher [20, Theorem 9.1.1].

Theorem 12.16 (KKT condition) *Consider the problem* (NLP) *with local minimizer $z^\star \in \Omega$. Moreover suppose the functions F, G, and H to be continuously differentiable and the (LICQ) to hold at $z^\star$. Then there exists a Lagrange multiplier $\lambda^\star \in \mathbb{R}^{r_g+r_h}$ such that the following conditions hold.*

$$\nabla_z L(z^\star, \lambda^\star) = 0 \tag{12.19}$$
$$G_i(z^\star) = 0 \quad \forall i \in \mathscr{E} \tag{12.20}$$
$$H_i(z^\star) \geq 0 \quad \forall i \in \mathscr{I} \tag{12.21}$$
$$\lambda_i^\star \geq 0 \quad \forall i \in \mathscr{I} \tag{12.22}$$
$$\lambda_i^\star G_i(z^\star) = 0 \quad \forall i \in \mathscr{E} \tag{12.23}$$
$$\lambda_i^\star H_i(z^\star) = 0 \quad \forall i \in \mathscr{I}. \tag{12.24}$$

The identity (12.24) is a so-called complementarity condition, which says that either $\lambda_i^\star = 0$ or $H_i(z^\star) = 0$ must hold. A special case which is important for nonlinear optimization algorithms is the following.

Definition 12.17 (*Strict complementarity*) Consider the problem (NLP) with local minimizer $z^\star \in \Omega$ and Lagrange multiplier $\lambda^\star \in \mathbb{R}^{r_g+r_h}$ satisfying (12.19)–(12.24). Then we say that the *strict complementarity condition* holds if $\lambda_i^\star > 0$ for all $i \in \mathscr{I} \cap \mathscr{A}(z^\star)$.

We will use this condition when discussing interior-point methods, below.

Interpreting the KKT conditions, we see that they connect the gradient of the cost function to active constraints. In particular, Theorem 12.16 states that for a given minimizer $z^\star$ moving along an arbitrary vector $v \in \mathscr{F}(z^\star)$ either increases the value of the first-order approximation of the cost function, i.e., $v^\top \nabla F(z^\star) > 0$, or keeps its value at the same level in the case $v^\top \nabla F(z^\star) = 0$.

In the second case—the so-called "critical" case $v^\top \nabla F(z^\star) = 0$—it is unknown if the cost function value is increasing or decreasing along v. Here second-order conditions come into play and the *curvature information* can be used to obtain more information about change of F along these directions, see Fletcher [20, Theorem 9.3.1] for a corresponding proof.

Theorem 12.18 (Second-order necessary conditions) *Consider the problem* (NLP) *with local minimizer* $z^\star \in \Omega$. *Suppose the functions* F, G *and* H *to be continuously differentiable and the (LICQ) to hold at* $z^\star$. *Let* $\lambda^* \in \mathbb{R}^{r_g + r_h}$ *be a Lagrange multiplier satisfying the KKT conditions* (12.19)–(12.24). *Then the inequality*

$$v^\top \nabla^2_{zz} L(z^\star, \lambda^\star) v \geq 0 \tag{12.25}$$

holds for all

$$v \in \mathscr{C}(z^\star, \lambda^\star) := \left\{ v \in \mathscr{F}(z^\star) \,\middle|\, \begin{array}{l} v^\top \nabla H_i(z^\star) = 0 \text{ for all} \\ i \in \mathscr{A}(z^\star) \cap \mathscr{I} \text{ with } \lambda_i^\star > 0 \end{array} \right\}. \tag{12.26}$$

The set $\mathscr{C}$ is also called the *critical cone*. It contains all directions which leave the active inequality constraints with $\lambda_i > 0$ as well as all equality constraints active if one moves a sufficiently small step along these directions. This, however, does not need to hold for those active inequality constraints with $\lambda_i = 0$. In particular, we have the equivalence

$$v \in \mathscr{C}(z^\star, \lambda^\star) \Longleftrightarrow \begin{cases} \nabla G_i(z^\star)^\top v = 0, & \text{for all } i \in \mathscr{E}, \\ \nabla H_i(z^\star)^\top v = 0, & \text{for all } i \in \mathscr{A}(z^\star) \cap \mathscr{I} \text{ with } \lambda_i^\star > 0, \\ \nabla H_i(z^\star)^\top v \geq 0, & \text{for all } i \in \mathscr{A}(z^\star) \cap \mathscr{I} \text{ with } \lambda_i^\star = 0. \end{cases}$$

Now, we want to get a converse result, i.e., we want to check whether a given feasible point is actually a local minimizer. As it turns out, the only differences between the previous necessary conditions and the sufficient conditions presented next is that the constraint qualification is not required whereas inequality (12.25) needs to be strengthened to a strict inequality, cf. Fletcher [20, Theorem 9.3.2]:

Theorem 12.19 (Second-order sufficient conditions) *Consider a feasible point* $z^\star \in \Omega$ *and suppose a Lagrange multiplier* $\lambda^\star \in \mathbb{R}^{r_g + r_h}$ *to exist satisfying* (12.19)–(12.24). *If we have*

$$v^\top \nabla^2_{zz} L(z^\star, \lambda^\star) v > 0 \tag{12.27}$$

for all $v \in \mathscr{C}(z^\star, \lambda^\star)$ *with* $v \neq 0$, *then* $z^\star$ *is a strict local minimizer of problem* (NLP).

We will now focus on the currently common approaches to solve nonlinear constrained optimization problems (NLP), these are the so-called *Sequential Quadratic Programming* (SQP) based on the active set $\mathscr{A}$ and the *Interior-Point Method* (IPM). In the remainder of this section, we describe the main ideas of these strategies. This description cannot replace a thorough treatment as provided, e.g., in the books by Bryson and Ho [8], Fletcher [20] or Nocedal and Wright [32], which were also our main sources for writing this section. Still, we decided to include this description because we consider it useful for the discussion of certain numerical aspects of NMPC algorithms in the subsequent Sects. 12.4–12.6.

Active Set SQP Methods

Following the approach outlined at the end of Sect. 12.2, the basic idea of solving (NLP) in the constrained case would be to iteratively determine search directions d_k by solving an auxiliary quadratic optimization problem which approximates F close to z_k, following either the line-search or the trust-region approach. Since we want to use the necessary conditions from Theorem 12.16, we will also construct a sequence λ_k, which is supposed to converge to the Lagrange multiplier λ^* in Theorem 12.16. However, when proceeding this way, the inequality constraints in (NLP) pose a severe problem because they are difficult to handle in a quadratic optimization problem. There are two ways to overcome this problem. One is to transform the inequality constraints into equality constraints via so-called *slack variables*; we will use this method later in the context of interior-point methods. The drawback of this approach is that each slack variable is an additional optimization variable and thus the dimension of the problem may grow significantly.

The alternative which we describe now is the so-called *active set methods*. Here, the fundamental idea is to consider a so-called *working set* $\mathcal{W}_k$ which contains some of the inequality and all equality constraints for the current iteration point z_k. All constraints within the working set are then treated as equality constraints and the resulting quadratic problem is solved. The working set $\mathcal{W}_k$ can be seen as an approximation to the active set $\mathcal{A}(z_k)$ and is updated in each iteration step. We require that the gradients of the constraints contained in the working set are linearly independent, even if the full set of active constraints at that point has linearly dependent gradients.

In this setting, the quadratic problem to be solved in each step for determining d_k is obtained from an approximation of the nonlinear problem

$$\boxed{\begin{array}{l} \text{minimize} \quad F(z) \\ \text{with respect to } z \in \mathbb{R}^{n_z} \\ \text{subject to } C_i(z) = 0 \text{ for all } i \in \mathcal{W}_k \end{array}} \qquad \text{(ECP)}$$

around the current iterate z_k. Here C_i denotes the components of the combined constraint function C in (12.18) and we assume F and C_i (i.e., G and H) to be twice continuously differentiable. The linear independence requirement for the gradients of the constraints in the working set implies that the (LICQ) condition holds for (ECP).

Note that we do not intend to solve (ECP) but rather use it for constructing an approximation in order to determine d_k. One way to obtain such an approximation is to apply Newton's method to the KKT conditions for (ECP). To formulate these conditions we use the notation $\tilde{\lambda}^{\mathcal{W}_k}$ in order to denote a Lagrange multiplier in $\mathbb{R}^{r_g+r_h}$ with $\tilde{\lambda}_i^{\mathcal{W}_k} = 0$ for $i \notin \mathcal{W}_k$. This way we can use the Lagrangian (12.18) from the original problem (NLP) for (ECP). With $\lambda^{\mathcal{W}_k}$ we denote the vector consisting of those components of $\tilde{\lambda}^{\mathcal{W}_k}$ corresponding to indices in $\mathcal{W}_k$, i.e.,

$$\lambda^{\mathcal{W}_k} = [(\tilde{\lambda}_i^{\mathcal{W}_k})_{i\in\mathcal{W}_k}].$$

Denoting the number of elements in $\mathcal{W}_k$ by $r_{\mathcal{W}_k}$, the vector $\lambda^{\mathcal{W}_k}$ is an element of $\mathbb{R}^{r_{\mathcal{W}_k}}$. Note that $\lambda^{\mathcal{W}_k}$ is uniquely determined by $\tilde{\lambda}^{\mathcal{W}_k}$ and vice versa. Furthermore, we use the notation

$$C^{\mathcal{W}_k}(z) = [(C_i(z))_{i\in\mathcal{W}_k}]$$

in order to denote the vector of constraints corresponding to the indices in $\mathcal{W}_k$. With this notation, the KKT conditions for (ECP) read

$$M(z, \tilde{\lambda}^{\mathcal{W}_k}) := \begin{pmatrix} \nabla_z L(z, \tilde{\lambda}^{\mathcal{W}_k}) \\ C^{\mathcal{W}_k}(z) \end{pmatrix} = 0. \tag{12.28}$$

If we apply Newton's method to this problem, then a step of the resulting iteration is given by

$$\begin{pmatrix} z^{new} \\ \lambda^{\mathcal{W}_k,new} \end{pmatrix} = \begin{pmatrix} z \\ \lambda^{\mathcal{W}_k} \end{pmatrix} - \left(\nabla_{z,\lambda^{\mathcal{W}_k}} M(z, \tilde{\lambda}^{\mathcal{W}_k})\right)^{-1} \begin{pmatrix} \nabla_z L(z, \tilde{\lambda}^{\mathcal{W}_k}) \\ C^{\mathcal{W}_k}(z) \end{pmatrix} \tag{12.29}$$

with

$$\nabla_{z,\lambda^{\mathcal{W}_k}} M(z, \tilde{\lambda}^{\mathcal{W}_k}) = \begin{bmatrix} \nabla_{zz} L(z, \tilde{\lambda}^{\mathcal{W}_k}) & -\nabla C^{\mathcal{W}_k}(z)^\top \\ \nabla C^{\mathcal{W}_k}(z) & 0 \end{bmatrix}.$$

Hence, if sufficient conditions for locally quadratic convergence of the resulting sequence are fulfilled, with this iteration we are able to numerically compute a solution of the problem (ECP). Since we assumed F, G, and H to be twice continuously differentiable, we only have to check whether the Jacobian of $M(z, \tilde{\lambda}^{\mathcal{W}_k})$ given in (12.28) is invertible. As we will see in Lemma 12.22, below, invertibility of this matrix follows at least locally around a minimizer $z^\star$ of (ECP) from the (LICQ) condition provided $z^\star$ satisfies the sufficient conditions from Theorem 12.18 for (ECP). Hence, under these conditions quadratic convergence can be guaranteed.

As already mentioned, we do not want to perform the iteration from Newton's method in order to solve (ECP). Rather, we would like to apply only one step of this iteration and then update the working set $\mathcal{W}_k$ if needed. While Newton's method is very useful for the convergence analysis, cf. Remark 12.23, it turns out that for the actual algorithm it is convenient to replace Newton's iteration by a different method, which is more closely related to the quadratic approximation idea outlined above.

To this end, let us assume that in the kth iteration of the iterative algorithm for solving (NLP) we are given iterates z_k, λ_k and a working set $\mathcal{W}_k$. Similar to the definition of $\tilde{\lambda}^{\mathcal{W}_k}$, above, we define $\tilde{\lambda}_k^{\mathcal{W}_k}$ to be the vector which coincides with λ_k for all components $i \in \mathcal{W}_k$ and whose components are zero for all $i \notin \mathcal{W}_k$. We call $\tilde{\lambda}_k^{\mathcal{W}_k}$ the full multiplier for working set $\mathcal{W}_k$.

Now we replace the cost function F in (ECP) by its Lagrangian L. The reason for using L instead of F will become clear in the discussion after Lemma 12.22, below. Then, we approximate the resulting optimization problem by a quadratic program with linear inequality constraints. This amounts to approximating the Lagrangian L close to z_k by the quadratic function

$$L(z_k + d_k, \tilde{\lambda}_k^{\mathscr{W}_k}) \approx L(z_k, \tilde{\lambda}_k^{\mathscr{W}_k}) + \nabla_z L(z_k, \tilde{\lambda}_k^{\mathscr{W}_k})^\top d_k + \frac{1}{2} d_k^\top \nabla_{zz}^2 L(z_k, \tilde{\lambda}_k^{\mathscr{W}_k}) d_k$$

and to replacing the equality constraints $C_i(z_k + d_k) = 0$ by their linearizations

$$C_i(z_k + d_k) \approx C_i(z_k) + \nabla C_i(z_k) d_k = 0$$

for all $i \in \mathscr{W}_k$. For all d_k satisfying these constraints a little computation shows the identity $L(z_k, \tilde{\lambda}_k^{\mathscr{W}_k}) + \nabla L(z_k, \tilde{\lambda}_k^{\mathscr{W}_k}) d_k = F(z_k) + \nabla F(z_k) d_k$, which we can insert into the approximation of $L(z_k + d_k, \tilde{\lambda}_k^{\mathscr{W}_k})$. This way, we arrive at the following quadratic optimization problem:

$$\boxed{\begin{array}{ll} \text{minimize} & F(z_k) + \nabla F(z_k)^\top d_k + \frac{1}{2} d_k^\top \nabla_{zz}^2 L(z_k, \tilde{\lambda}_k^{\mathscr{W}_k}) d_k \\ \multicolumn{2}{l}{\text{with respect to } d_k \in \mathbb{R}^{n_z}} \\ \multicolumn{2}{l}{\text{subject to } C_i(z_k) + \nabla C_i(z_k)^\top d_k = 0 \text{ for all } i \in \mathscr{W}_k} \end{array}} \tag{EQP}$$

The Lagrange multiplier for the optimal solution d_k of (EQP) according to Theorem 12.16 will be denoted by $\lambda_k^{(\mathrm{EQP})}$. The key idea of the (SQP) algorithm is to use these values in order to update z_k and λ_k according to[1]

$$z_{k+1} = z_k + d_k \quad \text{and} \quad \lambda_{k+1} = \tilde{\lambda}_k^{(\mathrm{EQP})}. \tag{12.30}$$

Here $\tilde{\lambda}_k^{(\mathrm{EQP})}$ denotes the vector in $\mathbb{R}^{r_g + r_h}$ whose components $(\tilde{\lambda}_k^{(\mathrm{EQP})})_i$ are defined by the relations

$$[(\tilde{\lambda}_k^{(\mathrm{EQP})})_{i,\, i \in \mathscr{W}_k}] = \lambda_k^{(\mathrm{EQP})} \quad \text{and} \quad (\tilde{\lambda}_k^{(\mathrm{EQP})})_i = 0 \ \text{ for } i \notin \mathscr{W}_k, \tag{12.31}$$

i.e., we extend the vector $\lambda_k^{(\mathrm{EQP})}$ whose dimension equals the number of indices in $\mathscr{W}_k$ to a vector in $\mathbb{R}^{r_g + r_h}$ by inserting zeros in the components corresponding to constraints which are not included in $\mathscr{W}_k$. A motivation for the choice of z_{k+1} and λ_{k+1} in (12.30) will be given in the discussion after Lemma 12.22, below, and the problem of determining $\mathscr{W}_{k+1}$ as well as suitable step lengths α_k will be considered after we have formulated the basic active set (SQP) algorithm in Algorithm 12.24.

Before we do this, we show how (EQP) can be solved by applying the necessary and sufficient conditions from Theorems 12.16 and 12.18, respectively, to (EQP).

Lemma 12.20 *If $\lambda_k^{(\mathrm{EQP})}$ denotes the Lagrange multiplier corresponding to the optimal solution d_k of* (EQP), *then we have*

---

[1] Appropriate step lengths α_k will be added to these updates, below.

$$\nabla^2_{zz} L(z_k, \tilde{\lambda}_k^{\mathscr{W}_k}) d_k + \nabla F(z_k) = \nabla C^{\mathscr{W}_k}(z_k)^\top \lambda_k^{(\text{EQP})}. \tag{12.32}$$

Proof This is an immediate conclusion of Theorem 12.16. □

Combining (12.32) with the constraints in (EQP) we arrive at the following characterization of the solution of (EQP).

Lemma 12.21 *Given the iterates z_k, λ_k, $\mathscr{W}_k$ and the corresponding full multiplier $\tilde{\lambda}_k^{\mathscr{W}_k}$ for working set $\mathscr{W}_k$, the optimal solution d_k of* (EQP) *and the corresponding Lagrange multiplier $\lambda^{(\text{EQP})}$ fulfill the linear equation system*

$$\begin{bmatrix} \nabla^2_{zz} L(z_k, \tilde{\lambda}_k^{\mathscr{W}_k}) & -\nabla C^{\mathscr{W}_k}(z_k)^\top \\ \nabla C^{\mathscr{W}_k}(z_k) & 0 \end{bmatrix} \begin{pmatrix} d_k \\ \lambda_k^{(\text{EQP})} \end{pmatrix} + \begin{pmatrix} \nabla F(z_k) \\ C^{\mathscr{W}_k}(z_k) \end{pmatrix} = 0. \tag{12.33}$$

Proof The stated linear equation system is a combination of the constraints of problem (EQP) and (12.32). □

For the numerical solution it would be more convenient if the matrix in (12.33) was symmetric. Since the Hessian $\nabla^2_{zz} L(z_k, \tilde{\lambda}_k^{\mathscr{W}_k})$ is symmetric, this can be easily achieved by multiplying $\nabla C^{\mathscr{W}_k}(z_k)$ and $C^{\mathscr{W}_k}(z_k)$ in (12.33) by -1.

The next lemma gives conditions under which (12.33) has a unique solution.

Lemma 12.22 *Consider a minimizer $z^\star$ with Lagrange multiplier $\tilde{\lambda}^{\mathscr{W}_k,\star}$ of* (ECP) *with working set $\mathscr{W}_k$. Assume that* (ECP) *satisfies the (LICQ) condition and that $z^\star$ satisfies the sufficient conditions from Theorem 12.18 for* (ECP). *Then there exist neighborhoods $\mathscr{N}_z$ of $z^\star$ and $\mathscr{N}_\lambda$ of $\tilde{\lambda}^{\mathscr{W}_k,\star}$ such that there exists a unique solution of* (12.33) *for all $z_k \in \mathscr{N}_z$ and $\tilde{\lambda}_k^{\mathscr{W}_k} \in \mathscr{N}_\lambda$.*

Proof In order to prove the assertion we show that the matrix

$$A_0 = \begin{bmatrix} \nabla^2_{zz} L(z^\star, \tilde{\lambda}^{\mathscr{W}_k,\star}) & -\nabla C^{\mathscr{W}_k}(z^\star)^\top \\ \nabla C^{\mathscr{W}_k}(z^\star) & 0 \end{bmatrix}$$

is invertible. This implies the assertion because all expressions in this matrix are continuous, hence the invertibility extends to neighborhoods of $z^\star$ and $\tilde{\lambda}^{\mathscr{W}_k,\star}$.

In order to prove invertibility of A_0 we show

$$A_0 \begin{pmatrix} v \\ w \end{pmatrix} = \begin{pmatrix} \nabla^2_{zz} L(z^\star, \tilde{\lambda}^{\mathscr{W}_k,\star}) v - \nabla C^{\mathscr{W}_k}(z^\star)^\top w \\ \nabla C^{\mathscr{W}_k}(z^\star) v \end{pmatrix} \neq 0,$$

for all vectors v, w of appropriate dimension with $(v^\top, w^\top) \neq 0$. If $\nabla C^{\mathscr{W}_k}(z^\star) v \neq 0$, then we are done. Otherwise, we have $\nabla C^{\mathscr{W}_k}(z^\star) v = 0$ and thus either $v = 0$ or $v \in \mathscr{C}(z^\star, \tilde{\lambda}^{\mathscr{W}_k,\star})$, where $\mathscr{C}$ is the critical cone for (ECP). If $v = 0$, then $w \neq 0$ must hold and since by (LICQ) the matrix $\nabla C^{\mathscr{W}_k}(z^\star)$ has full column rank we obtain

$$\nabla^2_{zz} L(z^\star, \tilde{\lambda}^{\mathscr{W}_k,\star}) v - \nabla C^{\mathscr{W}_k}(z^\star)^\top w = -\nabla C^{\mathscr{W}_k}(z^\star)^\top w \neq 0.$$

If $v \neq 0$ then $v \in \mathscr{C}(z^\star, \tilde{\lambda}^{\mathscr{W}_k,\star})$ must hold and by the positive definiteness (12.27) we obtain $v^\top \nabla^2_{zz} L(z^\star, \tilde{\lambda}^{\mathscr{W}_k,\star})v > 0$ which implies

$$v^\top \left(\nabla^2_{zz} L(z^\star, \tilde{\lambda}^{\mathscr{W}_k,\star})v - \nabla C^{\mathscr{W}_k}(z^\star)^\top w\right) = \underbrace{v^\top \nabla^2_{zz} L(z^\star, \tilde{\lambda}^{\mathscr{W}_k,\star})v}_{>0} - \underbrace{v^\top \nabla C^{\mathscr{W}_k}(z^\star) w}_{=w^\top \nabla C^{\mathscr{W}_k}(z^\star)v=0} > 0,$$

and thus

$$\nabla^2_{zz} L(z^\star, \tilde{\lambda}^{\mathscr{W}_k,\star})v - \nabla C^{\mathscr{W}_k}(z^\star)^\top w \neq 0.$$

Hence, in all cases we get $A_0(v^\top, w^\top)^\top \neq 0$, which shows the desired invertibility of A_0. □

Comparing the system of linear equations (12.33) derived from (EQP) and Newton's iteration (12.29) one makes the—at the first glance surprising—observation that both are identical if in (12.33) we set $z_k = z$, $d_k = (z^{new} - z)$, $\tilde{\lambda}_k^{\mathscr{W}_k} = \tilde{\lambda}^{\mathscr{W}_k}$ and $\lambda_k^{(\text{EQP})} = \lambda^{\mathscr{W}_k,new}$. However, this identity is not really a coincidence. In fact, the particular cost function in (EQP) was chosen exactly for the purpose to obtain this identity.

Remark 12.23 Interpreting the quadratic problem (EQP) as a Newton iteration is very useful for the analysis of the convergence speed of the algorithms. Furthermore, it provides the motivation for the choice $z_{k+1} = z_k + d_k$ and $\lambda_{k+1} = \tilde{\lambda}_k^{(\text{EQP})}$ in (12.30), as these are exactly the values z^{new} and $\tilde{\lambda}^{\mathscr{W}_k,new}$ from Newton's iteration.

Convergence issues will not be treated in detail here, but we would like to mention that the two properties one would like to have is that the sequences z_k, λ_k converge to a KKT point, which follows from Newton's method provided this method converges. Additionally, one would like to ensure that the cost function F decreases along the search direction d_k if $d_k \neq 0$. This decrease property allows to conclude that the limit is a candidate for a local minimizer without having to check second-order conditions.

The formulation of the iteration via (EQP) allows to obtain this decrease property under the conditions of Lemma 12.22 for z_k and $\tilde{\lambda}_k^{\mathscr{W}_k}$ in a neighborhood of $z^\star$ and $\tilde{\lambda}_k^{\mathscr{W}_k,\star}$. This is because $d_k \neq 0$ implies

$$\frac{1}{2} d_k^\top \nabla^2_{zz} L(z_k, \tilde{\lambda}_k^{\mathscr{W}_k}) d_k + \nabla F(z_k) d_k < 0 \tag{12.34}$$

since otherwise $d_k = 0$ would be the optimal solution. The positive definiteness of $\nabla^2_{zz} L(z^\star, \tilde{\lambda}_k^{\mathscr{W}_k,\star})$ for directions $d_k \in \mathscr{C}(z^\star, \tilde{\lambda}^{\mathscr{W}_k,\star})$ implies positivity of the first summand in (12.34) for z_k and $\tilde{\lambda}_k^{\mathscr{W}_k}$ close to these optimal values, hence we obtain $\nabla F(z_k) d_k < 0$. Thus, for $\alpha > 0$ sufficiently small we get

$$F(z_k + \alpha d_k) = F(z_k) + \nabla F(z_k)\alpha d_k + O(\alpha^2) < F(z_k),$$

i.e., decrease of F along d_k. This property is important because it shows that if we restrict the step $z_{k+1} := z_k + d_k$ in (12.30) to $z_{k+1} := z_k + \alpha_k d_k$ then we can still expect decrease of the cost function.

There are different ways to couple the solution of (EQP) with the update of the working set. The approach we explain here relies on the fact that what we actually want to solve in each iteration of the algorithms is the following inequality constrained quadratic program.

$$\begin{aligned} &\text{minimize} \quad F(z_k) + \nabla F(z_k)^\top d_k + \frac{1}{2} d_k^\top \nabla_{zz}^2 L(z_k, \lambda_k) d_k \\ &\text{with respect to } d_k \in \mathbb{R}^{n_z} \\ &\text{subject to } G_i(z_k) + \nabla_z G_i(z_k)^\top d_k = 0 \text{ for all } i \in \mathcal{E} \\ &\text{and } H_i(z_k) + \nabla H_i(z_k)^\top d_k \geq 0 \text{ for all } i \in \mathcal{I} \end{aligned} \tag{IQP}$$

We will utilize the solution of this program in order to determine the search direction d_k as well as the new working set $\mathcal{W}_{k+1}$. Before we explain the details of this procedure, we give the basic outline of an active set (SQP) algorithm. In this basic version we do not yet include the step length α_k, which will be discussed after Algorithm 12.25, below.

Algorithm 12.24 (*Basic active set* (SQP) *algorithm*) Suppose a pair of initial values (z_0, λ_0) and an initial working set $\mathcal{W}_0 \subseteq \mathcal{A}(z_0)$ to be given and set $k := 0$.
While convergence test is not satisfied do

1. Compute $F(z_k)$, $\nabla F(z_k)$, $\nabla_{zz}^2 L(z_k, \lambda_k)$, $C(z_k)$ and $\nabla C(z_k)$
2. Solve (IQP) using (EQP) and $\mathcal{W}_k$ and obtain d_k, $\lambda_k^{(\text{EQP})}$ and $\mathcal{W}_{k+1}$
3. Set $z_{k+1} := z_k + d_k$, $\lambda_{k+1} := \tilde{\lambda}_k^{(\text{EQP})}$ according to (12.31)
4. Set $k := k + 1$

As convergence test, here one typically checks whether the necessary conditions from Theorem 12.16 are satisfied up to some user-defined tolerance tol_{OPT}.

Furthermore, it should be noted that usually in (SQP) methods one avoids computing the Hessian $\nabla_{zz}^2 L(z^\star, \tilde{\lambda}_k^{\mathcal{W}_k})$ of the cost function and utilizes computationally cheaper update techniques for obtaining suitable approximations, instead. We will not discuss this topic here and refer to, e.g., Nocedal and Wright [32] for details.

Now we discuss the details of Step 2 of this algorithm. The solution of (IQP) is obtained by iteratively solving problems of type (EQP) and updating the working set $\mathcal{W}_k$. This defines another iteration inside the iteration of Algorithm 12.24 whose iteration index we denote by q. In this inner iteration, we iteratively determine vectors d_k^q and working sets $\mathcal{W}_k^q$, $q = 0, 1, \ldots$. We start this iteration with the working set $\mathcal{W}_k^0 = \mathcal{W}_k$ and initial value d_k^0 which we assume to be feasible for all constraints in (IQP). This iteration will terminate after a finite number of steps $q^\star$ with the optimal value $d_k = d_k^{q^\star}$ of (IQP). The corresponding working set $\mathcal{W}_k^{q^\star}$ will be the active set of problem (IQP) and will be used as $\mathcal{W}_{k+1}$ in Algorithm 12.24. The Lagrange multipliers $\lambda_k^{(\text{EQP})}$ needed in Algorithm 12.24 are obtained from the last solution of (EQP) in this iteration. The resulting algorithm will be given in Algorithm 12.25, below. Before we formulate this algorithm, we will now discuss the details of its different steps.

For its use within this iteration, we rewrite the problem (EQP) in order to take the previous iterate d_k^q into account.

$$\boxed{\begin{aligned} &\text{minimize} \quad F(z_k) + \nabla F(z_k)^\top (p^q + d_k^q) \\ &\qquad\qquad + \frac{1}{2}(p^q + d_k^q)^\top \nabla_{zz}^2 L(z_k, \tilde{\lambda}_k^{\mathcal{W}_k})(p^q + d_k^q) \\ &\text{with respect to } p^q \in \mathbb{R}^{n_z} \\ &\text{subject to } C_i(z_k) + \nabla C_i(z_k)^\top (p^q + d_k^q) = 0 \text{ for all } i \in \mathcal{W}_k^q \end{aligned}} \tag{EQP^q}$$

Here, for each q we assume that d_k^q is feasible for the constraints in (IQP) and for (EQP) with working set $\mathcal{W}_k = \mathcal{W}_k^q$. This implies that the constraints in (EQP^q) can equivalently be written as $\nabla C_i(z_k)^\top p^q = 0$. In order to simplify the presentation we assume that the problem (EQP^q) is strictly convex, i.e., that $(p^q)^\top \nabla_{zz}^2 L(z_k, \tilde{\lambda}_k^{\mathcal{W}_k}) p^q > 0$ holds for all $p^q \neq 0$ with $\nabla C_i(z_k)^\top p^q = 0$.

We denote the optimal solution of (EQP^q) by $p^{q,\star}$. From this optimal solution we want to construct d_k^{q+1} and a new working set $\mathcal{W}_k^{q+1}$, such that d_k^{q+1} is again feasible for (IQP) and feasible for (EQP) with $\mathcal{W}_k = \mathcal{W}_k^{q+1}$. To this end, we need to develop rules for adding and removing constraints from the working set $\mathcal{W}_k^q$ such that eventually (EQP^q) delivers an optimal solution of (IQP).

We first consider the problem of adding constraints which is closely related to the definition of d_k^{q+1}. Using the previous iterate d_k^q and the optimal solution $p^{q,\star}$ of (EQP^q) we define

$$d_k^{q+1} := d_k^q + \alpha^q p^{q,\star} \tag{12.35}$$

and compute the maximal $\alpha^q \in [0, 1]$ such that d_k^{q+1} is feasible for all constraints in (IQP). Note that if d_k^q satisfies the constraints contained in the working set $\mathcal{W}_k^q$, then these constraints are also satisfied for d_k^{q+1} for all $\alpha^q \in [0, 1]$. This is because $C_i(z_k) + \nabla C_i(z_k)^\top (p^{q,\star} + d_k^q) = 0$ and $C_i(z_k) + \nabla C_i(z_k)^\top d_k^q = 0$ imply $\nabla C_i(z_k)^\top p^{q,\star} = 0$ and thus

$$C_i(z_k) + \nabla C_i(z_k)^\top (d_k^q + \alpha^q p^{q,\star}) = C_i(z_k) + \nabla C_i(z_k)^\top d_k^q + \alpha^q \nabla C_i(z_k)^\top p^{q,\star} = 0.$$

Hence, for the computation of α^q we only need to consider the constraints H_i not contained in $\mathcal{W}_k^q$. If $\nabla H_i(z_k) p^{q,\star} \geq 0$ holds for $i \notin \mathcal{W}_k^q$, then we have

$$H_i(z_k) + \nabla H_i(z_k)^\top \cdot (d_k^q + \alpha^q p^{q,\star}) \geq H_i(z_k) + \nabla H_i(z_k)^\top d_k^q \geq 0$$

since we assumed d_k^q to be feasible for (IQP). Hence, $\alpha^q \geq 0$ can be chosen freely. In the case $\nabla H_i(z_k)^\top p^{q,\star} < 0$, we obtain $H_i(z_k) + \nabla H_i(z_k)^\top \cdot (d_k^q + \alpha^q p^{q,\star}) \geq 0$ only if

$$\alpha^q \leq \frac{-H_i(z_k) - \nabla H_i(z_k)^\top d_k^q}{\nabla H_i(z_k)^\top p^{q,\star}}$$

holds true. In order to maximize α^q we define

$$\alpha^q := \min\left\{1, \min_{i \notin \mathscr{W}_k^q,\, \nabla H_i(z_k)^\top d_k^q < 0} \frac{-H_i(z_k) - \nabla H_i(z_k)^\top d_k^q}{\nabla H_i(z_k)^\top p^{q,\star}}\right\} \tag{12.36}$$

Note that $\alpha^q = 0$ is possible since there might exist an active constraint which is not an element of the working set $\mathscr{W}_k^q$ and exhibits $\nabla H_i(z_k)^\top p^{q,\star} < 0$. We call the constraints $H_i(z_k)$ for which the minimum is achieved *blocking constraints* and denote the set of blocking constraints by

$$\mathscr{C}^q := \left\{ j \notin \mathscr{W}_k^q \;\middle|\; \alpha^q = \frac{-H_j(z_k) - \nabla H_j(z_k)^\top d_k^q}{\nabla H_j(z_k)^\top p^{q,\star}} \right\}. \tag{12.37}$$

If α^q can be chosen to 1, then none of the constraints not contained in $\mathscr{W}_k^q$ is active for d_k^{q+1} and we can set $\mathscr{W}_k^{q+1} := \mathscr{W}_k^q$. If $\alpha^q < 1$, however, then we know that at least one constraint is active for d_k^{q+1} which is not contained in $\mathscr{W}_k^q$. Hence, we construct $\mathscr{W}_k^{q+1}$ by adding one of the blocking constraints to $\mathscr{W}_k^q$.

We can iterate this procedure until the resulting algorithm reaches a point d_k^q which minimizes the quadratic objective function in (EQP^q) over its current working set $\mathscr{W}_k^q$. This situation can be easily identified since in the optimum we obtain $p^{q,\star} = 0$. Once such a point d_k^q is reached, we can decide which of the equality constraints in $\mathscr{W}_k^q \cap \mathscr{I}$ should be removed from $\mathscr{W}_k^q$ because we can only obtain a better solution by imposing the inequality constraints $H_i(z_k) + \nabla H_i(z_k)(p^q + d_k^q) \geq 0$ instead of the equality constraint $H_i(z_k) + \nabla H_i(z_k)(p^q + d_k^q) = 0$. Looking at the necessary conditions from Theorem 12.16, the constraints corresponding to negative components of the Lagrange multiplier $\lambda_k^{(\mathrm{EQP}^q)}$ are natural candidates for this. However, since a constraint is not checked anymore in the algorithm once the index i is removed from $\mathscr{W}_k^{q+1}$, we would also like to ensure that that it remains satisfied in the next step, i.e., that it is not immediately reinserted. Fortunately, as we will show next, the negativity of the respective component of the Lagrange multiplier $\lambda_k^{(\mathrm{EQP}^q)}$ guarantees that this indeed happens.

To this end, consider the iterate d_k^q yielding $p^{q,\star} = 0$ in (EQP^q) for working set $\mathscr{W}_k^q$ and the corresponding multiplier $\lambda_k^{(\mathrm{EQP}^q)}$. With $(\lambda_k^{(\mathrm{EQP}^q)})_{j_i}$, we denote the component of the multiplier corresponding to the constraint index $i \in \mathscr{W}_k^q$ and we assume $(\lambda_k^{(\mathrm{EQP}^q)})_{j_i} < 0$ for some $i \in \mathscr{W}_k^q \cap \mathscr{I}$.

Consider then the solution $p^{q+1,\star}$ of (EQP^{q+1}) for working set $\mathscr{W}_k^{q+1} = \mathscr{W}_k^q \setminus \{i\}$ with $d_k^{q+1} = d_k^q$ and corresponding multiplier $\lambda_k^{(\mathrm{EQP}^{q+1})}$. From Lemma 12.20 we then obtain

$$\nabla_{zz}^2 L(z_k, \tilde{\lambda}_k^{\mathscr{W}_k}) d_k^q + \nabla F(z_k) = \nabla C^{\mathscr{W}_k}(z_k)^\top \lambda_k^{(\mathrm{EQP}^q)}$$

and

$$\nabla^2_{zz} L(z_k, \tilde{\lambda}_k^{\mathscr{W}_k})(d_k^q + p^{q+1,\star}) + \nabla F(z_k) = \nabla C^{\mathscr{W}_k^{q+1}}(z_k)^\top \lambda_k^{(\mathrm{EQP}^{q+1})}$$

Subtracting the first from the second equation and using $C_i = H_i$ for $i \in \mathscr{I}$ we obtain

$$\begin{aligned}\nabla^2_{zz} L(z_k, \tilde{\lambda}_k^{\mathscr{W}_k}) p^{q+1,\star} &= \nabla C^{\mathscr{W}_k^{q+1}}(z_k)^\top (\lambda_k^{(\mathrm{EQP}^{q+1})} - [(\lambda_k^{(\mathrm{EQP}^q)})_{j \neq j_i}]) \\ &- \nabla H_i(z_k)(\lambda_k^{(\mathrm{EQP}^q)})_{j_i}.\end{aligned}$$

Multiplying from the left with $(p^{q+1,\star})^\top$ then yields

$$\begin{aligned}(p^{q+1,\star})^\top \nabla^2_{zz} L(z_k, \tilde{\lambda}_k^{\mathscr{W}_k}) p^{q+1,\star} &= (p^{q+1,\star})^\top \nabla C^{\mathscr{W}_k^{q+1}}(z_k)^\top (\lambda_k^{(\mathrm{EQP}^{q+1})} - [(\lambda_k^{(\mathrm{EQP}^q)})_{j \neq j_i}]) \\ &- (p^{q+1,\star})^\top \nabla H_i(z_k)(\lambda_k^{(\mathrm{EQP}^q)})_{j_i}.\end{aligned}$$

Since the constraints in (EQP^{q+1}) together with the feasibility of d_k^q imply the identity $(p^{q+1,\star})^\top \nabla C^{\mathscr{W}_k^{q+1}}(z_k) = 0$ we thus get

$$(p^{q+1,\star})^\top \nabla H_i(z_k)(\lambda_k^{(\mathrm{EQP}^q)})_{j_i} = -(p^{q+1,\star})^\top \nabla^2_{zz} L(z_k, \tilde{\lambda}_k^{\mathscr{W}_k}) p^{q+1,\star}.$$

Hence, the positive definiteness assumption on $\nabla^2_{zz} L(z_k, \tilde{\lambda}_k^{\mathscr{W}_k})$ together with the inequality $(\lambda_k^{(\mathrm{EQP}^q)})_{j_i} < 0$ yields

$$(p^{q+1,\star})^\top \nabla H_i(z_k) > 0.$$

Consequently, H_i grows along the search direction $p^{q+1,\star}$ and may hence be omitted in $\mathscr{W}_k^{q+1}$ without violating the corresponding inequality constraint in the next step.

In practice, not just any index i is chosen but the one corresponding to the most negative $(\lambda_k^{(\mathrm{EQP}^q)})_{j_i}$ which is motivated by sensitivity aspects.

Combining all we have derived so far, we end up with the following algorithm for solving (IQP) to be inserted in Step 2 of Algorithm 12.24.

Algorithm 12.25 (*Active set* (IQP) *algorithm*) Suppose a pair of values (z_k, λ_k) as well as the derivatives $\nabla F(z_k)$, $\nabla^2_{zz} L(z_k, \lambda_k)$, $\nabla C(z_k)$, the function values $F(z_k)$, $C(z_k)$ and a working set $\mathscr{W}_k \subseteq \mathscr{A}(z_k)$ to be given.

I. Set $q := 0$, $\mathscr{W}_k^0 := \mathscr{W}_k$ and find a starting point d_k^0 feasible for (IQP) and (EQP)
II. While not terminated
   a. Solve (EQP^q) to obtain $p^{q,\star}$ and $\lambda_k^{(\mathrm{EQP}^q)}$
   b. If $p^{q,\star} = 0$
      - If $(\lambda_k^{(\mathrm{EQP}^q)})_{j_i} \geq 0$ for all $i \in \mathscr{W}_k^q \cap \mathscr{I}$:
        terminate with $d_k = d_k^q$, $\lambda_k^{(\mathrm{EQP})} = \lambda_k^{(\mathrm{EQP}^q)}$ and $\mathscr{W}_{k+1} = \mathscr{W}_k^q$
        Else: Set $i := \operatorname*{argmin}_{i \in \mathscr{W}_k^q \cap \mathscr{I}} (\lambda_k^{(\mathrm{EQP}^q)})_{j_i}$, $d_k^{q+1} = d_k^q$, $\mathscr{W}_k^{q+1} := \mathscr{W}_k^q \setminus \{j\}$

      Else

- Compute α^q from (12.36) and $\mathscr{C}^q$ from (12.37)
- Set $d_k^{q+1} := d_k^q + \alpha^q p^{q,\star}$
- If $\mathscr{C}^q \neq \emptyset$: choose $i \in \mathscr{C}^q$ and set $\mathscr{W}_k^{q+1} := \mathscr{W}_k^q \cup \{i\}$
  Else: Set $\mathscr{W}_k^{q+1} := \mathscr{W}_k^q$

c. Set $k := k + 1$

One can show that, apart from certain exceptional cases, this algorithm terminates after a finite number of iterations if (IQP) is strictly convex, for details see [32, Sect. 16.4]. Since the solution satisfies the first-order necessary conditions for (IQP), under strict convexity it follows that upon termination the vector d_k is the optimal solution of (IQP). From our construction of the d_k^q it additionally follows that this d_k is the optimal solution of (EQP) with working set $\mathscr{W}_{k+1}$. Methods for finding the feasible starting point p_k^0 in Step I are discussed in [32, Sect. 16.4], too. Since d_{k-1} is feasible for z_{k-1} and $\mathscr{W}_k$ and since z_k is typically quite close to z_{k-1}, the vector $d_k^0 = d_{k-1}$ is usually a good initial guess.

Moreover, this algorithm allows us to maintain the linear independence property of constraints which are contained in $\mathscr{W}_k^q$. In particular, if the gradients of the active constraints of the initial value are linearly dependent, then we can consider a subset of linear independent constraints. During the iteration we have to add the blocking constraints. Since the normals of these constraints cannot be represented by a linear combination of the normals of the constraints contained in the working set $\mathscr{W}_k^j$, linear independence is preserved if one constraint is added. Last, the deletion of a constraint from the working set $\mathscr{W}_k^j$ clearly does not lead to linear dependency of the remaining constraint normals.

As already mentioned before, in general it is not a good choice to use $z_{k+1} := z_k + d_k$ in Algorithm 12.24. One reason for this is that d_k is obtained from minimizing a quadratic approximation of F near z_k, hence $F(z_k + d_k)$ may differ considerably from the value of this approximation if d_k is large. In unconstrained optimization, one would hence determine the new z_{k+1} by solving the one-dimensional minimization problem (12.13) in the line-search approach or one would restrict d_k by using the trust-region approach (12.14). In both cases, the decrease of the original nonlinear cost function F is used in order to measure the progress of the algorithm.

In constrained optimization, one also needs to take the constraints into account when measuring this progress. In order to achieve this, one does not use F for determining z_{k+1} but rather a function which defines a trade off between decrease of F and the violation of the constraints, the so-called *merit function*. In this function not only the constraints in the current working set $\mathscr{W}_{k+1}$ but all constraints in (NLP) are taken into account.

Merit functions are used in both the line-search and trust-region approach, however, the way this function is used is different in both approaches. For trust-region methods the merit function determines if the step is accepted or rejected and if the trust-region radius needs to be adapted. In contrast to that, in the line-search setting the merit function is used to control the step length itself.

In the following, we consider the merit function

$$\widetilde{L}(z,\mu) := F(z) + \mu\|A(z)\|_1 \tag{12.38}$$

with a positive parameter $\mu > 0$ and $A(z)$ defined by

$$A_i(z) = \begin{cases} G_i(z), & i \in \mathscr{E} \\ \min_{s_i \geq 0} H_i(z) - s_i, & i \in \mathscr{I}. \end{cases}$$

The variables s_i, $i \in \mathscr{I}$ are called *slack variables* and convert the inequality constraints $H_i(z) \geq 0$ into equality constraints $\min_{s_i \geq 0} H_i(z) - s_i = 0$. Indeed, using A the conditions $G_i(z) = 0, i \in \mathscr{E}$ and $H_i(z) \geq 0, i \in \mathscr{I}$ are now compactly expressed as $A(z) = 0$.

Note that this choice of a merit function is not necessarily the best and different function have been considered in the literature as well, see, e.g., the papers of Han [28], Powell [34] and Schittkowski [37, 38]. For simplicity of exposition, however, we exclusively consider (12.38). The merit function $\widetilde{L}$ from (12.38) has the important property that it decays along the search direction d_k from Algorithm 12.24 if the working set $\mathscr{W}_{k+1}$ coincides with the active set of (NLP) and the parameter μ is sufficiently large, see [32, Lemma 18.2]. This means that as long as there are no active constraints missing in the current working set, decrease of the merit function is guaranteed and the algorithm will show progress. Furthermore, one can show that even though $\widetilde{L}$ will in general not be differentiable due to the nonsmoothness of $\|A(z)\|_1$, the directional derivative $D(\widetilde{L}(z,\mu), d_k)$ along the search direction d_k in Algorithm 12.24 does exist.

For the line-search approach, this motivates the following strategy to determine the new iterate $z_{k+1} = z_k + \alpha_k d_k$ in Step 3 of Algorithm 12.24: Instead of solving a one-dimensional optimization problem, we use the information from the directional derivative in order to obtain a computationally less expensive criterion for computing α_k using a fixed parameter $\eta \in (0, 1/2)$

3a. Determine $\mu > 0$ such that $D(\widetilde{L}(z,\mu), d_k) < 0$
3b. Find α_k satisfying $\widetilde{L}(z_k + \alpha_k d_k, \mu) \leq \widetilde{L}(z_k, \mu) + \eta\alpha_k D\left(\widetilde{L}(z_k,\mu); d_k\right)$
3c. Set $z_{k+1} := z_k + \alpha_k d_k$ and update λ_{k+1}

Here, the update of λ_{k+1} can be done by using the step width α_k in order to define $\lambda_{k+1} = \lambda_k + \alpha_k(\tilde{\lambda}_k^{(EQP)} - \lambda_k)$. Alternatively, cf. [32, Sect. 18.3], one can determine λ_{k+1} by solving the least squares problem

$$\min_{\lambda} \|\nabla F(z_{k+1}) - C^{\mathscr{W}_{k+1}}(z_{k+1})\lambda\|_2^2. \tag{12.39}$$

The computation of μ in Step 3a. can be done in various ways, e.g., based on a quadratic approximation of $\widetilde{L}$ or by using the multiplier $\lambda_k^{(EQP)}$.

In the trust-region approach, one solves (IQP) under the additional constraint $\|d_k\|_2 \leq \Delta_k$. In the solution method for (IQP) discussed after Algorithm 12.24 this amounts to including the constraint $\|p^q + d_k^q\|_2 \leq \Delta_k$ in (EQP^q). Unfortunately, due to the additional constraint the problem may arise that the constraints $C_i(z_k) + \nabla C_i(z_k)^\top(p^q + d_k^q) = 0$ cannot be satisfied anymore.

There exists a wide variety of trust-region approaches, for sake of simplicity, however, we now only consider the so-called relaxation approach. In this approach, the original constraints in problem (EQP^q) are modified by a relaxation (or residual) vector r_k, that is,

$$C_i(z_k) + \nabla C_i(z_k)^\top(p^q + d_k^q) = r_{k,i}, \tag{12.40}$$

where r_k can be computed from

$$r_k := C(z_k) + \nabla C(z_k)^\top c_k \tag{12.41}$$

for c_k being the solution of

$$\begin{aligned} &\text{minimize} \quad \|C(z_k) + \nabla C(z_k) c_k\|_2^2 \\ &\text{with respect to} \quad c_k \in \mathbb{R}^d \quad \text{subject to} \quad \|c_k\|_2 \leq 0.8\Delta_k \end{aligned} \tag{12.42}$$

The safeguard factor 0.8 guarantees existence of a consistent solution of the relaxed problem. Since here we work with Euclidean norms, we modify the merit function to

$$\widetilde{L}(z_k, \mu) := F(z_k) + \mu\|A(z_k)\|_2$$

In order to determine whether the trust-region radius Δ_k should be reduced or not, one compares the decrease of the merit function $\widetilde{L}$ with the decrease of a quadratic approximation Q_μ of $\widetilde{L}$. This defines the ratio

$$\rho_k := \frac{\widetilde{L}(z_k, \mu) - \widetilde{L}(z_k + d_k, \mu)}{Q_\mu(0) - Q_\mu(d_k)} \tag{12.43}$$

which is large if the quadratic model Q_μ is close to the nonlinear function $\widetilde{L}$ and becomes the smaller the more these functions differ. Since a large difference indicates that the values of the nonlinear problem and its quadratic approximation differ significantly at $z_k + d_k$, this will be used as a criterion for reducing the trust-region radius Δ_k.

Hence, implementing the trust-region idea leads to the following modification of the Steps 2 and 3 in Algorithm 12.24, in which we use fixed parameters $\eta \in (0, 1/2)$ and $\gamma \in (0, 1)$ and an initial estimate $\Delta_0 > 0$ for the trust-region radius.

2a. Solve problem (12.42) for c_k and compute r_k from (12.41)
2b. Solve (IQP) using the modified constraints (12.40) and $\|d_k\| \leq \Delta_k$ in (EQP^q) to obtain d_k, $\lambda_k^{(\mathrm{EQP})}$ and $\mathscr{W}_{k+1}$

3a. Determine $\mu > 0$ such that $D(\widetilde{L}(z_k, \mu), d_k) < 0$
3b. Compute ρ_k according to (12.43)
3c. If $\rho_k > \eta$: Set $z_{k+1} := z_k + d_k$, $\lambda_{k+1} = \tilde{\lambda}_k^{(EQP)}$ and choose $\Delta_{k+1} \geq \Delta_k$
Else: Set $z_{k+1} := z_k$ and choose $\Delta_{k+1} \leq \gamma \|d_k\|$

Both for the line-search and the trust-region case, the algorithms just given mainly outline the main idea of these approaches and we refer to the optimization literature for all implementational details.

Interior-Point Methods

In contrast to active set methods, interior-point methods generate a sequence z_k which always lies in the interior of the feasible set Ω. For generating this sequence, in each iterate the entire set of inequality constraints H is used. To this end, these inequality constraints are transformed into equality constraints using *slack variables*, similar to the definition of A in the merit function (12.38). This way the number of constraints to be considered in each iteration may become considerably larger and thus the computational effort in each iteration grows. On the other hand, one avoids the potentially time-consuming identification of the working set. Which of the two advantages is predominant crucially depends on the problem to be solved and can only be assessed on a case by case basis.

Once the constraints are reformulated, the fundamental idea of interior point algorithms is to modify the problem under consideration such that all inequality constraints are always active. In the literature, there are two main ways to achieve this goal, the so-called *continuation method* and the *barrier method*. These approaches, however, lead to very similar KKT equation systems.

In the continuation (or homotopy) method, the problem

$$\begin{array}{l} \text{minimize} \quad F(z) \\ \text{with respect to } z \in \mathbb{R}^{n_z}, s \in \mathbb{R}^{n_s} \\ \text{subject to } G_i(z) = 0 \text{ for all } i \in \mathcal{E}, \\ H_i(z) - s_i = 0 \text{ for all } i \in \mathcal{I} \text{ and } s_i \geq 0 \end{array} \tag{IPM}$$

is considered where $s \in \mathbb{R}^{n_s}$ are slack variables filling the gap to modify inactive constraints to be active. The Lagrangian for the problem (IPM) is given by

$$L(z, s, v, w) = F(z) - v^\top G(z) - w^\top (H(z) - s). \tag{12.44}$$

Using that at the minimum the equality $H(z) = s$ must hold, for this problem the KKT conditions from Theorem 12.16 can be written as

$$\nabla F(z) - \nabla G(z)^\top v - \nabla H(z)^\top w = 0 \tag{12.45}$$

$$Sw - \mu e = 0 \tag{12.46}$$

$$G(z) = 0 \tag{12.47}$$

$$H(z) - s = 0 \tag{12.48}$$

with $s \geq 0$, $w \geq 0$ and $\mu = 0$, using the vector $e := (1, 1, \ldots, 1)^\top$ and the matrices $S := \operatorname{diag}(s)$ and $W := \operatorname{diag}(w)$ to simplify the notation.

Here, the additional parameter μ is introduced as a perturbation parameter in order to enforce the solution to stay away from the boundary of Ω by setting $\mu > 0$. More precisely, we consider a sequence of (perturbed) KKT conditions with positive perturbation parameters μ_j where $\mu_j \to 0$ as $j \to \infty$. The strict positivity of μ_j in each iterate forces the slack variable s and the multiplier w to be positive and thus the iterates to stay in the interior of Ω. Similar to the previously mentioned (SQP) methods, the hope is that the limit of the generated sequence satisfies the KKT conditions for problem (IPM) and, if a merit function decreases along the iterates, that the limit is actually a minimizer.

The second, so-called barrier approach consists of a so-called self concordant barrier function used to encode the feasible set $\{s \geq 0\}$ via

$$\begin{aligned}
&\text{minimize} \quad F(z) - \mu \sum_{i=1}^{n_s} \log s_i \\
&\text{with respect to } z \in \mathbb{R}^{n_z},\ s \in \mathbb{R}^{n_s} \\
&\text{subject to } G_i(z) = 0 \text{ for all } i \in \mathcal{E} \\
&\text{and } H_i(z) - s = 0 \text{ for all } i \in \mathcal{I}
\end{aligned}$$

again with parameter $\mu > 0$. In contrast to the homotopy approach where we explicitly stated the constraint $s \geq 0$, here it is not necessary to add this constraint to the barrier approach problem since minimization of the barrier term $-\mu \sum_{i=1}^{n_s} \log s_i$ prevents components of s from becoming too close to zero. As for the continuation method, the idea of the barrier approach is to generate approximate solutions for a sequence of positive parameters μ which converges to zero.

Applying Theorem 12.16 to the barrier problem, we obtain the conditions

$$\nabla F(z) - \nabla G(z)^\top v - \nabla H(z)^\top w = 0 \tag{12.49}$$

$$-\mu S^{-1} e + w = 0 \tag{12.50}$$

$$G(z) = 0 \tag{12.51}$$

$$H(z) - s = 0 \tag{12.52}$$

which are identical to (12.45)–(12.48) except for (12.50) which is moreover nonlinear in s. However, multiplying (12.50) by S we can easily convert this condition into (12.46).

Similar to the (SQP) problem, the ideas of line-search and the trust-region approach can be applied to the (IPM) problem as well. Before discussing these methods, we first look at the application of Newton's method for solving (IPM). Applying Newton's method to (12.45)–(12.48) we obtain the system of equations

$$\begin{pmatrix} \nabla^2_{zz} L(z,s,v,w) & 0 & -\nabla G(z)^\top & -\nabla H(z)^\top \\ 0 & W & 0 & S \\ \nabla G(z) & 0 & 0 & 0 \\ \nabla H(z) & -\mathrm{Id} & 0 & 0 \end{pmatrix} \begin{pmatrix} d_z \\ d_s \\ d_v \\ d_w \end{pmatrix} = - \begin{pmatrix} \nabla F(z) - \nabla G(z)^\top v - \nabla H(z)^\top w \\ Sw - \mu e \\ G(z) \\ H(z) - s \end{pmatrix} \tag{12.53}$$

which is also called primal-dual system (the so-called primal system arises from (12.49)–(12.52) in a similar manner). After computing the step (d_z, d_s, d_v, d_w) the new state iterate can be obtained via

$$\begin{aligned} z_{k+1} &= z_k + \alpha_s^{\max} d_z & s_{k+1} &= s_k + \alpha_s^{\max} d_s \\ v_{k+1} &= v_k + \alpha_w^{\max} d_v & w_{k+1} &= w_k + \alpha_w^{\max} d_w \end{aligned} \tag{12.54}$$

where

$$\begin{aligned} \alpha_s^{\max} &= \max\{\alpha \in (0,1] \mid s + \alpha d_s \geq (1-\tau)s\} \\ \alpha_w^{\max} &= \max\{\alpha \in (0,1] \mid w + \alpha d_w \geq (1-\tau)w\} \end{aligned} \tag{12.55}$$

with $\tau \in (0, 1)$, typically 0.995. The latter condition is called *fraction to boundary rule* since it prevents the state and slack variables z and s to reach their lower bounds too fast. Note that the matrix in (12.53) is regular throughout the iteration for z_k in a neighborhood of an optimal solution which satisfies the second-order sufficient conditions from Theorem 12.19 and the strict complementarity condition from Definition 12.17. In particular, if the strict complementarity condition holds at a solution $z^\star$, then for every index i we see that either z_i or s_i remains bounded away from zero as the iterates approach $z^\star$, which guarantees that the second block row of the matrix in (12.53) has full row rank. Hence, the interior-point approach itself is not ill conditioned and will not show singularities.

In a practical implementation, the matrix in (12.53) is not used. This is due to the fact that transforming the equation system (12.53) into

$$\begin{pmatrix} \nabla^2_{zz} L(z,s,v,w) & 0 & -\nabla G(z)^\top & -\nabla H(z)^\top \\ 0 & \Sigma & 0 & \mathrm{Id} \\ -\nabla G(z) & 0 & 0 & 0 \\ -\nabla H(z) & \mathrm{Id} & 0 & 0 \end{pmatrix} \begin{pmatrix} d_z \\ d_s \\ d_v \\ d_w \end{pmatrix} = - \begin{pmatrix} \nabla F(z) - \nabla G(z)^\top v - \nabla H(z)^\top w \\ w - \mu S^{-1} e \\ -G(z) \\ -H(z) + s \end{pmatrix} \tag{12.56}$$

for $\Sigma = S^{-1}W$ (or $\Sigma = \mu S^{-2}$ for the primal case), we obtain a symmetric form which can be dealt with much more effectively.

Moreover, the nonconvex case and the fact that the matrix in (12.56) may be singular—if the conditions discussed after (12.55) are not satisfied—have to be treated in order to approximate minimizers instead of mere KKT points and to render the approach applicable, i.e., the matrix to be invertible.

In order to ensure the approximation of a minimizer, as for the (SQP) algorithm (cf. Remark 12.23) we want to ensure that the value of the merit function is decreasing during the iteration. To this end, one can show that (d_z, d_s, d_v, d_w) is a descent direction of (12.56) if

$$\begin{pmatrix} \nabla^2_{zz}L(z,s,v,w) & 0 \\ 0 & \Sigma \end{pmatrix} \text{ is positive definite on the null space of } \begin{pmatrix} -\nabla G(z) & 0 \\ -\nabla H(z) & \mathrm{Id} \end{pmatrix}.$$

Here, Σ is positive definite by construction, but $\nabla^2_{zz}L(z,s,v,w)$ may be indefinite. To compensate for this deficiency, one can replace the Hessian by $\nabla^2_{zz}L(z,s,v,w) + \delta\mathrm{Id}$ where $\delta > 0$ is sufficiently large to ensure positive definiteness. The size of this modification is a priori unknown but can be obtained by successively enlarging δ. Additionally, a possible rank deficiency of $\nabla G(z)$ must be considered. In the primal-dual matrix

$$\begin{pmatrix} \nabla^2_{zz}L(z,s,v,w) + \delta\mathrm{Id} & 0 & -\nabla G(z)^\top & -\nabla H(z)^\top \\ 0 & \Sigma & 0 & \mathrm{Id} \\ -\nabla G(z) & 0 & \gamma\mathrm{Id} & 0 \\ -\nabla H(z) & \mathrm{Id} & 0 & 0 \end{pmatrix} \begin{pmatrix} d_z \\ d_s \\ d_v \\ d_w \end{pmatrix} = -\begin{pmatrix} \nabla F(z) - \nabla G(z)^\top v - \nabla H(z)^\top w \\ w - \mu S^{-1}e \\ -G(z) \\ -H(z) + s \end{pmatrix} \tag{12.57}$$

this is done by including a regularization parameter $\gamma > 0$.

Since the iteration (12.54) does not terminate in finite time, we impose the following error function

$$E(z,s,v,w;\mu) = \max\{\|\nabla F(z) - \nabla G(z)^\top v - \nabla H(z)^\top w\|, \tag{12.58}$$
$$\|Sw - \mu e\|, \|G(z)\|, \|H(z) - s\|\} \tag{12.59}$$

for some vector norm $\|\cdot\|$ which will be used for defining a termination criterion. Then, we obtain the following algorithm.

Algorithm 12.26 (*Basic interior-point algorithm*) Suppose a pair of initial values (z_0, s_0) to be given and set $k := 0$.
Compute multipliers v_0 and w_0, define parameters $\mu_0 > 0$, $\sigma, \tau \in (0,1)$

While convergence test not satisfied

1. While $E(z_k, s_k, v_k, w_k; \mu_k) \geq \mu_k$

    a. Compute search direction $d = (d_z, d_s, d_v, d_w)$ by solving (12.57)
    b. Determine $\alpha_s^{\max}$, $\alpha_w^{\max}$ via (12.55)
    c. Obtain new iterate $(z_{k+1}, s_{k+1}, v_{k+1}, w_{k+1})$ from (12.54)
    d. Set $\mu_{k+1} := \mu_k$ and $k := k + 1$

2. Choose $\mu_k \in (0, \sigma \mu_k)$

Again, for the convergence test it is checked whether the KKT conditions from Theorem 12.16 are satisfied up to some user defined tolerance tol_{OPT}.

For this algorithm, we can show the following result:

Theorem 12.27 (Convergence of interior-point algorithm) *Suppose F, G and H to be continuously differentiable functions and Algorithm 12.26 to generate a sequence $(z_k)_{k=0,\ldots,\infty}$ for a parameter sequence $\mu_k \to 0$ as $k \to \infty$. Then all points $z^\star$ with $\lim_{l\to\infty} z_{k_j} = z^\star$ for some subsequence k_l are feasible. If additionally the (LICQ) condition holds at $z^\star$, then the KKT conditions from Theorem 12.16 hold at $z^\star$.*

Proof Let $\hat{z}$ be the limit point of the sequence z_{k_l} from Algorithm 12.26. Since $\mu_k \to 0$ holds, the error function (12.58) converges to zero, and we have that $G_i(z_k) \to 0$ and $H_i(z_k) - s_k \to 0$. Since G and H are continuous, it follows that $G(\hat{z}) = 0$, $H(\hat{z}) \geq 0$ and $\hat{s} = H(\hat{z}) \geq 0$ showing the first assertion.

Since the error function (12.58) converges to zero and $\hat{s}_i = H_i(\hat{z}) > 0$ for all $i \notin \mathscr{A}(\hat{z})$, we obtain $[w_{k_l}]_i \to 0$ for all $i \notin \mathscr{A}(\hat{z})$. Again using that the error function converges to 0 this implies

$$\nabla F(z_{k_l}) - \sum_{i\in\mathscr{E}} [v_{k_l}]_i \nabla G_i(z_{k_l}) - \sum_{i\in\mathscr{A}(\hat{z})} [w_{k_l}]_i \nabla H_i(z_{k_l}) \to 0 \tag{12.60}$$

Since by (LICQ) the vectors $\nabla G_i(\hat{z})$, $i \in \mathscr{E}$ and $\nabla H_i(\hat{z})$, $i \in \mathscr{A}(\hat{z})$ are linearly independent and ∇F and ∇H are continuous, the vectors $\nabla G_i(z_{k_l})$ and $\nabla H_i(z_{k_l})$ are linearly independent for all sufficiently large k_l, too. It follows that $[v_{k_l}]_i$ and $[w_{k_l}]_i$ in (12.60) are unique and depend continuously on z_{k_l}, hence they converge to some $(\hat{v}, \hat{w}) \geq 0$. Using continuity of the expressions in the error function, it follows that the KKT conditions are satisfied for $\hat{z}$ with $\hat{\lambda} = (\hat{v}^\top, \hat{w}^\top)^\top$ as Lagrange multiplier. This proves the second assertion. □

Similar to the Line-Search (SQP) Algorithm, a merit function is imposed in the (IPM) variant of a line-search which may be of the form

$$\widetilde{L}(z, s, \nu) = F(z) - \mu \sum_{i=1}^{n_s} \log s_i + \nu \|G(z)\| + \nu \|H(z) - s\| \tag{12.61}$$

where the penalty parameter ν needs to be updated throughout the iteration. Note that we do not have to reprove results concerning descent of the cost function along the computed direction since results shown for the (SQP) case carry over to the (IPM) case. The line-search iteration is given by

$$z_{k+1} = z_k + \alpha_s d_z \qquad s_{k+1} = s_k + \alpha_s d_s$$
$$v_{k+1} = v_k + \alpha_w d_v \qquad w_{k+1} = w_k + \alpha_w d_w \tag{12.62}$$

where the step lengths are chosen to satisfy

$$\alpha_s \in (0, \alpha_s^{\max}], \qquad \alpha_w \in (0, \alpha_w^{\max}] \tag{12.63}$$

and

$$\widetilde{L}(z_{k+1}, s_{k+1}, \nu) \leq \widetilde{L}(z_k, s_k, \nu) + \eta \alpha_s D\widetilde{L}(z_k, s_k, \nu; (d_z, d_s)) \tag{12.64}$$

for some $\eta \in (0, 1)$. To incorporate the changes due to the introduction of the merit function (12.61), we modify Algorithm 12.26 by defining the fixed parameter $\eta \in (0, 1)$ and changing Steps 1b to 1d as follows.

1b. Compute $\alpha_s^{\max}$, $\alpha_w^{\max}$ via (12.55)
1c. Determine α_s, α_w such that (12.63) and (12.64) hold
1d. Obtain new iterate $(z_{k+1}, s_{k+1}, v_{k+1}, w_{k+1})$ from (12.62)

Similar to the (SQP) method we can also formulate a trust-region version of the (IPM) algorithm in which a trust-region ball constraint $\|(d_z, \tilde{d}_s)\|_2 \leq \Delta_k$ is added with $\tilde{d}_s = S_k^{-1} d_s$. This constraint ensures that the quadratic approximation is not treated as a global model, but only in a small neighborhood of z_k. The model problem itself is given by

$$\begin{aligned}
&\text{minimize} \quad \nabla F(z_k)^\top d_z + \frac{1}{2} d_z^\top \nabla_{zz}^2 L(z_k, s, v, w) d_z - \mu e^\top \tilde{d}_s + \frac{1}{2} \tilde{d}_s^\top S_k \Sigma_k S_k \tilde{d}_s \\
&\text{with respect to } d_z \in \mathbb{R}^{n_z}, d_s \in \mathbb{R}^{n_s} \\
&\text{subject to } \nabla G(z_k) d_z + G(z_k) = r_{G_k}, \\
&\nabla H(z_k) d_z - S_k \tilde{d}_s + (H(z_k) - s_k) = r_{H_k}, \\
&\|(d_z, \tilde{d}_s)\|_2 \leq \Delta_k \text{ and } \tilde{d}_s \geq \tau e.
\end{aligned}$$

where the descent direction $\tilde{d}_s$ is bounded away from zero by a parameter $\tau \in (0, 1)$. To compute the residuals r_{G_k}, r_{H_k} a subproblem similar to (12.42) is derived, but now the variable $c_k = (c_z, c_s)$ is considered:

$$\text{Minimize} \quad \|\nabla G(z_k) c_z + G(z_k)\|_2^2 + \|\nabla H(z_k) c_z - S_k c_s + (H(z_k) - s)\|_2^2 \tag{12.65}$$
$$\text{with respect to} \quad c_k \in \mathbb{R}^{n_z + n_s} \quad \text{subject to} \quad \|c_k\|_2 \leq 0.8\Delta \text{ and } c_s \geq -\frac{\tau}{2} e$$

Then, the radii r_G and r_H are given by

$$r_{G_k} = \nabla G(z_k) c_z + G(z_k), \qquad r_{H_k} = \nabla H(z_k) c_z - S_k c_s + (H(z_k) - s_k). \tag{12.66}$$

Again similar to the (SQP) case, cf. (12.43), a step is accepted if for the direction (d_z, d_s) the ratio ρ of the reduction of the merit function $\widetilde{L}$ with respect to the predicted reduction using a quadratic model Q_ν of the merit function $\widetilde{L}$

$$\rho = \frac{\widetilde{L}(z, s, \nu) - \widetilde{L}(z + d_z, s + d_s, \nu)}{Q_\nu(0) - Q_\nu(d_z, d_s)} \tag{12.67}$$

is at least $\rho \geq \eta$ where ν needs to be chosen sufficiently large.

To incorporate the trust-region idea into Algorithm 12.26, we introduce the fixed parameter $\eta \in (0, 1)$ and modify Steps (1a) to (1c):

1a. Evaluate normal step c_k from (12.65) and radii r_G, r_H via (12.66)
1b. Compute solution $d_z, \tilde{d}_s$ for quadratic model of (IPM) and retrieve $d_s = S\tilde{d}_s$
1c. Obtain ρ from (12.67)
    If $\rho \geq \eta$: Set $z_{k+1} = z_k + d_z$, $s_{k+1} = s_k + d_s$, update multipliers v_k, w_k and choose $\Delta_{k+1} \geq \Delta_k$
    Else: Set $z_{k+1} = z_k$, $s_{k+1} = s_k$, $v_{k+1} = v_k$, $w_{k+1} = w_k$ and choose $\Delta_{k+1} < \Delta_k$

Here, the update of the Lagrange multipliers can be done similar to the least squares approach outlined in (12.39).

Note that there also exist other types of solvers for the general (NLP) problem like penalty and augmented Lagrangian methods which we do not discuss here. Additionally, there exists a wide variety of solvers for quadratic problems which we did not mention here but which are of particular interest: on the one hand, quadratic problem solvers are required within (SQP) and (IPM) algorithms. On the other hand, many practical problems directly lead to a quadratic cost function.

Additionally, many extensions, modifications, and improvements for the presented algorithms have been developed whose explanation is beyond the scope of this section; some of these are given in a short overview in Sect. 12.7.

12.4 Implementation Issues in NMPC

So far, we discussed the numerical components which are typically part of an NMPC algorithm and whose interplay is sketched in Fig. 12.10. Our aim now is to combine these components, i.e., the differential equation solver from Chap. 11, the discretization technique from Sect. 12.1 and the optimizer from Sect. 12.3, and to analyze the interactions and effects between them.

Since the interaction heavily depends on the used implementation, one cannot give a complete analysis. Therefore, the aim of this section is to give some guidelines regarding the interplay of the components with respect to certain key parameters such as the tolerance values for the optimization and integration, the horizon length or the number of multiple shooting nodes.

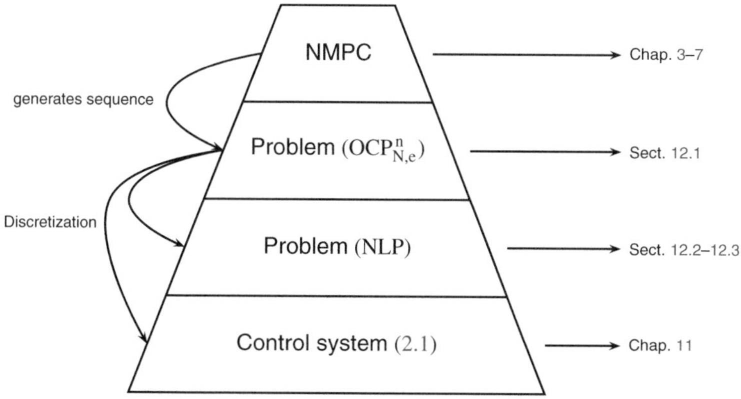

Fig. 12.10 Architecture of NMPC solver

Structure of the Derivatives

We start by analyzing the connection between the discretization and the optimization method. As we have seen in Sect. 12.3, both the (SQP) and the (IPM) algorithms require derivatives of the cost function and the Jacobian of the constraints in order to compute a search direction d_k. Since for many examples algebraic expressions for these function are not at hand, these derivatives need to be computed numerically. To this end, in the continuous time case one may numerically solve the variational equation (cf. Hairer and Wanner [27, Sect. I.14]). Alternatively, one may directly approximate all derivatives by difference quotients with respect to every component of the optimization variable z. Here we follow this second approach which approximates the directional derivative of a function $g : \mathbb{R}^{n_z} \to \mathbb{R}^p$ at a point z in direction v via

$$\nabla g(z)^\top v = \lim_{t \to 0} \frac{g(z + t \cdot v) - g(z)}{t} \approx \frac{g(z + t \cdot v) - g(z)}{t} \tag{12.68}$$

for small $t > 0$. Here, the step length t should be chosen depending on the used computer, i.e., its floating point accuracy. Using the ith unit vector $v = e_i$, we thus obtain an approximation of the ith row of the Jacobi matrix $\nabla g(z)$.

Note that if g involves the solution of a continuous time optimal control problem using the methods from Chap. 11, then the computation of the difference quotient (12.68) may depend sensitively on the time grids used to compute $g(z + t \cdot v)$ and $g(z)$. To circumvent this problem, a synchronized evaluation of f for the nominal vector z and for the disturbed vectors $z + t \cdot v$ using the same grid for both computations can be used.

A similar problem arises from a possible integral type cost function (3.4) with running cost L, which has to be evaluated along the trajectory. To this end, we use the technique described in Sect. 11.4, i.e., we include the integral as an additional component in the ordinary differential equation and use the same discretization for

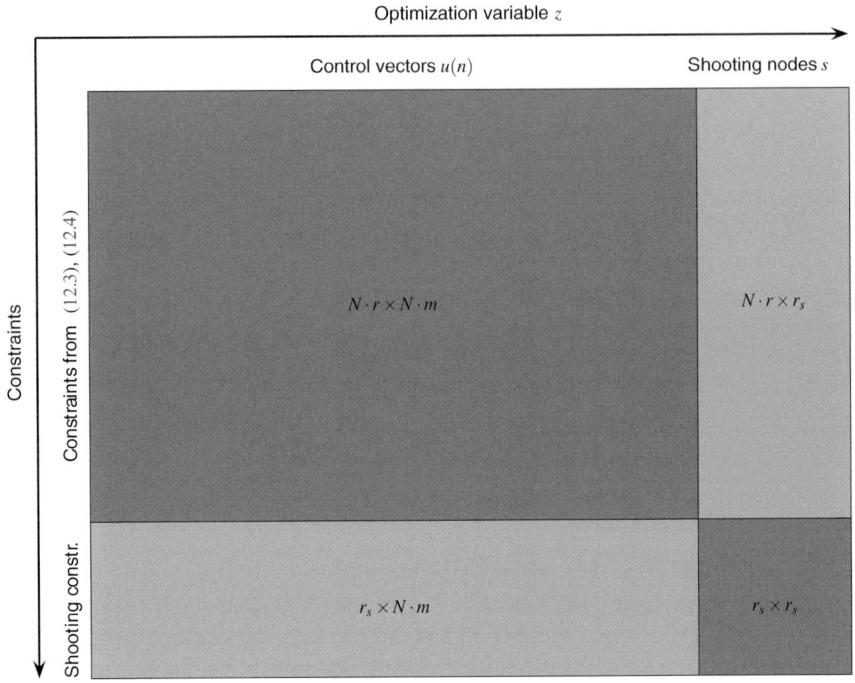

Fig. 12.11 Structure of the Jacobian of the constraints $\nabla_z C(z)^\top$

the evaluation of this integral and the dynamics of the problem, see also the paper of Gyurkovics and Elaiw [26] for a detailed analysis in an NMPC context.

Most if not all common optimization algorithms use an approximation $B(z)$ of the Hessian $\nabla_{zz} L$ of the cost function which can be obtained, e.g., by so-called DFP (Davidon-Fletcher-Powell) [11, 21] or BFGS (Broydon-Fletcher-Goldfarb-Shanno) updates [7, 19, 22, 39]. For this reason, the computationally most expensive part within each step of the optimization algorithm is the evaluation of the Jacobian $\nabla C(z)$ of the constraints. We now consider this problem in some more depth for the multiple shooting method. In the following, $r = r_g + r_h$ denotes the number of constraints and—as before—r_s the number of shooting nodes. Using that the optimization variable $z \in \mathbb{R}^{n_z}$ is of the form $z = (u(0)^\top, \ldots, u(N-1)^\top, s^\top)^\top$, this Jacobi matrix has the structure sketched in Fig. 12.11.

This figure indicates that the number of function evaluations is growing quadratically in the horizon length N, which is therefore the dominating parameter for the computing time of the optimization algorithm. For this reason, setting up an NMPC scheme which is stable for small optimization horizon N—for instance by choosing a good cost function in the sense of Sect. 6.6, by adding terminal weights as described in Sect. 10.2 or by using the adaptation techniques of Sect. 10.8—will in general significantly reduce the computational effort.

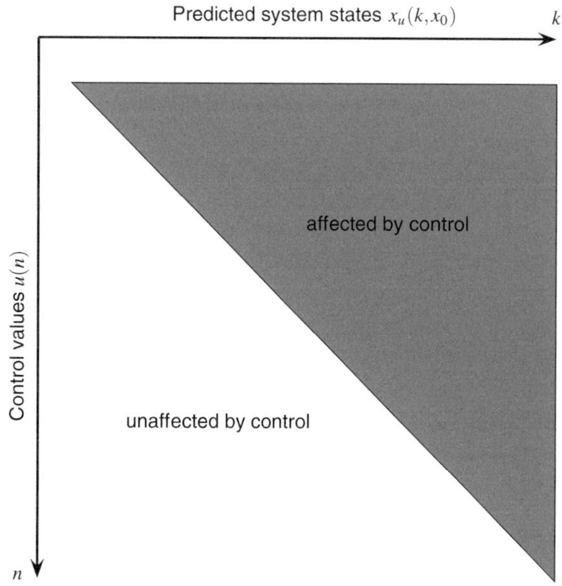

Fig. 12.12 Dependence of $x_u(k, x_0)$ on $u(n)$

Regardless of the length of the horizon, a recomputation of the Jacobian $\nabla C(z)$ has to be performed for a large number of the iteration steps within any of the presented minimization routines. Hence, an efficient implementation to evaluate this matrix is required. Additionally, efficient (SQP) and (IPM) implementations usually update only a few entries of the Jacobian whereas for the other entries the linearization is considered to be sufficiently accurate. This implies that an efficient implementation of the evaluation of the Jacobian should perform only those evaluations of the dynamics of the system needed for the entries of the Jacobian which are actually requested by the optimization algorithm.

We now sketch such an efficient evaluation which is based on the following observation: if we look at the way the predicted states $x_u(k, x_0)$ depend on the control values $u(n)$, one observes the triangular structure sketched in Fig. 12.12 in which the arrows indicate increasing indices k and n.

This structure is due to that fact that changing $u(n)$ does not affect the state trajectory values $x_u(k, x_0)$ for $k \in \{0, \ldots, n\}$.

Our proposed efficient implementation builds upon this observation. Assume that the columns of $\nabla C(z)^\top$ colored in dark gray in Fig. 12.13 have to be recomputed. To this end, we need to evaluate the nominal function $C(z)$ as well as the perturbed values $C(z + te_i)$ for those i corresponding to the dark gray columns. Each of these evaluations requires the computation of a trajectory following the rules described before (12.8). Since $z = (u(0)^\top, \ldots, u(N-1)^\top, s^\top)^\top$, each e_i either corresponds to an entry of a control vector $u(n)$ or to a shooting node s_k. Let us first consider those e_i which correspond to entries of control vectors and denote the time index of the corresponding control vectors by n_i. Then, the definition of z yields that $i_1 \geq i_2$ implies $n_{i_1} \geq n_{i_2}$.

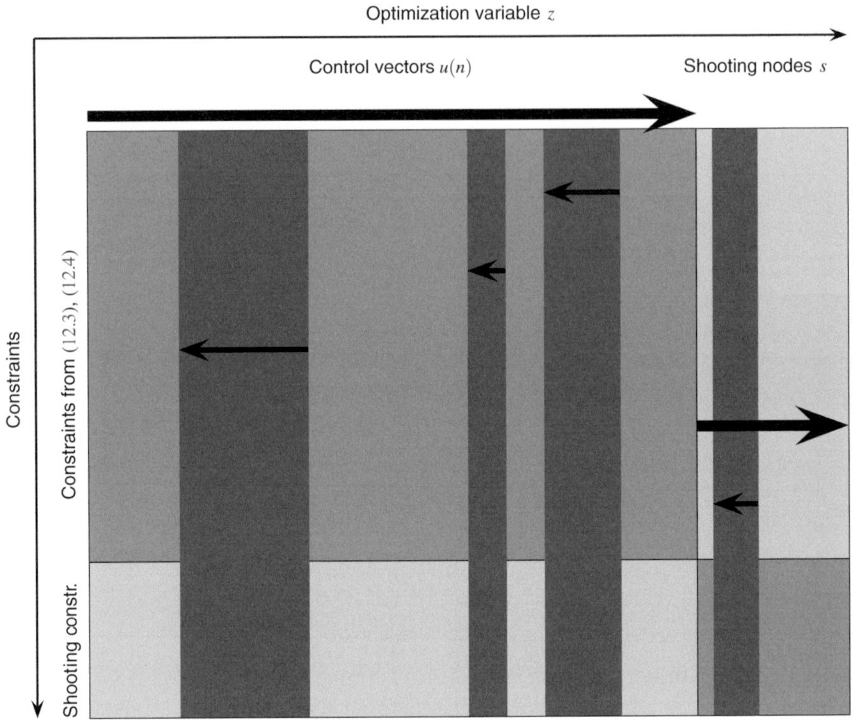

Fig. 12.13 Efficient implementation for computing the Jacobian $\nabla_z C(z)^\top$

The structure from Fig. 12.12 now implies that the nominal trajectory x_u corresponding to z and the perturbed trajectory $x_{u'}$ corresponding to $z + te_i$ satisfy $x_u(k) = x_{u'}(k)$ for $k = 0, \ldots, n_i$. This means that these values can be reused for $x_{u'}$ and do not need to be recomputed. Hence, for each perturbed trajectory to be computed one could copy the nominal trajectory and then change only those entries which do not coincide. While this approach is already very efficient in terms of the number of function evaluations, it is not efficient in terms of memory access because we need to make many copies of the nominal trajectory. Since memory access is one of the bottle necks in modern computers, this procedure will be quite time consuming.

To solve this problem we use a second implication of Fig. 12.12: consider two perturbed trajectories x_{u_1} and x_{u_2} corresponding to $z + te_{i_1}$ and $z + te_{i_2}$ with $i_1 \geq i_2$. Then the time indices of the corresponding control vectors satisfy $n_{i_1} \geq n_{i_2}$ and consequently the triangular structure from Fig. 12.12 yields the identity $x_{u_1}(k) = x_{u_2}(k)$ holds for $k = 0, \ldots, n_i$. This means that when computing $x_{u_2}(k)$, instead of using the values of the nominal trajectory x_u we may also use the values of any perturbed trajectory x_{u_1} with $i_1 \geq i_2$. Thus, if we schedule the computations of $C(z + te_i)$ such that the indices i are monotone decreasing, then we can compute each perturbed trajectory by modifying the perturbed trajectory computed before, i.e., we can compute all entries in the Jacobian by working on a single copy of the nominal trajectory instead of having to make many copies.

The indices e_i corresponding to shooting nodes can be treated in the same manner observing that again the change of a shooting node only affects the solution at times greater or equal than the shooting time.

The resulting scheduling of the computations is indicated by the arrows in Fig. 12.13. The two thicker arrows pointing to the right correspond to the computation of the nominal trajectory for the control and the shooting values, respectively, and the thinner arrows pointing to the left indicate the order of the computations of the perturbed trajectories w.r.t. the indices of the corresponding e_i.

An alternative to this efficient sequential scheduling is given by parallelization of the computations of the perturbed trajectories. Except for the evaluation of the nominal trajectory, all evaluations of the dynamics (2.1) (or (2.8) in the continuous time case) required to compute $\nabla C(z)$ and $\nabla F(z)$ are fully decoupled and may be executed in parallel. Hence, if the problem is large and many cores can be used to parallelize these computations, a significant speedup can be expected.

Condensing

As outlined in the discussion of Fig. 12.11, the computation of the Jacobian of the constraints massively depends on the number of optimization variables of the discretized optimal control problem. Hence, according to Sect. 12.1 the minimal number of optimization variables is $N \cdot m$ for control values $u \in \mathbb{R}^m$. Adding multiple shooting nodes increases the number of optimization variables but may also improve the numerical solutions, cf. Example 12.2. In the following, we present an approach which allows us to reduce the additional numerical effort induced by using shooting nodes and which has become quite popular in the NMPC community, the so-called *condensing* of constraints, see, e.g., the PhD thesis of Diehl [12] or the paper by Bock and Plitt [6].

In order to simplify the exposition, here we apply the method to the full discretization technique which can be regarded as a special case of the multiple shooting technique in which each component of the solution vector at each sampling instant is a shooting node. To simplify notation, we now define s_j to be a vector of dimension d, which allows us to get rid of the index function $\iota(\cdot)$, and we set the index function $\varsigma(\cdot)$ to be the identity function. According to this change, the equality constraints induced by the continuity conditions (12.9) take the form

$$S(z) := \Big[\big[s_{j+1} - f\left(s_j, u(j)\right)\big]_{j\in\{0,\ldots,N-1\}} \Big] = 0.$$

Using a Newton like approach for the KKT conditions from Theorem 12.16, as outlined for the active set and interior-point methods in Sect. 12.3, we obtain the linearized continuity condition

$$S(z) + \nabla_z S(z)^\top \Delta s = 0 \tag{12.69}$$

where Δs denotes the part of the search direction d corresponding to the shooting nodes. Within this equation, the Jacobian of $S(z)$ takes the sparse form

$$\nabla_z S(z)^\top = \begin{pmatrix} -\frac{\partial f(s_0,u(0))}{\partial s_0} & 1 & 0 & \cdots & \cdots & 0 \\ 0 & -\frac{\partial f(s_1,u(1))}{\partial s_1} & 1 & 0 & \cdots & \vdots \\ \vdots & \ddots & \ddots & \ddots & 1 & 0 \\ 0 & \cdots & \cdots & 0 & -\frac{\partial f(s_{N-1},u(N-1))}{\partial s_{N-1}} & 1 \end{pmatrix}.$$

This allows us to compute Δs from (12.69) via

$$\begin{pmatrix} \Delta s_1 \\ \vdots \\ \Delta s_N \end{pmatrix} = \begin{pmatrix} \frac{\partial f(s_0,u(0))}{\partial s_0} & 0 & \cdots & 0 \\ 0 & \frac{\partial f(s_1,u(1))}{\partial s_1} & 0 & \vdots \\ \vdots & \ddots & \ddots & 0 \\ 0 & \cdots & 0 & \frac{\partial f(s_{N-1},u(N-1))}{\partial s_{N-1}} \end{pmatrix} \begin{pmatrix} \Delta s_0 \\ \vdots \\ \Delta s_{N-1} \end{pmatrix} - S(z),$$

i.e., $\Delta s_1, \ldots, \Delta s_N$ can be computed from Δs_0 which is in the search direction for the shooting node corresponding to the initial value. Hence, if we solve (12.69) outside the discretized optimal control problem, then the fully discretized problem reduces to a shooting discretization in which only the components of the initial value are considered to be shooting nodes, i.e., $r_s = d$. Note that the dynamics of the system still have to be evaluated using the initial values given by the shooting nodes of the original problem. However, apart from this technical difference, the condensing technique allows us to use a full discretization of the optimal control problem while the complexity of the problem handed over to the optimization algorithm is reduced to a multiple shooting approach with only d instead of $d \cdot (N + 1)$ shooting nodes.

A useful side effect of using the full discretization we would like to mention is that all evaluations of the dynamics (2.1) (or (2.8) in the continuous time case) are fully decoupled. Hence, computing all components of the (discretized) optimization problem can be done in parallel, which may lead to a significant speedup in generating the iterates of the optimization problem.

Optimality and Computing Tolerances

Let us now turn to the investigation of the interplay of the different numerical accuracies in the optimization method and the differential equation solver. In particular, we analyze the impact of the user dependent choices of tolerance values for these two components on the stability of the closed-loop system on the one hand, and on the resulting computing time on the other hand. For the differential equations solver, the tolerance tol_{ODE} is the parameter discussed in Sect. 11.3. For the optimization routine, the parameter tol_{OPT} is the accuracy up to which the sufficient KKT conditions of Theorem 12.16 are satisfied. Both tol_{OPT} and tol_{ODE} are assumed to be small in this section. The situation in which the optimization is terminated before the necessary conditions are approximately satisfied, which was theoretically investigated in Sect. 10.9 will be discussed in Sect. 12.6.

To illustrate effects of different choices for the tolerance levels, let us first give the following example.

Example 12.28 Consider the inverted pendulum Example 12.1. As tolerances for the differential equation solver and the optimization algorithm we consider the grid of parameters

$$(\text{tol}_{\text{OPT}}, \text{tol}_{\text{ODE}}) \in \{10^{-8}, 10^{-7}, \ldots, 10^{-1}\}^2$$

in order to evaluate the NMPC controller μ_N. The resulting closed-loop trajectory $x_{\mu_N}(\cdot, x_0)$ was computed on the interval $[0, 20]$ with sampling period $T = 0.1$, initial value $x_0 = (2, 2, 0, 0)$ and numerical accuracy $\text{tol}_{\text{ODE}} = 10^{-10}$. By using a much smaller tolerance for computing $x_{\mu_N}(\cdot, x_0)$ we "emulate" the setting of Sect. 11.5 in which the computation of μ_N is affected by numerical errors and the closed-loop solution is obtained by using the numerically computed μ_N and the exact dynamics.

Let us first consider one of the critical cases mentioned in Example 12.1 and set the horizon length to $N = 53$. In this case, the numerical results in Fig. 12.14 illustrate that the parameters $(\text{tol}_{\text{OPT}}, \text{tol}_{\text{ODE}})$ should be chosen carefully due to their effects on the closed loop. The closed-loop costs in the following figures are computed as in Example 12.2.

Here, Fig. 12.14a shows that for some pairs $(\text{tol}_{\text{OPT}}, \text{tol}_{\text{ODE}})$ the closed-loop costs are significantly higher. If we analyze the corresponding closed-loop trajectories, these costs correspond to solutions which are driven to the downward position $(-2\pi, 0, 0, 0)$ for $\text{tol}_{\text{OPT}} = 10^{-1}$ and several values considered for tol_{ODE}, cf. the dotted lines in Fig. 12.14b. All other pairs, however, do "mainly" cause the pendulum to tend toward an upright position. However, these positions vary in two ways.

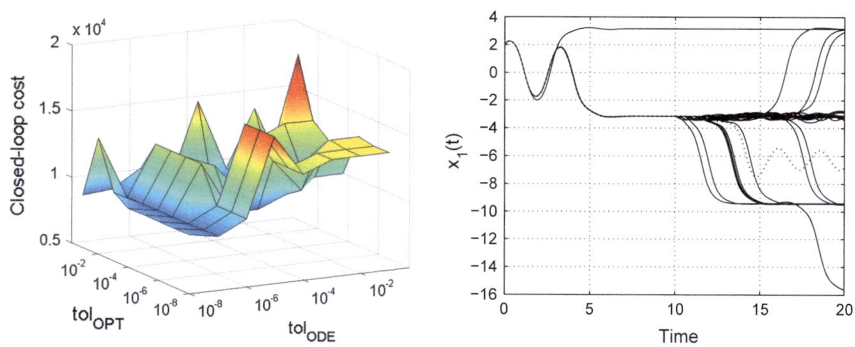

(a) Closed-loop cost (b) $x(\cdot)$ trajectories

Fig. 12.14 Exceptional closed-loop costs for $N = 53$ and various optimization and differential equation solver tolerances

For one, the x_1-component of the closed-loop trajectory does not converge to the same upright position, similar to what we observed in Example 12.2. And, more importantly, once an upright position is reached, the trajectory may become unstable and move to another upright position, an effect we already observed in Example 12.1. In this case, the observed closed-loop costs increase due to the repeated transitions between different upright positions, see again Fig. 12.14a.

The reasons for this unpredictable behavior can be found in the implementation and the unstable nature of the equilibrium which we want to stabilize. One source of errors is the tolerance tol_{ODE} used in computing μ_N. Since the closed-loop trajectory is always computed with accuracy $\text{tol}_{\text{ODE}} = 10^{-10}$ and the open-loop predictions are computed with different tolerance levels of the differential equation solver, the solutions deviate slightly. Consequently, according to Theorem 11.10, we can only expect practical asymptotic stability and the radius δ of the neighborhood to which the solutions converge according to Definition 11.9 will grow with the tolerance tol_{ODE} used in the computation of μ_N (and most likely also with tol_{OPT}, even though this was not rigorously analyzed in Sect. 11.5). Once δ becomes too large, the pendulum may leave a neighborhood of the upright equilibrium and move to another upright position.

Another source of errors stems from the fact that the optimization algorithm does not yield a globally optimal solution. In fact, since the problem is nonconvex we can in general never guarantee to obtain a globally optimal solution, but the growth of the closed-loop costs in Fig. 12.14a for increasing tol_{OPT} indicates that this is more likely for larger tol_{OPT}. In contrast to numerical errors in the solution of the ordinary differential equation, this problem can be dealt with in various ways.

For example, additional constraints can be added once a trajectory is close to the equilibrium in order to exclude the open-loop trajectory from moving to other equilibria. Alternatively, the initial guess of the control may be changed when starting of the optimization such that the initial solution stays close to the equilibrium. A third possibility is to add shooting nodes and set their values to the desired equilibrium, as already discussed in the introduction to multiple shooting in Sect. 12.1. This is less restrictive than adding constraints and incorporates the aspect of modifying the initial guess at the same time. As discussed in Example 12.2(ii), introducing shooting nodes may obstruct the optimization routine. Close to the equilibrium, however, we have never experienced such effects during our numerical experiments.

Although too large choices of the tolerance levels may have destabilizing effects, from a computation time point of view it is in general desirable to choose these values as large as possible in order to speed up the computation required for each NMPC iterate. In particular, for larger tolerance levels less steps within the differential equation solver as well as the optimization method are required to satisfy these tolerance levels. Unfortunately, inaccuracies in the differential equation solver may obstruct the optimization method and vice versa. Still, if we want to accelerate the NMPC algorithm, this tuning parameters should be considered closely. To illustrate these interactions, we continue with Example 12.28.

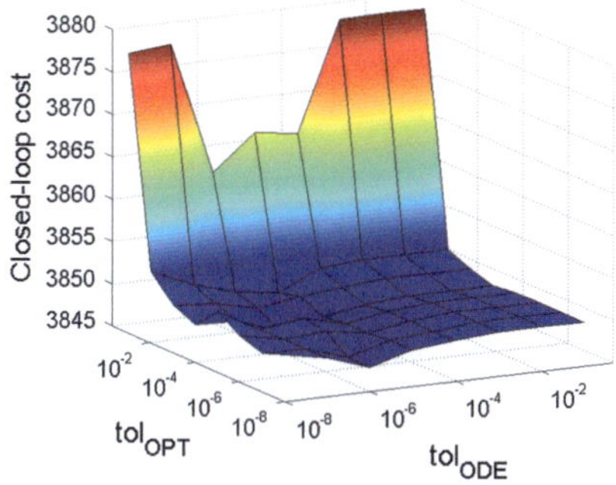

(a) Closed-loop costs

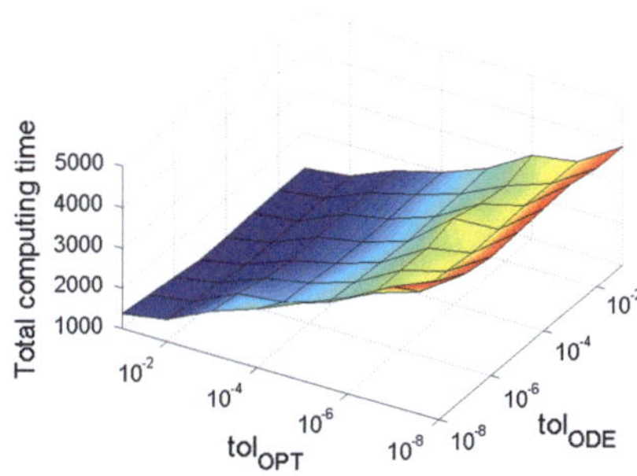

(b) Computing times

Fig. 12.15 Comparison of closed-loop costs and computing times for $N = 15$ and various optimization and differential equation solver tolerances

Example 12.29 In the following, we compare the closed-loop behavior and the computing time for various tolerance levels. In Fig. 12.15, we show the limit case $N = 15$ for the optimization horizon. Figure 12.15a, b allow us to conclude that tolerance level tol_{OPT} may be eased to 10^{-2} without increasing the closed-loop cost significantly or affecting the stability of the closed loop while the computing times are lowered. Our default initial setting for $(\text{tol}_{\text{OPT}}, \text{tol}_{\text{ODE}})$ is the pair $(10^{-6}, 10^{-6})$, but already for $(\text{tol}_{\text{OPT}}, \text{tol}_{\text{ODE}}) = (10^{-3}, 10^{-4})$ the computing time can be reduced by approximately 35 %.

There are several guidelines which can be drawn from this example, and which have shown to be reasonable in our numerical experiments.

First, the horizon length should always be chosen to be as small as possible before different tolerance levels are considered. The reasons for this are twofold: For one, the additional complexity of an enlarged horizon always leads to an increase in the computing time to solve the optimal control problem. While this may be compensated by increasing the tolerance levels of the optimizer and the differential equation solver, another aspect comes into play which we have not yet discussed: increasing the tolerance level of the differential equation solver may heavily corrupt the numerical evaluation of the derivatives needed in the optimization algorithm. For this reason, the optimization iteration may be unable to converge toward a point satisfying the KKT conditions of Theorem 12.16 with accuracy less than tol_{OPT}.

As a consequence, the tolerance levels should always satisfy $\text{tol}_{\text{OPT}} \gtrsim \text{tol}_{\text{ODE}}$. Here, we use this imprecise formulation on purpose since properties of the differential equation like its rate of change or simple its scale influence the "best" relation between these two tolerance parameters. Yet, it is usually reasonable to set the tolerance level of the differential equation solver to be smaller than the tolerance level of the optimization method.

12.5 Warm Start of the NMPC Optimization

In the previous section, we concentrated on solving Step (2) of Algorithm 3.11 which is computationally quite demanding. Although it is not obvious from the algorithm itself, the consecutive Step (3) can be used to significantly reduce the time required to solve the subsequent problem of type $(\mathrm{OCP}^{\mathrm{n}}_{\mathrm{N,e}})$. More precisely, the optimal solution $u^\star$ of $(\mathrm{OCP}^{\mathrm{n}}_{\mathrm{N,e}})$ at time n can be efficiently used in order to construct an initial guess for $(\mathrm{OCP}^{\mathrm{n}}_{\mathrm{N,e}})$ at time $n+1$. In order to see how this can be done, one needs to take a closer look to the time evolution of the NMPC problem itself as displayed in Figs. 12.16 and 12.17 for a sampled data continuous time system with sampling period T. For better visibility, in these figures we have plotted the whole continuous time trajectory instead of only the discrete values of $x_u(\cdot, x_0)$.

Figure 12.16 schematically sketches the step of solving the problem $(\mathrm{OCP}^{\mathrm{n}}_{\mathrm{N,e}})$ for a given initial value x_0 for the nth iterate of the NMPC procedure. As a result, a prediction of the trajectory $x_u(\cdot, x_0)$ on the considered time horizon and a minimizing control sequence $u^\star(\cdot) \in \mathbb{U}^N(x_0)$ are obtained.

Now, according to Step (3) of Algorithm 3.11, the first element of the control sequence $u^\star(\cdot)$ is used as control value in the next sampling period, see Fig. 12.17 for a schematic sketch. However, the algorithm does not say what should be done with the remainder of the control sequence. The simplest approach is to discard these values and go to the next iterate. Yet, as obvious from Fig. 12.17, the sequence contains a lot of information on the prediction at time $n+1$.

Hence, this information should be reused in the subsequent $(n+1)$st NMPC step. In particular, one has to keep in mind that in most cases an iterative solver is used to compute the optimal control $u^\star(\cdot)$ for problem $(\mathrm{OCP}^{\mathrm{n}}_{\mathrm{N,e}})$. As mentioned in the previous Sect. 12.3, one can only guarantee local convergence for such solvers, i.e., a good initial guess is required. Furthermore, supplying a good initial guess will

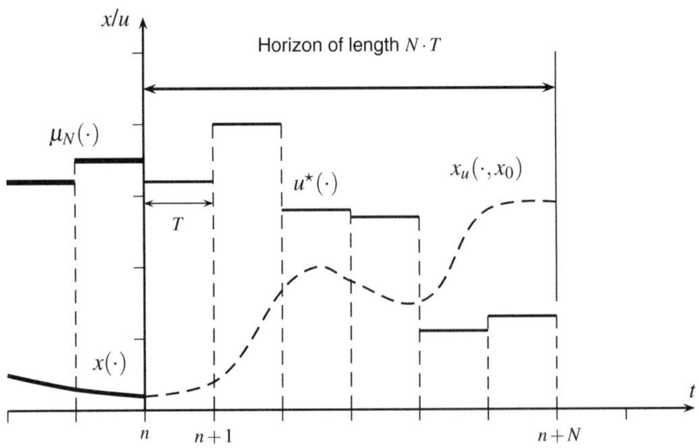

Fig. 12.16 Step (2) of Algorithm 3.11

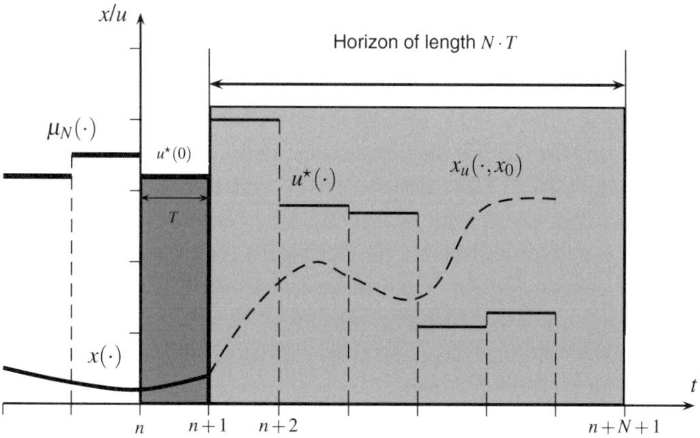

Fig. 12.17 Step (3) of Algorithm 3.11

reduce the number of iterations to be performed by both the (IPM) and the (SQP) method.

In what follows we will explain three methods to address the "*good initial guess*" problem.

Initial Value Embedding

The *initial value embedding* technique proposed by Diehl, Bock, Schlöder, Findeisen, Nagy and Allgöwer [13] is used for optimal control problems ($\text{OCP}_{\text{N,e}}$) discretized with full discretization or with recursive elimination and solved using (SQP) methods. The approach utilizes the fact that the subsequent problem differs only in the initial value x_0. In particular, taking a closer look at the problem (NLP) using full discretization, one can see that the problem depends linearly on the x_0 in the equality constraint

$$x_u(0, x_0) - x_0 = 0 \tag{12.70}$$

where $x_u(0, x_0)$ is the first entry of the optimization vector z as displayed in (12.5). Now, if the (SQP) solver handling the subsequent problem ($\text{OCP}_{\text{N,e}}$) with time index $n + 1$ is initialized via

$$z_0 := z^\star = (x_{u^\star}(0, x_0)^\top, \ldots, x_{u^\star}(N, x_0)^\top, u^\star(0)^\top, \ldots, u^\star(N-1)^\top)^\top$$

where $z^\star$ is the optimal solution of the nth problem ($\text{OCP}_{\text{N,e}}$), then most of the data of the previous linear problem

$$\begin{bmatrix} \nabla^2_{zz} L(z^\star, \tilde{\lambda}^{\mathscr{W}_k,\star}) & -\nabla C^{\mathscr{W}_k}(z^\star)^\top \\ \nabla C^{\mathscr{W}_k}(z^\star) & 0 \end{bmatrix} \begin{pmatrix} d^\star \\ \tilde{\lambda}^{(\text{EQP}),\star} \end{pmatrix} + \begin{pmatrix} \nabla F(z^\star) \\ C^{\mathscr{W}_k}(z^\star) \end{pmatrix} = 0 \tag{12.71}$$

solved at time n, cf. Algorithm 12.25, can be reused for the first (SQP) step at time $n + 1$. This is because only those very few entries in (12.71) which depend on the new initial value x_0^{new} at time $n + 1$ need to be updated. Hence, the approximation in the first (SQP) step at time $n + 1$ is almost readily computed. Note, however that this advantage only holds for the full discretization, since for the problem with recursive elimination all values in (12.71) depend on x_0 and thus all gradients need to be recomputed. The impact of the latter issue can be reduced if the entire sensitivity matrix in (12.71) is precomputed and the derivative regarding x_0 is added online as the measurement is received, or if the latter is included in the quadratic problem (EQP). Furthermore, this procedure becomes less efficient in the case of nonconstant reference x^{ref} because in this case ℓ and thus F change with time and the respective gradients have to be recomputed, too.

If, moreover, x_0 and x_0^{new} differ only slightly, as it is the case for small sampling periods, then the remaining entries in (12.71) may still be sufficiently accurate in order to allow for an accurate solution of $(\text{OCP}_{\text{N,e}})$ for the new initial value x_0^{new}. In this case, a single (SQP) iteration is sufficient in order to deliver a sufficiently accurate solution at time $n + 1$. Furthermore, since the constraint (12.70) is linear in $x_0 = x_0^{\text{new}}$, the infeasibility introduced by inserting x_0^{new} will be removed after this single iteration.

An interesting application of the initial value embedding technique is the so-called *real-time iteration* introduced by Diehl [12]. While real-time iterations were used to update solutions, the explicit usage of initial value embedding was proposed later in in the articles of Bock, Diehl, Kühl, Kostina, Schlöder and Wirsching [4] and Ferreau, Bock and Diehl [17]. In this approach one updates the control sequence several times in one sampling interval, assuming, of course that the used hardware (sensors, actuators, etc.) allows for a faster sampling of the control variable u. The structure of the resulting scheme does not fully comply with the theoretical analysis in the previous chapters, but we conjecture that it can be analyzed by suitably extending the multirate sampling approach discussed before Remark 2.9.

In the real-time iteration, one considers the initial condition x_0 as a parameter of the nonlinear optimization problem resulting from a full discretization. In addition, a scalar parameter $\tau \in [0, 1]$ representing the time between two NMPC sampling instants is introduced to obtain

$$
\begin{aligned}
&\text{minimize} \quad F(\tau, \hat{z}) \\
&\text{with respect to } \hat{z} \in \mathbb{R}^{n_z} \\
&\text{subject to } G_i(\tau, \hat{z}) = 0 \text{ for all } i \in \mathscr{E} \\
&\text{and } H_i(\tau, \hat{z}) \geq 0 \text{ for all } i \in \mathscr{I}
\end{aligned}
$$

Here, $\tau = 0$ represents the discretized minimization problem of the nth problem $(\text{OCP}_{\text{N,e}})$ and $\tau = 1$ the problem at time instant $n + 1$. The parametric formulation now allows for different ways of updating the control μ_N in between the standard NMPC time instants n and $n + 1$ by repeatedly solving (SQP) subproblems with partially updated data. These updating strategies can ordered in terms of the relative time τ as indicated in Table 12.1.

Table 12.1 Multilevel NMPC control update possibilities using parametric quadratic programming

	Step size	Required updates	
A	$\Delta\tau_A \lll 1$		Linear MPC
B	$\Delta\tau_B \ll 1$	$C^{\mathcal{W}_k}(\tau + \Delta\tau_B, \hat{z})$, $\nabla F(\tau + \Delta\tau_B, \hat{z})$	Feasibility improvement
C	$\Delta\tau_C < 1$	$C^{\mathcal{W}_k}(\tau + \Delta\tau_C, \hat{z})$, $\nabla F(\tau + \Delta\tau_C, \hat{z}), \tilde{\lambda}^{\mathcal{W}_k,\star}$	Optimality improvement
D	$\Delta\tau_D = 1$	$C^{\mathcal{W}_k}(\tau + \Delta\tau_D, \hat{z})$, $\nabla F(\tau + \Delta\tau_D, \hat{z}), \tilde{\lambda}^{\mathcal{W}_k,\star}$, $\nabla C^{\mathcal{W}_k}(\tau + \Delta\tau_D, \hat{z})$ $\nabla^2_{zz} L(\tau + \Delta\tau, z^\star, \tilde{\lambda}^{\mathcal{W}_k,\star})$	Single SQP step

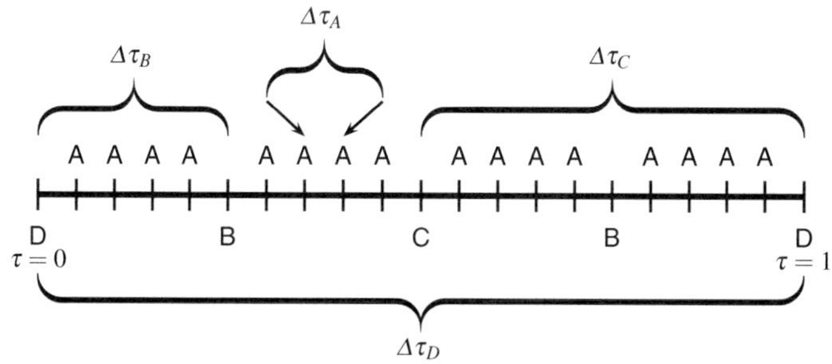

Fig. 12.18 Example of a multilevel scheme

The different *update modes* A, B, C, and D are then performed according to a schedule like, e.g., the one in Fig. 12.18. The time $\Delta\tau$ between two updates needs to be large enough such that the required updates can be performed.

The different *modes* A, B, C, and D can also be interpreted in terms of the kind of improvement they induce for the optimization problem.

In Mode A, a new measurement x_0^{new} is incorporated into the optimization problem but the remaining data in the discretized problem are kept entirely frozen. This corresponds to a linear MPC setup in which the linearization is inherited from the last (SQP) step.

Mode B aims at improving feasibility of a solution by evaluating/approximating local copies of the gradient of the cost function and the vector of constraints using the new measurement x_0^{new}. Note that $\nabla F(\tau + \Delta\tau_B, \hat{z})$ does not necessarily have to be reevaluated but can be approximated using the Hessian $\nabla^2_{zz} L(0, \hat{z}, \tilde{\lambda}^{\mathcal{W}_k,\star})$. For this setting one can show that the optimization variable converges locally to a suboptimal but feasible point of the discretized optimal control problem.

The third Mode C introduces a new local Lagrange multiplier $\tilde{\lambda}^{\mathcal{W}_k,\star}$ of the active constraints. Since the Lagrange multipliers can be interpreted as a penalization of the constraints, updating the Lagrange multiplier leads to a more adequate penalization and thus a repeated evaluation of (12.71) given the new measurement x_0^{new} and the new $\tilde{\lambda}^{\mathscr{W}_k,\star}$ leads to an improvement in terms of optimality. Here, the gradient $\nabla F(\tau + \Delta\tau_C, \hat{z})$ and the Jacobian $\nabla C^{\mathscr{W}_k}(\tau + \Delta\tau_C, \hat{z})$ are updated, too, in order to incorporate the effect of the changing multiplier. This update can be computed efficiently, e.g., via the reverse mode of the automatic differentiation, see the work of Griewank and Walther [25]. The updated $\nabla C^{\mathscr{W}_k}$ is, however, not inserted into equation (12.71) because otherwise the matrix in (12.71) would change and would have to be factorized causing significant computational effort. Still, it can be shown that a sequence generated in the manner described above locally converges to a KKT point of the discretized optimal control problem.

Last, the Mode D corresponds to a single (SQP) step which is in most cases sufficient to generate an approximately optimal solution of the discretized optimal control problem ($\text{OCP}_{\text{N,e}}$). If, however, the linearization happens to be not accurate enough, this deficiency can be overcome easily via refining the hierarchical scheme shown in Fig. 12.18 by inserting more Mode D steps.

We like to point out that in Modes A, B, and C it is not necessary to factorize the matrix in (12.71) again. Hence, the solution of (12.71) requires only a single backsolve which is orders of magnitudes faster than the factorization which needs to be performed in Mode D.

On a lower level, a parametrization of the quadratic approximation of nonlinear optimization problem (NLP) can be considered, which allows the usage of parametric active set methods, cf. Best [3]. In that case, the parameter τ introduces a linear affine homotopy into the problem (IQP) giving

$$\begin{array}{l} \text{minimize} \quad \nabla F(\tau, \hat{z})^\top d + \frac{1}{2} d^\top \nabla^2_{zz} L(\tau, \hat{z}, \lambda) d \\ \text{with respect to } \hat{z} \in \mathbb{R}^{n_z} \\ \text{subject to } G_i(\tau, \hat{z}) + \nabla_z G_i(\tau, \hat{z})^\top d = 0 \text{ for all } i \in \mathscr{E} \\ \text{and } H_i(\tau, \hat{z}) + \nabla H_i(\tau, \hat{z})^\top d \geq 0 \text{ for all } i \in \mathscr{I}. \end{array}$$

The main difference between these two approaches is that real-time iterations are allocated on the level of the constrained nonlinear problem, whereas the parametric active set methods act on the linear quadratic approximation of the latter. Since in real-time iterations typically only one linear quadratic approximation is solved, these problems are rather similar. However, real-time iterations allow for further iteration steps with different linearization points, which is not possible for parametric active set methods.

Sensitivity-Based Warm Start

The second technique applies to (IPM) and is called *sensitivity-based warm start*, see, e.g., the article of Zavala and Biegler [41]. Similar to the initial value embedding described before, it focuses on a fast approximation for neighboring problems around the nominal solution $x_u(\cdot, x_0)$ used for the optimization at time instant n. Since the problem is parametric in the initial value x_0, a standard sensitivity result can be applied. This result guarantees—under suitable conditions—the existence of Lipschitz constants L_x, L_F such that

$$\begin{aligned} \|x_{u^\star}(N, x_0) - x_{\tilde{u}^\star}(N, \tilde{x}_0)\| &\leq L_x \|x_0 - \tilde{x}_0\| \\ \|F(z^\star) - F(\tilde{z}^\star)\| &\leq L_F \|x_0 - \tilde{x}_0\| \end{aligned}$$

holds, cf. the work of Fiacco [18] for further details. In particular, this results guarantees the existence of a neighboring solution $x_{\tilde{u}^\star}(\cdot, \tilde{x}_0)$ for $\tilde{x}_0 \approx x_0$.

If additionally the penalty parameter μ of the (IPM) in the current NMPC step is sufficiently small, then the KKT conditions (12.45)–(12.48) can be expressed in a parameter dependent form by $\phi(z^\star, x_0) = 0$. Now we freeze the primal-dual iteration matrix from (12.57) to

$$K^\star(x_0) := \begin{pmatrix} \nabla^2_{zz} L(z^\star, s, v, w) & 0 & -\nabla G(z^\star)^\top & -\nabla H(z^\star)^\top \\ 0 & \Sigma & 0 & \mathrm{Id} \\ -\nabla G(z^\star) & 0 & \gamma \mathrm{Id} & 0 \\ -\nabla H(z^\star) & \mathrm{Id} & 0 & 0 \end{pmatrix}.$$

Utilizing the existence result, we compute the solution of the frozen primal-dual system (12.53)

$$K^\star(x_0)\Delta z = -(\phi(z^\star, x_0^{\mathrm{new}}) - \phi(z^\star, x_0)) = -\phi(z^\star, x_0^{\mathrm{new}}). \qquad (12.72)$$

Since the primal-dual system (12.57) arose from applying Newton's method to the KKT conditions, solving (12.72) corresponds to a Newton step from the optimal solution $z^\star$ toward the neighboring solution $\tilde{z}$ for $\tilde{x}_0 = x_0^{\mathrm{new}}$. Hence, for the new iterate $\hat{z} := z^\star + \Delta z$ we have

$$\|\hat{z} - \hat{z}^\star\| \leq L_z \|x_0 - x_0^{\mathrm{new}}\|^2$$

for some constant $L_z > 0$ where $\hat{z}^\star$ represents the optimal solution for the neighboring problem ($\mathrm{OCP}_{\mathrm{N,e}}$) with initial value x_0^{new}. Furthermore, since the KKT matrix $K^\star(x_0)$ is already available and its factorization has been computed for the current NMPC step n, only a single backsolve needs to be performed to determine $\hat{z}$. In fact, this approach is similar to the technique for compensating measurement errors mentioned at the end of Sect. 10.6. Here, however, we use the first-order approximation $\hat{z}$, as an initial guess for the fully discretized problem ($\mathrm{OCP}_{\mathrm{N,e}}$).

We can use even more "old" information from the previous time step if we consider the problem with new initial value as a perturbation of the old problem with perturbation $(x_0^{\mathrm{new}} - x_0)$. If the active set does not change under the perturbation $(x_0^{\mathrm{new}} - x_0)$, then the penalty parameter μ can be fixed to a small value and the KKT matrix $K^\star(x_0)$ can be reused to perform fixed-point iterations on the system

$$K^\star(x_0)\Delta z_i = -\phi(z_i, x_0^{\mathrm{new}})$$

with initial value $z_0 = z^\star$. These iterations not only reduce the primal and dual infeasibility of the perturbed problem, but if the perturbation $(x_0^{\mathrm{new}} - x_0)$ is sufficiently small, the iteration converges to the solution of the perturbed problem. For large perturbations, however, the KKT matrix needs to be reevaluated and re-factorized.

If the perturbation $(x_0^{\text{new}} - x_0)$ leads to a change in the active set, then the linearization of the complementarity relaxation (12.46) which is contained in the frozen KKT matrix $K^\star(x_0)$ will drive the first Newton iteration outside of the feasible set and the sensitivity approximation is inconsistent. To avoid this problem, on could reuse the factorization of $K^\star(x_0)$ using a Schur complement scheme to correct the active set which is equivalent to an active set (SQP) iteration.

Shift Method

Both initial value embedding technique and sensitivity method require the optimal control problem to be fully discretized and are restricted to problems of the form ($\text{OCP}_{\text{N,e}}$) whereas the last approach we present now, the so-called *shift method*, can be combined with any discretization method and also with the time varying problem ($\text{OCP}^{\text{n}}_{\text{N,e}}$). The method utilizes the time similarity of two consecutive NMPC steps instead of the state similarity of x_0 and x_0^{new} exploited by the other approaches. In particular, we do not need to assume that x_0^{new} is close to x_0, hence the method also works for large sampling periods.

In the shift approach, the internally computed optimal open-loop trajectory $x_{u^\star}(\cdot, x_0)$ and the control $u^\star(\cdot)$ are shifted in time by removing the first entries $x_{u^\star}(0, x_0)$ and $u^\star(0)$. The initial guess for the subsequent NMPC step is then given by

$$z_0 := (x_{u^\star}(1, x_0), \ldots, x_{u^\star}(N, x_0)^\top, \overline{x}^\top, u^\star(1)^\top, \ldots, u^\star(N-1)^\top, \overline{u}^\top)^\top \quad (12.73)$$

in the case of the full discretization, or by

$$z_0 := (u^\star(1)^\top, \ldots, u^\star(N-1)^\top, \overline{u}^\top)^\top \quad \text{and} \quad x_0 := x_{u^\star}(1, x_0) \quad (12.74)$$

in case of the recursive elimination technique. Since the original optimal sequences contain $(N+1)$ state and N control vectors, we have to add one additional element at the end of each sequence after the shift. In the sequences above, these are denoted by $\overline{x} \in \mathbb{R}^d$ and $\overline{u} \in \mathbb{R}^m$ and we will discuss different ways for choosing these values below.

Since the approach does not rely on first-order information available from the previous step but rather on the nonlinear prediction of the model itself, its advantage is that it is indeed a nonlinear update. Moreover, the shift operation can easily be extended to cover the Lagrange multipliers $\lambda^\star$ as well as the vectors of constraints G, H, the gradient of the cost function ∇F, the Jacobian of the constraints ∇G, ∇H and the Hessian of the cost function $\nabla^2_{zz} L$. Similar to the state and the control trajectories these components of the discretized optimal control problem require a new last component.

In order to obtain values for $\overline{x}$ and $\overline{u}$ we sketch three approaches: The first approach simply copies the last entry of the trajectories and sets

$$\overline{x} := x_{u^\star}(N, x_0) \qquad \text{and} \qquad \overline{u} := u^\star(N-1).$$

This choice is reasonable if the process is close to its steady state. In general, however, it will result in an infeasible initial guess z_0.

The second approach computes the values $\overline{x}$ and $\overline{u}$ by solving the problem ($\mathrm{OCP}^{\mathrm{n}}_{\mathrm{N,e}}$) with horizon $N = 1$ and initial value $x_0 = x_{u^\star}(N, x_0)$. Implicitly, the constraints of the additional optimal control problem will force its solution to be feasible, hence also the new initial guess z_0 for the subsequent NMPC step is feasible.

The third approach can be used in the case of terminal conditioned schemes if Assumption 5.9(i) holds and a formula $u_x = \kappa(x)$ for the control value u_x from this assumption is available to the NMPC algorithm. In this case, we can proceed as in Theorem 10.26 and set $\bar{u} = \kappa(x_{u_n^0}(N-1, x_0))$ and $\overline{x} := f(x_{u^\star}(N, x_0), \bar{u})$.

Here, we like to stress the fact that this update can be processed prior to obtaining a new measurement x_0^{new}. Hence, this technique can also be combined with the initial value embedding technique and sensitivity method. To this end, first the shift method is applied and once the new measurement is available one defines

$$z_0 := (x_0^{\mathrm{new}}, x_{u^\star}(2, x_0), \ldots, x_{u^\star}(N, x_0)^\top, \overline{x}^\top, u^\star(1)^\top, \ldots, u^\star(N-1)^\top, \overline{u}^\top)^\top.$$

or

$$z_0 := (u^\star(1)^\top, \ldots, u^\star(N-1)^\top, \overline{u}^\top)^\top \quad \text{and} \quad x_0 := x_0^{\mathrm{new}}$$

respectively and proceeds with one of the previously mentioned strategies.

If the sampling time is sufficiently short, one can expect that the shifted control sequence is very close to the optimal control in the next sampling interval. In this case, a single SQP iteration may be sufficient in order to obtain a new optimal control with sufficient accuracy. This idea can indeed be made rigorous, as shown by Diehl, Findeisen, Allgöwer, Bock and Schlöder in [14]. In fact, this approach is not restricted to SQP-type optimization algorithms. Conceptually similar results have been obtained for general classes of optimization method including gradient-based steepest descent methods. For details we refer to, e.g., Graichen and Kugi [24], see also the discussion in Sect. 10.9.

In this context, we recall the multistep update technique described at the end of Sect. 10.4, which can also be implemented using sensitivity techniques or with a limited number of optimization iterations. Since in the multistep update technique the optimization is performed only for the remaining part of the optimal control, i.e., on a shortened horizon, a shift is not necessary. Hence, there is no error from shifting the initial guess which is why this method also works well for larger sampling times and allows for a simpler analysis. A disadvantage of the updated multistep approach on the other hand is that the full reoptimization after the end of the m steps can be computationally expensive.

12.6 Nonoptimal NMPC

In Sect. 10.9 we introduced Algorithms 10.22 and 10.28 which use nonoptimal open-loop predictions instead of optimal ones within the NMPC Algorithm 3.1. While in theory it is much more convenient to use optimal controls, from a numerical point of view we can only expect the solution computed by one of the algorithms in Sect. 12.3 to be locally optimal with a predefined optimality tolerance. Therefore, (almost) all numerical results are nonoptimal. While small deviations from optimality can be considered as perturbations, cf. Examples 12.28 and 12.29, large deviations need to be analyzed in a different way.

Common optimization routines check (at least) two termination criteria, that is, a KKT-based optimality criterion and a maximal number of allowable iterations. The latter condition is similar to terminating the iteration after j^* steps in Algorithm 10.22. Since an optimization algorithm will not make further progress once the optimality criterion is satisfied, setting the maximal number of iterations to j^* can be regarded as equivalent to Algorithm 10.22, even though the maximal number of iterations only affects the solution if the optimization routine has not been terminated by the KKT criterion before. This is the algorithm we use in Example 12.30, below.

Compared to the termination criterion via decrease of the value function (10.25), this is a much less sophisticated approach. Yet, we like to emphasize that such a "brute force" restriction of the number of iterations might be necessary from a practical point of view in order to bound the computing times within each iterate of the NMPC algorithm. Such bounds may, in turn, be needed in order to guarantee that the computation is finished before the control must be implemented. Note that even if a compensation technique for the computation times as discussed in Sect. 10.6 is used, the optimization must be finished within an a priori fixed amount of time.

As the following numerical example will show, in the NMPC setting it is quite likely that an optimization algorithm is able to find an optimal solution within only a few iterations. Yet, it also shows that the number of iterations j^* in Algorithm 10.22 should be set carefully.

Example 12.30 Consider once more the inverted pendulum on a cart problem from Example 12.1. We analyze the impact of terminating the optimization algorithm after a maximal number of iterations j^*. The following Fig. 12.19 shows results for the set of horizon lengths $N \in \{15, \ldots, 28\}$ and the set of maximal numbers of iterations $\{1, \ldots, 100\}$. Again, the closed-loop costs in the following figures are computed as in Example 12.2. For this setting, we obtain that if the maximal number of iterations is chosen to be small, then the closed-loop system does not converge toward an upright position which corresponds to the large closed-loop costs displayed in Fig. 12.19a. Moreover, if the horizon N is chosen large, then the maximal number of iterations is required to be slightly larger than for small horizons. For $N = 15$, the receding horizon controller using up to two optimization iterations per NMPC iterate stabilizes the closed loop in an upright position. For $N = 28$, however, seven optimization iterations per NMPC iterate are required.

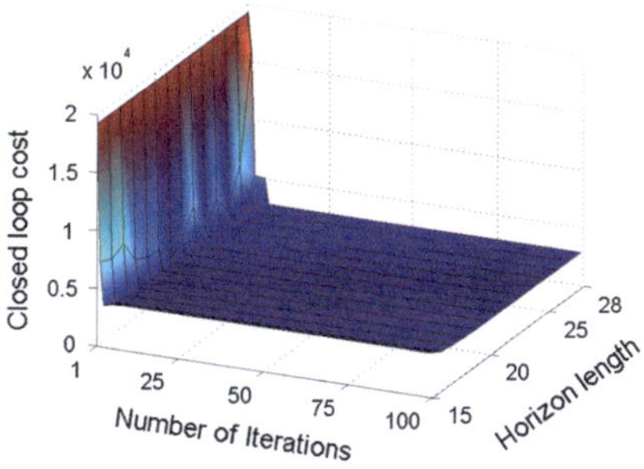

(a) Closed-loop costs

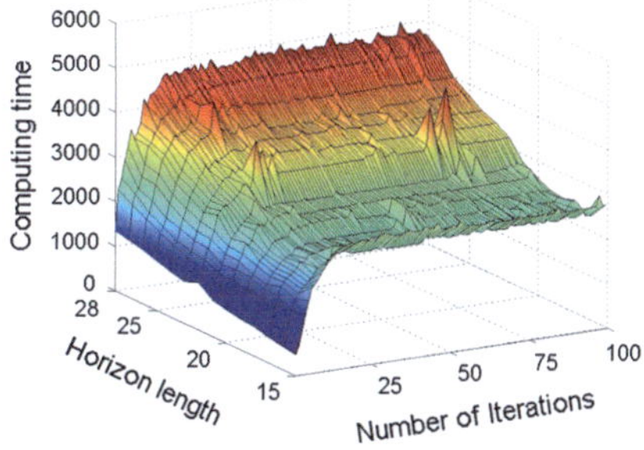

(b) Computing times

Fig. 12.19 Closed-loop costs and computing times for various maximal numbers of iterations of the minimization routine and horizon lengths N

As a consequence, we obtain an additional indicator why the parameter N should be chosen as small as possible to accelerate the receding horizon control algorithm: For one, small horizon lengths require less steps of the optimization method and therefore the computing time is reduced, see Fig. 12.19b. Additionally, each step in the iterative optimization can be performed faster as discussed after Fig. 12.11 which further reduces the computing time, again see Fig. 12.19b.

When choosing the maximal number of iterations j^* one has to take an algorithmic detail of the optimization algorithm into account which we have not yet discussed. Often, optimization routines allow for increases in the cost function to compensate for the so-called Maratos effect, cf. the PhD thesis of Maratos [31] and the article of Powell [35], which describes the possible obstruction of the optimization algorithm by the merit function (12.38). If no measures are taken, the merit function may slow down the convergence by rejecting steps which else make good progress toward a solution. To avoid the Maratos effect, typically either a second-order correction of the search direction as presented by Coleman and Conn [10] or a nonmonotone strategy, given by Chamberlain, Powell, Lemarechal and Pedersen [9], which allows for an increase in the merit function are used. These two approaches differ significantly: The second-order correction aims at improving the merit function and hence, the satisfaction of the constraints of the original problem. The nonmonotone strategy tries to enhance feasibility and optimality at the same time by allowing for temporary increases in the merit function. Allowing for an increase in the cost function may lead to the situation that even if a number of iterations has been performed, the currently best solution obtained so far may still be the initial guess. Thus, the maximal number of iterations j^* should be chosen larger than the allowed number of trial steps used in the Maratos compensation algorithms.

As an alternative to terminating the optimization after j^* iterations steps one could also use a time dependent termination criterion. Such an approach, however, may be critical since a priori it is not clear how many iterations can be executed by the optimization routine in the given time. In particular, the resulting number of iterations may be too small in order to cover the trial steps just discussed. Hence, a time-dependent termination criterion needs to be tuned carefully.

Regardless of whether a fixed number of iterations j^* or a more sophisticated termination criterion like (10.25) is used, apart from suboptimality affecting stability of the closed loop, the feasibility of the nonoptimal open-loop solutions is an important issue. Recall that in Algorithm 10.22 we require the open-loop control $u_n^{j^\star}(\cdot)$ and the corresponding trajectory $x_{u_n^{j^\star}}(\cdot, x(n))$ to satisfy all constraints of the optimal control problem and in Theorem 10.26 we assumed $u_0^0(\cdot) \in \mathbb{U}^N_{\mathbb{X}_0}(x_0)$ to hold which is a quite strong assumption. This is due to the fact that most systems are far too complicated to intuitively come up with such an initial guess.

Example 12.31 Again consider the inverted pendulum Example 12.1 and the initial value to be set to the stable downward position $x_0 = (0, 0, 0, 0)$. While for small j it is clear that $u_N(j, x_0) = \pm 5$ is optimal in order to swing up the pendulum with maximal energy, at some time instant j_0 a switching of the control value has to occur in order to swing the pendulum into the opposite direction. However, both the determination of the switching time and the values of the open-loop control after the switching is a difficult task if feasibility of the given state constraints is to be guaranteed. Thus, obtaining an initial guess for optimization horizons $N > j_0$ is in general a difficult task.

In case a heuristic derivation of u_0^0 is not possible, one may generate a feasible initial control u_0^0 is by using the optimization problem, itself. This is due to the fact that the nonlinear optimization methods presented in Sect. 12.3 can be started with an infeasible initial guess z_0 and still deliver a feasible solution $z^\star$. Hence, before starting Algorithm 10.22 we can solve ($\mathrm{OCP}^{\mathrm{n}}_{\mathrm{N,e}}$) in order to determine u_0^0, provided, of course that the optimization algorithm is able to find a feasible solution.

It may, however, happen that even when starting from a feasible initial guess z_0 the optimization algorithm may compute intermediate iterates z_k which are infeasible. In order to understand why this may happen, let us look once again at the merit function

$$\widetilde{L}(z, \mu) := F(z) + \mu \|A(z)\|_1$$

defined in (12.38). By the definition of this function the violation of one or more constraints corresponds to $\widetilde{L}(z, \mu) > F(z)$ for $\mu > 0$. Therefore, the decrease condition of the merit function used in both the line-search and the trust-region algorithms ensures that the search direction d_k is supposed to lead to an improvement of optimality in terms of the cost function F, but also to an improvement of feasibility. Note, however that it is quite possible that a search direction d_k causes a reduction of $\widetilde{L}(z_{k+1}, \mu)$ while $\|A(z_{k+1})\|_1 \geq \|A(z_k)\|_1$ holds. To prevent this, the choice of μ, in particular its initial value μ_0 becomes the critical component. Within the presented (SQP) and (IPM) algorithms of Sect. 12.3, μ is chosen such that a decrease in $\widetilde{L}$ can be guaranteed. As a consequence, intermediate solutions may improve optimality in expense of larger violations of constraints before (eventually) tending back to the feasible set. For this reason, even if previous iterates are feasible, the current step may generate an infeasible iterate. This may happen for both the line-search and the trust-region algorithms unless specialized algorithms are used, as, e.g., the method proposed by Panier and Tits [33].

Common implementations of optimization algorithms do often not use such specialized methods. Instead, a rather simple mechanism is employed in order to solve this infeasibility issue: during the iteration, feasibility of an iterate and the corresponding cost function value are compared. If more than one feasible iterate is found, then the one with the minimal cost function value is returned. If no feasible iterate is found, the method returns an error message, and in most cases the iterate which causes the least constraint violation. Hence, if we terminate one such algorithm before the KKT-based optimality conditions are satisfied, then a feasible solution is returned if a feasible iterate has been encountered during the run of the optimization routine.

The easiest way to ensure the existence of such a solution surely is to start the optimization using a feasible initial guess z_0. As outlined before, this may be a difficult task when starting the NMPC scheme at time $n = 0$; hence if no good heuristic guess for u_0^0 is available at startup then in general one has to hope that the optimization routine is able to find a feasible solution.

For $n \geq 1$, however, we can exploit the fact that the optimal control problems in two consecutive NMPC steps are quite similar. We can use this similarity to obtain a feasible initial guess via the shift methods presented in Sect. 12.5. In particular, as discussed in the paragraph on shift methods in Sect. 12.5, the second and third approach presented in this paragraph will always provide a feasible initial guess z_0 if the shifted initial value $x_{u^\star}(1, x_0)$ and the new measurement x_0^{new} are identical. Note that this is exactly the argument used in the proof of Theorem 10.26 in which the third shift strategy from Sect. 12.5 was applied.

Unfortunately, for all other described restarting methods in Sect. 12.5 feasibility of the initial guess cannot be guaranteed, in general. However, the initial value embedding and the sensitivity-based warm start technique only cause the d constraints corresponding to the initial value shooting node

$$x_u(0, x_0) - x_0 = 0$$

to be violated. Since this condition depends linearly on x_0, the infeasibility problem is resolved after only one iterate provided the linearization of the discretized problem is accurate enough, i.e., the difference between x_0 and x_0^{new} is sufficiently small. If this is not the case, then one may still expect the infeasibility problem to be resolved in only a handful of iteration steps.

Summarizing, once a feasible initial solution is found, there are ways to implement the optimization iteration such that the two termination criteria discussed in Sect. 10.9, i.e., terminating after j^* iteration steps, see Algorithm 10.22, or terminating after condition

$$J_N(x(n), u_n^{j^*}(\cdot) \leq \widetilde{V}_N(n-1) - \alpha\ell(x(n-1), u_{n-1}(0))$$

as in Algorithm 10.28, can be expected to be reached with feasible control sequences.

12.7 Notes and Extensions

In Sect. 12.1 we have shown three discretization techniques which allow us to convert an optimal control problem into a optimization problem in standard form (NLP). These methods are more or less standard in NMPC and are well suited for the case of zero-order hold control since the error induced by the discretization of the control is zero due to the knowledge of the switching points of the control. If other types of control functions are considered, one can show convergence of the discretized optimal control toward the continuous optimal control for the Euler discretization, cf. Malanowski, Büskens and Maurer [29]. Alternatively, higher order discretization techniques can be used, see, e.g., Dontchev, Hager and Veliov [16].

For the nonlinear optimization problem in standard form (NLP), we sketched the analytical background of nonlinear optimization and presented both most common algorithms for nonlinear optimization problems—(SQP) and (IPM)—in both the line-search and the trust-region setting in Sects. 12.2 and 12.3. Yet, since we only want to point out the special properties of the NMPC setting for nonlinear optimization algorithm, we skipped all modifications which can be used to speed up the iteration of such methods. Amongst the most popular modifications we mentioned the DFP (Davidon-Fletcher-Powell) [11, 21] or BFGS (Broydon-Fletcher-Goldfarb-Shanno) updates [7, 19, 22, 39] methods to update the Hessian in Sect. 12.4 as well as the second-order correction method of the search direction as presented by Coleman and Conn [10] or the nonmonotone strategy given by Chamberlain, Powell, Lemarechal and Pedersen [9] from Sect. 12.6 to compensate for the Maratos effect, cf. the PhD thesis of Maratos [31] and the article of Powell [35]. A good survey for these and other methods be found in, e.g., the monograph of Nocedal and Wright [32]. For simplicity of exposition, we also did not discuss convergence theory of such algorithms for the more general nonconvex case, see, e.g., Gould, Orban and Toint [23], which can be handled by common implementations as well. Since there exists a wide range of available implementations, we only state a short and incomplete list here:

Name	Website
SNOPT	http://www.sbsi-sol-optimize.com/asp/sol_product_snopt.htm
NLPQLP	http://www.ai7.uni-bayreuth.de/nlpqlp.htm
HQP	http://hqp.sourceforge.net
IPOPT	http://www.coin-or.org/ipopt
LOQO	http://www.princeton.edu/~rvdb/loqomenu.html
KNITRO	http://www.ziena.com
WORHP	http://www.worhp.de

Additionally, we like to note that apart from the first-discretize-then-optimize (direct) approach featured in this chapter, there also exist so-called indirect methods to solve optimal control problems. These methods are based on Pontryagin's maximum principle and compute the optimal control from the resulting analytical boundary value problem via shooting or collocation methods, see, e.g., Bryson and Ho [8].

This typically requires a good initial guess for the Lagrange multipliers which is why these methods are rarely used for NMPC. An exception to this rule is Bell, Limebeer and Sargent [2], in which the authors investigate such a method for an NMPC problem governed by a differential algebraic equation (DAE).

In the remaining Sects. 12.4 to 12.6 we discussed several issues of NMPC from both the implementation as well as the analytical side. In particular, we on the one hand focused on the internal structure of each discretized optimal control problem which led us to the idea of condensing described in Diehl [12] and Bock and Plitt [6] and the triangular structure of the Jacobian of the constraints. These observations are particularly useful if one considers parallel algorithms to evaluate these matrices. On the other hand, we have shown how the relationship between two consecutive NMPC iterates can be used to generate a good initial guess. The idea of the initial value embedding and the hierarchical NMPC approach have been taken from Diehl, Bock, Schlöder, Findeisen, Nagy and Allgöwer [13] and Albersmeyer, Beigel, Kirches, Wirsching, Bock and Schlöder [1] respectively, whereas a detailed analysis of the sensitivity-based warm start can be found in Zavala and Biegler [41]. A classification of the respective methods including numerical tricks can be found in Diehl, Ferreau and Haverbeke [15]. Stability for the combination of the shift method and the initial value embedding has been shown in [14], yet we are not aware of respective results regaring the sensitivity-based warm start. Considering implementations, there exist several commercial and academic NMPC packages. In the following table, we summarize several examples:

Name	Website
Acado	`http://www.acadotoolkit.org`
GRAMPC	`https://sourceforge.net/projects/grampc`
Jmodelica	`http://jmodelica.org`
MPC Tools	`http://jbrwww.che.wisc.edu/software/mpctools`
μAO-MPC	`http://ifatwww.et.uni-magdeburg.de/syst/muAO-MPC`
BLOM	`http://www.mpclab.net/trac`
jMPC Toolbox	`http://www.i2c2.aut.ac.nz/Resources/Software/jMPCToolbox.html`
PlantPAx MPC	`http://discover.rockwellautomation.com/plantpaxmpc`
BrainWave	`http://www.andritz.com/aa-brainwave`
DMCplus	`http://www.aspentech.com/products/advanced-process-control/DMCplus`
SORTiA-MPC	`http://www.azbil.com`
PlantTriage	`http://www.expertune.com/PlantTriage.aspx`
Predict and Control	`http://www.abb.com`
Matlab MPC Toolbox	`http://www.mathworks.com/products/mpc`

A list of commercial packages is also given in Qin and Badgwell [36].

Problems

1. Consider the constraint set $\Omega = \{(z_1, z_2, z_3) \mid z_3^2 \geq z_1^2 + z_2^2\}$. Show that while Ω is a set with nonsmooth boundary, the set itself is described by a smooth function.
2. One can trivially construct an example of a feasible set Ω and a feasible point $z^\star$ at which the LICQ (see Definition 12.14) is satisfied but the constraints are nonlinear. Prove whether or not the reverse situation holds, i.e., the active constraints are linear but the LICQ is not satisfied.
3. Consider the constraints
$$z_2 \leq z_1^3 \quad \text{and} \quad z_2 \geq 0.$$
Show that at $z = (0, 0)$ we have $T_\Omega(z) \subsetneq \mathscr{F}(z)$.
4. Similar to the idea of Condensing in Sect. 12.4 suppose that s_j is a vector of dimension d allowing to get rid of the index function $\iota(\cdot)$ and that the index function $\varsigma(\cdot) : \{0, \ldots, r_s - 1\} \to \{0, \ldots, N\}$ is strictly monotonically increasing with $\varsigma(0) = 0$. Consider the corresponding linearized continuity condition
$$S(z) + \nabla_z S(z)^\top \Delta s = 0 \tag{12.75}$$
where again Δs denotes the part of the search direction d corresponding to the shooting nodes. Show that $\Delta s_1, \ldots, \Delta s_{r_s - 1}$ can be computed from Δs_0 which is the search direction for the shooting node corresponding to the initial value.
Hint: Reformulate the continuity condition
$$\left[s_{j+1} - f\left(x_u(\varsigma(j+1) - 1, x_0), u(\varsigma(j+1) - 1)\right)\right]_{j \in \{0, \ldots, r_s - 1\}} = 0, \tag{12.76}$$
to get rid of all state vectors $x_u(k, x_0)$ which are not shooting node vectors, i.e., for which $\varsigma(j) \neq k$.
5. Formulate the optimization problem in standard form for problem ($\mathrm{OCP}_{\mathrm{N,e}}^{\mathrm{n}}$) using
   (a) recursive elimination and an additional endpoint constraint $\mathbb{X}_0(n) = \{x_*\}$.
   (b) multiple shooting with shooting vector $s_N \in \mathbb{R}^d$ with initial value $s_N = x_*$ inducing the constraints
$$S(z) = [s_N - f\left(x_u(N - 1, x_0), u(N - 1)\right)] = 0.$$
Using either of the presented optimization methods, show that the iterates will not coincide in general.

References

1. Albersmeyer, J., Beigel, D., Kirches, C., Wirsching, L., Bock, H.G., Schlöder, J.: Fast nonlinear model predictive control with an application in automotive engineering. In: Magni, L.,

Raimondo, D.M., Allgöwer, F. (eds.) Nonlinear Model Predictive Control. Lecture Notes in Control and Information Sciences, vol. 384, pp. 471–480. Springer, Berlin (2009)
2. Bell, M.L., Limebeer, D.J.N., Sargent, R.W.H.: Robust receding horizon optimal control. Comput. Chem. Eng. **20**, 781–786 (1996) (Supplement 2, European Symposium on Computer Aided Process Engineering 6)
3. Best, M.J.: An algorithm for the solution of the parametric quadratic programming problem. Applied Mathematics and Parallel Computing, pp. 57–76 (1996)
4. Bock, H., Diehl, M., Kühl, P., Kostina, E., Schlöder, J., Wirsching, L.: Numerical methods for efficient and fast nonlinear model predictive control. In: Findeisen, R., Allgöwer, F., Biegler, L.T. (eds.) Assessment and Future Directions of Nonlinear Model Predictive Control. Lecture Notes in Control and Information Sciences, vol. 358, pp. 163–179. Springer, Berlin (2007)
5. Bock, H.G.: Numerical treatment of inverse problems in chemical reaction kinetics. In: Ebert, K.H., Deuflhard, P., Jäger, W. (eds.) Modelling of Chemical Reaction Systems. Springer Series in Chemical Physics, vol. 18, pp. 102–125. Springer, Heidelberg (1981)
6. Bock, H.G., Plitt, K.: A multiple shooting algorithm for direct solution of optimal control problems. In: Proceedings of the 9th IFAC World Congress Budapest, Pergamon, Oxford, pp. 242–247 (1984)
7. Broyden, C.G.: The convergence of a class of double-rank minimization algorithms. II. The new algorithm. J. Inst Math. Appl. **6**, 222–231 (1970)
8. Bryson, A.E., Ho, Y.C.: Applied optimal control. Hemisphere Publishing Corp. Washington, D. C. Optimization, estimation, and control, Revised printing(1975)
9. Chamberlain, R.M., Powell, M.J.D., Lemarechal, C., Pedersen, H.C.: The watchdog technique for forcing convergence in algorithms for constrained optimization. Math. Program. Stud. **16**, 1–17 (1982) (Algorithms for constrained minimization of smooth nonlinear functions)
10. Coleman, T.F., Conn, A.R.: Second-order conditions for an exact penalty function. Math. Program. **19**(2), 178–185 (1980)
11. Davidon, W.C.: Variable metric method for minimization. Research and Dev. Rep. ANL-5990 (Rev.), Argonne Nat. Lab. (1959)
12. Diehl, M.: Real-time optimization for large scale nonlinear processes. PhD thesis, University of Heidelberg (2001)
13. Diehl, M., Bock, H.G., Schlöder, J.P., Findeisen, R., Nagy, Z., Allgöwer, F.: Real-time optimization and nonlinear model predictive control of processes governed by differential-algebraic equations. Journal of Process Control **12**(4), 577–585 (2002)
14. Diehl, M., Findeisen, R., Allgöwer, F., Bock, H.G., Schlöder, J.P.: Nominal stability of the real-time iteration scheme for nonlinear model predictive control. IEE Proc. Control Theory Appl. **152**, 296–308 (2005)
15. Diehl, M., Ferreau, H., Haverbeke, N.: Efficient Numerical Methods for Nonlinear MPC and Moving Horizon Estimation, vol. 384, pp. 391–417. Springer, Heidelberg (2009)
16. Dontchev, A.L., Hager, W.W., Veliov, V.M.: Second-order Runge–Kutta approximations in control constrained optimal control. SIAM J. Numer. Anal. **38**(1), 202–226 (2000) (electronic)
17. Ferreau, H.J., Bock, H.G., Diehl, M.: An online active set strategy to overcome the limitations of explicit MPC. Int. J. Robust Nonlinear Control **18**(8), 816–830 (2008)
18. Fiacco, A.V.: Introduction to sensitivity and stability analysis in nonlinear programming. Mathematics in Science and Engineering, vol. 165. Academic Press Inc., Orlando (1983)
19. Fletcher, R.: A new approach to variable metric algorithms. Comput. J. **13**, 317–322 (1970)
20. Fletcher, R.: Practical Methods of Optimization, 2nd edn. Wiley-Interscience, New York (2001)
21. Fletcher, R., Powell, M.J.D.: A rapidly convergent descent method for minimization. Comput. J. **6**, 163–168 (1963)
22. Goldfarb, D.: A family of variable metric methods derived by variational means. Math. Comput. **24**, 23–26 (1970)
23. Gould, N.I.M., Orban, D., Toint, P.L.: Numerical methods for large-scale nonlinear optimization. Acta Numer. **14**, 299–361 (2005)
24. Graichen, K., Kugi, A.: Stability and incremental improvement of suboptimal MPC without terminal constraints. IEEE Trans. Autom. Control **55**, 2576–2580 (2010)

25. Griewank, A., Walther, A.: Evaluating derivatives: principles and techniques of algorithmic differentiation, 2nd edn. Society for Industrial and Applied Mathematics (SIAM), Philadelphia (2008)
26. Gyurkovics, E., Elaiw, A.M.: Conditions for MPC based stabilization of sampled-data nonlinear systems via discrete-time approximations. In: Findeisen, R., Allgöwer, F., Biegler, L.T. (eds.) Assessment and Future Directions of Nonlinear Model Predictive Control. Lecture Notes in Control and Information Sciences, vol. 358, pp. 35–48. Springer, Berlin (2007)
27. Hairer, E., Wanner, G.: Solving Ordinary Differential Equations. II, Springer Series in Computational Mathematics, vol. 14, 2 edn. Springer, Berlin (1996)
28. Han, S.P.: A globally convergent method for nonlinear programming. J. Optim. Theory Appl. **22**(3), 297–309 (1977)
29. Malanowski, K., Büskens, C., Maurer, H.: Convergence of approximations to nonlinear optimal control problems. In: Mathematical Programming with Data Perturbations. Lecture Notes in Pure and Applied Mathematics, vol. 195, pp. 253–284. Dekker, New York (1998)
30. Mangasarian, O.L.: Nonlinear Programming. McGraw-Hill Book Co., New York (1969) (Reprinted by SIAM Publications 1995)
31. Maratos, N.: Exact penalty function algorithms for finite dimensional and control optimization problems. PhD thesis, University of London (1978)
32. Nocedal, J., Wright, S.J.: Numerical Optimization, 2nd edn. Springer Series in Operations Research and Financial Engineering. Springer, New York (2006)
33. Panier, E.R., Tits, A.L.: A superlinearly convergent feasible method for the solution of inequality constrained optimization problems. SIAM J. Control Optim. **25**(4), 934–950 (1987)
34. Powell, M.J.D.: A fast algorithm for nonlinearly constrained optimization calculations. In: Numerical Analysis (Proceedings of the 7th Biennial Conference, University of Dundee, Dundee, 1977), pp. 144–157. Lecture Notes in Mathematics, vol. 630. Springer, Berlin (1978)
35. Powell, M.J.D.: Convergence properties of algorithms for nonlinear optimization. SIAM Rev. **28**(4), 487–500 (1986)
36. Qin, S.J., Badgwell, T.A.: A survey of industrial model predictive control technology. Control Eng. Pract. **11**, 733–764 (2003)
37. Schittkowski, K.: The nonlinear programming method of Wilson, Han, and Powell with an augmented Lagrangian type line search function. I. Convergence analysis. Numer. Math. **38**(1), 83–114 (1981/82)
38. Schittkowski, K.: On the convergence of a sequential quadratic programming method with an augmented Lagrangian line search function. Math. Op. Stat. Ser. Optim. **14**(2), 197–216 (1983)
39. Shanno, D.F.: Conditioning of quasi-Newton methods for function minimization. Math. Comput. **24**, 647–656 (1970)
40. Stoer, J., Bulirsch, R.: Introduction to numerical analysis. Texts in Applied Mathematics, vol. 12, 3rd edn. Springer, New York (2002) (Translated from the German by R. Bartels, W. Gautschi and C. Witzgall)
41. Zavala, V.M., Biegler, L.T.: Nonlinear programming strategies for state estimation and model predictive control. In: Magni, L., Raimondo, D.M., Allgöwer, F. (eds.) Nonlinear Model Predictive Control, Lecture Notes in Control and Information Sciences, vol. 384, pp. 419–432. Springer, Berlin (2009)

Appendix A
NMPC Software Supporting This Book

The NMPC simulations in this book were computed with two different software packages. For the computationally less demanding examples, we have developed a MATLAB routine *nmpc.m* which provides a straightforward implementation of the NMPC Algorithm 3.11. For the more complicated Examples 2.10, 2.11 and 2.12 we have used the C++ software YANE by Jürgen Pannek and Thomas Jahn, cf. www.nonlinearmpc.com. This appendix provides a brief introduction to these software packages. Moreover, we briefly explain additional MATLAB® and MAPLE™ code which has been used for several computations and for generating the figures in this book. All the software described in this appendix is available from the webpage

www.nmpc-book.com

A.1 The MATLAB NMPC Routine

The MATLAB M-File *nmpc.m* provides a straightforward implementation of the NMPC Algorithm 3.11. Upon call of

```
function [t, x, u] = nmpc(runningcosts, terminalcosts, ...
              constraints, terminalconstraints, ...
              linearconstraints, system, ...
              mpciterations, N, T, tmeasure, xmeasure, u0, ...
              varargin)
```

the function executes a simulation of the NMPC closed loop for a user-defined number of `mpciterations` closed-loop steps. Internally, the steps Algorithm 3.11 and its variants are implemented in different auxiliary function which are easily extended to more complex settings.

(1) In function *measureInitialValue*, a new measurement of the initial value $x_0 = x(n)$ is obtained by copying the value $x_{u^\star}(1, x(n-1))$ of the optimal predicted trajectory computed at time $n-1$. This means that the routine simulates the

© Springer International Publishing Switzerland 2017
L. Grüne and J. Pannek, *Nonlinear Model Predictive Control*,
Communications and Control Engineering, DOI 10.1007/978-3-319-46024-6

nominal closed-loop system. If measurement errors are to be simulated then this routine can be modified accordingly.

(2) The solution of the problem ($\text{OCP}^{\text{n}}_{\text{N,e}}$) is computed in the function *solveOptimalControlProblem*. Here, the problem ($\text{OCP}^{\text{n}}_{\text{N,e}}$) is transformed to a static nonlinear constrained optimization problem (NLP) via a recursive discretization, see Sect. 12.1 for details. The transformed problem is then solved via the MATLAB optimization method *fmincon*, which—depending on the chosen options—uses one of the algorithms described in Sect. 12.3. The initial guess of the control sequence is computed in the function *shiftHorizon* via the shift technique from Sect. 12.5.

(3) Lastly, the NMPC feedback $\mu_N(n, x(n))$ is implemented via the function *applyControl* which evaluates the dynamics of the system for the control computed by the optimization method in Step (2). Here the same dynamics as in the computation of the predictions is used. This could be changed if modeling errors or external perturbations shall be simulated.

For all simulations in this book which were carried out with *nmpc.m* we provide ready to use M-Files on our webpage whose names correspond to the number of the example in this book, e.g., the M-File *example_6_26.m* contains the simulation in Example 6.31. These M-Files may also be used as templates for building new examples by suitably adjusting the functions defining problem ($\text{OCP}^{\text{n}}_{\text{N,e}}$). The following table provides a list of the ingredients of problem ($\text{OCP}^{\text{n}}_{\text{N,e}}$) and the corresponding functions in the MATLAB implementation.

Ingredient in ($\text{OCP}^{\text{n}}_{\text{N,e}}$)	Function in implementation
$\ell(n, x, u)$	`runningcosts(t, x, u)`
$F(n, x)$	`terminalcosts(t, x)`
$\mathbb{X}$, $\mathbb{U}(x)$	`linearconstraints(t, x, u)`
	`constraints(t, x, u)`
$\mathbb{X}_0$	`terminalconstraints(t, x)`
$f(n, x, u)$	`system(t, x, u, T)`

Together with the length of the horizon N, the sampling interval T and initial values for the time t_0, the (closed-loop) state x_0 and the control u, the functions in the right column of this table are supplied as arguments to *nmpc*. During the run of *nmpc*, these functions are called by the optimization routine once for each summand in J_N in ($\text{OCP}_{\text{N,e}}$) with `t` containing the current sampling time—`t` $= (n+k)T$ in the notation of ($\text{OCP}^{\text{n}}_{\text{N,e}}$)—and `x`, `u` containing the corresponding state and control vectors $x_u(k)$ and $u(k)$.

Observe that in the implementation the sets of admissible control sequences $\mathbb{U}^N(x_0)$ or $\mathbb{U}^N_{\mathbb{X}_0}(x_0)$ do not need to be provided explicitly. Rather, the state and control constraint sets $\mathbb{X}$, $\mathbb{U}(x)$ and—if applicable—the terminal constraints set $\mathbb{X}_0$ are provided in the implicit form of Definition 3.6. Here MATLAB's optimization routine *fmincon* distinguishes between linear and nonlinear equality and inequality constraints; the declaration of the respective functions *linearconstraints* and *constraints* follows the usual MATLAB convention. For instance, for adding an equilibrium terminal constraint of the form $\mathbb{X}_0 = \{x_*\}$ one may use the function $G_i^S(x) = \|x - x_*\|_2^2$. For, e.g., $x_* = (1, 2)^\top$ this amounts to defining the array `ceq` in *terminalconstraints* as `ceq = [norm(x-[1, 2],2)^2];`. More details can be found in the MATLAB help for *fmincon*.

The NMPC implementation given in *nmpc.m* can also be configured more specifically to match the needs of a certain example. To this end, additional inputs can be used to specify, e.g., the optimization method and its optimality tolerance or output functions. Additionally, the core routine *nmpc* can also deal with continuous time dynamics and the user may set tolerances for the differential equation solver differently for the open- and the closed-loop dynamics. Last, example specific output can be generated by suppliable output functions. For details we refer to the comments and the help text in the file.

A.2 Additional MATLAB and MAPLE Routines

In addition to *nmpc.m* and the example files using this routine, we provide several other MATLAB routines which we used for computations as well as for several figures. Most of these M-Files are related to the Optimization Problem (6.16). The file *mpcalpha_cn.m* solves this problem and thus computes the suboptimality index α for an NMPC problem satisfying the Controllability Assumption 6.5 with

$$\beta(r, n) = c_n r \tag{A.1}$$

for some real sequence $(c_n)_{n\in\mathbb{N}_0}$ with $c_n \geq 0$ and $c_n = 0$ for all $n \geq n_0$, i.e., *finite time controllability* with linear overshoot bound. The case of *exponential controllability*

$$\beta(r, n) = C\sigma^n r \tag{A.2}$$

for real constants $C \geq 1$ and $\sigma \in (0, 1)$ can be handled via this function as well by setting $c_n = C\sigma^n$. To make evaluations for the exponential case more convenient, the function `mpcalpha_Csigma` in file *mpcalpha_Csigma.m* automatically computes the required sequence c_n from the input data C and σ and calls `mpcalpha_cn` to derive α.

Upon call of

```
function [alpha, lambda] = mpcalpha_cn(c,N,varargin)
```

the linear Optimization Problem (6.16) is constructed from the input data `c` and the horizon length N for the case of finite time controllability. Here, `c` is expected to be a vector containing the values c_n from (6.5). If the supplied vector `c` is shorter than the horizon N, it is automatically filled up with zeros. In a second step, the MATLAB routine *linprog* is applied to compute the desired values of α and $\lambda_0, \ldots, \lambda_{N-1}$ for the constructed linear optimization problem. For exponential controllability, the respective call is

```
function [alpha, lambda] = mpcalpha_Csigma(C,sigma,N,varargin)
```

with C and σ from (6.4).

In both functions, as an additional argument the user may supply an integer m denoting the number of control elements of the open-loop control to be implemented, cf. Sect. 10.4. Without defining m the case of classical MPC ($m = 1$) is imposed. Similarly, an endweight different from $\omega = 1$ as outlined in Sect. 10.2 can be defined. Lastly, different levels of output of the function can be chosen. In particular, with parameter `output` ≥ 1 the optimal sequence $\lambda_0, \ldots, \lambda_N$ is displayed graphically.

For creating figures, we additionally provide the files *alpha_m_plot.m* and *alpha_omega_plot.m*. Upon call of

```
function alpha_m_plot(C,sigma,N,m,varargin)
```

```
function alpha_omega_plot(C,sigma,N,m,endweight,varargin)
```

these functions generate plots of the suboptimality index α with respect to the number of control horizons m or with respect to the endweight ω, respectively. For examples of how to use these functions we refer to the M-Files *example_fig_7_1.m* and *example_fig_7_3.m* which were used in order to compute Figs. 10.1 and 10.3.

Figure 10.4 was computed for a linear inverted pendulum using Algorithm 3.1 with running cost

$$\ell(x, u) = 2\|x\|_1 + 4\|u\|_1,$$

sampling period $T = 0.5$, optimization horizon $N = 10$ and linear constraints. Such an optimal control problem can be transformed into a static linear optimization problem and can thus be solved with a linear optimization routine which is considerably faster than a nonlinear optimization method. This is what is done in *example_fig_7_4.m*, where MATLAB's linear optimization routine *linprog* is used for solving the resulting linear optimization problem. While we encourage testing this program, we like to note that the runtime of the program may easily exceed one day for a large number of initial values.

Finally, our webpage also provides a number of MAPLE worksheets (for MAPLE 12 or newer), which were used for several numeric and symbolic computations throughout this book. Like our M-Files, the worksheets are named according to the example or figure they refer to. Each of these worksheets is comprehensively explained by comments in the files which is why we refrain from giving further explanation here.

A.3 The C++ NMPC Software

While the MATLAB implementation is quite nice for tutorial purposes, we solved the more complicated Examples 2.10, 2.11 and 2.12 via the C++ NMPC software YANE which can be downloaded from www.nonlinearmpc.com.

Unpacking any of the files via

```
tar -xvf"package_filename".tar.gz
```

will create a new folder containing the source code files. Within this folder, a new subfolder *build* for compiling the source code should be generated to avoid overwriting the CMake compilation and installation routines. Now, the configuration file for CMake needs to be generated from within the created subfolder *build*. Here, user-specific options can be supplied, e.g., a local installation path:

```
cmake -DCMAKE_INSTALL_PREFIX="installation_path"../
```

Once the configuration is complete, the package can be compiled and installed via

```
make
make install
```

Note that depending on the chosen *installation_path* the install command may require superuser rights. Moreover, the environment variables used by the C++ compiler of the system must contain the installation path which can be added via

```
export LD_LIBRARY_PATH="installation_path"/lib:$LD_LIBRARY_PATH
export LIBRARY_PATH="installation_path"/lib:$LIBRARY_PATH
export CPLUS_INCLUDE_PATH="installation_path"/include:
    $CPLUS_INCLUDE_PATH
```

A tutorial of the C++ implementation as well as explanations of the classes and methods can be found at www.nonlinearmpc.com.

In a similar manner the example archive which can be downloaded from our homepage www.nmpc-book.com can be unpacked and compiled but does not have to be installed. The archive is structured as follows:

./
- `Examples` ... Within this directory, the C++ files of the NMPC problems are defined using the models from subdirectory `./Models`. Similar to the examples solved using Matlab and MAPLE, the examples files are named according to the example they refer to.
- `Models` ... This directory holds the models defined in Examples 2.10, 2.11 and 2.12 as well as the respective NMPC problem related components such as the cost functional or the shooting nodes.
- `cmake` ... This directory contains auxiliary modules required by CMake.

Apart from the C++ files, each subdirectory contains a file *CMakeLists.txt* which provide information required by CMake to compile the package. If additional examples shall be implemented, these files need to be adapted accordingly, see, e.g., www.cmake.org for further information.

Compiling the package generates several executables which can be found in the subdirectory *build/Examples*—assuming that CMake is called within the subdirectory *build*. Upon execution, each file generates a problem-specific screen output. For additional file outputs of the computed trajectories, we again refer to the documentation of the YANE software.

Appendix B
Glossary

The following table gives an overview of the notation we used throughout this book. Note that auxiliary notations introduced within proofs or used only within a single example are not displayed here.

Acronyms

(ECP) Equality constrained nonlinear optimization problem, p. 390
(EQP) Equality constrained quadratic optimization problem, p. 395
(EQP^q) Equality constrained quadratic subproblem of (IQP), p. 395
IOSS Input/output-to-state stability, p. 302
(IPM) Interior point method, p. 380
(IQP) Inequality constrained quadratic optimization problem, p. 394
ISS Input-to-state stability, p. 193
LICQ Linear independent constraint qualification, p. 386
LQR Linear-quadratic regulator, p. 104
MPC Model predictive control, p. 4
(NLP) Nonlinear optimization problem, p. 371
NMPC Nonlinear model predictive control, p. iii
($\mathrm{OCP_N}$) Finite horizon optimal control problem, p. 46
($\mathrm{OCP_N^n}$) Finite horizon time-varying optimal control problem, p. 53
($\mathrm{OCP_\infty}$) Infinite horizon optimal control problem, p. 71
($\mathrm{OCP_\infty^n}$) Infinite horizon time-varying optimal control problem, p. 71
($\mathrm{OCP_{N,e}}$) Extended finite horizon optimal control problem, p. 56
($\mathrm{OCP_N^{n,e}}$) Extended finite horizon time-varying optimal control problem, p. 56
($\mathrm{OCP_N^p}$) Local finite horizon optimal control problem, p. 272
(SQP) Sequential quadratic programming, p. 380

Sets and Spaces

$A \subseteq \mathbb{X}$ Subset of the state space in Definition 7.24, p. 193
$\mathscr{A}(z) \subseteq \mathscr{E} \cup \mathscr{I}$ Active set of constraints, p. 386
$\mathscr{B}_r(x)$ Open ball centered at x with radius r, p. 29

L. Grüne and J. Pannek, *Nonlinear Model Predictive Control*,
Communications and Control Engineering, DOI 10.1007/978-3-319-46024-6

$\overline{\mathscr{B}}_r(x)$ Closed ball centered at x with radius r, p. 97
$\mathscr{C}(z^*, \lambda^*)$ Critical cone, p. 388
$E_k \subseteq X$ Exit set, p. 179
$\mathscr{E} \subset \mathbb{N}$ Index set of equality constraints of an optimization problem, p. 373
$\mathscr{E}^S \subset \mathbb{N}$ Index set of equality constraints describing sets, p. 52
$\mathscr{F}_N \subseteq \mathbb{X}$ Feasible set without terminal constraints for horizon N, p. 178
$\mathscr{F}_\infty \subseteq \mathbb{X}$ Infinite horizon feasible set, also called viability kernel, p. 178
$\mathscr{F}(z) \subseteq \mathbb{R}^{n_z}$ Linearized feasible directions, p. 386
$\mathscr{G} \subseteq \mathbb{R}$ Time grid, p. 345
$I \subseteq \mathbb{R}$ Open interval, p. 16
I^p Set of neighboring data of subsystem p, p. 259
$\mathscr{I} \subset \mathbb{N}$ Index set of inequality constraints of an optimization problem, p. 373
$\mathscr{I}^s \subset \mathbb{N}$ Index set of inequality constraints describing sets, p. 52
$\mathscr{I}^p = (p_1, \ldots, p_m) \subset \mathscr{P}\backslash\{p\}$ Set of neighbors of subsystem p, p. 261
$\mathscr{K}$ Class of continuous functions $\alpha:\mathbb{R}_0^+ \to \mathbb{R}_0^+$which are strictly increasing with $\alpha(0) = 0$, p. 29
$\mathscr{K}_\infty$ Class of functions $\alpha \in \mathscr{K}$ which are unbounded, p. 29
$\mathscr{KL}$ Class of continuous functions β: $\mathbb{R}_0^+ \times \mathbb{R}_0^+ \to \mathbb{R}_0^+$ with $\beta(., t) \in \mathscr{K}$ and $\beta(r, .) \in \mathscr{L}$, p. 29
$\mathscr{KL}_0$ Class of continuous functions $\beta : \mathbb{R}_0^+ \times \mathbb{R}_0^+ \to \mathbb{R}_0^+$ with $\lim_{t\to\infty}\beta(r, t) = 0$ for each $r > 0$ and $\beta(., t) \in \mathscr{K}_\infty$ or $\beta(., t) \equiv 0$, p. 125
$\mathscr{L}$ Class of continuous functions $\delta : \mathbb{R}_0^+ \to \mathbb{R}_0^+$ which are strictly decreasing with $\lim_{t\to\infty}\delta(t) = 0$, p. 29
$L^\infty(\mathbb{R}, \mathbb{R}^m)$ Space of locally Lebesgue integrable functions from $\mathbb{R}$ to $\mathbb{R}^m$, p. 17
$\mathbb{N}$ Natural numbers, p. 13
$\mathbb{N}_0$ Natural numbers including zero, p. 13
$\mathbb{N}_\infty$ Natural numbers including ∞, p. 13
$P \subset X$ Practical stability region, p. 31
$P(n) \subset X$ Time-varying practical stability region, p. 32
$\mathscr{P}$ Index set of subsystems p, p. 259
$\mathscr{Q} := \{p\} \cup \mathscr{I}^p \subset \mathscr{P}$ Partial index set for subsystem p, p. 265
$\mathbb{R}$ Real numbers, p. 13
$\mathbb{R}_0^+$ Nonnegative real numbers, p. 29
$S \subseteq X$ Domain of a Lyapunov function, p. 33
$S(n) \subseteq X$ State space component of the domain of a time-varying Lyapunov function, p. 36
$\mathscr{S} \subseteq \mathbb{N}_0 \times X$ Domain of a time-varying Lyapunov function, p. 36
$S_{(\bar{d},\bar{e})}(x_0)$ Set of all perturbed trajectories with bounded perturbation and measurement errors, p. 192
$(S^p)_{p\in\mathscr{P}}$ Decomposition of S for index set $\mathscr{P}$, p. 261
$T_\Omega(z) \subseteq \mathbb{R}^{n_z}$ Tangent cone, p. 385
U Control value space p, p. 13
U^N Set of finite horizon control sequences, p. 13
U^∞ Set of infinite horizon control sequences, p. 13
$\mathbb{U}(x) \subseteq U$ Control constraint set, p. 48

$\mathbb{U}^N(x) \subseteq^N$ Set of admissible finite horizon control sequences, p. 47
$\mathbb{U}^\infty(X) \subseteq \mathbb{U}^\infty$ Set of admissible infinite horizon contro sequences, p. 48
$\mathbb{U}^N_{\mathbb{X}_0}(x), \mathbb{U}^N_{\mathbb{X}_0}(n,x) \subseteq U^N$ Set of admissible finite horizon control sequences for terminal constraint set $\mathbb{X}_0$, p. 55
U^p Control value space of subsystem p, p. 259
$\widetilde{U}^p$ Neighboring control value space of subsystem p required for evaluation, p. 261
$\overline{U}^p$ Neighboring control value space of subsystem p not required for evaluation, p. 261
$\mathbb{U}^p \subseteq U^p$ Control constraint set for subsystem p, p. 277
$\mathbb{U}^{p,N}(x_0^p, i^p) \subseteq U^{p,N}$ Set of admissible finite horizon control sequences for subsystem p, p. 268
$\mathbb{U}^{p,N}_{\mathbb{X}_0^p}(x_0^p, i^p) \subseteq U^{p,N}$ Set of admissible finite horizon control sequences for terminal constraint set $\mathbb{X}_0^P$ for subsystem p, p. 309
$\mathbb{V}^\tau(x) \subseteq L^\infty([0,\tau], \mathbb{R}^m)$ Set of admissible continuous time control functions, p. 309
V_N^{-1} $([0, L])$ sublevel set of $V_N, V_N^{-1}([0, L]) := \{x \in \mathbb{X} | V_N(x) \in [0, L]\}$, p. 389
$\mathscr{W}_K \subseteq \mathscr{E} \cup \mathscr{I}$ Working set of optimization algorithm, p. 396
$\mathscr{W}_K^q \subseteq \mathscr{E} \cup \mathscr{I}$ Working set of problem (EQP^q), p. 13
X State space, p. 48
$\mathbb{X} \subseteq X$ State constraint set, p. 54
$\mathbb{X}_0, \mathbb{X}_N(n) \subset X$ Terminal constraint set, p. 55
$\mathbb{X}_N, \mathbb{X}_N(n) \subset X$ Feasible set for terminal constraint set $\mathbb{X}_0$ and horizon N, p. 55
$\mathbb{X}^k \subseteq X$ Time dependent state constraint set p. 204
X^p State space of subsystem p, p. 259
$\mathbb{X}^P$ State constraint set for subsystem p, p. 260
$\widetilde{X}_P$ Neighboring state space of subsystem p required for evaluation, p. 261
$\mathbb{X}_0^P \subset X^P$ Terminal constraint set for subsystem p, p. 286
$\overline{X}^P$ Neighboring state space of subsystem p not required for evaluation, p. 261
$\mathbb{X}_N^P(i^P) \subset X^P$ Feasible set for terminal constraint set $\mathbb{X}_0^P$ for subsystem p, p. 268
$Y \subseteq X$ Forward invariant subset of the state space, p. 30
$Y(n) \subseteq X$ Forward invariant family of subsets of the state space, p. 32
Y Output space, p. 304
$\mathbb{Y} \subseteq X \times U$ Set of admissible pairs, p. 48
$\Omega \subset \mathbb{R}^{n_z}$ Feasible set of an optimization problem, p. 385

Variables

$\alpha_k \in [0, 1]$ Step length in optimization algorithm, p. 384
$\alpha \in (0, 1]$ Suboptimality parameter, p. 80
$\overline{\alpha} \in (0, 1)$ Suboptimality threshold, p. 325
$\gamma_k \in \mathbb{R}_0^+$ Auxiliary values in Formula 6.18, $\gamma_k = B_K(r)/r$, p. 133
$\Delta \in \mathbb{R}_0^+$ Radius of semi-global asymptotic stability region, p. 154
$\delta \in \mathbb{R}_0^+$ Radius of practical asymptotic stability region, p. 154
$\delta \in \mathbb{R}_0^+$ Radius of feasible ball around x_*, p. 185

$\lambda \in \mathbb{R}_0^+$ Weight of control penalization in stage cost, p. 46
$\lambda_k \in \mathbb{R}_0^+$ Stage cost values along an optimal trajectory, p. 13
$\lambda \in \mathbb{R}^{d_c}$ Lagrange multiplier in optimization problem, p. 287
$\lambda \in \mathbb{R}^{r_g+r_h}$ Lagrange multiplier in optimization problem, p. 387
$\lambda^* \in \mathbb{R}^{r_g+r_h}$ Optimal Lagrange multiplier in optimization problem, p. 387
$\lambda^{\mathscr{W}_k} \in \mathbb{R}^{r_{\mathscr{W}_K}}$ Lagrange multiplier for working set $\mathscr{W}_k$, p. 390
$\widetilde{\lambda}^{\mathscr{W}_k} \in \mathbb{R}^{r_g+r_h}$ Full Lagrange multiplier for working set $\mathscr{W}_k$, p. 390
$\widetilde{\lambda}^{\mathscr{W}_{k,*}} \in \mathbb{R}^{r_g+r_h}$ Full optimal Lagrange multiplier of problem (ECP), p. 392
$\lambda_k^{(\mathrm{EQP})} \in \mathbb{R}^{r_{\mathscr{W}_K}}$ Optimal Lagrange multiplier of problem (EQP), p. 392
$\widetilde{\lambda}_k^{(\mathrm{EQP})} \in \mathbb{R}^{r_g+r_h}$ Full optimal Lagrange multiplier of problem (EQP), p. 392
$\nu \in \mathbb{R}_0^+$ Value of the optimal value function $V_N(x_{u*}(1, x))$, p. 157
$\sigma \in (0, 1)$ Decay rate in exponential controllability, p. 126
$\tau_c \in \mathbb{R}_0^+$ Computing time required to solve an optimal control problem, p. 47
$\tau_c^{\max} \in \mathbb{R}_0^+$ Maximal allowable computing time, p. 313
$\omega_{N-k} \in \mathbb{R}_0^+$ Weights in cost functional, p. 55
$Q \in \mathbb{R}^{d \times d_c}$ Linearization for local constraints regarding states $x_{u(\cdot, x_0)}$, p. 289
$R \in \mathbb{R}^{m \times d_c}$ Linearization for local constraints regarding controls $u(\cdot)$, p. 289
$c \in \mathbb{R}^{d_c}$ Right hand side for local constraints, p. 289
$C \in \mathbb{R}_0^+$ Overshoot parameter in exponential controllability, p. 126
$c_n \in \mathbb{R}_0^+$ Coefficients for finite time controllability, p. 126
$\overline{d} \in \mathbb{R}_0^+$ Upper bound of additive perturbation sequence $d : \mathbb{N}_0 \to X$, p. 192
$d_k \in \mathbb{R}^{n_z}$ Search direction in optimization algorithm, p. 383
$d_k^q \in \mathbb{R}^{n_z}$ Iterates for computing d_k via problem (EQP^q), p. 395
$\overline{e} \in \mathbb{R}_0^+$ Upper bound of measurement error sequence $e : \mathbb{N}_0 \to X$, p. 192
$h_i \in \mathbb{R}^+$ Step size for time grid and one step method, p. 345
$h_i \in \mathbb{R}^+$ Maximal step size for time grid and one step method, p. 344
$i^P(n) \in I^P$ Neighboring data of subsystem, p. 262
$N \in \mathbb{N}$ Neighboring data of subsystem, p. 46
$N_n \in \mathbb{N}$ Adapted optimization horizon, p. 326
$n_z \in \mathbb{N}$ Dimension of optimization variable of an optimization problem, p. 367
$p^q \in \mathbb{R}^{n_z}$ Optimization variable in problem (EQP^q), p. 396
$p \in \mathscr{P}$ Index of subsystems, p. 259
$p_g \in \mathbb{N}_0$ Number of equality constraints describing a set, p. 52
$p \in \mathbb{N}$ Number of subsystems $p \in \mathscr{P}$, p. 261
$p_h \in \mathbb{N}_0$ Number of inequality constraints describing a set, p. 52
$r_g \in \mathbb{N}_0$ Number of equality constraints of an optimization problem, p. 265
$r_h \in \mathbb{N}_0$ Number of inequality constraints of an optimization problem, p. 265
$r_s \in \mathbb{N}_0$ Number of shooting nodes, p. 376
$r_{\mathscr{W}_k} \in \mathbb{N}_0$ Number of elements in the working set $\mathscr{W}_k$, p. 390
$s \in \mathbb{R}^{r_s}$ Shooting node values, p. 376
$t_0 \in \mathbb{R}$ Initial time of a trajectory, p. 16
$t_n \in \mathbb{R}$ Sampling times, p. 17
$T \in \mathbb{R}^+$ Sampling period, p. 17
$T_{\mathrm{opt}} \in \mathbb{R}^+$ Optimization horizon in continuous time, p. 64

$u \in U$ Control value, p. 13
$u_* \in U$ Control value in equilibrium, p. 46
$u^e \in \mathbb{U}$ Control value in optimal equilibrium, p. 220
$u^p \in U$ Control value of subsystem, p. 262
$\widetilde{u}^p \in \widetilde{U}^p$ Neighboring control of the subsystem p required for evaluation, p. 261
$\overline{u}^p \in \overline{U}^p$ Neighboring control of the subsystem p not required for evaluation, p. 261
$x \in X$ State of the system, p. 13
$x^+ \in X$ State at the next time instant, p. 13
$x_0 \in X$ Initial value of a trajectory, p. 13
$x_*^p \in X^p$ Equilibrium for subsystem p, to be stabilized, p. 269
$x^{p,\mathrm{ref}} : \mathbb{N}_0 \to X^p$ Reference trajectory for subsystem p, to be stabilized, p. 286
$x_* \in X$ Equilibrium, to be stabilized, p. 28
$x^e \in \mathbb{X}$ Optimal equilibrium, p. 220
$x^p \in X^p$ State of the subsystem p, p. 262
$x_0^p \in X^p$ Initial value of a trajectory of subsystem p, p. 271
$\tilde{x}^p \in \tilde{X}^p$ Neighboring states of the subsystem p required for evaluation, p. 261
$\bar{x}^p \in \bar{X}^p$ Neighboring states of the subsystem p not required for evaluation, p. 261
$z \in \mathbb{R}^{n_z}$ Optimization variable of the optimization problem, p. 37
$z^\star \in \mathbb{R}^{n_z}$ Optimal solution of optimization problem, p. 374
$z_k \in \mathbb{R}^{n_z}$ Iterates of optimization variable,z_0 is the initial guess, p. 374

Functions

$|z_1|_{z_2}$ Distance between $z_1, z_2 \in Z$, brief notation for $|z_1|_{z_2} = d_Z(z_1, z_2)$, p. 30
$||x||$ Norm of x in a vector space, p. 2
$\alpha_1 \in \mathcal{K}_\infty$ Lower bound of a Lyapunov function V, p. 33
$\alpha_2 \in \mathcal{K}_\infty$ Upper bound of a Lyapunov function V, p. 33
$\alpha_3 \in \mathcal{K}_\infty$ Lower bound of the minimal stage cost function l^*, p. 135
$\alpha_4 \in \mathcal{K}_\infty$ Upper bound of the minimal stage cost function l^*, p. 135
$\alpha_V \in \mathcal{K}_\infty$ Bound of the decrease of a Lyapunov function V, p. 33
$\bar{\alpha}_W \in \mathcal{K}_\infty$ Upper bound of W in detectability condition, p. 304
$\alpha_W \in \mathcal{K}_\infty$ Bound for the decrease of W in detectability condition, p. 304
$\beta \in \mathcal{KL}$ Comparison function used for stability analysis, p. 30
$\gamma_W \in \mathcal{K}_\infty$ Bound for the increase of W in detectability condition, p. 304
$i : \{1, \ldots, r_s\} \to \{1, \ldots, d\}$ Shooting state index function, p. 376
$\kappa : \mathbb{X}_0 \to \mathbb{U}$ Local feedback map on terminal constraint set $\mathbb{X}_0$, p. 334
$\lambda : \mathbb{X} \to \mathbb{R}$ Storage function, p. 223
$\mu : X \to U$ State feedback law, p. 15
$\mu : \mathbb{N}_0 \times X \to U$ Time-varying state feedback law, p. 32
$\mu_N : X \to U$ NMPC-feedback law, p. 48
$\mu_N : \mathbb{N}_0 \times X \to U$ Time-varying NMPC-feedback law, p. 55
$\mu_N^\varepsilon : X \to U$ NMPC-feedback law computed from numerical model f^ε, p. 360
$\mu_\infty : \mathbb{N}_0 \times X \to U$ Infinite horizon optimal feedback law, p. 77
$\mu_N^p : X^p \times I^p \to U^p$ Local NMPC-feedback law, p. 261

$\mu_\alpha : \mathbb{N}_0 \times X \to U$ Suboptimal asymptotically stabilizing feedback law, p. 84
$\pi : S \to S$ Idempotent linear map or projection, p. 261
$\varphi(\cdot, t_0, x_0, v) : \mathbb{R} \to \mathbb{R}^d$ Continuous time open-loop trajectory, p. 16
$(\pi^p)_{p\in\mathscr{P}}$ with $\pi^p : S \to S$ Set of idempotent linear maps or projections, p. 261
$\pi_X^P : X \to X$ Projection with such that $\mathrm{Im}(\pi_X^P) = X^p$, p. 261
$\pi_X^P : X \to X$ Projection with such that $\mathrm{Im}(\pi_U^P) = U^p$, p. 261
$\tilde{\varphi}(\cdot, t_0, x_0, v) : \mathscr{G} \to \mathbb{R}^d$ Numerical approximation of $\varphi(\cdot, t_0, x_0, v)$, p. 344
$\Phi : \mathbb{R}^d \times \mathbb{U} \times \mathbb{R} \to \mathbb{R}^d$ Numerical one step method, p. 345
$\zeta : \{1, \ldots, r_s\} \to \{0, \ldots, N\}$ Shooting time index function, p. 376
$\omega \in \mathscr{K}$ Modulus of continuity, p. 194
$B_K : \mathbb{R}_0^+ \to \mathbb{R}_0^+$ Upper bound of the cost functional J_N, p. 126
$C : \mathbb{R}^{n_z} \to \mathbb{R}^{r_g+r_h}$ Constraint function of an optimization problem, p. 387
$C^{\mathscr{W}_k} : \mathbb{R}^{n_z} \to \mathbb{R}^{r_k}$ Constraint function for working set $\mathscr{W}_k$, p. 390
$d_Z : Z \times Z \to \mathbb{R}_0^+$ Metric on a metric space Z, p. 13
$d : \mathbb{N}_0 \to X$ Perturbation sequence, p. 192
$e : \mathbb{N}_0 \to X$ Measurement error sequence, p. 192
$e : \mathscr{G} \to \mathbb{R}_0^+$ Approximation error of one step method, p. 348
$F : X \to \mathbb{R}_0^+$, $F : \mathbb{N}_0 \times X \to \mathbb{R}_0^+$ Terminal cost function, also denoted F_J in Chap. 12, p. 55
$F : \mathbb{R}^{n_z} \to \mathbb{R}_0^+$ Cost function of an optimization problem, p. 372
$\tilde{F} : X \times U \to \mathbb{R}_0^+$ Rotated or modified terminal cost function, p. 226
$F^p : X^p \times I^p \to \mathbb{R}_0^+$ Terminal cost function for subsystem, p. 268
$f : X \times U \to X$ Transition map of a discrete time control system, p. 13
$f^\varepsilon : X \times U \to X$ Numerically approximated transition map, p. 363
$f_c : \mathbb{R}^d \times \mathbb{R}^m \to \mathbb{R}^d$ Vector field of a continuous time control system, p. 16
$f^p : \mathbb{R}^{d^p} \times \mathbb{R}^{m^p} \to \mathbb{R}^{m^p}$ Transition map of th pth discrete time control subsystem, p. 262
$g : X \to X$ Transition map of a discrete time system, p. 29
$g : \mathbb{N}_0 \times X \to X$ Transition map of a time-varying discrete time system, p. 33
$G_i^S : X \times U \to \mathbb{R}$ Equality constraint function of a set, p. 53
$H_i^S : X \times U \to \mathbb{R}$ Inequality constraint function of an optimization problem, p. 53
$h : X \to Y$ Output function, p. 305
$J_N : X \times U^N \to \mathbb{R}_0^+$ Finite horizon cost functional, p. 47
$\tilde{J}_N : X \times U^N \to \mathbb{R}_0^+$ Rotated or modified finite horizon cost functional, p. 226
$J_\infty : \mathbb{N}_0 \times X \times U^\infty \to \mathbb{R}_0^+$ Infinite horizon cost functional, p. 72
$J_\infty^{cl}(\cdot, \cdot, \mu) : \mathbb{N}_0 \times X \to \mathbb{R}_0^+$ Infinite horizon cost of closed-loop trajectory, p. 80
$\bar{J}_\infty^{cl}(\cdot, \mu) : X \to \mathbb{R}_0^+$ Averaged infinite horizon cost of closed-loop trajectory, p. 223
$J_N^{cl}(\cdot, \mu) : X \to \mathbb{R}_0^+$ Finite horizon cost of closed-loop trajectory, p. 223
$J_N^{uc} : X \times U^N \to \mathbb{R}_0^+$, $J_\infty^{uc} : X \times U^N \to \mathbb{R}_0^+$ Cost functionals without terminal conditions (Sect. 8.4 only), p. 229
$J_N^p : X^p \times U^{p,N} \times I^p \to \mathbb{R}_0^+$ Finite horizon cost functional for subsystem, p. 268
$\ell : X \times U \to \mathbb{R}_0^+$ Stage cost function, p. 47

$\ell : \mathbb{N}_0 \times X \times U \to \mathbb{R}_0^+$ Time-varying stage cost function, p. 53
$\ell^\star : X \to \mathbb{R}_0^+$ Minimal stage cost function, p. 125
$\ell^\star : \mathbb{N}_0 \times X \to \mathbb{R}_0^+$ Minimal time-varying stage cost function, p. 143
$\tilde{\ell} : X \times U \to \mathbb{R}_0^+$ Rotated or modified stage cost function, p. 225
$\tilde{\ell} : X \times U \to \mathbb{R}_0^+$ Stage cost function for inverse optimality, p. 113
$\ell^p : X^p \times U^p \times I^p \to \mathbb{R}_0^+$ Stage cost function for subsystem, p. 272
$L : X \times U \to \mathbb{R}_0^+$ Running cost function of an integral form stage cost, p. 46
$L : \mathbb{R}^{n_z} \times \mathbb{R}^{r_g+r_h} \to \mathbb{R}$ Lagrangian of an optimization problem, p. 387
$L : X^{N+1} \times U^N \to \mathbb{R}$ Lagrangian of an optimization problem, p. 288
$M : \mathbb{R}^{n_z} \times \mathbb{R}^{n_z+} \to \mathbb{R}^{2n_z}$ KKT condition vector of problem (ECP), p. 391
$S : \mathbb{R}^{n_z} \to \mathbb{R}^{nN}$ Equality constraint function in multiple shooting nodes, p. 395
$u : \{0, \ldots, N-1\} \to U$ Finite horizon control sequence, p. 14
$u : \mathbb{N}_0 \to U$ Infinite horizon control sequence, p. 13
$u^\star : \{0, \ldots, N-1\} \to U$ Finite horizon optimal control sequence, p. 48
$u^\star : \mathbb{N}_0 \to U$ Infinite horizon optimal control sequence, p. 70
$u^{\mathrm{ref}} : \mathbb{N}_0 \to U$ Reference control sequence, p. 70
$u_x : \{0, \ldots, N-1\} \to U$ Control sequence in controllability assumption, p. 125
$u_n^j : \{0, \ldots, N-1\} \to U$ Control sequence provided by iterative optimization method, p. 331
$u^{p,\star} : \{0, \ldots, N-1\} \to U^p$ Finite horizon optimal control sequence of subsystem, p. 261
$v : \mathbb{R} \to \mathbb{R}^m$ Control function in continuous time, p. 16
$V : S \to \mathbb{R}_0^+$ Lyapunov function, p. 33
$V : \mathcal{S} \to \mathbb{R}_0^+$ Time-varying Lyapunov function, p. 36
$V_N : \mathbb{N}_0 \times X \to \mathbb{R}_0^+$ Finite horizon optimal value function, p. 59
$\tilde{V}_N : X \to \mathbb{R}_0^+$ Rotated or modified finite horizon optimal value function, p. 226
$V\infty : \mathbb{N}_0 \times X \to \mathbb{R}_0^+$ Infinite horizon optimal value function, p. 70
$V_N^{uc} : X \to \mathbb{R}_0^+$, $V_\infty^{uc} : X \to \mathbb{R}_0^+$ Optimal value functions without terminal conditions (Sect. 8.4 only), p. 229
$W : X \to \mathbb{R}_0^+$ Auxiliary function in detectability condition, p. 304
$x_u(\cdot, x_0), x_u : \{0, \ldots, K-1\} \to X$ Predicted or open-loop trajectory, p. 14
$x_{\mu_N}(\cdot, x_0), x_{\mu_N} : \mathbb{N}_0 \to X$ Nominal NMPC closed-loop trajectory, p. 48
$\tilde{x}_{\mu_N}(\cdot, x_0), \tilde{x}_{\mu_N} : \mathbb{N}_0 \to X$ Perturbed NMPC closed-loop trajectory, p. 192
$x^\varepsilon_{\mu_n^\varepsilon}(\cdot, x_0), x^\varepsilon_{\mu_n^\varepsilon} : \mathbb{N}_0 \to X$ Numerical NMPC closed-loop trajectory, p. 360
$\tilde{x}^\varepsilon_{\mu_n^\varepsilon}(\cdot, x_0), \tilde{x}^\varepsilon_{\mu_n^\varepsilon} : \mathbb{N}_0 \to X$ Perturbed numerical NMPC closed-loop trajectory, p. 360
$x^{ex}_{\mu_n^\varepsilon}(\cdot, x_0), x^{ex}_{\mu_n^\varepsilon} : \mathbb{N}_0 \to X$ Exact closed-loop trajectory with numerical NMPC-feedback law μ_N^ε, p. 360
$x^{\mathrm{ref}} : \mathbb{N}_0 \to X$ Reference trajectory, to be stabilized, p. 28
$x^p_{u^p}(\cdot, x_0^p, i^p), x^p_{u^p} : \{0, \ldots, K-1\} \to \mathbb{X}^p$ Predicted or open-loop trajectory of subsystem, p. 268

Index

L. Grüne and J. Pannek, *Nonlinear Model Predictive Control*,
Communications and Control Engineering, DOI 10.1007/978-3-319-46024-6

MIX
Papier aus verantwortungsvollen Quellen
Paper from responsible sources
FSC® C105338

If you have any concerns about our products,
you can contact us on
ProductSafety@springernature.com

In case Publisher is established outside the EU,
the EU authorized representative is:
Springer Nature Customer Service Center GmbH
Europaplatz 3, 69115 Heidelberg, Germany

Printed by Libri Plureos GmbH
in Hamburg, Germany